SDGs and Textiles

Series Editor

Hafeezullah Memon, International Institute of Silk, Zhejiang Sci-Tech University, Hangzhou, China

The book series "SDGs and Textiles" addresses the strategies to achieve sustainable development goals (SDGs) in the present, past, and future. It presents books about the present and future policies of textile ministries of different countries, and books related to sustainability education around different parts of the world in the textile sector. Moreover, it would welcome the conference proceeding related to SDGs and Textiles. The series would cover books comparing the sustainability and SDGs of different institutions and countries. The individual book volumes in the series are thematic. The goal of each book is to give readers a comprehensive overview of a different area of sustainability in the textile sector. As a collection, the series provides valuable resources to a broad audience in academia, the research community, industry, and anyone looking to expand their knowledge of SDGs and Textiles.

Textiles and life are together – life cannot be separated from textiles as it is the most important need for human beings after food. In 2015, the United Nations General Assembly proposed 17 interlinked global goals to be achieved by 2030. Since then, academia and industry have paid much attention to achieving these goals. Textile found its close relation with almost all of these 17 goals.

SDG 1 - No Poverty: Poverty would never be overcome by a charity only; it is essential to develop people's skills to have a better and wealthy life. Thus, the textile can be considered an excellent discipline to achieve this goal by creating jobs and small and medium businesses.

SDG 2 - Zero Hunger: Through the effective utilization of advanced application of Agrotech Textiles, it is possible to have higher crop yields and save crops from rough weather, unexpected rains, floods, insects, etc.; thus, geotextiles play an essential in achieving this goal of sustainable development.

SDG 3 - Good Health & well-being: There has been much health consciousness after Covid19, and medical textiles assist in getting good health and well-being.

SDG 4 - Learning & Education: Textile or fashion has remained a significant discipline for societies for ages, and there has always remained much to explore in this field. Textile-related universities may play a vital role by offering free access to their education resources, training and spreading information among the locals.

SDG 5 - Gender Equality: The textile sector is one of the industrial sectors that accepted gender equality long ago; in particular, the garment sector has more females than males. Thus, the textile sector has been doing gender equality. Moreover, there has been a recent trend for Gender Neutral Clothing, which need worth studying and may further assist gender equality.

SDG 6 - Clean Water & Sanitation: Textiles could be achieved through filtration, and of course, textile is one of the critical materials for filtration.

SDG 7 - Affordable & Clean Energy: With the recent advancement in material science and engineering, the textile sector has come on the front for, not only by using this clean energy during textile production but also by assisting the production of this clean energy, either in the form of wind turbines blades made of textile composites or by energy harvesting from T-Shirts, etc.

SDG 8 - Decent Work: Recently, there has been much attention that the textile workers are not paid well, labor rights are not cared about, etc.

SDG 9 - Industry and innovation: Textile Industry always follows innovation; the textile companies that do not chase innovation cannot survive in the market.

SDG 10 - Reduced Inequalities: Getting better life and well-being would help reduce inequalities in the textile industry.

SDG 11 - Sustainable Cities: Sustainable Textile Cities through Buildtech and transport textiles.

SDG 12 - Consumption and Production: Textile and garment consumption and production all come under.

SDG 13 - Climate Action: Oekotech or Ecotech Textile, waste management of textiles are upfront to achieve this goal of sustainable development.

SDG 14 - Life Below Water: Mitigating microfiber waste in rivers and oceans may come under the context of it. There has been much attention on this subject after passing the bill at the parliament level of the UK.

SDG 15 - Life on Land: Geotech or Geotextiles studies life on land.

SDG 16 - Peace, Justice, and Strong Institutions: Protective textiles are doing their best to achieve peace, justice, and strong institutions.

SDG 17 - Partnerships for the Goals: The application of textiles to achieve sustainable development goals is only an example. In all textiles sectors, combined efforts of all the goals are essential to achieve true sustainability.

Muhammad Arshad
Editor

Enzymes in Textile Processing: A Climate Changes Mitigation Approach

Textile Industry, Enzymes, and SDGs

Springer

Editor
Muhammad Arshad
Department of Basic Sciences
Jhang-campus
University of Veterinary and Animal
Sciences Lahore
Lahore, Pakistan

ISSN 2948-1236 ISSN 2948-1244 (electronic)
SDGs and Textiles
ISBN 978-981-97-8057-0 ISBN 978-981-97-8058-7 (eBook)
https://doi.org/10.1007/978-981-97-8058-7

This Springer imprint is published by the registered company Springer Nature Singapore Pte Ltd.
The registered company address is: 152 Beach Road, #21-01/04 Gateway East, Singapore 189721, Singapore

To my bundles of joy, the juvenile characters Muhammad Talha Arshad and Muhammad Taha Arshad

Preface

Sustainability in the textile sector is crucial to guarantee its long-term survival, reduction of its hurtful environmental impact, and satisfy customers with regulatory expectations. In the textile business, wet processing is a significant part of the industry; has adverse ecological implications and health problems because it uses a lot of chemicals, energy, and water. In response to these issues, the book intends to highlight the green approach of utilizing enzymes in the textile sector and the applications of different enzymes in textile production processes. It also provides several ways to treat effluent in the textile industry with bioenergy production to mitigate climate change.

Overall, the employment of enzymes in the textile sector can help to reduce harsh chemicals use with the minimization of water and energy consumption to achieve environmental sustainability. Environment-friendly processes with green norms and laws are essential for the reduction of chemical use in wet processing to save the environment and human health.

Competent writers and researchers with textile processing and enzyme science backgrounds have accomplished the book. I am grateful for their expertise and dedication in completing this book as their insights and research have greatly enriched this publication.

Jhang, Pakistan
May 2024

Dr. Muhammad Arshad

Acknowledgments

I would like to express my sincere gratitude to every contributor to this book, whose knowledge and combined efforts have enabled this publication. I sincerely thank the writers for their commitment and knowledge. I deeply appreciate the reviewers, whose insightful comments and helpful critiques enabled this book to take on its current form. Its information has been substantially enhanced by their knowledge and perception. I would especially want to thank the Series Editor for all of his help, advice, and dedication to keeping this series current and of high quality. His assistance has been crucial in making this endeavor a success.

I express our gratitude to the students who contributed to this initiative at different phases. Their energy, diligence, and novel viewpoints have been tremendously helpful.

Jhang, Pakistan
May 2024

Dr. Muhammad Arshad

Contents

Distinctive Role of Enzymes in Textile Industry

Maha Gul Zafar, Aqsa Mumtaz, Ali Akbar, Iram Liaqat, Mubbashar Hassan, Mubarik Mahmood, Ghulam Mustafa, Sher Zaman Safi, Syed Shatir A. Syed-Hassan, Abdur Rahman Ansari, Mazhar Abbas, and Muhammad Arshad

Abstract Enzymes being biological catalysts play a pivotal role in accelerating the chemical reaction. The rise in environmental pollution due to the use of hazardous chemicals in textiles leads to environmental unsustainability affecting the ecosystem. Replacement of these chemicals with biocatalysts that are biodegradable, reliable, stable, environment-friendly, and renewable resources has revolutionized the textile industry. These can be easily obtained from fungi, algae, bacteria, plants, and microorganisms. Hydrolases and oxidoreductases are two important enzymes used in the

M. G. Zafar · A. Mumtaz · M. Abbas · M. Arshad (✉)
Biochemistry section, Jhang-campus, University of Veterinary and Animal Science Lahore, Lahore, Pakistan
e-mail: muhammad.arshad@uvas.edu.pk

A. Akbar
Centre for Biotechnology and Microbiology, University of Swat, 19120 Khyber Pakhtunkhwa, Swat, Pakistan

I. Liaqat
Department of Zoology, Government College University Lahore, Lahore, Pakistan

M. Hassan
Department of Clinical Sciences, Jhang-campus, University of Veterinary and Animal Science Lahore, Lahore, Pakistan

M. Mahmood
Department of Animal Sciences, Jhang-campus, University of Veterinary and Animal Science Lahore , Lahore, Pakistan

G. Mustafa
Department of Biochemistry, Government College University Faisalabad, Faisalabad, Pakistan

S. Z. Safi
Bioscience and Nursing, MAHSA University, George, Selangor Malaysia, Malaysia

S. S. A. Syed-Hassan
School of Chemical Engineering, College of Engineering, Universiti Teknologi MARA, 40450 Shah Alam, Malaysia

A. R. Ansari
Anatomy section, Jhang-campus, University of Veterinary and Animal Science Lahore, Lahore, Pakistan

M. Arshad (ed.), *Enzymes in Textile Processing: A Climate Changes Mitigation Approach*, SDGs and Textiles, https://doi.org/10.1007/978-981-97-8058-7_1

textile industry, i.e., lipases, amylase, cellulases, pectinase, laccases, and catalases. These are used for bio-sourcing, de-sizing, enhancing dye uptake, enhancing the leather quality, biopolishing, improving yarn lubricity, denim stone washing, defibrillation of lyocell, wool sourcing, bleaching agent, dye decolorization, and laundry detergent. These viable sources make sure the quality of fabric whether it is denim, cotton, yarn, and polyester is not compromised. It plays a key role in improving air quality, the health of workers dealing with these-biocatalyst and most importantly the stability of the ecosystem.

Graphical Abstract

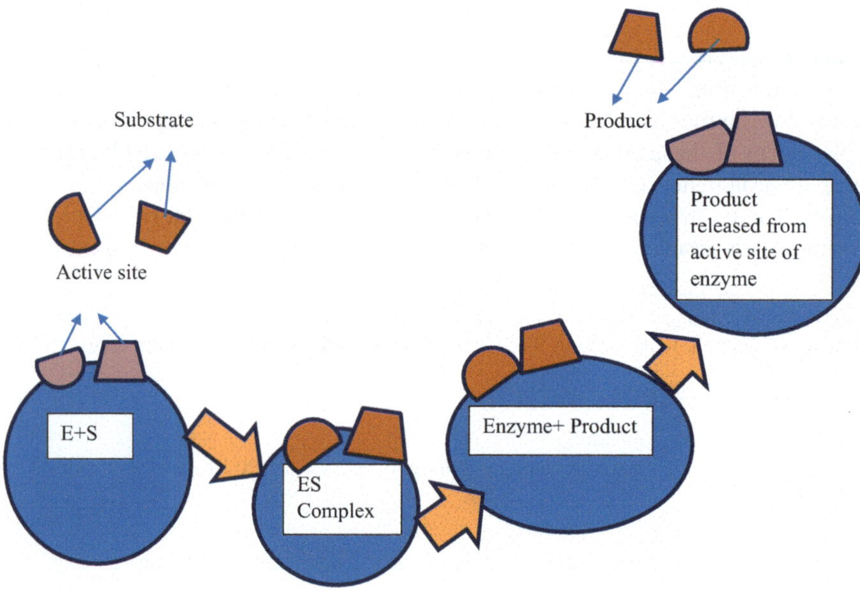

Keywords Enzymes · Environment friendly · Hydrolases · Oxidoreductases · Denim

1 Introduction

All chemical reactions in living things are driven by enzymes, which are specialized, versatile, and efficient catalysts to regulate various biological functions in organisms from bacteria to humans. The enzymes show exceptional accuracy, specificity, and efficiency during biological transformations. Their extraordinary active sites, which properly orient the catalytic residues to effectively stabilize the transition state (TS), are the source of their remarkable catalytic power [12, 47, 60]. Chemical processes can be sped up by enzymes to occur in milliseconds instead of billions of years, as they work as dynamic ensembles of interconverting conformations [1]. Since the known history, enzymes have been employed to produce products like linen, leather, and indigo. Such natural phenomena depend on the enzymes derived from microbes or other preparations such as papaya fruit or rumen from calves. The majority of industrial enzymes in use today are hydrolytic, suggesting they help break down different natural materials [22]. Enzymes are vital, potent catalysts that can transform various compounds, including environmental pollutants [21]. Enzymes are crucial in organic matter transformation, nutrient cycling, food quality, and quantity. In the pharmaceuticals and food industries, they are considered a safer alternative to hazardous inorganic catalysts. A huge number of enzymes are part of the textile manufacturing process including pectinases, amylases, laccases, cellulases, proteases, lipases, and catalase [5, 14, 61].

2 Enzyme Kinetics

Enzymes can selectively catalyze particular substrates to alter the speeds of biological reactions.

The Michaelis–Menten equation (MM equation), which was created by Michaelis and Menten and subsequently more thoroughly determined by Briggs and Haldane using the standard quasi-steady-state approximation (sQSSA), has served as the foundation for a conventional method used to explain enzyme kinetics for a century.

The catalytic constant, k cat, and the Michaelis–Menten constant, Km, are the two constants used in the equation to describe how the concentration of substrate affects the rates of enzyme-catalyzed reactions. V max = k cat E T, where E T is the total enzyme concentration and Km is the substrate concentration at which the reaction rate is half of the V max, is the maximum rate of the reaction at saturating substrate concentrations determined by the k cat [13].

To evaluate the parameters of enzyme kinetics, two main assays are used: k cat and K M, which are derived from the product buildup over time using the reaction progress curve assay and the starting velocity assay. Initial velocity assays assess initial reaction rates at different substrate concentrations, and the results are usually analyzed using Lineweaver–Burk plots to determine parameters. As an alternative, the response progress curve assay makes better use of computer resources than the starting velocity assay by fitting the whole time course to a differential equation [16, 64].

The Michaelis–Menten (MM) equation is the basis for both tests, which should be carried out in circumstances when the equation provides valid, i.e., when the enzyme concentration is much less than the total of the substrate concentration and K. M. There is a chance that utilizing in vitro experiments to forecast in vivo enzyme activity won't be possible in this situation because of greater endogenous enzyme concentrations.

An alternative method based on the total quasi-steady-state approximation (tQ model) built from Bayesian inference is offered to overcome the shortcomings of the

MM equation and identification problems with parameters. The traditional QSSA (sQ) model has biases that this model addresses, especially in situations, where enzyme concentrations are not low. The tQ model improves parameter estimate accuracy and precision by enabling unrestricted pooling of experimental data under different situations [70].

When paired with Bayesian inference, the tQ model provides a strong foundation for precise estimation of the kinetics parameters of enzymes, circumventing identification problems and the limits of the MM equation. The method allows for ideal experimental design without requiring prior knowledge of enzyme kinetics parameters, while simultaneously improving the accuracy of parameter estimation [10].

3 Role of Enzymes in Textile Industries

The textile processing business is expanding, but historically it has required a lot of energy, water, and harsh chemicals, which have led to waste streams that have a significant negative environmental impact [30]. The textile sector releases harmful compounds into the environment, it contributes to over 20% of the cause of water pollution [42]. The substances utilized in conventional textile production processes, such as heavy metals, dyes, bleaches, and detergents (Cutting out textile pollution. Chem Eng News. https://cen.acs.org/articles/93/i41/Cutting-Textile-Pollution.html). Chemicals such as urea, caustic soda, acids, bases, salts, and bleach were once used in the textile production processes of desizing, scouring, and bleaching. These substances are hazardous to humans and can contaminate the environment when disposed of after usage [65]. Because enzymes are non-toxic and environmentally friendly, their usage in the chemical processing of fibres and textiles is fast receiving wider attention as the textile industries encounter an increasingly pressing need to eliminate pollution in the textile production process [52]. Enzymes are biocatalysts that accelerate a reaction's rate by lowering its activation energy [65]. Enzymes are being used in various industrial processes (Chauhan and Gupta 2017; Sharma et al. 2020). Enzymes are frequently employed in industry to substitute harsh chemicals since they are safe, biodegradable, stable, and reliable under mild circumstances [52].

Because enzyme reactions involve rapid reaction rates and are very selective, they can frequently be conducted in mild circumstances. Industrial enzymes are derived from biological systems and are separated from microorganisms by fermentation utilizing mostly renewable resources, which promotes sustainable development [25]. Hydrolases and oxidoreductases are two important enzymes used in the textile industry [42]. For a number of reasons, microbial enzymes are favoured by those sourced from plants or animals. First of all, a variety of catalytic activities are provided by microbial enzymes, giving textile processes flexibility. Second, because genetic manipulation is simple and microorganisms grow quickly in cheap conditions, they may be created in large yields. Furthermore, because microbial enzymes are not impacted by seasonal variations, they provide a steady and dependable supply all year round [72].

In addition, enzymes derived from microorganisms are more stable than those derived from plants and mammals. Because of its stability, enzyme activity remains intact for an extended period of time, which improves textile operations' consistency and efficiency. Microbial enzymes are the favored option for industrial applications since they are safer and easier to produce. Overall, microbial enzymes are invaluable for a variety of enzymatic applications in the textile industry due to their advantages, which include their broad catalytic activities, high yields, stability, and ease of manufacture [4, 45, 51].

A large range of complicated chemicals and auxiliaries are used in the chemical processing of textiles, which leads to the production of highly hazardous and polluting effluent. Acids, alkalis, detergents, grease, oils, sulphates, solvents, heavy metals, inorganic salts, fibres, and leftover colourants (dyes) are among the materials found in this effluent. In addition to discolouring water, these contaminants seriously endanger aquatic life and other kinds of life with their toxicity. Therefore, before discharging the effluent into the environment, it must be treated, with special attention paid to the removal of any remaining colouring agents [4, 63]. Amylases were the main enzymes used in textile processing until the 1980s, mostly for the post-weaving removal of starch-based sizing from fabrics. Since then, however, cellulases have been developed and have shown to be successful in eliminating fuzz and fibrils, especially in the cotton textile industry. Furthermore, cellulases are being used to give denim and other clothing an aged look, extending their usefulness beyond conventional textile techniques.

Proteolytic enzymes have also been tested for the possibility of eliminating wool fibre scales, which has improved the material's anti-felting qualities [3]. From the 7000 enzymes known, only about 75 are commonly used in textile industry processes [59].

The enzymes of the hydrolase class, which are commonly utilized in the desizing, biopolishing, bioscouring, and bleaching of textiles, include amylase, cellulase, catalase, lipase, and pectinase [49]. The transferases, such as transglutaminases, which modify the surface of synthetic fibres, and laccases, which decolorize dyes are members of the oxidoreductase class of enzymes [67].

Enzymes are used in textile chemical processing at several stages to meet the demands of the fourth industrial revolution through green technology initiatives. Eco-friendly techniques are aided by the use of peroxidases in bleaching processes. The use of cellulases in denim finishing and bio-polishing highlights the need for sustainable methods in the textile industry. The enzyme amylase is frequently used to desize woven fabrics. After bleaching, the catalase enzyme can be used to remove hydrogen peroxide.

These uses of enzymes show how the industry is moving towards more ecologically friendly practices that are in line with the Fourth Industrial Revolution's fundamentals [25, 29].

4 Lipases Role in Textile Industry

Lipases (triacylglycerol lipase EC 3.1.1.3) are enzymes that catalyze the hydrolysis of lipid esters into fatty acids and glycerol. They are essential for many biological functions, such as metabolism, digestion, and cellular signaling. Lipases can be combined with other units or enzymes to increase their specificity and efficiency. Lipases, for instance, can be made more stable, active, and reusable by immobilizing them on solid substrates or combining them with other enzymes in multienzyme systems. Lipases are thought to be among the third largest classes of commercial enzymes.

Lipases have been widely employed in a variety of sectors, such as food, medicine, and biofuels. Lipases have a number of important uses, including biosourcing, where they facilitate effective production procedures [24].

Lipases are incredibly adaptable; they work well in non-aqueous conditions and are stable in organic solvents. They are active at interfaces in micro-aqueous settings. These characteristics highlight their importance as catalysts in a range of industrial processes, including synthesis and bioremediation, providing effective and long-lasting solutions in a number of industries [2].

Lipases are active on amphiphilic compounds and form micelles in the water state and their hydrolysis involves an enzyme having active functioning at the water/lipid interface. They are present in a wide variety of microbes, mammals, and plants [20]. Lipases play a vital role in enzymatic degradation processes, providing persistent solutions for industrial environmental remediation [57].

Lipases meet the increasing demand for sustainable industrial processes because they are ecologically friendly and biodegradable. Microbial lipases are commercially feasible for use in large-scale applications because they can be manufactured in huge quantities using fermentation methods. Numerous microbial lipases exhibit stability throughout an extensive range of temperatures and pH levels, rendering them appropriate for use in a variety of industrial processes under varying circumstances [24].

Lipases are mostly employed in the textile industry for bioscouring, desizing cotton fabrics, and surface modification of synthetic fibres. They can be used alone or in conjunction with other enzymes including xylanases, pectate lyase, and proteases [32]. lipases are essential for the processing and modification of synthetic fibres at different phases of the production process of yarn, fabric, and textiles. Their activity promotes sustainable manufacturing techniques and enables improved performance, durability, and functional qualities [65]. Fabrics treated with detergent plus lipase, demonstrated improved cleaning efficacy compared to detergent alone [71].

The enzyme lipase improves polyester fabrics' capacity to be dyed while causing less surface abrasion and weight loss [20]. In the textile business, lipases are used to help remove size lubricants, which improves the absorbency of the cloth and allows for better dyeing consistency. Additionally, using it lessens the potential of cracks and streaks in denim fabric abrasion systems [24, 43]. A lipase treatment can increase a polyester fabric's wettability and absorbency (PCT Publication No. 97/33001,

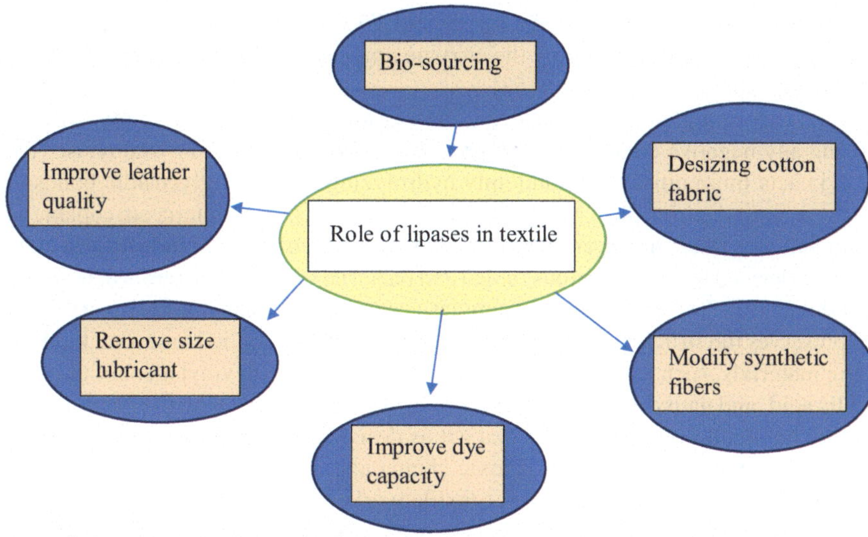

Fig. 1 Role of lipases in textile

Genencor International, Inc.) [34]. The research indicated that treated polyester fabrics (bioactive and basic)with lipases dyed better than untreated ones in terms of dye uptake, colour fastness, and overall dyeing performance [19]. while rinsing it breaks down oils and fats into fatty acids and glycerol, removing the most stubborn lipid stains. As an alternative to alkaline hydrolysis, it makes polyester hydrophilic.

One can accomplish this by adding a detergent component [11]. When preparing leather, enzymes such as lipase and protease offer numerous advantages. When raw hides are rehydrated, they use less water and generate less waste [20].

Using lipases results in leather that is of higher quality, with a clearer surface, precise coloring, and increased water resistance. This highlights the significance of lipases in the leather processing industry, since they improve the finished product's functionality and aesthetic appeal while supporting eco-friendly manufacturing approaches [2] (Fig. 1).

5 Amylases Role in Textile Industry

Amylases belong to the hydrolases class of enzymes [67]. The textile industry's primary use of amylase is still enzymatic desizing. The application of enzymes, specifically amylases, in the textile sector dates back to the early 1900s. During this period, greige fabrics were treated with barley extracts, which were rich in active enzymes [32].

Starch and its derivatives make up around 75% of size agents. Dietary starch can be hydrolyzed more easily by the enzyme amylase, which splits it into maltose, dextrin, and shorter-chain sugars [53].

Amylases completely eliminate the size components from the cloth without causing any negative effects. In contrast to β-amylase (EC 3.2.1.2), α-amylase (EC 3.2.1.1) acts more quickly by randomly hydrolyzing the α-1,4 glycosidic bonds to release α-limit dextrins [42]. Many types of fungi, yeasts, and bacteria are capable of producing amylases; however, the most often utilized enzymes in industrial sectors are those derived from filamentous fungal and bacterial sources. The removal of sizing ingredients from woven fabrics with amylase enzymatic desizing usually works well and preserves the structural integrity of the fabric. It can be applied to several kinds of sizing materials, such as blends with other polymers like polyvinyl alcohol or polyacrylic acid, and native or modified starches [25]. It can be used with the combination of other hemicellulolytic enzymes that act as desizing agent removing harsh sizing agents without the involvement of toxic chemicals ensuring not only high-quality fabric but also a safer environment for textile workers [52].

In the textile industry, desizing processes are often used to remove the adhesive coating, or "size," that is applied to warp threads during weaving. This size lubricates and protects threads from friction, minimizing breakage during weaving. It is typically composed of starch and its derivatives due to its reasonable and good film-forming ability. After weaving, cotton fibers must be desized in order to remove non-cellulosic materials, which will facilitate the dyeing and finishing processes. pH levels between 5.5 and 6.5 and temperatures between 30 and 60°C are excellent for amylase desizing. These ideal optimum conditions not only boost enzymatic activity but also reduce water consumption because it does not require repeated washings to get rid of chemical residue after desizing [65].

α-amylase enzymes are usually used for desizing because they cleave starch more quickly than β-amylase and at random sites. α-amylases are commonly utilized in the enzymatic stone washing of denim clothing due to their increased stability and activity at higher pH values. For these α-amylases to remain stable during heat, calcium ions are usually required. To maximize their effectiveness, they are frequently used in conjunction with metal chelators in textile applications [20].

A number of variables, including pH, oxidative stability, chelator resistance, and temperature behavior, affect how effective amylase is. Protein engineering methods including homology considerations, random mutagenesis, and site-directed mutagenesis are used to create redesigned amylases with improved properties, like thermostability, in order to increase their performance [11].

To improve yarn lubricity, a mixture of enzymes is preferred in some situations. To accomplish the intended results, a combination of enzymes, including lipase, protease, and amylase, may be used. Laundry detergents contain amylases to help remove a wide range of stains, such as fatty, protein, and starch-based stains, among others [31]. Along with amylase, neutral cellulase improves fabric wettability, dye absorption, and minimizes stiffness. In addition, this combined treatment has no negative effects on machines, safer enzyme handling, and softer fabrics as a result.

The production of amylases by bacteria such as B. licheniformis, B. amyloliquefaciens, and B. stearothermophilus is a typical method of desizing in the textile industry. Furthermore, Aspergillus species fungal amylases can be employed for the same function [6].

The use of amylase as an ingredient in laundry detergent formulas has grown in the past several years. Improving α-amylase's oxidative stability is necessary to do this. This is frequently achieved by replacing no oxidizable residues with oxidation-sensitive amino acids, such as cysteines and methionine. Notably, Purastar OxAm (created by Genencor) and Duramyl (manufactured by Novozymes) are two of the most commonly utilized oxidative-stable α-amylases available on the market [11] (Fig. 2).

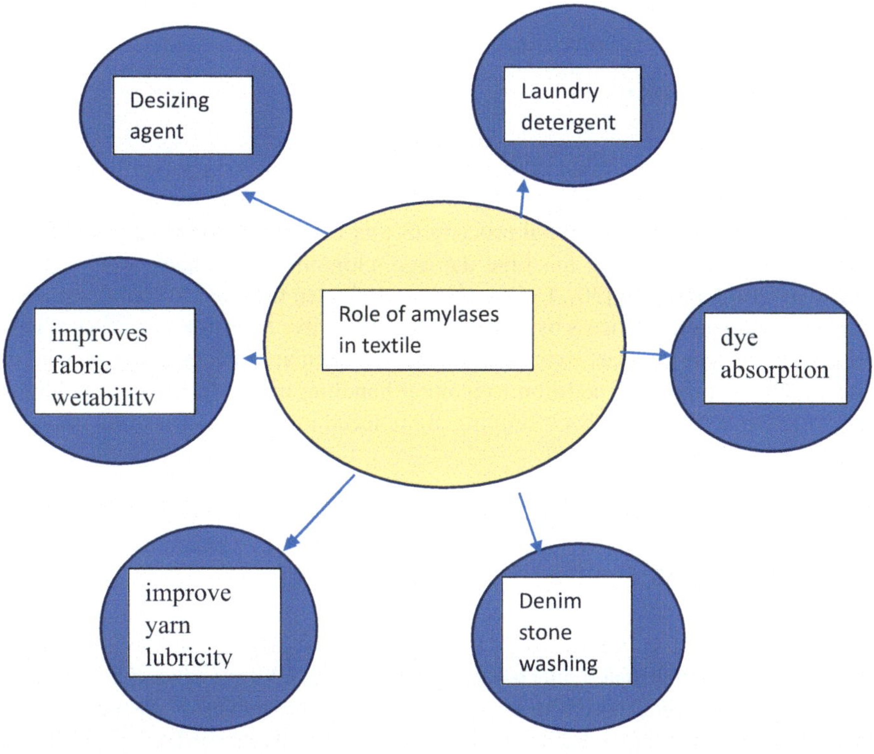

Fig. 2 Role of amylases in textile

6 Cellulase Role in Textile Industry

A key component of lignocellulosic materials like plant cell walls, cellulose is broken down by cellulases. These enzymes break down cellulose into simpler sugars, mainly glucose, by hydrolyzing the β-1,4-glycosidic linkages in the fibres. This procedure is important because it makes it possible to ferment cellulose and turn it into useful products like cellulosic biofuels.

They are effective at catalyzing the breakdown of cellulose, cellulases have been used in a wide range of industrial applications for more than thirty years [17]. It plays numerous roles in textile industries, i.e., bio-stone washing, bio-carbonization and wool scouring, biopolishing and finishing, bio-scouring, and defibrillation of lyocell [40].

6.1 Stone Washing

It is one of the denim washing procedures that has been used historically to obtain the desired faded appearance and softer feel. On the other hand, there are a number of practical issues with traditional procedures that use pumice stones and oxidative bleaching agents, including machine damage, clogging of drainage systems, and excessive fabric abrasion [50]. Cellulases help to lessen the need for pumice stones during the bio-stoning process by releasing the indigo dye from denim surfaces. This reduces fabric deterioration, wear on machinery, and environmental issues related to pumice dust formation in addition to making handling easier. Moreover, problems such as tissue stiffness and back staining are reduced in cellulase-based bio-stoning [41].

6.2 Biopolishing

During biopolishing, cellulases target the tiny fibers that protrude on fabric surfaces. These fibers are eliminated mechanically, giving the cloth a smoother, glossier appearance. The colour brightness, hydrophilicity, and moisture absorbance are improved by this technique. It's an environmentally friendly method that consistently enhances fabric finishing [15].

Commercially available endoglucanase II can often be combined with other cellulase components, particularly endoglucanase I, during the finishing and biostoning processes. The garments lose weight as a result of the hydrolysis of this mixture [54].

6.3 *Bio-Sourcing*

Cellulase enzymes are environmentally friendly and great alternatives to hazardous chemicals as these specifically target non-cellulosic components. In order to target non-cellulosic sticky molecules including pectin, natural waxes, esters, grease, dirt, and oil without affecting the integrity of the cellulose component in the fabric, cellulase enzymes are essential [66]. Cellulases enable the elimination of these substances, which improves the fabric's wettability and, consequently, its quality and functionality [27, 54].

6.4 *Defibrillation of Lyocell*

When the fibrillated lyocell fabric is treated with enzymes, the surface fibrils that were released during the fibrillation process are eliminated. Due to the particular hydrolysis that happens during enzymatic treatment, leading to a faster breakdown of the mechanical properties [7]. While the mechanical defibrillation treatment just affects the surface fibres of the fabric, the enzyme treatment can impact all of the fibres within it [26].

6.5 *Bio-Carbonization and Wool Scouring*

Wool dyeing frequently calls for an acidic bath to optimize dye absorption; a high pH is necessary to get good results. The use of low pH values is known to improve dye exhaustion, or increased dye uptake [18]. Sulphuric acid carbonizing, which is the conventional method, raises concerns about the environment and wool quality. By facilitating the breakdown of plant residue in wool, cellulase enzymes may reduce the need for sulfuric acid and their negative effects on the environment. Enzymes such as cellulase also improve wool scouring process efficiency by lowering water consumption and environmental issues related to conventional methods [46].

7 Pectinase Role in Textile Industry

Pectinases being complex enzymes hydrolyze pectic compounds; they were first identified as virulence factors in the breakdown of plant cell walls. However, these have currently gained significance across a number of industries, including food, textile, and biofuel [68]. It is produced by a variety of living organisms, including plants, microorganisms (such as fungi and bacteria), insects, nematodes, and protozoa [62]. These are commonly sourced from various natural materials such as citrus peel, apple pomace, sugar beet, and sunflower. Their increasing demand is a result of their adaptability, wide substrate selectivity, stability, inducibility, and ability to function on a variety of pectic materials [48]. Textile processing has benefited greatly in both environmental and product quality aspects through the use of enzymes. Prior to weaving of yarn into fabric, the warps yarns are coated with a sizing agent to lubricate and protect the yarn from abrasion during weaving [8]. Pectinases, along with enzymes such as lipases, amylases, cellulases, and hemicellulases, have completely changed the process of removing sizing agents from cotton. They provide a more environmentally friendly and safe substitute for hazardous caustic soda. This novel technique, called bio-scouring, uses certain enzymes to efficiently extract non-cellulosic contaminants from fibres. Pectinases are essential to this process because they guarantee complete cleaning without interfering with the breakdown of cellulose [56]. The use of Bacillus subtilis pectinase in cotton textile manufacturing has shown to be economical. A new process has been developed recently to produce premium cotton and blended fabrics in just one step. Results from this technology have shown to be reliable and consistent, making it appropriate for experiments on an industrial scale [38, 58]. Pectate lyase sourced from Bacillus pumilus BK2 has demonstrated significant efficacy in cotton scouring. This enzyme has successfully attained the objective of achieving 100% chemical scouring while preserving the integrity of cellulose, thus avoiding any destructive effects on the cotton fibers [37]. This results in an improvement in both the safety of working conditions for textile workers and the overall quality of the fabric [23].

8 Laccases Role in Textile Industry

Benzenediol: oxygen oxidoreductase (EC 1.10.3.2), formerly known as laccase, is one the oldest and most common enzymes in nature. The blue oxidase family, which includes these enzymes, is thought to be the largest subgroup of multinuclear copper-containing oxidases. All of the major biological kingdoms plants, fungi, insects, bacteria, lichens, and sponges have a plethora of laccases [36]. In the textile industry, between 1 and 20% of the colors produced worldwide are lost during the dyeing process. Unfortunately, the high concentration of organic substances, particularly carcinogenic waste compounds, in the colored effluents produced during the textile production process is a major cause of ecosystem disruption [39]. Industrial dyes frequently have complicated structures that provide them resistance to conventional processes like exposure to light, water, and chemicals. Using laccases has become the best method for color removal because these conventional methods have proven to be useless.

When combined with particular stabilizers, laccases can efficiently eliminate waste. These enzymes are also used to bleach different kinds of fabric and clothes [73]. Their main use in the production of textiles is in the bleaching of denim [28]. Notably, laccases can also aid in the formation of color by oxidizing aromatic molecules.

Because laccases can remove odors during the fabric-making process, they are often added to laundry detergents [44]. The potential for minimizing the shrinkage of wool exists through the implementation of laccase-mediated technologies [67].

Laccases can break down textile dyes that are not bonded to the fibres during the dyeing process, which frequently results in visual pollution. Chlorine dioxide is commonly used as a bleaching agent in the conventional bleaching of cellulose pulp in order to oxidize residual lignin that gives the pulp its brown color. Hazardous dyes found in wastewater released into the environment can cause allergies, rashes, respiratory illnesses, and systemic problems in living things.

However, when organic biomass reacts with chlorine in this process, it creates organochlorine chemicals that are harmful to both human and animal health.

Despite these benefits, the large-scale application of enzymatic bleaching is still in the beginning stages due to high operating costs. Laccases provide an alternative by oxidizing phenolic and amine compounds found in lignin, helping to remove them from cellulose pulp and lowering the need for chlorine-based bleaching agents. However, the addition of mediators opens up the possibility of developing chlorine-free bleaching sequences, offering a viable path towards ecological bleaching methods [36, 55].

9 Catalase Role in Textile Industry

Catalase (EC 1.11.1.6) is an oxidoreductase enzyme more precisely, a peroxidase. It effectively suppresses the reactive oxygen species (ROS) generated in cells, especially as a byproduct of aerobic metabolism, by catalyzing the breakdown of hydrogen peroxide (H_2O_2) into water (H_2O) and oxygen (O_2). Hydrogen peroxide found in bleaching effluents is broken down by catalase into oxygen and water, which are safe byproducts. Textile wastewater is less harmful to the environment after this treatment, making it safer to release or maybe reuse.

It is viable to recycle bleaching effluents for use in textile dyeing by effectively removing hydrogen peroxide from them. This lessens the use of chemicals, energy, and the overall environmental impact of the textile manufacturing process in addition to conserving water resources [33].

9.1 Dye Decolorization

Given the potential hazards of synthetic dyes and their effects on the environment, catalase is being used more and more in the treatment of dye effluents from sectors like textiles [9]. These dyes have poisonous, carcinogenic, and mutagenic qualities, which make them dangerous for ecosystems and human health when they are frequently released into aquatic habitats.

Conventional dye remediation techniques can be expensive and produce hazardous consequences. However, methods for breaking down enzymes, especially those that use catalases that have been isolated from a variety of sources, such as bacteria, plants, algae, and mammalian cells, present a viable option for treating dye effluents in a way that is both efficient and sustainable.

Significant increases in the activity and biodegradability of azo dyes have been seen when plant catalases, such as those from Salvinia molesta and Lemna minor L., are employed for the degradation of azo dyes. Additionally, catalase-mediated procedures, such as ozonation with catalase treatment, have demonstrated encouraging outcomes in terms of improving wastewater's organic component biodegradability and dye degradation.

By reducing pH, increasing the efficiency of chemical oxygen demand (COD) removal, and improving biodegradability ratios, this method helps to break down dyes into simpler organic acids and makes it easier to remove them from wastewater [35, 69].

10 Conclusion

The textile business is increasing day by day with the recent demand, so the release of toxic chemicals and gases is growing. The use of eco-friendly substituent enzymes has revolutionized this field. Enzymes are biocatalysts that lower the activation energy and carry out reactions at a high rate. These biological sources are more liable, sustainable, nontoxic, and renewable. The inclusion of enzymes has offered long-lasting solutions for the cleanup of industrial environmental contamination. Bleaching agents, dyes, and organic substances that are carcinogenic in nature affecting not only the environment but also potential risk to industrial workers should be replaced by biocatalysts. Amylases, laccase, catalase, cellulases, pectinases, and catalases are proven to be effective in bio-sourcing, biopolishing, detergent, dye decolorization, releasing the crack and streak in denim, enhancing the fabric sustainability (cotton, polyester, leather, yarn) and stone washing of denim. Currently, these enzymes are used in textiles. Two of the most widely used oxidative-stable α-amylases on the market are Purastar OxAm (developed by Genencor) and Duramyl (produced by Novozymes).

References

1. Agarwal PK, Bernard DN, Bafna K, Doucet N (2020) Enzyme dynamics: Looking beyond a single structure. Chem Cat Chem 12(19):4704–4720
2. Akram F, Mir AS, Haq I, u., & Roohi, A. (2023) An appraisal on prominent industrial and biotechnological applications of bacterial lipases. Mol Biotechnol 65(4):521–543
3. Araújo R, Casal M, Cavaco-Paulo A (2008) Application of enzymes for textile fibres processing. Biocatal Biotransform 26(5):332–349. https://doi.org/10.1080/10242420802390457

4. Archna LK, Siva K (2012) Biological methods of dye removal from textile effluents-A review. J Biochem Tech, 3(5)
5. Arya PS, Yagnik SM, Raval VH (2023) Role of microbial enzymes in agricultural industry. In Biotechnol Microb Enzym (pp 525–550). Elsevier
6. Aydemir D, Gecili F, Özdemir N, Ulusu NN (2020) Synthesis and characterization of a triple enzyme-inorganic hybrid nanoflower (TrpE@ ihNF) as a combination of three pancreatic digestive enzymes amylase, protease and lipase. J Biosci Bioeng 129(6):679–686
7. Berto GL, Mattos BD, Rojas OJ, Arantes V (2021) Single-step fiber pretreatment with monocomponent endoglucanase: defibrillation energy and cellulose nanofibril quality. ACS Sustain Chem & Eng 9(5):2260–2270
8. Bhardwaj V, Degrassi G, Bhardwaj RK (2017) Microbial pectinases and their applications in industries: a review. polymer, 4(8), 829–836
9. Blackburn R (2009) Sustainable textiles: life cycle and environmental impact. Elsevier
10. Choi B, Rempala GA, Kim JK (2017) Beyond the Michaelis-Menten equation: Accurate and efficient estimation of enzyme kinetic parameters. Sci Rep 7(1):17018
11. Choudhury AKR (2020) Enzyme applications in textile chemical processing. In Sustainable Technologies for Fashion and Textiles (pp 91–115). Elsevier
12. Copeland RA (2023) Enzymes: a practical introduction to structure, mechanism, and data analysis. John Wiley & Sons
13. Cornish-Bowden A (2015) One hundred years of Michaelis-Menten kinetics. Perspect Sci 4:3–9
14. Das D, Goyal A (2013) Pharmaceutical enzymes. In Biotransformation Waste Biomass High Value Biochem (pp 367–387). Springer
15. de Souza Lima J, Immich APS, de Araújo PHH, de Oliveira D (2022) Cellulase immobilized on kaolin as a potential approach to improve the quality of knitted fabric. Bioprocess Biosyst Eng 45(4):679–688
16. Duggleby RG, Wood C (1989) Analysis of progress curves for enzyme-catalysed reactions. Automatic construction of computer programs for fitting integrated rate equations. Biochem J, 258(2), 397–402.
17. Ejaz U, Sohail M, Ghanemi A (2021) Cellulases: from bioactivity to a variety of industrial applications. Biomimetics 6(3):44
18. El-Sayed H, Abou Taleb M, Mowafi S (2021) Potential applications of textile wastes and by-products in preparation of textile auxiliaries. Egypt J Chem 64(8):4433–4447
19. El-Shemy N, El-Hawary N, El-Sayed H (2016) Basic and reactive-dyeable polyester fabrics using lipase enzymes. J Chem Eng Process Technol 7(1):1000271
20. Galante YM, Formantici C (2003) Enzyme applications in detergency and in manufacturing industries. Curr Org Chem 7(13):1399–1422
21. Gianfreda L, Rao M, Scelza R, De la Luz Mora M (2016) Role of enzymes in environment cleanup/remediation. In Agro-Ind Wastes Feed Enzym Prod (pp. 133–155). Elsevier
22. Gurung N, Ray S, Bose S, Rai V (2013) A broader view: microbial enzymes and their relevance in industries, medicine, and beyond. Biomed Res Int 2013(1):329121
23. Haile S, Ayele A (2022) Pectinase from microorganisms and its industrial applications. Sci World J
24. Hasan F, Shah AA, Hameed A (2006) Industrial applications of microbial lipases. Enzyme Microb Technol 39(2):235–251
25. Hasan MM, Nabi F, Mahmud R (2015) Benefits of enzymatic process in textile wet processing. Int J Fiber TextE Res 5(2):16–19
26. Ibbett R, Gaddipati S, Hill S, Tucker G (2013) Structural reorganisation of cellulose fibrils in hydrothermally deconstructed lignocellulosic biomass and relationships with enzyme digestibility. Biotechnol Biofuels 6:1–15
27. Jagajanantha P, Morey M, Satankar V, Mageshwaran V (2022) Bio-scouring of non-spinnable cotton by a crude enzyme of a new fungal strain Aspergillus sp. VM-1, isolated from banana pseudostem waste. Waste Biomass Valorization 1–10

28. Jennifer C, Stephie C, Abhishri S, Shalini B (2012) A review on skin whitening property of plant extracts. Int J Pharm Bio Sci 3(4):332–347
29. Kabir SMM, Koh J (2021) Sustainable textile processing by enzyme applications. Biodegrad Technol Org Inorg Pollut, 339–366
30. Kadam SK, Tamboli AS, Sambhare SB, Jeon B-H, Govindwar SP (2018) Enzymatic analysis, structural study and molecular docking of laccase and catalase from B. subtilis SK1 after textile dye exposure. Eco Inform 48:269–280
31. Kakkar P, Wadhwa N (2022) Extremozymes used in textile industry. J TextE Inst 113(9):2007–2015
32. Kalantzi S, Kekos D, Mamma D (2019) Bioscouring of cotton fabrics by multienzyme combinations: application of Box-Behnken design and desirability function. Cellulose 26:2771–2790
33. Kaushal J, Mehandia S, Singh G, Raina A, Arya SK (2018) Catalase enzyme: Application in bioremediation and food industry. Biocatal Agric Biotechnol 16:192–199
34. Kellis Jr JT, Poulose AJ, Yoon M-Y (2001) Enzymatic modification of the surface of a polyester fiber or article. In: Google Patents
35. Khataee A, Movafeghi A, Torbati S, Lisar SS, Zarei M (2012) Phytoremediation potential of duckweed (Lemna minor L.) in degradation of CI Acid Blue 92: Artificial neural network modeling. Ecotoxicol Environ Saf 80:291–298
36. Khatami SH, Vakili O, Movahedpour A, Ghesmati Z, Ghasemi H, Taheri-Anganeh M (2022) Laccase: Various types and applications. Biotechnol Appl Biochem 69(6):2658–2672
37. Klug-Santner BG, Schnitzhofer W, Vršanská M, Weber J, Agrawal PB, Nierstrasz VA, Guebitz GM (2006) Purification and characterization of a new bioscouring pectate lyase from Bacillus pumilus BK2. J Biotechnol 121(3):390–401
38. Kohli P, Gupta R (2015) Alkaline pectinases: a review. Biocatal Agric Biotechnol 4(3):279–285
39. Konstantinou IK, Albanis TA (2004) TiO2-assisted photocatalytic degradation of azo dyes in aqueous solution: kinetic and mechanistic investigations: a review. Appl Catal B 49(1):1–14
40. Korsa G, Konwarh R, Masi C, Ayele A, Haile S (2023) Microbial cellulase production and its potential application for textile industries. Annals of Microbiology 73(1):13
41. Korsa G, Masi C, Konwarh R, Tafesse M (2022) Harnessing the potential use of cellulolytic Klebsiella oxytoca M
42. Kumar D, Bhardwaj R, Jassal S, Goyal T, Khullar A, Gupta N (2021) Application of enzymes for an eco-friendly approach to textile processing. Environ Sci Pollut Res, 1–11
43. Kumar JA, Senthil Kumar M (2020) A study on improving dyeability of polyester fabric using lipase enzyme. Autex Res J 20(3):243–249
44. Kunamneni A, Plou FJ, Ballesteros A, Alcalde M (2008) Laccases and their applications: a patent review. Recent Pat Biotechnol 2(1):10–24
45. Madhu A, Chakraborty JN (2017) Developments in application of enzymes for textile processing. J Clean Prod 145:114–133
46. Maiti S, Kulkarni K, Adivarekar R (2018) Biotechnology in textile wet processing. Glo J Biomed Sci 2:7–13
47. Maria-Solano MA, Serrano-Hervás E, Romero-Rivera A, Iglesias-Fernández J, Osuna S (2018) Role of conformational dynamics in the evolution of novel enzyme function. Chem Commun 54(50):6622–6634
48. Martín MC, López OV, Ciolino AE, Morata VI, Villar MA, Ninago MD (2019) Immobilization of enological pectinase in calcium alginate hydrogels: A potential biocatalyst for winemaking. Biocatal Agric Biotechnol 18:101091
49. Mate DM, Alcalde M (2015) Laccase engineering: from rational design to directed evolution. Biotechnol Adv 33(1):25–40
50. Mazotto AM, de Ramos Silva J, de Brito LAA, Rocha NU, de Souza Soares A (2021) How can microbiology help to improve sustainability in the fashion industry? Environ Technol Innov 23:101760
51. Mojsov K (2011) Application of enzymes in the textile industry: a review
52. Mojsov K (2012) Biotechnological applications of pectinases in textile processing and bioscouring of cotton fibers

53. Mojsov K (2014) Industrial enzymes in textile processing and the healty environment: A review. Tekstilna industrija 61(1):12–16
54. Pandit P, Maiti S, Singha K, Maity S, Shanmugasundaram O (2022) Biotechnological and nano-biotechnological approaches in finishing of textile materials. In Appl Biotechnol Sustain TextE Prod (pp 173–186). Elsevier
55. Paraschiv G, Ferdes M, Ionescu M, Moiceanu G, Zabava BS, Dinca MN (2022) Laccases—versatile enzymes used to reduce environmental pollution. Energies 15(5):1835
56. Pasha KM, Anuradha P, Subbarao D (2013) Applications of pectinases in industrial sector. Int J Pure Appl Sci Technol 16(1):89
57. Patel GB, Rakholiya P, Shindhal T, Varjani S, Tabhani N, Shah KR (2021) Lipolytic Nocardiopsis for reduction of pollution load in textile industry effluent and SWISS model for structural study of lipase. Biores Technol 341:125673
58. Pilnik W, Voragen AG (1993) Pectic enzymes in fruit and vegetable juice manufacture. Enzym Food Process 3:363–399
59. Quandt C, Kuhl B (2001) Enzymatic processes: Operational possibilities and optimization (Enzymes Possibilités et perspectives). L'Industrie Textile (1334Á1335)
60. Rao MA, Scelza R, Acevedo F, Diez M, Gianfreda L (2014) Enzymes as useful tools for environmental purposes. Chemosphere 107:145–162
61. Raveendran S, Parameswaran B, Ummalyma SB, Abraham A, Mathew AK, Madhavan A, Rebello S, Pandey A (2018) Applications of microbial enzymes in food industry. Food Technol Biotechnol 56(1):16
62. Rehman H, Baloch AH, Nawaz MA (2021) Pectinase: immobilization and applications. A review. Trends Pept Protein Sci, *6*, 1–16 (e11)
63. Sarayu K, Sandhya S (2012) Current technologies for biological treatment of textile wastewater–a review. Appl Biochem Biotechnol 167:645–661
64. Schnell S, Maini P (2003) A century of enzyme kinetics. Should we believe in the Km and vmax estimates?
65. Sen A, Kapila R, Chaudhary S, Nigam A (2021) Biotechnological applications of microbial enzymes to replace chemicals in the textile industry-a review. TextE Assoc 82(2):68–73
66. Sharma S, Yazdani SS (2016) Diversity of microbial cellulase system. In New Futur Dev Microb Biotechnol Bioeng (pp. 49–64). Elsevier
67. Shen J (2019) Enzymatic treatment of wool and silk fibers. In Adv TextE Biotechnol (pp. 77–105). Elsevier
68. Shrestha S, Rahman MS, Qin W (2021) New insights in pectinase production development and industrial applications. Appl Microbiol Biotechnol 105:9069–9087
69. Takio N, Yadav M, Yadav HS (2021) Catalase-mediated remediation of environmental pollutants and potential application–a review. Biocatal Biotransform 39(6):389–407
70. van Eunen K, Bakker BM (2014) The importance and challenges of in vivo-like enzyme kinetics. Perspectives in Science 1(1–6):126–130
71. Varanasi A, Obendorf SK, Pedersen LS, Mejldal R (2001) Lipid distribution on textiles in relation to washing with lipases. J Surfactants Deterg 4(2):135–146
72. Wavhal S, Balasubramanya R (2011) Role of biotechnology in the treatment of polyester fabric. Indian J Microbiol 51(2):117–123
73. Yavuz M, Kaya G, Aytekin Ç (2014) Using Ceriporiopsis subvermispora CZ-3 laccase for indigo carmine decolourization and denim bleaching. Int Biodeterior Biodegradation 88:199–205

Enzyme Applications in Textile Industry: A Step Toward Sustainable Development Goals

Hamid Masood, Samiullah Khan, Shujaul Mulk Khan, Ali Akbar, Atiq Ur Rehman, Ubaid Ur Rahman, Nauman Khan, Khurshid Ahmad, Gulnar K. Tulindinova, and Abdullah Abdullah

Abstract A significant portion of developing nations' national economies is derived from the textile and apparel industries. With 110 million metric tons of fiber produced globally in 2018, this industry ranked fourth in the globe. These textile producers provide goods valued at millions of dollars to industrialized nations annually. However, harsh and costly chemicals are used by textile firms all over the world to process fibers, endangering both the environment and public health. As a result, ecologically friendly alternatives must be incorporated into the manufacturing process right away. Utilizing biotechnological products such as different enzymes in textile production as a sustainable substitute for conventionally hazardous chemicals. Enzymes have the potential to significantly impact the textile industry's transition into a completely sustainable market that produces no waste or pollution. Globally recognized companies in the enzyme production industry include DSM, Novozymes, Amano Enzymes Inc., and DuPont. The textile industry will benefit from the use of

H. Masood · S. Khan (✉) · A. U. Rehman · U. U. Rahman · N. Khan
Department of Microbiology, Faculty of Biological Sciences, Quaid-I-Azam University, Islamabad 45320, Pakistan
e-mail: samikhan@qau.edu.pk

S. M. Khan (✉) · A. Abdullah
Department of Plant Sciences, Faculty of Biological Sciences, Quaid-I-Azam University, Islamabad 45320, Pakistan
e-mail: shuja@qau.edu.pk

S. M. Khan
Pakistan Academy of Sciences, Islamabad, Pakistan

A. Akbar
Center for Microbiology and Biotechnology, University of Sawat, Swat, Pakistan

K. Ahmad
State Key Laboratory of Marine Food Processing and Safety Control, College of Food Science and Engineering, Ocean University of China, No. 1299, Sansha Road, Qingdao, Shandong Province 266404, P.R. China

G. K. Tulindinova
Department of Biology, Pavlodar Pedagogical University Named After. A. Margulan, Pavlodar, Kazakhstan

M. Arshad (ed.), *Enzymes in Textile Processing: A Climate Changes Mitigation Approach*, SDGs and Textiles, https://doi.org/10.1007/978-981-97-8058-7_2

enzymes to keep straggling clothing out of landfills. In this case, textile materials might also be biodegraded by enzymes, producing raw materials that could be utilized to recycle textile goods. We can protect our world from the buildup of harmful and toxic waste in the environment by using enzymes.

Graphical Abstract: Enzymatic Sustainability in Textile Manufacturing

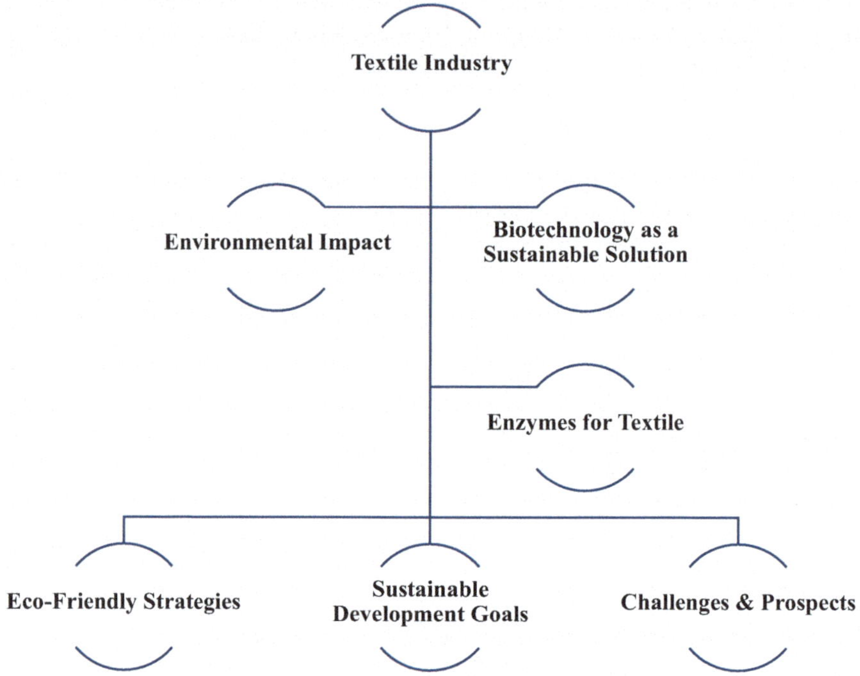

Keywords Textile · Pollution · Enzymes · Zero waste · Zero pollution · Sustainable market

1 Introduction

The human population is predicted to increase dramatically and reach approximately 8.5 billion people by 2030 [44]. Demand for necessities including food, health care, education, and other necessities rises. Humans employ a variety of methods and approaches to meet needs, provide for sustainability, and complete requirements. Among those tactics, supply chain management techniques that are environmentally friendly and sustainable are vital for commercial operations [11]. The major 17 sustainable development goals (SDGs) were the first initiative of the United Nations' Agenda for Sustainable Development 2030, which was introduced in 2016. The SDGs outline the most recent objectives for social, environmental, and economic advancements, including environmental management, economic growth, and the eradication of poverty [4]. The Sustainable Development Goals (SDGs) encourage everyone on the planet to contribute to the goals, including public and commercial organizations, governments, and civil society. Up to 12 trillion dollars in sustainable commercial and economic capacities can come from private ownership. Two large fashion firms, Inditex and H&M, responded to the UN SDGs quite effectively by adopting strict measures for each objective. Three H&M intended to limit its manufacturing to using recycled or sustainable materials, while Inditex closely adhered to the code of conduct for the sake of the health and welfare of its employees.

When compared to other industries, the textile sector has a highly specialized supply chain that is marked by a lengthy value chain, labor-intensive production, and a high level of environmental pollution [11]. When it comes to ecology, the textile sector releases dangerous contaminants into the air. However, it uses a lot of resources, such as water, electricity, wool, cotton, and linen. The textile sector, which uses 72 different types of chemically hazardous pollutants during the dyeing process, is responsible for between 17 and 20% of water contamination, according to World Bank statistics.

Depending on the kind of textile produced, different textile industries have different effects on climate change. Nonetheless, the frequent requirement for final checks and changes poses a considerable environmental concern within the textile production chain. These corrective cycles entail labor-chemical-intensive procedures and have caused the textile sector to become one of the major producers of greenhouse gas emissions globally, making approximately as much as 10% of total emissions. Because 119 million tons of textiles are produced worldwide, each kilogram of plastic-based fabric emits roughly 11.9 kg of CO_2, or 1291 million tons of CO_2 overall (or 7.3 trillion kilometers driven by car, to put this into perspective) [19, 57].

Textile and fashion sustainability has become a critical factor for both consumers and industry due to growing awareness of the impact purchases have on the environment [10, 21]. Many eco-friendly strategies are being researched and developed to make the fashion and textile industries more ecologically friendly. These consist of the following:

1. Using other sources of fiber staple, such as bamboo, to mitigate the disadvantages of conventional cotton farming.
2. Making use of natural colors and dyes.
3. Replacing water in cleaning and reduction procedures with supercritical CO_2.
4. Manufacturing high-quality, long-lasting textiles.
5. Putting in place heat recovery systems, in which steam boilers or stentering frames provide the energy needed to heat water throughout subsequent stages of the process.
6. Reusing groundwater of the same quality after using it for cooling processes.

2 Textile Biotechnology

Textile biotechnology is an emerging field, it utilizes advancements in biotechnology to transform the textile industry by offering eco-friendly solutions, innovative materials, and improved performance [9]. In response to the growing global focus on sustainable practices, textile biotechnology tackles urgent issues like resource depletion and pollution through innovative approaches to material development and manufacturing. This interdisciplinary field includes bioengineered fibers, fabrics, and dyes, leading to a new era of textile production that is environmentally sustainable. Textile biotechnology will certainly influence the future of the industry because of consumers' growing demand for high-quality, sustainable products [13, 52].

These days, the widespread use of cutting-edge and novel technical applications is changing our lives more and more. The textile industry may save costs, safeguard the environment, address health and safety issues, and enhance quality and functionality with the help of biotechnology, a cutting-edge and rapidly evolving field of science. Biotechnology has a great chance to replace conventional textile manufacturing, particularly with the establishment and implementation of ever-stricter laws and regulations on wastewater disposal. Current advancements in textile biotechnology are highlighted, along with the following areas where biotechnology may become more significant in the textile industry: The creation of novel textile fibers and polymers for use in the textile industry; the use of enzyme technology in textile processes; and the use of biotechnology to remediate textile effluents [55].

3 Microbial Enzymes in the Textile Industry

In textile industries, enzymes are used to facilitate the creation of fiber processing technologies that are environmentally benign and to improve the quality of the final product [15]. Due to their many benefits over the chemicals that were previously utilized, enzymes are the most ideal chemical substitutes for usage in the textile industry. Enzymes are safer than harmful chemicals, simpler to manage, and biodegradable, speeding up the tempo of the reaction in addition to being highly substrate specific [38, 45]. Microbial enzymes offer a far superior and more ecologically beneficial choice. The two main uses of enzymes in the textile industry are the softening and bio-stoning of cotton products (cellulases) and the de-sizing (amylases) of textile products. Lipases, pectinases, catalases, proteases, xylanases, and other enzymes are utilized in the textile processing industry (Fig. 2.1). These enzymes are employed in processes such as peroxide removal, wool finishing, fading of denim and non-denim, bio-scouring, and bio-polishing [8].

The two main types of enzymes employed in the pre-treatment and finishing of cotton are hydrolase and oxidoreductase. Hydrolases such as amylase, cellulase, cutinase, protease, pectinase, and lipase/esterase are employed in various

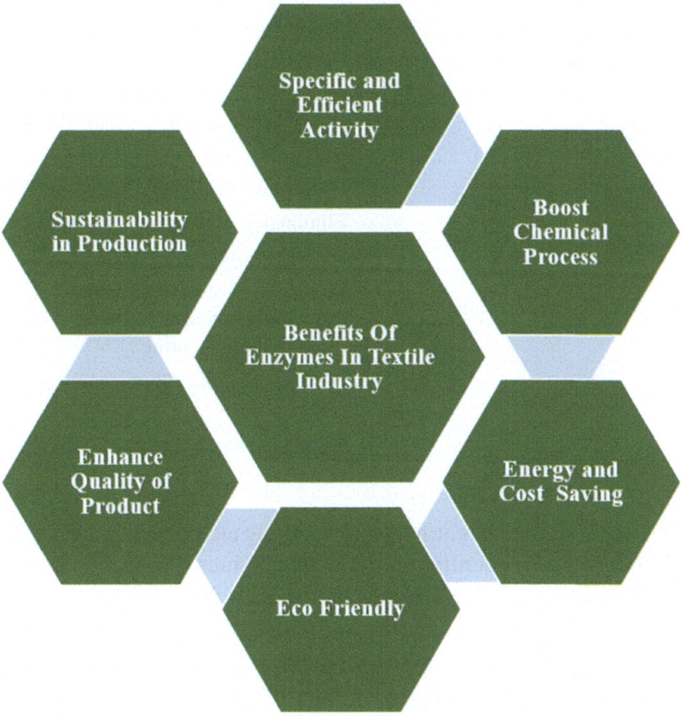

Fig. 2.1 Applications of enzymes in the textile industry

processes such as fabric biopolishing and bioscouring, wool anti-felting, cotton softening, denim finishing, wool desizing, and synthetic fiber finishing [14]. Textile wastewater pretreatment methods, whether physical or chemical, are costly, highly energy-intensive, and inefficient for the environment [2]. They also make dangerous mud. Therefore, as opposed to physicochemical processes, employing microbial approaches to destroy textile dyes will be advantageous for the environment and may be economical. Two examples of cheap, easily obtainable, easily processed, and easily mobilizable microbial enzymes are laccase and azoreductase [50, 53]. The search for bacteria that may degrade the azo dye began in 1970 with the finding of three strains: Bacillus subtilis, Aeromonas hydrophila, and Bacillus cereus [16].

4 Applications of Microbial Enzymes in the Textile Industry

The textile industry currently uses hydrolases and, to a lesser extent, oxidoreductase. The use of alpha-amylases in cotton de-sizing has long been regarded as highly developed. Furthermore, cellulases, hemicellulases, lipases, pectinases, and catalases are

Table 2.1 Major enzymes used in the fashion and textile industry

Important enzymes used in the textile industry	Uses
Amylase	Desizing
Cellulases and hemicellulases	Desizing of CMC Biostoning of jeans For the stylish effect on cellulose fibers
Lipases	Elimination of waxes and fat particles
Laccases	Bio-bleaching
Proteases	Modification of wool properties Degumming of silk Animal fibers scouring
Pectinases	Bast and vegetable fibers scouring, e.g., jute and cotton
Peroxidases	Wood pulp bio-bleaching

used in a variety of cotton pre-treatment and finishing techniques [33, 35, 37]. Furthermore, natural fibers are frequently treated with enzymes. Proteases are employed in felt-free wool finishing [20], sericinases are used in the enzymatic degumming of silk [23], and cellulases and xylanases are in charge of softening jute [32]. When compared to more conventional, non-enzymatic methods, the use of enzymes has various advantages. Following the typical pre-treatment of amylases to desize raw cotton textiles enzymatically, these chemicals are removed with a strong alkaline treatment conducted at high temperatures (Table 2.1). This general alkaline scouring process can damage cellulosic material and consume a lot of energy, water, and alkali [28, 39].

4.1 Microbial Enzymatic Desizing

To lubricate and save the yarn from friction and stop the fiber from breaking in the time of weaving, the warp fibers covered with an adhesive material for textiles composed of cotton or mixes are known as "size." After weaving, the cotton's natural sizing agent and other non-cellulosic components must be taken out to make the cloth ready for dyeing and finishing. Desizing is the process of eliminating the starch-origin shielded coating (size) [3, 18, 59]. To provide more effective and suitable wet processing, amylases are employed in the fashion and textile industries to remove size from starch material. The hydrolytic enzyme amylase breaks down nutritional starch into two short-chain sugars, maltose and dextrin. Because of their starch-specificity, these enzymes can remove starch from materials without causing any harm to the support fabric [17]. An amylase can be employed for the desizing process at low temperatures (30–60 °C) and an optimum pH range (5.5–6.5) [12]. Neutral cellulase is added to amylase to improve wettability, decrease stiffness, and improve dye absorption. According to [34], the cloth is softer, handling enzymes is safer, and there are no adverse effects on machinery.

4.2 Degumming of Silk

Papain is used to degum silk and boil off cocoons. To get rid of a proteinaceous covering sericin on the fiber in raw silk, it must be degummed. An alkaline solution including soap is mostly used for degumming, which is a severe procedure that also destroys the arrangement of the fibrin. As degumming agents, several acidic, alkaline, and neutral proteases have been investigated because they can dissolve sericin but not the protein in silk fiber. Although this is not in common usage, alkaline proteases are the most effective in eliminating sericin and enhancing silk exterior qualities like smoothness, shine, and handle [7].

4.3 Bioscouring

The main cell walls of cellulosic fibers used to create textiles contain non-cellulosic impurities like natural colorants, pectins, proteins, lipids, waxes, and minerals. These decrease finishing and dyeing productivity, which leads to a soiled look and poor wettability [47, 60]. Therefore, cotton yarn or fabric must be prepared to eliminate substances that prevent dye from binding before it can be colored. Scouring is a process that increases the fabric's wettability and often makes use of alkalis like sodium hydroxide. These compounds also destroy the sensitive cellulose material, which reduces the strength and lightens the weight of the cloth. Enzymatic or bioscouring preserves practically all of the cellulose structure, preserving the weight and strength of the material. The cellulose structures under the cuticle membrane are broken down by the enzyme cellulases produced by *Trichoderma reesei* and *Clostridium thermocellum* [31]. While pectinase can harm cotton cuticle structure by dissolving the link between the cotton fiber body and the cuticle and breaking down pectin, cellulase can harm cotton cuticle structure by digesting the main wall cellulose directly beneath the cotton cuticle [41].

Wax, pectin, and non-cellulosic contaminants are broken down and eliminated from cotton by pectinases made by Tetracladium species and Aspergillus niger. This is done by accelerating the 1,4-D-galactosiduronic unsymmetrically hydrolysis connection in pectin compounds. This breaks down the cotton fibers' cuticle layer. Since the overall process is substrate-specific, cellulosic components are unaffected [54]. Proteinaceous impurities from the cellulose fibers are hydrolyzed by proteases obtained from *Pseudomonas aeruginosa, Proteus, Bacteroides fragilis, Propionibacterium acnes* and *Staphylococcus aureus* [6]. Cutin and wax are broken down by hydrolytic enzymes called cutinases, which are generated by *Pseudomonas putida, Streptomyces scabies* and *Fusarium solani* [24].

Compared to conventional scouring, bioscouring offers many benefits, such as being more affordable and safe for the environment, as well as requiring fewer chemicals overall, and posing less of a danger of accidents. Compared to alkaline scouring (100%) bioscouring has much lower chemical oxygen demand (COD), biological oxygen demand (BOD), and total dissolved solids (TDS) [47]. Additionally, the finish is much softer after bioscouring than it is after alkaline scouring, which has a harsher texture. Additionally, bioscouring has no detrimental effects on the environment, the fabric, or the employees' health [41].

4.4 Biobleaching (Enzymatic Bleaching)

Cotton fibers are bleached to remove the flavanoids' inherent coloration and turn them white so that they may subsequently be colored to match consumer preferences. The most popular textile-bleaching agent is hydrogen peroxide [1, 46]. The hydrogen peroxide must then be removed from the clothes using a lot of water since the high range of temperature and an alkaline pH can extremely damage the fibers. Therefore, using an enzymatic bleaching technique (also known as biobleaching) instead of hydrogen peroxide would provide a higher-quality product. A suitable combination of enzyme systems can be used in place of this process. The pH and temperature ranges of the selected amyloglucosidases, pectinases, and glucose oxidases are suitable for their active states. Laccases oxidize flavonoids, which makes cotton whiter.

Originally, [42] showed that T. hirsuta laccase was responsible for improving cotton's whiteness, possibly through the oxidation of flavonoids. More recently, a combination ultrasoundlaccase therapy for cotton bleaching was proposed by [26]. The advantage of laccase is its reaction specificity with the indigo pigment used in denim, preserving the original raw white of the weft yarn while imparting a deeper hue to the woven fabric that bleaching with hypochlorite cannot. Other dyes are unaffected by this enzyme. Enzymes are used in the process, therefore there is little chance of environmental damage or hazardous waste discharge [29].

A lot of research has been done on laccases and their ability to decolorize textile effluents. Because laccases can break down colors with a range of structures, including synthetic dyes, they offer an environmentally benign way to remediate dye effluent [29].

4.5 Biowashing of Denims

Most clothing is given a wash treatment to give it a worn-in appearance, such as stone-washing denim jeans. The abrasive action of pumice stones during denim stone washing causes blue denim to fade in jeans. Cellulases have reduced or eliminated stone usage in jeans production, reducing harm to machines and clothing, and reducing pumice dust in the laundry environment. This also enhances productivity by reducing stone presence in laundry machines [43].

Denims were traditionally treated using pumice stones made of sodium hypochlorite and potassium permanganate [46]. These techniques have several drawbacks, such as fabric wear and tear and back-staining. These factors led to the usage of cellulases in denim washing, combined with the enormous amounts of pumice stones required. The indigo dye is loosened by cellulases during bio-stone washing. The small amount of enzyme needed is a benefit. Denim washing can also be done using laccase. More recently, several writers demonstrated that employing a mediator or not, laccase which are efficient agent for the stonewashing result of denim fabric. [29].

4.6 Biopolishing

Biopolishing is the fuzz fibers and pilling up of threads removal. The final procedure is to enhance the quality of cloth polishing. Synthetic cellulose fibers, natural and cotton can benefit from the 'biostoning' procedure as well as an enzymatic process termed 'biopolishing'. The key benefit of this method is that pilling is avoided. In the textile industry, a fuzz ball is referred to as a "pill." These have an impact on garment quality because they make the cloth seem ugly and tangled.

Cellulases are most used and obtained from *Aspergillus, Trichoderma* species. and *Penicillium* [51]. The small fibrils (fuzz or hairs) that protrude from the exterior of yarn are particularly vulnerable to enzyme assault, therefore cellulases hydrolyze them. The microfibrils become weaker as a result, and they are more likely to separate from the fiber's main structure and create a smoother exterior on the yarn (Yu et al. 2022).

The polishing of cotton textiles is not necessary to improve the fabric. For the fiber lyocell, developed in 1991, this phase is crucial. It is constructed of wood pulp and has a propensity to quickly fibrillate when moist (the fiber's surface fibers peel apart). Finished lyocell clothing will become coated in pills if they are not taken out. When Lyocell fabric is finished, cellulases are applied to prevent fibrillation as well as to improve its smooth look.

5 Treatment of Wool and Leather

According to studies conducted on wool textiles, the use of Transglutaminases (TGs) for treatment increases dye uptake, handle, softness, tensile strength retention, wettability, resistance, shrink, and protection from common detergent-related damage [48]. The following dyeing and color qualities of leather are improved when leather is treated with TG in combination with keratin or casein [58]. Although it is still in the research stage, the use of TG for wool treatment and leather seems to be the best method.

6 Effluent Treatment

The expansion of the fashion and textile industry also increases the aquatic pollution level due to uncontrollable chemicals and dye contaminants at the dye concentration level of (10–200 mg/L) and non-biodegradable and additional pollutants such as detergents, surfactants, dispersants, acids, bases, salts, oxidants, and humectants. Thus, it is imperative to effectively treat textiles before releasing them into the environment [3] (Yaun et al. 2020).

Many of them are synthetic pigments such as reactive orange-16, which can be decolored by ligninases like peroxidases and laccases (Manganese&Lignin Peroxidase) from white rot fungi like *Chrysosporium*, *Phanerochate,* and *Serratia marcescens* bacteria [22].

7 Enzyme's Role in Sustainable Development Goals

Notably, the use of enzymes has not yet been properly integrated into these methods. However, enzymes have a great deal of potential to contribute significantly to the transition of the textile industry into a zero-waste, zero-pollution, totally sustainable market. Sustainable enzyme applications used throughout the manufacturing process have the potential to supplant conventional textiles [30, 36].

Enzymes can reduce the gross carbon footprint of fiber spinning, solvent cleaning, dying, washing, and fabric finishing, according to life cycle assessments. A single gram of enzyme can reduce carbon dioxide emissions by 0.42 to 1.0 kg per kg of dry matter basis of yarn. This can be applied to the 119 million tons of textiles produced globally, which means that between 50 and 119 million tons of CO_2 emissions can be avoided [19, 27, 40, 57].

The amount that enzyme synthesis contributes to the overall carbon footprint of textile bioprocessing depends on several variables, including transportation, raw material sources, and manufacturing techniques. Recent life cycle studies, however, have discovered low values of at least 8.9 g of CO_2 equivalent for each gram of enzyme, from cradle to gate in the life cycle (Set al. 2021). These results highlight the potential benefits of enzymes in the bioprocessing of fashion and textiles, considering their remarkably low environmental impact.

The solution to the problem of non-biodegradable clothing in landfills lies in enzymes as well. Existing textile materials can be biodegraded by enzymatic processes, allowing for their repurposing in the creation of new recycled textiles. In contrast to chemical procedures, recycling can be accomplished with a notable reduction of 5.5 kg of CO_2 per kg of textile material by using enzymes. Enzymes may be used to recycle all fashion and textiles, which weigh 119 million tons. This would save about 655 million tons of CO_2 emissions or 3.7 trillion kilometers driven by cars.

Developing new enzymes that can alter textile processes is the key to making sustainable clothing a commercial product rather than merely a commitment to lessening environmental effects and minimizing climate change. The most recently created enzymes can specifically target important processes that lead to the manufacture of fabric, such as the dyeing process optimization and the elimination of leftover spinning oils and sizing products. Water-based, low-temperature processes that use less energy and water discharge are made easier by enzymes.

Enzymes have broader positive effects on biodiversity and the health of the planet beyond their impact on GHG emissions. Decarbonization requires methods that support the recycling of carbon from renewable sources, and they help to reduce the consumption of chemicals in these processes. Moreover, enzymes facilitate biological degradation processes that lessen pollutants and restore contaminated areas. This is important because environmental pollution has a detrimental impact on ecosystems and biodiversity, both of which are essential to preserving the health of the planet. Enzymes are a component of natural remedies and closed-loop biological systems that can aid in reducing atmospheric pollution and chemical discharge, thereby enhancing biodiversity. This strategy is in line with initiatives to improve environmental management, such as the Natura 2000 network. The search for novel enzymes that can generate biobased compounds and help remove pollutants is crucial to protecting biodiversity and the health of the earth.

8 Challenges and Obstacles Faced by the Textile Biotechnology Sector

Textile biotechnology is gaining ground in many areas of the textile industry but there are still some obstacles and bottlenecks which need particular attention to address and come up with practical strategies and approaches for its long-term solutions. Thus, there has been limited success in utilizing available commercial enzyme preparations. The method has been noted to be extremely slow, which makes it economically unviable for modern applications. This is mostly related to the persistent properties of several components.

9 Expert Viewpoint and Perspective in Textile Biotechnology

In the Fashion and textile industry biotechnology, especially enzyme technology provides sustainable and environmentally friendly alternatives to chemical processes, especially in the desizing and contaminant removal processes. In the conversion of starch into sugars many commercial enzymes are in practice, but they can also increase pH levels, making it difficult to maintain a consistent temperature. It is desirable to use such enzymes that can show excellent activity throughout a larger pH and temperature range.

To improve fiber preparation, the textile industry has optimized the combinations of enzymes (amylase, pectinase, cellulase, and lipases). Hydrogen peroxide (H_2O_2) and sodium hydroxide (NaOH) are both frequently employed for bleaching, even though the concentration of H_2O_2 may not be enough to attain the specified whiteness levels. Gox enzyme and chemical H_2O_2 may function well together and are acceptable for the environment.

Another expert perspective on the benefits of using enzymes in the textile sector says that while there is little knowledge on how to use immobilized enzymes in the textile industry, this limits their potential. The difficulty of recycling enzymes increases prices. Enzymatic techniques are less harmful to the environment than traditional chemical processes, but further research, including lifecycle analyses, is required to fully understand their true environmental impact on the textile industry. So, lifecycle assessment is another important method of quantitative analysis that gives full information for making decisions [56]. Moreover, traditional plant-based items are suitable alternatives to various products causing environmental issues [5].

10 Prospects in Textile Sustainability by Using Biotechnology

In the twenty-first century, textile biotechnology particularly enzyme technology—has received increased interest. Through genetic engineering, new or altered enzymes with bitter and better qualities are created. Genetically modified enzymes of commercial significance are undoubtedly predicted to proliferate in many textile industry areas, eventually displacing current chemical or mechanical processes. To sum up, the application of biotechnology in the textile industry presents a workable way to achieve greener and more sustainable production techniques, especially when contrasted with extremely harmful traditional methods. Future predictions indicate that the textile industry will gain even more from biotechnology.

References

1. Arshad M (2010) Bioethanol: a sustainable and environment friendly solution for Pakistan. A Sci J COMSATS–Science Vis 2011:16–7
2. Arshad M, Adil M, Sikandar A, Hussain T (2014) Exploitation of meat industry by-products for biodiesel production: Pakistan's perspective. Pakistan J Life Soc Sci 12:120–125
3. Arshad M (2017) Clean and sustainable energy technologies. In: Clean energy for sustainable development. Academic Press, pp 73–89
4. Arshad M, Ansari AR, Qadir R, Tahir MH, Nadeem A, Mehmood T, Alhumade H, Khan N (2022) Green electricity generation from biogas of cattle manure: an assessment of potential and feasibility in Pakistan. Front Energy Res 25(10):911485
5. Abdullah Khan SM, Haq ZU, Khalid N, Ahmad Z, Ejaz U (2022) Utilization of three indigenous plant species as alternative to plastic can reduce pollution and bring sustainability in the environment. In: Natural resources conservation and advances for sustainability. Elsevier, pp 533–544
6. Al-Dhabi NA, Esmail GA, Ghilan AKM, Arasu MV, Duraipandiyan V, Ponmurugan K (2020) Characterization and fermentation optimization of novel thermo stable alkaline protease from Streptomyces sp. Al-Dhabi-82 from the Saudi Arabian environment for eco-friendly and industrial applications. J King Saud Univ-Sci 32(1):1258–1264
7. Arami M, Rahimi S, Mivehie L, Mazaheri F, Mahmoodi NM (2007) Degumming of Persian silk with mixed proteolytic enzymes. J Appl Polym Sci 106(1):267–275
8. Barrett AJ, Woessner JF, Rawlings ND (eds) (2012) Handbook of proteolytic enzymes, vol 1. Elsevier
9. Bano I, Ghauri MA, Arshad M, Yasin T, Younus M (2017) Bioactivity of variant molecular weight chitosan against drug-resistant bacteria isolated from human wounds. Microb Drug Resist 1;23(8):958–65
10. Bano I, Arshad M (2018) Climatic changes impact on water availability. Perspect Water Usage Biofuels Prod: Aquat Contam Clim Chang 39–54
11. Cai YJ, Choi TM (2020) A United Nations' sustainable development goals perspective for sustainable textile and apparel supply chain management. Transp Res E Logist Transp Rev 141:102010
12. Cavaco-Paulo A, Gubitz G (eds) (2003) Textile processing with enzymes, vol 29. Woodhead publishing
13. Chatha SAS, Asgher M, Iqbal HM (2017) Enzyme-based solutions for textile processing and dye contaminant biodegradation—a review. Environ Sci Pollut Res 24:14005–14018
14. Chen S, Su L, Chen J, Wu J (2013) Cutinase: characteristics, preparation, and application. Biotechnol Adv 31(8):1754–1767
15. Choi JM, Han SS, Kim HS (2015) Industrial applications of enzyme biocatalysis: current status and future aspects. Biotechnol Adv 33(7):1443–1454
16. Dave SR, Patel TL, Tipre DR (2014) Bacterial degradation of azo dye containing wastes. Microbial degradation of synthetic dyes in wastewaters. Springer International Publishing, Cham, pp 57–83
17. Etters JN, Annis PA (1998) Textile enzyme use: a developing technology. Am Dye Report 87(5):18–23
18. Feitkenhauer H, Fischer D, Fäh D (2003) Microbial desizing using starch as model compound: enzyme properties and desizing efficiency. Biotechnol Prog 19(3):874–879
19. Filho WL, Ellams D, Han S et al (2019) A review of the socio-economic advantages of textile recycling. J Clean Prod 218:10–20
20. Fornelli S (1994) Enzymatic treatments of proteic fibers. Melliand Textilber 75:120–125
21. Gazzola P, Pavione E, Pezzetti R et al (2020) The perception of sustainabilityand circular economy: a gender/generation quantitativeapproach. Sustain 12:2809
22. Gondaliya N, Parikh SC (2014) Statistical analysis of variable parameters and its effect on reactive dyes decolorization. Microbes 8

23. Gulrajani ML (1992) Degumming of silk. Rev Prog Color Relat Top 22(1):79–89
24. Gururaj P, Khushbu S, Monisha B, Selvakumar N, Chakravarthy M, Gautam P, Nandhini Devi G (2021) Production, purification and application of Cutinase in enzymatic scouring of cotton fabric isolated from acinetobacter baumannii AU10. Prep Biochem Biotechnol 51(6):550–561
25. Hao J, Song F, Huang F, Yang C, Zhang Z, Zheng Y, Tian X (2007) Production of laccase by a newly isolated deuteromycete fungus Pestalotiopsis sp. and its decolorization of azo dye. J Ind Microbiol Biotechnol 34(3):233
26. Hassan MM, Saifullah K (2023) Ultrasound-assisted sustainable and energy efficient pretreatments, dyeing, and finishing of textiles–a comprehensive review. Sustain Chem Pharm 33:101109
27. Jegannathana KR, Nielsen PN (2013) Environmental assessment of enzyme use in industrial production-a literature review. JClean Prod 42:228–240
28. Kakkar P, Wadhwa N (2022) Extremozymes used in the textile industry. J Text Inst 113(9):2007–2015
29. Khaled JM, Alyahya SA, Govindan R, Chelliah CK, Maruthupandy M, Alharbi NS, Li WJ (2022) Laccase-producing bacteria influenced the high decolorization of textile azo dyes with advanced study. Environ Res 207:112211
30. Khan R, Jan HA, Shaheen B, Abidin SZU, Ullah A, Basit A, Khan SM (2023) Biofertiliser from animal wastes. In: Climate changes mitigation and sustainable bioenergy harvest through animal waste: sustainable environmental implications of animal waste. Springer Nature Switzerland, Cham, pp 413–429
31. Kuhad RC, Gupta R, Singh A (2011) Microbial cellulases and their industrial applications. Enzym Res
32. Kundu AB, Ghosh BS, Chakrabarti SK, Ghosh BL (1991) Enhanced bleaching and softening of jute by pretreatment with polysaccharide degrading enzymes. Text Res J 61(12):720–723
33. Liu X, Kokare C (2023) Microbial enzymes of use in industry. In: Biotechnology of microbial enzymes. Academic Press, pp 405–444
34. Madhu A, Chakraborty JN (2017) Developments in application of enzymes for textile processing. J Clean Prod 145:114–133
35. Marcher D, Hagen HA, Castelli S (1993) Entschlichten mit enzymen. ITB Veredlung 39(3):20−32
36. Masood H, Khan SU, Khan SM, Nawaz A, Wajid SH, Rehman AU, Abdullah A (2023) Lifecycle and risk assessment of animal manure utilization. In: Climate changes mitigation and sustainable bioenergy harvest through animal waste: sustainable environmental implications of animal waste. Springer Nature Switzerland, Cham, pp 241–262
37. Meyer-Stork LS (2002) Enzymanwendungen in der Textilindustrie: Möglichkeiten,-Grenzen—Potenziale. Maschen-Industrie 52(5):32–35
38. Mojsov K (2011) Application of enzymes in the textile industry: a review
39. N N (2002) Reference document on best available techniques for the textiles industry. In: European integrated pollution prevention and control (IPPC) Bureau. European Commission
40. Nielsen PH, Kuilderd H, Zhou W et al (2009) Enzyme biotechnology for sustainable textiles. In: Blackburn RS (ed) SustainableTextiles. Elsevier, Woodhead Publishing Series in Textiles, Amsterdam, pp 113–138
41. Pawar SB, Shah HD, Andhorikar GR (2002) Enzymatic processing of cotton: a bio-technical approach. Man Made Text India 45(4):133–138
42. Pereira L, Bastos C, Tzanov T, Cavaco-Paulo A, Gübitz GM (2005) Environmentally friendly bleaching of cotton using laccases. Environ Chem Lett 3:66–69
43. Periyasamy AP, Periyasami S (2023) Critical review on sustainability in denim: a step toward sustainable production and consumption of denim. ACS Omega 8(5):4472–4490
44. Priya, Berry S, Agrawal PR (2024) Sustainable textile production in the developing countries: Major challenges or concerns and their possible solutions. In: Dye pollution from textile industry: challenges and opportunities for sustainable development (pp 263-278). Singapore, Springer Nature Singapore

45. Qasim M, Bukhari SA, Ghani MJ, Masoud MS, Huma T, Arshad M, Haque A, Ibrahim Z, Javed S, Rajoka MI (2016) Relationship of oxidative stress with elevated level of DNA damage and homocysteine in cardiovascular disease patients. Pak J Pharm Sci 2:29
46. Rahman M, Hack-Polay D, Billah MM, Un Nabi MN (2020) Bio-based textile processing through the application of enzymes for environmental sustainability. Int J Technol Manag Sustain Dev 19(1):87–106
47. Rajulapati V, Dhillon A, Kumar Gali K, Katiyar V, Goyal A (2020) Green bioprocess of degumming of jute fibers and bioscouring of cotton fabric by recombinant pectin methylesterase and pectate lyases from Clostridium thermocellum. Process Biochem 92:93−104
48. Rather LJ, Akhter S, Hassan QP, Mohammad F (2018) Biofunctionalization of various textile materials using enzyme biotechnology as a green chemistry alternative. Enzym Food Technol: Improv Innov 263–276
49. Salman M, Adeel S, Batool F, Usama M, Hassan A, Ozomay M, Hosseinnezhad M (2022) Sustainable application of flavonoid-based natural colorants. Text Dye Pigment: Green Chem Approach 295–321
50. Sarkar S, Banerjee A, Halder U, Biswas R, Bandopadhyay R (2017) Degradation of synthetic azo dyes of textile industry: a sustainable approach using microbial enzymes. Water Conserv Sci Eng 2:121–131
51. Shabbir M, Ahmed S, Sheikh JN (eds) (2020) Frontiers of textile materials: polymers, nanomaterials, enzymes, and advanced modification techniques. John Wiley & Sons
52. Shanmugasundaram OL, Pandit P (2022) Conclusion on biotechnology in textile production and future scope. In: Applications of biotechnology for sustainable textile production. Woodhead Publishing, pp 241–249
53. Sharif N, Munir N, Hasnain M, Naz S, Arshad M (2021) Environmental impacts of ethanol production system. Sustain Ethanol Clim Chang: Sustain Assess Ethanol Distilleries 2021:205–223
54. Singh A, Varghese LM, Battan B, Patra AK, Mandhan RP, Mahajan R (2020) Eco-friendly scouring of ramie fibers using crude xylano-pectinolytic enzymes for textile purpose. Environ Sci Pollut Res 27:6701–6710
55. Singh AK, Iqbal HM, Cardullo N, Muccilli V, Fern'andez-Lucas J, Schmidt JE, Bilal M (2023) Structural insights, biocatalytic characteristics, and application prospects of lignin-modifying enzymes for sustainable biotechnology—a review. Int J Biol Macromol 124968
56. Tochetto GA, Aragão AMI, de Oliveira D, Immich APS (2022) Can enzymatic processes transform textile processes? a critical analysis of the industrial application. Process Biochemistry
57. United Nations Economic Commission for Europe/Food andAgriculture Organization of the United Nations (UNECE/FAO).Forests for Fashion (2022). https://unece.org/forests/forests-fashnion. Accessed 2 Dec 2022
58. Wu X, Liu A, Wang W, Ye R (2018) Improved mechanical properties and thermal-stability of collagen fiber based film by crosslinking with casein, keratin or SPI: effect of crosslinking process and concentrations of proteins. Int J Biol Macromol 109:1319–1328
59. Wu XT, Ansari AR, Pang XX, Li HZ, Zhang ZW, Luo Y, Arshad M, Song H (2019) Visfatin plays a significant role in alleviating lipopolysaccharide-induced apoptosis and autophagy through PI3K/AKT signaling pathway during acute lung injury in mice. Arch Immunol Ther Exp 1(67):249–261
60. Zaini MS, Arshad M, Syed-Hassan SS (2023) Adsorption isotherm and kinetic study of methane on palm kernel shell-derived activated carbon. J Bioresour Bioprod 1;8(1):66–77

Enzymes in Textile—A Step Towards Sustainability

Uzma Hameed, Muhammad Tahir Hayat, Isha Fatima, and Makumburage Rusiru Chadeeswara Makumbura

Abstract The textile industry, one of the oldest sectors, has long been associated with significant environmental concerns, contributing to approximately 20% of water pollution. The textile industry is striving to mitigate its environmental impact. Thus, there is a growing industrial shift towards adopting environmentally friendly and sustainable technologies. A noteworthy development in this regard is the emergence of enzymes as a greener alternative to traditional chemical-based methods. This chapter provides an in-depth exploration of applications of key enzymes such as amylases, cellulases, laccases, lipases, pectinases, and proteases in the textile industry. These enzymes, alone or together, serve as sustainable alternatives in almost all the sectors of the textile industry as their applications span across dry and wet processes, finishing, and bioremediation of textile effluent. Ongoing trends, current challenges, and prospects surrounding the integration of enzymes as a sustainable approach within the textile industry are also reviewed. The chapter sheds light on how enzymes offer a promising avenue to achieve economical and eco-friendly processes. By delving into the nuances of enzyme applications and their potential, this chapter contributes to the broader understanding of sustainable practices in the textile industry.

Keywords Enzymes · Green catalysts · Textile manufacturing process · Wet processing · Advantages

1 Introduction

Enzymes are biocatalysts that enhance the rate of reaction by lowering the energy of activation. Enzymes govern all the metabolic reactions in living organisms. According to the International Commission of Enzymes (EC) established in 1956, enzymes are divided into 7 classes, i.e., oxidoreductases, transferases, hydrolases, lyases, isomerases, lignases, and translocases [32]. Noteworthy attributes of enzymes

U. Hameed (✉) · M. T. Hayat · I. Fatima · M. R. C. Makumbura
Institute of Industrial Biotechnology, Government College University Lahore, Lahore, Pakistan
e-mail: uzmahameed@gcu.edu.pk

M. Arshad (ed.), *Enzymes in Textile Processing: A Climate Changes Mitigation Approach*, SDGs and Textiles, https://doi.org/10.1007/978-981-97-8058-7_3

include high activity, specificity, and enantio-selectivity towards a wide variety of non-native substrates. Since the discovery of the very first enzyme "diastase" in 1883, by the French chemist Anselme Payen [57, 171], enzymes have emerged as robust and sustainable biocatalysts. Enzymes have been proven to be of utmost importance for diverse industries and processes such as food, feed, paper & pulp, pharmaceutical, organic synthesis, detergent, leather, textiles, bioremediation, cosmetics, and personal care. [29, 32]. The global industrial enzyme market is continuously expanding and is expected to reach 90,028.41 million US$ by 2028.

The global textile industry, one of the oldest industries in human history, is a key player in international trade and has witnessed significant growth attributed to the escalating population [139]. The textile industry stands on the cusp of a transformative era powered by biotechnological innovations, particularly for the use of enzymes. Enzymatic biotechnology heralds a new epoch of eco-friendly, cost-effective, and energy-efficient textile manufacturing, signaling a departure from traditional methods laden with hazardous chemicals. Enzymes such as amylases, lipases, cellulases, laccases, catalases, proteases, and pectinases are at the forefront, facilitating processes that once relied heavily on water, energy, and time-intensive practices [82].

Enzymes are being actively utilized or under exploration in nearly every significant segment of the textile industry with a current global textile market of 488.8 million USD that is forecasted to grow at a CAGR (compound annual growth rate) of 5.2% by 2032 [51]. Cellulases and amylases have already established commercial roles in textiles. Amylases are specifically employed for desizing processes, which involve the removal of starch-based sizing agents applied during yarn weaving [45, 70]. Lipases, known to enhance fiber properties like microstructure, gloss, dyeing, and fabric wettability, have also been investigated for their potential contribution to textile desizing along with amylases. Cellulases are increasingly being used in the textile as a sustainable alternative for fabric softening, polishing, and denim stone-washing. Similarly, pectinases have the potential to perform scouring. Catalase and xylanases have been reported for bleaching [9, 32, 74, 85]. In addition to detergent formulation, proteases are used in silk degumming. Along with pectinases, these can be employed for scouring during wet processing [9, 120]. In textile, laccases have the potential to be used for cotton bleaching, dyeing, coating, denim finishing, fabric surface modification, treatment of synthetic dyes, removal of detergents and odors during cloth-washing, and the bioremediation of effluents and degradation of natural dyes [36, 106]. Glucose oxidase, known for its capability to produce H_2O_2, can function as an eco-friendly bleaching agent [44]. Transglutaminases enhance the wool shrink resistance and improve the tensile strength [20, 155]). Other enzymes, such as nitrilases, lignases and cutinases have textile applications such as modification of textile chemicals, dyes, and textile waste treatment. These enzymes are also being investigated to develop novel processes for the textile [7, 136, 159].

Overall, enzymes, either alone or in combination, find extensive applications from dry processes to wet processes such as in silk degumming, wool and ramie fibers processing, modification of synthetic fiber, synthesis of dyes, and effluent treatment. The eco-conscious wave is also propelling the textile sector to embrac natural dyes and enzymes-based processes which play a pivotal role in enhancing sustainability and reducing the ecological footprint of synthetic dyes. This shift not only satisfies the increasing consumer demand for environmentally friendly products but also enriches textiles with additional qualities such as antioxidant properties, UV protection, and antimicrobial potential provided by plant-derived dyes [58].

2 Enzymes

Enzymes are remarkable and ubiquitous biocatalysts that orchestrate all intricate metabolic pathways in all living organisms [27]. Enzymes reduce activation energy to accelerate the rate of chemical reactions and remaining intact during that process. Most enzymes are proteins in nature and are highly substrate specific. Enzymes contribute to the maintenance of homeostasis, metabolism, and overall cellular functions and thus are indispensable to sustaining life. Enzymes are omnipotent and can be obtained from diverse living sources including microbes, plants, and animals [62, 92, 143]. The term enzyme, first coined by Kuhne in 1878, is derived from the Greek word "enzumé" meaning "in (en) yeast (zumé)". Enzymes are either proteins or catalytic RNA molecules known as ribozymes. Predominantly proteinaceous enzymes are high molecular weight three-dimensional protein complexes composed of single (monomer) or multiple (multimer) polypeptide chains.

Enzymes differ from each other depending upon specific protein sequences, substrate specificity, type of bond cleaved, and mechanism of cleavage. Enzymes may or may not require small protein and/or nonprotein molecules. For example, certain enzymes use bivalent metal ions (e.g., Mg^{2+}, Ca^{2+}, Cu^{2+}, Mn^{2+}, Fe^{2+}, etc.) to bind with the substrate or stabilize the enzyme–substrate complex. Enzymes are designed to work optimally at specific pH, temperature, and salt concentrations. These are susceptible to denaturation by high temperature, pH, radiations, and other factors that can disrupt the 3D protein structure [32, 163]. Enzymes are categorized into seven groups based on the type of reactions they catalyze and each group is assigned an EC (Enzyme Commission) number (Table 1) according to the International Commission of Enzymes [32, 164].

Enzymes are outstanding in terms of high activity, specificity, and enantioselectivity, towards a wide variety of non-native substrates. The specific substrate, which is usually much smaller than the size of the enzyme, binds with the help of a binding cleft to a specific site known as the "active site" of the enzymes containing catalytic residues (amino acid). The number and types of catalytic residues vary for each class of enzyme. The substrate binding at the active site induces conformational changes leading to swift conversion of substrate to the product due to the inherent ability of enzymes to lower the activation energy of the reaction (Fig. 1) [32, 54].

Table 1 Seven classes of enzymes with their catalyzed reactions

Enzyme Class	EC	Type of reaction catalyzed
Oxidoreductases	EC 1	Catalyze oxidation reduction reactions
Transferases	EC 2	Transfer a functional group from one substrate to another (methyl- acyl- amino- or phosphate group
Hydrolases	EC 3	Catalyze the hydrolysis of the bond
Lyases	EC 4	Cleave various bonds by means other than hydrolysis and oxidation
Isomerase	EC 5	Catalyze isomerization changes within a single molecule
Ligases	EC 6	Join two molecules with covalent bonds usually at the expense of an energy source (usually ATP)
Translocases	EC 7	Catalyze the movement of ions or molecules across membrane or their separation within membrane

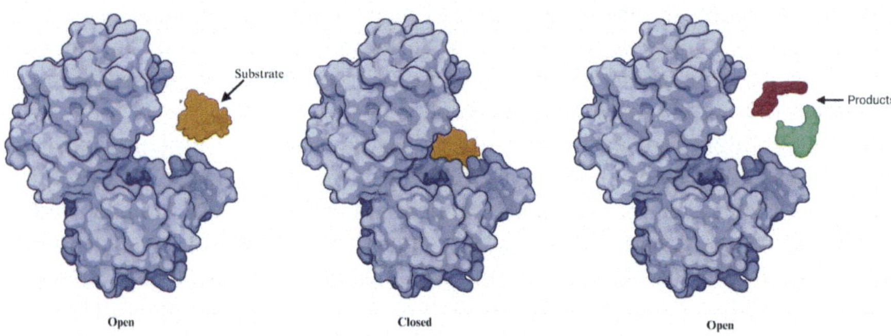

Fig. 1 Enzyme substrate binding and catalysis: Phosphoglycerate kinase-induced fit. (PDB ID: 2xe6, 2wzb)

2.1 *Enzymes as Green Catalysts*

The term *Green Chemistry* was first described by Paul Anastas in 1991 to describe, "the design of sustainable chemical products and processes that reduce or eliminate the use or generation of hazardous substances" [10]. Enzymes are cleaner and greener alternatives for harsh and toxic chemical catalysis for manufacturing high-quality products. Thus, enzymes are preferred for the development of sustainable and ecofriendly industrial processes. These biocatalysts can contribute to reducing carbon footprint leading to an improved environment by reducing CO_2 emissions reduction [128].

The applications of enzymes in industrial processes also present challenges such as reusability, and sensitivity to pH, temperature, salt, and other additives. But despite these are favored for industrial uses owing to the following advantages [32]:

- High specificity, precision reaction rate, and process control
- Ability to work in milder conditions unlike chemical catalyst
- Safe to handle, non-corrosive
- Biodegradable and eco-friendly
- Less harmful byproduct
- Energy efficient
- Reduced pollution
- Low water input
- low emission of greenhouse gasses and reduced global warming

More than 3000 enzymes from different have been isolated so far from different domains of life. The importance of enzymes extends beyond cells and living organisms to bustling industrial processes, where they turn ordinary into extraordinary. Nevertheless, the availability of enzymes for commercial application is restricted to a finite number of available for industrial uses. Despite this constrain, enzymes, as a green approach, have been replacing the toxic catalysts and chemicals required in many industries and processes such as food, feed, textile, paper & pulp, pharmaceutical, organic synthesis, leather, textile, bioremediation, cosmetics, detergent, and personal care [32, 50, 92, 148].

The global industrial enzyme market size is continuously growing. It was 6409.48 million USD in 2022 and is forecasted to be 90,028.41 million US$ by 2028 with a CAGR of 5.88% (360 Industry [1]). Notable corporations engaged in the production and promotion of enzymes include Novozymes, Bestzyme Bio-Engineering Co., Ltd, AB Enzymes, Infinita Biotech Pvt. Ltd., DSM, Röhm and Haas and Genencor among others [32, 52].

3 Enzymes-Based Textile Manufacturing Processes

The textile industry involves an intricate series of steps from raw materials to the final marketing of the product (Fig. 2). Over the years, the industry has evolved, incorporating innovative techniques and materials to improve the quality and efficiency of textile products. However, with these advancements came the environmental challenges. Traditional textile manufacturing processes have been a significant source of pollution, contributing to nearly 20% of global water pollution due to the release of toxic substances, heavy metals, dyes, bleaches, and detergents [140]. To address these environmental challenges, textile industry is increasingly turning to biological methods, particularly the use of enzymes. Enzymes, due to their specificity, environmentally friendly nature, and ability to work under mild conditions, have emerged as a promising alternative to harsh chemicals traditionally used in textile processing [153]. Enzymes are employed in the textile industry at different steps (Fig. 3) such

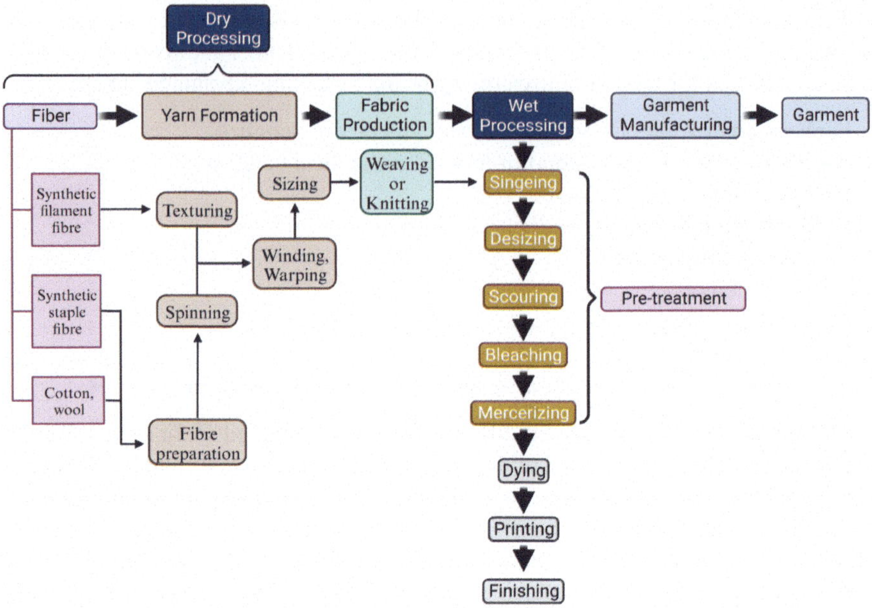

Fig. 2 Overview of textile manufacturing process

as wet-processing and treatment of effluents, processing of fibers (silk, wool, ramie), modification of synthetic fibers, and textile dye synthesis. The inclination of textile industry towards enzyme-based technologies epitomizes the integration of environmentally amicable processes. This approach is reshaping the textile landscape by mitigating harmful waste production and introducing less harmful technology into the various stages of textile finishing and waste degradation. The role of important enzymes in different textile processes is discussed here in detail. (Fig. 3).

3.1 Enzyme Based Wet-Processing

Wet-processing involves the gray fabric coloration and/or chemical treatments involving pretreatments (singeing, desizing, scouring, bleaching, and mercerizing), dyeing, printing, finishing and laundry, etc. These steps involve the utilization of harsh chemicals and polluted effluents (Fig. 4).

3.1.1 Desizing

In the realm of textile engineering, both sizing and desizing are essential processes that contribute to fabric refinement. Sizing of yarn enhances the weaveability of

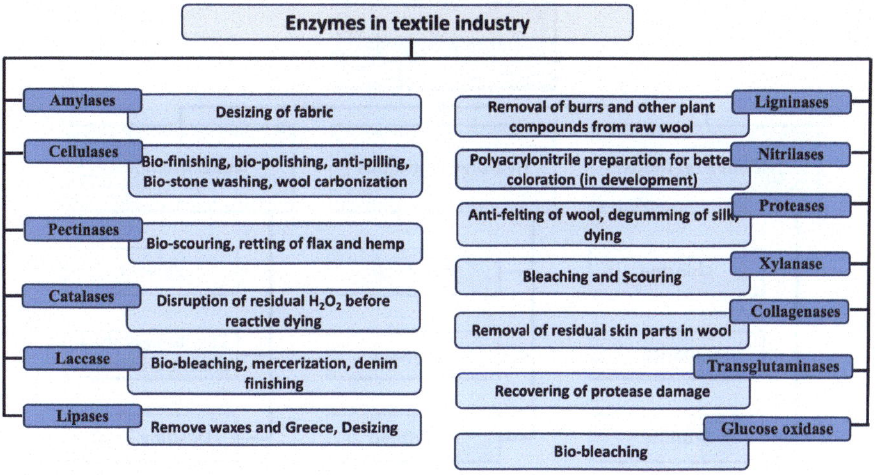

Fig. 3 Enzymes with their application in the textile industry

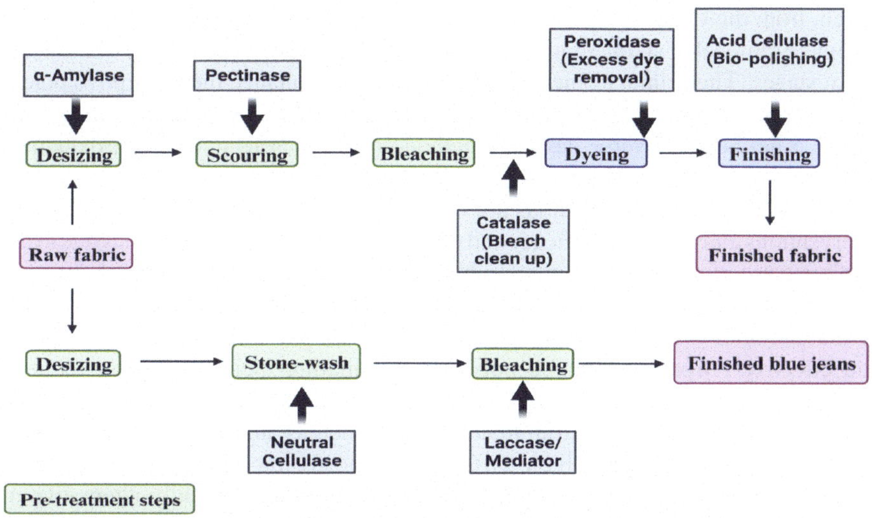

Fig. 4 Enzymes involved in different steps of textile wet processing

the warp yarn [154]. After sizing, the yarn exhibits enhanced properties, including increased resistance to absorption, friction, mechanical stresses, tensile strength and elasticity [6, 135]. Various sizing agents including modified starch, polyvinyl alcohol (PVA), carboxymethyl cellulose (CMC) and acrylates are in use [6]. Starch, a primary sizing agent, not only enhances the fabric's structural integrity but also provides a protective barrier against external adversities [80]. Desizing, a preliminary phase in fabric finishing, requires the methodical elimination of sizing agents, predominantly

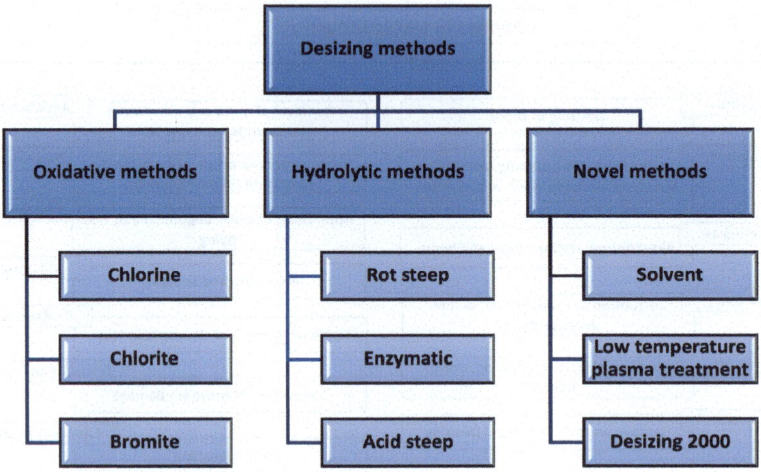

Fig. 5 Desizing method used in the textile industry

starch, from the surface of yarn. During wet textile processing, desizing ensures the complete removal of sizing remnants, thereby facilitating optimal outcomes in subsequent stages. The primary objective of desizing is to enhance the absorptive potential of yarn, optimizing its chemical reactivity during the dyeing and printing [55]. Figure 5 delineates diverse desizing methodologies such as hydrolytic, oxidative, and contemporary techniques available [86].

Conventional desizing methods are contingent on rigorous washing or chemical interventions and often lack efficacy and raise environmental apprehensions due to the profligate use of water and chemicals. Nonetheless, biotechnological breakthroughs have facilitated the incorporation of enzymes, particularly amylases, in desizing. These enzymes are adept at targeting and degrading starch-based sizing agents and effectively hydrolyze starch polymers, facilitating their subsequent removal during the washing phase [5]. Alpha amylases are commonly used enzymes in the desizing process. Amylases from malt, animal pancreas, and bacteria are commercially available for desizing (Table 2). Currently, mostly microbial enzymes are used for textile processes. Rehman et al. [127] explored the eco-friendly alternative of using amylase from *Bacillus cereus AS2*. Optimal desizing occurred with 1000 IU/ml of amylase at 60°C in 15 h, especially when combined with 0.5% Triton-X. This enzymatic method was comparable to commercial desizing in terms of efficiency, with a notable 22% desizing efficiency observed over water-desized fabric. The enzymatic approach offers a more sustainable and eco-friendly alternative to chemical methods.

Table 2 Amylases and their optimum working conditions for textile desizing

Amylase source	Optimum pH	Optimum temperature (°C)	Required concentration (g/L)	Effect of $CaCl_2$
Bacterial	5.0–7.5	60–70*	0.1–1.0	+
Pancreas	6.8–7.0	40–55	1.0–5.0	+
Malt	4.6–5.2	55–65	3.0–20	+

* Thermostable bacterial enzymes are able to work near 100 °C

Enzymatic Desizing Process of Grey Cotton Fabric

The enzymatic desizing process is initiated by immersing the grey fabric in hot water (Fig. 6). This thermal treatment is instrumental in weakening the hydrogen bonds within the size material, thereby making it more susceptible to enzymatic degradation. Following this thermal treatment, the fabric is introduced to a desizing solution. This aqueous concoction comprises specific hydrolytic enzymes, electrolytes in the form of salts, and pH modulators such as acids or alkalis. Among these, enzymes like amylase, derived from malt extracts, are predominantly preferred for their efficacy in liquefying the starch under neutral pH conditions. They are activated over a broad spectrum of temperatures and pH levels, with optimal activity observed between 60-80°C [157].

To ensure uniform exposure of the fabric to the desizing solution, a mechanized guide roller system is employed (Fig. 7). This system adeptly channels the fabric over a winch, directing it into a specialized dwelling pit. Within this pit, the fabric naturally assumes a sinuous or wavy morphology, which effectively augments the surface area exposed to the desizing solution. The inherent gradient or slope of the pit is meticulously calibrated to provide a predetermined dwell time, ensuring the comprehensive enzymatic breakdown of the size material.

Post enzymatic treatment, the fabric undergoes rigorous washing cycles, including multiple hot washes at 95°C, followed by a cold wash or an oxalic acid wash. This phase is paramount for the elimination of enzymatic residues and any solubilized size material, culminating in a fabric that is impeccably primed for subsequent textile operations. The desizing process is carried out using specialized machinery, such as the Menzel desizing machine, which operates at speeds ranging from 70m/min to 100m/min. The machine's length spans 220m, with a desizing chamber capacity of 1000L. During the desizing process, the temperature is maintained between 65°C and 70°C, with a pH range of 5.5–6.5. The dwelling time of fabric in the desizing chamber varies between 8 to 12 min, depending on the specific requirements. The desizing recipe for cotton typically includes bacterial enzymes (3120DE) at concentrations of 5–7 gpl, wetting agents (3160CE) at 3–4 gpl, sequestering agents (3120CI) at 2–3 gpl, and anti-tinting agents (3199FI) for yarn-dyed fabric at 1–2 gpl [28].

After the desizing process, it is imperative to conduct a systematic analysis of the fabric to ascertain the uniformity, effectiveness, and completeness of the treatment. The prevalent test methods employed include assessment of weight loss, TEGEWA

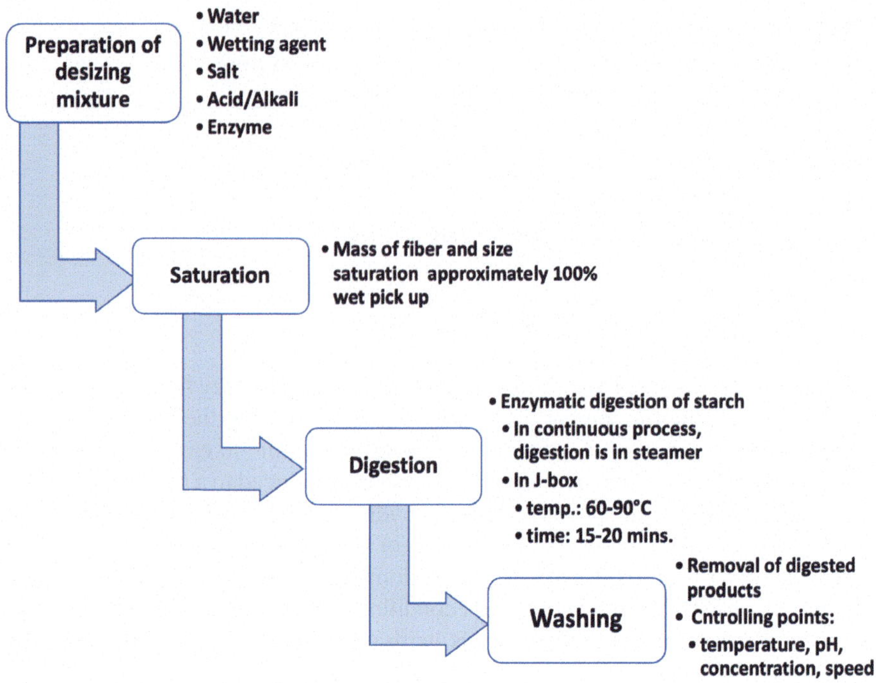

Fig. 6 Steps of enzyme-based desizing process

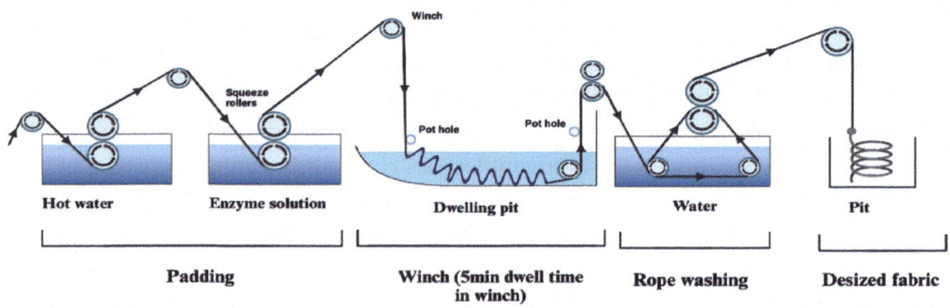

Fig. 7 Enzymatic process of desizing

scale rating, Water and Colour Solution Absorbency, Hand Feel, Bending Length, Effect on Tensile Strength, and Tearing Strength.

Advantages of Enzymatic Desizing

A salient advantage of utilizing amylase enzymes in desizing is their efficacy, which translates to comprehensive removal of size materials and a reduced environmental

Table 3 Comparison of conventional and enzymatic desizing

Criteria	Conventional desizing	Enzymatic desizing
Chemical usage	Reliance on chemicals (like caustic soda, H_2O_2)	Utilizes enzymes such as amylase to process starch and size agents
Energy	Higher energy needs	Energy-efficient
Temperature	Needs elevated temperatures	Operates at cooler temperatures
Washing steps	Several washes to clear chemical remnants	Reduced washes due to no chemical usage
Water use	Increased water use	Conserves water
Waste production	Production of chemical waste; proper disposal essential	Limited chemical residue due to enzyme action
Environmental Impact	Potential harm to the environment from chemicals	Eco-friendly and degradable

impact. This enzymatic approach requires less water and negates the use of detrimental chemicals. Furthermore, recent studies have underscored the versatility of amylase enzymes in the desizing process. These enzymes have showcased their capability to operate effectively across a range of pH levels, thereby offering enhanced adaptability in textile processing conditions. The enzymatic approach ensures fabric integrity, efficient starch removal, and environmental sustainability [96]. Table 3 presents the comparison of conventional desizing and enzymatic desizing.

3.1.2 Scouring

Scouring is crucial in preparing fabrics for dyeing and ensuring the desired qulaity and comfort of end product. It is a critical pretreatment process in textile manufacturing, especially for cotton fabrics, which are predominantly composed of cellulose. Traditional chemical-based scouring involves boiling of cotton boiling in sodium hydroxide (up to 5%). While this method is effective, it presents several challenges. This process consumes a considerable amount of energy, and the alkaline treatment of cotton demands intensive rinsing and neutralization, leading to a large water footprint. Additionally, the caustic chemicals often result in damage to the cotton fibers, compromising the integrity of fabric [87].

The introduction of enzymatic scouring emerged as a solution to the environmental and quality concerns associated with traditional scouring. Enzyme-based scouring is leveraged, improving both the environmental sustainability and economics of the textile pretreatment process. Notably, enzymes such as endo-polygalacturonase (EC 3.2.1.15) and pectate lyase (EC 4.2.2.2) are employed for scouring. Pectate lyase has gained popularity as a bioscouring agent for cotton and other natural bast fibers like hemp and flax [151]. Cellulases are particularly well-suited for the scouring of cotton materials. Certain enzyme preparations include cellulases along with pectinases. The collective effects of both types of enzymes result in more significant reductions in weight and strength compared to the use of pectinase alone [32].

The initial phase involves the removal of superficial dirt and any foreign material from the fabric. This is achieved by subjecting the fabric to a light wash or mild mechanical agitation to shake off dust particles. After this, the fabric undergoes an inspection to ensure it is devoid of any residual contaminants and is primed for the enzymatic treatment. The enzymatic bath step is critical as during this stage, specific enzymes, such as pectinase and cellulase, are selected based on their affinity for the impurities in the cotton. Typically, the enzyme concentrations in the bath range from 0.5–2.0% owg (on weight of the goods). To enhance the wetting and penetration of the solution into the fabric, nonionic surfactants are added, usually in the range of 0.1–0.5% owg. The pH of the bath is meticulously adjusted, typically between 4.5 and 7.5, using precise amounts of acid or alkali to cater to the requirements of the enzyme. The enzyme solution is then prepared in water, maintaining a specific liquor ratio, which is the ratio of water to fabric weight, typically in range of 1:10 to 1:20. The fabric is immersed in the prepared enzyme solution. The process temperature and duration are crucial parameters that are set based on the optimal activity range of the enzyme and are usually between 50 - 60°C and 30–90 min, respectively. To ensure even treatment and prevent fabric damage, continuous agitation is maintained throughout the process. Furthermore, the process is fully monitored and controlled to maintain the pH and temperature.

Upon the completion of the scouring process, it is imperative to halt the enzymatic reaction to prevent the over-processing of the fabric. This is achieved by either raising the bath temperature to about 80 °C or introducing an alkaline solution to shift the pH beyond the active range of the enzyme [20]. After the enzymatic action is terminated, the fabric is thoroughly rinsed with water, typically at temperatures ranging from room temperature to 40°C to ensure the removal of any residual enzymes and loosened impurities. If the pH is significantly altered, a neutralization step is incorporated, treating the fabric with a mild neutralizing agent to restore the pH to a neutral range. The treated fabric is then dried at controlled temperatures, to prevent over-drying, which can lead to brittleness. In some cases, to enhance the fabric feel, a softening agent might be applied before the fabric undergoes further processing stages in the textile manufacturing pipeline. The general outline of the bioscouring process is shown in Fig. 8.

It is essential to evaluate the fabric properties to ensure the effectiveness of treatment. Technical methods such as absorbency tests, whiteness index measurements, and scanning electron microscopy are employed for this purpose. Additionally, physical attributes like tensile strength, weight loss percentage, and fabric hand feel are assessed to ensure uncompromised fabric quality [21, 161].

Advantages of Enzymatic Scouring

The transition to enzymatic scouring in textile is a pivotal shift towards sustainable practices, addressing several environmental and operational challenges inherent

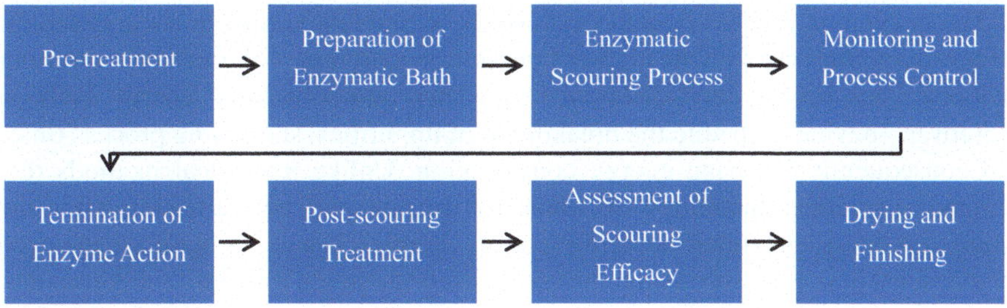

Fig. 8 Steps of the bioscouring process

Table 4 Comparison of traditional vs enzymatic scouring process

Traditional scouring	Enzymatic scouring
High pollution levels	Eco-friendly
Fabric damage	Fabric integrity preservation
Extensive energy use	Energy-efficient
Large water consumption	Water-saving
Hazardous work conditions	Safer work environment
Resource-intensive	Cost-effective
Regulatory challenges	Easier compliance

in traditional methods (Table 4). This advanced approach harnesses the catalytic prowess of enzymes, offering a myriad of advantages that are both eco-friendly and economically beneficial [15, 47]. Enzymatic scouring significantly diminishes the reliance on harsh chemicals commonly used in traditional scouring, such as sodium hydroxide, complexing agents, and surfactants. These conventional agents, while effective, often contribute significantly to total dissolved solids (TDS) and a high chemical oxygen demand (COD) in the wastewater, necessitating extensive treatment before disposal or recycling. Enzymes, being biodegradable proteins, reduce the chemical load in effluents. They specifically target the non-cellulosic impurities in cotton, such as pectin, waxes, and fats, without adversely affecting the cellulose backbone. This specificity ensures that the process is environmentally benign, aligning with global standards for sustainable textile production.

Enzymatic scouring stands out for its gentle yet effective interaction with textile fibers. Traditional scouring, particularly alkaline treatments, can severely compromise the structural integrity of cellulose fibers, leading to weight loss, reduced tensile strength, and an overall degradation in fabric quality. Enzymes offer a milder alternative. By operating under less aggressive conditions (neutral pH and lower temperatures), they preserve the natural strength and luster of the fibers. Furthermore, enzymes like cellulases aid in the removal of surface fuzz, enhancing the appearance and feel of fabric without adverse impacts on its durability. This aspect is particularly crucial for high-value cotton textiles, where quality cannot be compromised [13].

The enzymatic process operates efficiently at ambient temperatures and neutral pH levels, contrasting with the high-temperature requirements of traditional scouring methods. This fundamental difference translates into substantial energy savings. Moreover, enzymes expedite the breakdown of impurities, shortening process times and consequently lowering energy consumption. Unlike traditional methods that demand substantial quantities of water for rinsing and neutralization after the scouring process, enzymatic scouring reduces chemical usage. It minimizes the need for extensive post-process rinsing, thereby conserving water, a resource of paramount importance in textile manufacturing regions often plagued by water scarcity. Furthermore, reduced energy, water, and chemical usage contribute to lower operational costs. By reducing effluent toxicity and contributing to safer working conditions, this method aligns with the industry's commitment to environmental stewardship and social responsibility [14, 161].

3.1.3 Bleaching

It is the last step of the pretreatment process in wet processing. Bleaching is aimed at achieving maximum whiteness and hydrophilicity by removing the natural coloration of the fabric [115]. This procedure not only elevates the visual appeal of the fabric but also optimizes its dye uptake, ensuring a uniform coloration.

The conventional bleaching system comprises H_2O_2 (oxidizing agent), NaOH (activator for H_2O_2), and additional substances such as sequestering agents and stabilizers for H_2O_2. The material is submerged in an alkaline solution with a pH of 10–12 and commercially available H_2O_2 at temperatures and durations, ranging from 30–120°C and 30–60 min, respectively. The precise conditions are meticulously selected based on the desired level of whiteness and the material. Following this procedure, the fabric undergoes a washing step, after which it is acidified to neutralize the sodium hydroxide [34, 110, 115].

The textile industry has shown an increasing interest in enzymatic bleaching as an alternative to traditional methods. This innovative approach employs enzymes, notably glucose oxidase (GOx), to produce H_2O_2 in situ. Here, glucose acts as the substrate, and when combined with the enzyme GOx, it facilitates the generation of H_2O_2, which subsequently serves as the bleaching agent. A typical method involves treating D-glucose in a pH-adjusted medium at pH 5.0 with sodium acetate and glucose oxidase at 37°C [12, 44]. Figure 9 shows the detailed bleaching process [101]. During this process, the fabric is immersed in the enzyme solution that generates H_2O_2 which acts as a bleaching agent. The H_2O_2 is responsible for the decolorization of the fabric, and works effectively as a mild and environmentally friendly bleaching agent. The process is generally slower compared to conventional methods, allowing for a controlled and gradual bleaching effect. This slower process is beneficial as it is less harsh on the fibers of fabric, and reduces the risk of damage that is often associated with traditional bleaching methods.

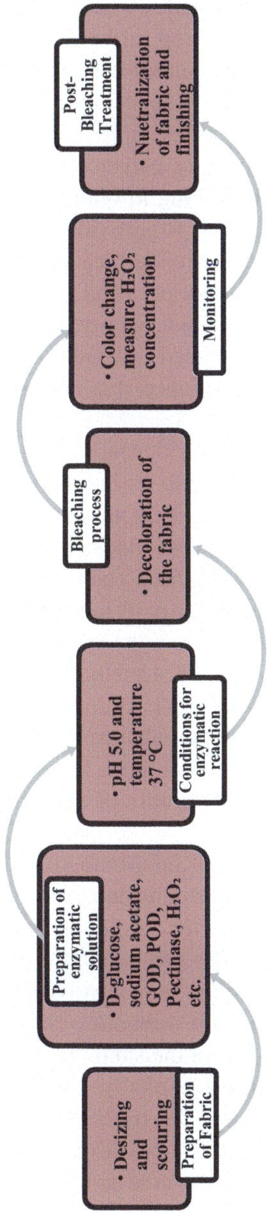

Fig. 9 Enzymatic bleaching process

Throughout the bleaching process, the reaction progress can be monitored by assessing the color change of fabric or using specific assays to measure the concentration of H_2O_2. Once the desired level of whiteness or decoloration is achieved, the fabric is removed from the solution and rinsed throughly to remove any residual enzymes or H_2O_2. This step ensures to stop further bleaching or degradation. After bleaching, the fabric is neutralized and conditioned with appropriate finishing treatments to achieve the desired hand feel, appearance, and performance characteristics. In one study, when protease preparation Bactosol SI (Clariant) was used for biobleaching of wool. It significantly enhanced whiteness and hydrophilicity compared to conventional H_2O_2 bleaching in both oxidative and reductive conditions. However, due to protease sensitivity to peroxides, its treatment is recommended to be carried out before exposure to peroxide bleaching as a pretreatment step for one hour at a temperature of 50°C and a pH level of 9.0. Some proteases such as Subtilisins from *Bacillus* spp. break down peptide and ester bonds. Unlike other proteases, alkaline proteases with enhanced peroxide resistance, such as Durazyme 16.0L (Novo Nordisk), have been developed that are promising for the bleaching process [32].

The assessment of enzymatic bleaching depends on carefully examining the level of whiteness added to the fabric relative to its initial state. This critical evaluation usually makes use of accepted benchmarks. One common metric is the degree of whiteness also known as the CIE whiteness index, which offers a numerical representation of the degree of whiteness attained. The success of the process is closely related to maintaining the structural integrity of fabric after bleaching, even beyond just changing its color. Minimal damage to the fibers is indicative of an effective enzymatic bleaching process. As a result, a method that preserves the strength of fabric and indicates less damage to the fibers is considered more effective, striking a balance between the intended aesthetic enhancements and material preservation. This two-pronged evaluation approach highlights the delicate equilibrium between maintaining the inherent quality and durability of the fabric and accomplishing the desired bleaching effect [26].

Despite the innovative approach and environmental benefits of enzyme-based bleaching, it faces several challenges and considerations that necessitate attention. A paramount concern is the enzyme stability during the process, especially how GOx are rapidly inactivated by the H_2O_2 they generate. This delicate paradox is partially addressed by employing a synergistic system involving GOx and glucose, which collaboratively sustains the reaction while mitigating enzyme deactivation. However, another hurdle emerges when comparing the subtle bleaching effect achieved enzymatic methods with more pronounced results of conventional chemical methods. This disparity highlights an imperative for further optimization of the enzymatic process. Potential refinements could encompass exploring alternative enzymes with more robust characteristics, fine-tuning enzyme and substrate concentrations, or even recalibrating the overall reaction conditions to enhance efficacy. These considerations highlight the ongoing need to balance enzymatic bleaching's eco-friendly allure with performance competencies that meet or exceed those of traditional chemical counterparts [56].

Recent studies in enzymatic bleaching underscore the advancements in the domain of environmental sustainability and efficiency, particularly in the context of cotton fabric bleaching and the extraction of nanocrystalline cellulose. Wang et al. [162] introduced an innovative UV/H_2O_2 system for cotton fabric bleaching, highlighting its environmental sustainability and efficiency in achieving significant whiteness improvement. The research quantitatively establishes that optimal UV irradiation and H_2O_2 concentration are crucial for maximizing bleaching efficacy while preserving fabric integrity, evidenced by a whiteness index increase to 81.23% and considerations of fabric strength loss. Fast-forwarding to 2023, [38] quantitatively evaluated the enzymatic bleaching process, revealed that optimized enzyme formulations can decrease chemical and energy consumption by 20 and approximately 30%, respectively. Furthermore, the research establishes a direct correlation between enzyme concentrations and the reduction of harmful effluents, positioning enzymatic bleaching as a quantifiably more sustainable and efficient process.

Peroxide Removal

Bleaching of natural fiber by H_2O_2 in textiles is followed by dyeing and printing. However, any residues of leftover H_2O_2 on the fabric can interrupt the dyeing process. Hence, the complete elimination of peroxide is imperative. The conventional bleaching system is comprised of H_2O_2 (oxidizing agent), NaOH (activator for H_2O_2), and additional substances such as stabilizers and sequestering agents for H_2O_2. The chemical bleaching involves immersing the fabric in an alkaline solution with a pH ranged from 10 to 12 at a temperature as high as 120 °C for a duration of 10 to 20 min. Following this procedure, the fabric undergoes a washing step, after which it is acidified to neutralize the NaOH. The presence of NaOH residues on the fabric leads to yellowing upon drying.

Both processes require significant quantities of water. Alternatively, the use of enzymes such as catalases facilitates the quick and thorough decomposition of H_2O_2 into non-harmful H_2O and O_2. This approach is more convenient and expeditious. Catalases for industrial applications can easily be produced by various microbes through fermentation [108, 152]. Enzymatic H_2O_2 elimination through the use of catalase has been extensively employed since the late 1990s in various textile industrial setups, including both new and existing installations. This approach can be adapted for continuous, semi-continuous, and batch operations [115]. These enzymes have a very high turnover number and can decompose 500 million moles of H_2O_2 per minute under moderate optimum working conditions such as, a temperature range of 20–50°C and a neutral pH. Neutralizing the solution before adding the dye is advantageous since catalase exhibits its highest level of activity within the pH range of 6–8. Instead of pre-treating with catalase before dyeing, it can be introduced directly to the dye bath 5–10 min before adding dyes and chemicals, as catalase does not exhibit any catalytic activity toward the dyes [32].

The water requirement depends on the fiber type being processed in the textile industry. However, the number of rinses required when utilizing catalase can be significantly reduced. Another study reported that the use of a commercial catalase to remove H_2O_2 remnants from cotton textiles enhanced the dyeing performance and significantly augmented the color output while using a bifunctional monofluortriazinyl dye [32, 74, 152]. In summary, the enzymatic removal of H_2O_2 is a sustainable process that results in energy and water conservation, less polluted effluent, and recycling of bleach effluent with no impact on the subsequent dyeing process [32, 74, 115]

New ecological regulations, the issues related to ROS species as water pollutents of residual waters, and the increased process costs have become the focus of attention leading to process development for the industrial application of catalases [74]. Peroxidases efficiently break down H_2O_2 within a pH range of 3 to 9 and at temperatures ranging from 30 to 80 °C [32]. The integration of various enzymatic processes in a single solution, such as the simultaneous application of dyeing and biopolishing, bioscouring, and desizing would contribute to water conservation, time efficiency, and cost reduction.

3.1.4 Dyeing and Printing

Dyeing and printing are fundamental aspects of the textile manufacturing process, dedicated to bestowing fabrics with desired colors and patterns. Historically, methods such as block printing and screen printing have been predominant. Block printing, an age-old technique, involves imprinting designs onto fabric using intricately carved blocks. While it produces unique and intricate designs, it has faced challenges like poor appearance, inappropriate patterns, and color loss [114]. On the other hand, screen printing, an evolution of the stenciling process, transfers color through cuts in a screen placed on the fabric. The design is printed by using traditional agents including various dyes and thickeners [121]. The dyeing and printing processes not only add aesthetic appeal but also functional value to textiles. They enable customization, differentiation, and can even bestow specific properties like UV protection or antimicrobial characteristics to fabrics [137]. However, these methods not only involve use of hazardous chemicals and dyes, but also produce highly polluted textile effluents, posing serious environmental and health problems. In recent times, the shift towards more sustainable methods has led to the rise of enzyme-based dyeing and printing methods. These methods employ enzymes as eco-friendly agents to achieve the desired colors and patterns on textiles. Examples include the use of the laccase enzyme for eco-friendly discharge printing on cotton fabrics, and protease extracted from *Moringa Oleifera* leaves, and lipase from sunflower seeds [19].

The process of coloring and grafting functional molecules (such as antibacterial agents and hydrophobic enhancing compounds) using laccase is being extensively pursued. Several investigations have examined the simultaneous treatment, dyeing, and finishing of cotton, wool, and silk using natural colorants and employing different enzymes. For example, laccases oxidize a variety of aromatic compounds such as aminonaphthols, aminophenols, diamines, phenols, and flavonoids, leading to the formation of radicals that interact with the fiber, resulting in the production of various shades. By manipulating the reaction time, and concentrations of modifiers and laccases, diverse tones and intensities on the fabric can be achieved. Thus, laccases have been used to color cotton fabric using natural polyphenols and to facilitate the linkage of hydrophobic phenolic chemicals, like lauryl gallate to wool fibers [63, 78, 105, 119, 166]. Similarly, other enzymes such as cellulases, α-amylase, amyloglucosidase, trypsin, and transglutaminase have effectively been employed in the preparation of materials for dyeing with saffron, curcumin, and sappan [115]. The fabric pretreatment with cellulases, which efficiently degrade surface of cellulose fibers to produce a more absorbent surface, improves dye-receptivity and whiteness.

The protocol for the dyeing and printing of a fabric involves pretreatment of the fabric, enzyme application, dye bath preparation, and finishing (Fig. 10). Each stage of the pretreatment process (desizing, scouring, mercerizing, and bleaching) influences the quality of the dyeing outcomes. The order in which the preparatory pretreatment procedures are performed varies based on the kind of textile fabric and dye used, as well as the amount of water and chemicals needed for each step. The fabric is submerged in a dye bath following an incubation period and a thorough rinse to eliminate any leftover enzymes. In dye baths high textile dyeing rates and leveling properties are achieved near boiling points. Discontinuous dyeing machines, like those using infrared for polyester fibers, can even dye at high temperatures and pressure. Textiles with brilliant and uniform colors are obtained after the cellulases treatment of fabrics due to enhanced dye penetration and fixation. This enzymatic method reduces the requirement for harsh chemical treatments during the pretreatment stage, which not only improves dyeing efficiency and color fastness but also aligns with sustainable textile practices. As part of the after-treatment process or finishing, chemical residues are removed from the fabric, and a series of steps are taken to ensure colorfastness. Additional treatments like softening, conditioning and rinsing maintain sustainability by removing excess dye and chemicals, and enhance dye binding. Fibers can be treated in the finishing stage to repel water, soften, shield ultraviolet light, protect against static electricity, self-clean, be fire retardant, and resist soil rot.

Prajapati and his collegues developed a laccase-based dyeing method for wool and nylon 6,6 fibers to attain a wide range of colors, including pinks, greens, and blues. The dyed fabrics showed reasonable wash fastness but poor resistance to light. Enzymatic dyeing allowed for easy color variations, compared to conventional methods. Laccase also showed potential for creating surface patterns, indicating opportunities for sustainable textile practices [119]. In another study, Zhang and his colleagues devised a novel strategy to use laccases from *Trametes versicolor* for coloration in conjunction with 1-ethyl-3-(3-dimethylaminopropyl) carbodiimide hydrochloride (EDC) for the pretreatment of wool fabric. Results showed that, upon pretreatment, the color difference of wool fabrics dyed with poly(syringic acid) and the K/S values significantly improved when compared to untreated control samples. Furthermore, pretreated wool fabrics exhibited superior color fastness and uniformity [174].

The advantages of enzymatic dyeing and printing methods are manifold over the traditional ones (Table 5). Enzymes minimize the use of harmful chemicals, ensuring

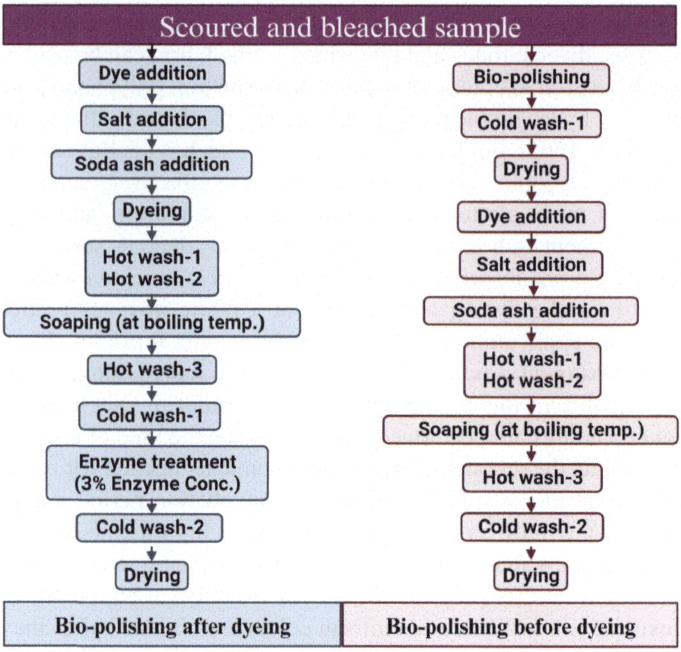

Fig. 10 Dyeing step in the textile wet process

safety for both the environment and the end-users. Additionally, enzymatic methods can offer superior dye penetration and fixation, resulting in vibrant and long-lasting colors [123] (Table 5).

Table 5 Comparison of Conventional and enzymatic dyeing and printing

Criteria	Conventional dyeing and printing	Enzymatic dyeing and printing
Chemical use	Reliance on chemicals like Procion, Rhodamine B, wax, and starch	Utilizes enzymes such as cellulase for dying and printing
Waste production	Wastewater containing residual dyes, chemicals, and impurities	Limited chemical residues
Washing steps	Consumes substantial water, contributing to pollution, and the potential for color fading poses durability challenges	Requires less water, reduces energy, and improves color fastness due to its efficient removal of excess dyes
Water usage	Increase water use	Conserves water
Environmental Impact	Potential harm to the environment	Eco-friendly and degradable

3.1.5 Finishing

Finishing in the textile manufacturing process refers to a series of operations that improve the appearance, texture, and performance of fabrics. Traditionally, the finishing process employes various chemicals and agents to achieve desired fabric properties. These agents, often synthetic, are used to impart specific characteristics to the fabric, such as softness, wrinkle resistance, and water repellency. The significance of finishing lies in its ability to improve the functional and aesthetic attributes of textiles, making them more appealing and suitable for specific uses [84]. Enzyme-based finishing, also known as bio-finishing, bio-polishing, or bio-washing, involves utilization of enzymes for the modification of the surface and properties of fabrics. It is commonly applied to cotton, cotton blends, viscose rayon, and Tencel fabrics. It has emerged as an eco-friendly and sustainable alternative to traditional finishing methods.

Bio-finishing can also be used on denim jeans, shirts, and jackets [95]. This process involves the use of acidic textile cellulases to eliminate tiny hairs and fuzzy fibrils from the surface of fabrics and the subsequent dyeing process. It is crucial in preventing fibrillation, as it significantly enhances various fabric aspects such as appearance, color vibrancy, texture, water absorbency, degree of crystallinity, softness, drapability, and luster [23, 113, 118, 147]. Biopolishing can be performed at various stages of wet processing, with post-bleaching being the most common. It can be carried out in both continuous and batch processes, with continuous processes requiring an incubation period for enzymatic degradation [134].

Biopolishing is best achieved by endoglucanase-rich formulations. Neutral cellulases with a working temperature and pH range of 50–60 °C and 5.5–8, respectively, are reported to be more desirable compared to acidic cellulases with a working pH and temperature of 4.5–5.5 and 45–55 °C, respectively [59]. The processes are performed under stringent control to prevent fabric degradation. The enzymatic activities are susceptible to concentrations of salts, ionic surfactants, and reactive dyes. High agitation is provided at the industrial level for better adsorption and desorption of cellulases on the fiber and the separation of partially hydrolyzed fibrils. Biopolishing can be performed after or before dyeing (Fig. 10) but the latter is more common. However, synergistic dyeing and finishing in a single step is more economical and gaining interest [115].

Denim, a jeans fabric, is primarily composed of woven cotton threads dyed with indigo to achieve a distinctive blue color. However, these cotton fibers contain about 10% non-cellulosic impurities, diminishing their commercial value. To enhance the quality of jeans, microbial enzymes can be employed during the cleaning and desizing processes, leading to improvements in color, texture, and covering ability. A recombinant Xylanase II (XynA2), derived from *Caulobacter crescentus* has been reported to clean and refine jeans fabric. The effective treatment involved using 1 unit/mg of XynA2 at 60 °C and pH 8 for 12 h. The results indicated that the XynA2 from *C. crescentus* could successfully improve denim jeans in a more eco-friendly way compared to other chemical-based or abrasive methods. [65]

The enzymatic treatments align with the principles of green chemistry, leading to reduced utilization of harsh chemicals, conservation of energy and water, and minimal waste generation. Furthermore, enzymatic processes are often more fiber-specific and gentle, ensuring enhanced fabric quality without compromising the integrity of the material. Applications of Enzymes in textiles hold a promising future, with the potential for complete replacement of conventional chemicals, from pretreatment to final finishing processes, especially for cellulosic fabrics [40].

4 Other Enzyme-Based Textile Processes

4.1 Bioprocessing of Wool

Wool is primarily comprised of cystine-rich keratinous proteins (approximately 82%), non-cysteine-rich nonkeratinous proteins (roughly 17%), and non-protein materials (1%) that include waxy and a small quantity of polysaccharides [131]. Wool is esteemed for being natural, renewable, biodegradable, having a low carbon impact, and being energy-efficient with numerous inherent benefits. Owing to resiliency, insulation against cold, comfort, and aesthetic look, wool has established a worldwide reputation for excellence among manufacturers and consumers [81].

4.1.1 Bio-Carbonization of Woo

Carbonization, the removal of cellulosic or vegetable material from the wool surface, is carried out at high temperatures and acidic conditions. These conditions can also reduce fiber length and wool strength. Therefore, researchers aim to develop an enzyme-based process to solve this problem. For this purpose, various mixtures of oxidoreductases (ligninolytic enzymes), pectinases, lyases, and cellulases are being investigated as potential candidates [4, 115].

4.1.2 Shrink Proofing of Wool

Wool clothing and items are prone to felting and shrinking washing, impacting the appearance and size to the extent that they may become unusable. Woolen garments, on average, shrink to about 55% of their original volume after the initial wash. Consequently, preventing shrinkage is crucial for using wool [59]. The wool shrinkage is of two types i.e., relaxation and felting. Relaxation arises due to the removal of internal strain accrued during textile fiber processing. It is reversible once removed from machines [98]. Whereas, the felting is irreversible and caused due to the scales present on the surface of wool fibers. These scales become interlocked with other scales like ratchet while washing, tumble drying, etc. Felting is problematic as it changes the size and texture of the apparels [59]. Traditional methods to prevent felting involve chlorination followed by an antichlorination step to remove chlorine entailing the use of harsh chemicals such as permonosulphuric acid (PMS), dichloroisocyanuric acid (DCCA), sodium hypochlorite, and sodium sulfite [76, 130]. While these processes are commercially appealing, they pose significant environmental concerns, notably in the form of wastewater contamination with organic chlorides that can be absorbed, a problem evident in the commonly used Chlorine-Hercosett process [89].

Proteases alone or in combination are the most extensively investigated enzymes for imparting shrink-resistance to wool fiber (Fig. 11) that also help to remove impurities, and increase dye affinity [37, 59, 115]. The sequential treatment of wool with lipase, glutathione reductase, and papain resulted in the removal of lipids from the wool surface, reduction of disulfide bonds, and smoothing and softening of wool, respectively [115]. Proteases, being smaller molecules, can penetrate deeper into the wool and destroy the inner structure of wool. However, this issue can be addressed by enzyme immobilization or by recombinant technology to increase the size of proteases and restrict them to the surface of wool. A thermozyme from *Bacillus safensis* FO-36bMZ836779, immobilized on a sericin carrier, was used for the treatment of wool. It partially removed the scales from the fiber surface without an aggressive impact on the bulk of the fibers thereby leading to machine-washable wool [53].

The H_2O_2 exposure in combination with protease can induce modification of the cysteine residues on the surface [115]. Lenting and colleagues developed a method to prevent full wool shrinkage and tensile strength loss. Pretreatment of wool with H_2O_2 in high salt concentration and alkaline conditions modified the surface and reduced the penetration of the protease enzyme [88]. Researchers extracted polypeptides from wool protein fibers using Esperase (a serine-type protease) from *Bacillus* sp. These polypeptides were used to modify the wool fiber surface using glycerol diglycidyl ether (a cross-linking agent) and subsequent curing steps for binding the polypeptides. The resulting knitted fabric without excessive fiber damage exhibited high shrinkage resistance during machine washing. Only 1–2% shrinkage was observed after 5 cycles of 5A washes with $< 1\%$ weight loss due to washing, and a burst strength of 317 kPa [149].

Kadam and colleagues devised an innovative "trifunctional protease" to enable the production of machine-washable wool textiles. Their approach involved linking protease with L-cysteine by carbodiimide/N-hydroxysuccinimide (EDC/NHS) and

Fig. 11 Electron microscopic image of wool fiber, **a** Wool fiber without any treatment, **b** Wool fiber after enzyme treatment showing smooth and antifelt resistance

poly(ethylene glycol) bis(carboxymethyl) ether (HOOC-PEG-COOH) coupling. The proteases, when applied to wool fabrics displayed the qualities such as reducibility, hydrolytic capability, and precise localization. The modified protease exhibited significant treatment benefits, enhancing dimensional stability by reducing felting (6.12%), managing scale dislodgement and minimizing strength loss (11.7%) as affirmed by scanning electron microscopy. This versatile enzyme was intricately designed to modify wool surfaces specifically, potentially paving the way for more sustainable and enhanced textile treatment processes [99].

Another study demonstrated the protease-chitosan treatment of wool exhibited the lowest shrinkage at 3.57% ($p < 0.05$), followed by the laccase-chitosan treatment at 4.67%. However, when employing enzyme-biopolymer combinations, the shrink resistance aligns with sustainability, as it avoids the release of AOX (adsorbable organic halides), preserves the inherent chemical structure of wool, improves tensile strength, enhances whiteness, reduces yellowness, and maintains comparable bending properties [69]. Laccase application on wool demonstrates potential as an alternative to proteases by effectively reducing felting without adversely affecting other properties of wool fabric. The use of 5% laccase at a liquor-to-goods ratio of 40:1, pH 4.5, and 60 °C, notably enhanced fabric strength. Additionally, shrinkage, water absorption time, and weight reduction percentages decreased, leading to increased resistance to felting [105]. Treating wool textiles with transglutaminases unveils potential for diverse and innovative finishes, and promising novel functional possibilities for wool textiles.

Enzymatic grafting of proteins or amines onto wool fibers enhances both tensile strength and shrink resistance. Protein-crosslinking enzymes like transglutaminases (TGases) can repair wool and yarn after chemical and enzymatic treatment restoring the fiber strength to 25% compared to the control. Combining these enzymes with keratin or casein improves the dyeing process and color characteristics of leather [35]. Overall, TGases enhance tensile strength, shrink resistance, wettability, handling, softness, dye uptake, minimize felting, and safeguard wool from detergent degradation.

In a distinct approach, a separate study employed two enzymes, a keratinase from *Bacillus subtilis* and a cutinase from *Thermobifida fusca*, in a sequential treatment applied to wool fabrics for durations of 4 h and 24 h, respectively. The primary goal was to reduce the shrinkage of wool fabrics. The significantly reduced shrinkage from 12.4% to 5.86% by partially disrupting the cuticle layer was obtained. This decrease in shrinkage aligns with the requirements for machine-washable fabrics with a strength loss of 7.46% which was at an acceptable level. Moreover, the combined enzymatic treatments not only aid in minimizing shrinkage but also enhance the wettability and dyeability of the resultant fabric due to enzymatic breakdown of the lipid layer and keratin. The collective effects of these enzymatic treatments effectively eliminate scales from wool fabrics and disintegrate cuticles, presenting an environmentally friendly and efficient alternative to traditional anti-felting procedures [173].

4.2 Degumming of Silk

Silk, a long natural protein fiber, is a biocompatible material with a role in medical treatments since ancient times. Recently, it has been the focus of attention for its applications in wound healing and tissue engineering [160]. Silk is produced by several different insects including silkworms, spiders, flies, mites, and scorpions. Each of these produces a unique variety of silk [33]. The silkworms (*Bombyx mori*) are domesticated for silk production which is the most widely used silk material in textiles [25]. The cocoon silk fiber is composed of two protein components, i.e., fibroin and sericin. Fibroin is the main fibrous structure that constitutes 75% of silk fiber. Whereas sericin constitutes 25% of the silk fiber and is a glue-like protein that holds the cocoon together [4, 104].

Degumming is the removal of sericin (silk gum) from fibroin. Silk gum gives a stiff and harsh feeling and dull appearance to the fiber. It also hides the whiteness, softness, and typical shiny aspect of silk. In the textile industry, fibroin is converted into raw silk after the degumming process, which is subsequently used in the manufacturing of a variety of yarns and silk fabrics [18]. The conventional degumming process involves treatment with alkali, oxidizing agents, and soap for extended time at high temperatures (Fig). Nevertheless, the retrieved sericin to undergoes a significant degradation, leading to the loss of certain functional properties and a decrease in its molecular weight. Additionally, separating the soap from sericin poses considerable complexity, potentially leaving residual traces of soap in the sericin. As a result, its suitability for biomedical and pharmaceutical applications is limited [124, 144].

Bio-degumming of silk (Fig. 12) is not only an eco-friendly method but also provides better results for the sheen, softness, color, and texture of the silk [112, 177]. Thus, the exploration of the potential of enzymes for silk degumming is a currently active area of research. Several neutral, acidic, and alkaline proteases are under investigation. Alkaline proteases exhibit significant potential in enhancing the features of silk surfaces, such as texture, luster, and sleekness [115]. An in-silico approach has revealed the interplay of the trypsin enzyme, from *Diatraea saccharalis* with the sericin protein, revealing resemblances between trypsin and cocoonase, and implying the potential use of trypsin in cocoon degumming processes (Sneha et al. 2022). Another study incorporated both in silico and in vitro methods to assess the degumming effect of cocoonase. Although the degumming by Na_2CO_3 treatment results in more pronounced cocoon sheet weight loss compared to cocoonase, it is still widely used due to its efficiency. However, cocoonase-treated silk cocoon sheet retains more smoothness, luster, and the natural color of tasar silk [9].

Recombinant cocoonase, a serine protease from silk moths, has also been explored for silk degumming and bleaching. Mixtures of proteases and lipases have also been reported for effective removal of the gum from silk without any damage. Additionally, these mixtures have the added advantage of making the silk supple and improving the efficiency of the dyeing process. The application of MTGase (Microbial transglutaminase), followed by H_2O_2, protease, and ultrasound showed enhanced crease resistance and tensile strength of the silk fabric [115].

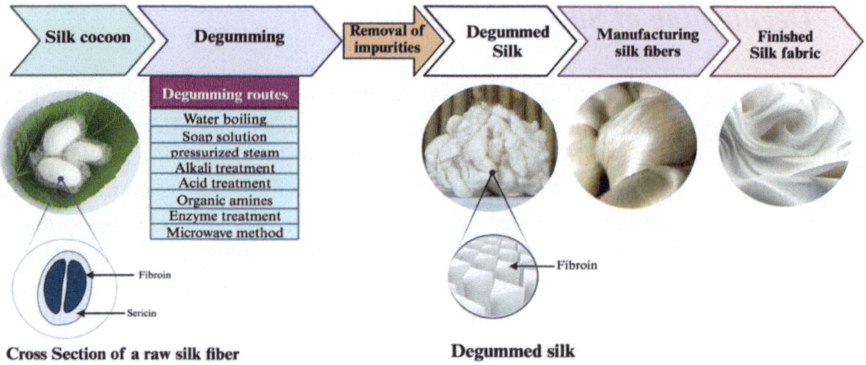

Fig. 12 The degumming process of silk

4.3 Degumming of Ramie

Ramiez is the fiber that is obtained from the ramie plant (*Boehmeria nivea* (L.) Gaud, *Boehmeria nivea* var. tenacissima). It is also known as white ramie or Chinese grass and belongs to the Urticaceae family [79, 165]. Raw ramie fibers contain 20–35% gum primarily composed of pectin and hemicellulose which are removed during degumming. To meet textile standards, the residual gum content should be between 1.5% to 2.5% [175]. Currently, degumming is constrained due to adverse environmental impact, elevated energy consumption, increased production expenses, and fiber damage caused by traditional chemical-based degumming [93]. Thus, biodegumming is sought after. Ramie is expected to gain visibility in the global textile market through this approach, thereby leading to economic benefits for ramie cultivators [126].

Direct use of pectate lyases and xylanases-producing bacterial strains (such as *Pectobacterium* sp., *Bacillus thuringiensis* MCC2138, *B. subtilis*, *B. pumilus*, *B. megaterium,* etc.) or their enzymes have been reported for degumming of ramie fiber [22, 93, 109, 126, 175, 176]. The direct use of *Pectobacterium* sp. CXJZU-120 strain is reported to eliminate over 90% of the gum present in raw ramie within 6 h. This method proved effective in extracting ramie fibers from various grades of raw materials while preserving their inherent structural and textile properties. It also reduced production cost and pollution by 20.5% and 80%, respectively, and 50% enhanced resource utilization compared to conventional methods [93]. In another study, *B. thuringiensis* MCC2138 and *B. subtilis* ABDR01 were both used to prepare novel microbial formulations for degumming, resulting in a soft, durable, and lustrous fiber with higher tensile strength with less chemicals consumption [126]. Pectate lyase and xylanase-producing *Bacillus* sp. can decrease residual gum to 9.4% within 5 h [175]. Thermostable alkaline pectate lyase from *Bacillus* sp. RN1, cloned and expressed in *Pichia pastoris* showed potential as a compelling candidate for ramie degumming [176]. These findings affirm the substantial advantages of degumming ramie fibers using bacteria and their enzymes. Enzymatic degumming notably increases the fiber brightness while causing less reduction in bundle-breaking tenacity [175].

4.4 Bio-Stone Washing

Denim items including blue jeans are popular, with approximately 800 million pairs producion per year [24]. Traditionally, pumice stone washing is used to give a stylish worn effect to denim. It involves nonhomogenous indigo dye removal to get the worn look of blue jeans. Traditional stone washing involves the use of 1–2 kg of stones per kg of jeans in 1 h. However, the effect of pumice stone washing is severe for clothing, equipment, and machines. Moreover, large amounts of sand due to stone erosion and sludge (18 metric tons per week) are produced leading to a serious environmental impact [4, 111]. During the mid-1980s use of microbial cellulases emerged as a prominent alternative for pumice-based denim washing. Enzymatic aging has been shown to increase productivity and reduce time, strength loss, machine damage, waste, and water issues. The stonewashing effect is due to the strong ability of endoglucanases to remove indigo dye [4, 115]. For example, *Trichoderma reesei* endoglucanase II effectively removed denim color, resulting in a satisfactory stone-washing effect with minimal hydrolysis [60]. Enzyme-based stone washing has not completely replaced pumice stones. Concomitant use of mechanical abrasion and cellulases leads to the defibrillation of surface fibers, generation of cotton powder, and dye release from dyed yarn resulting in the ideal old and worn aspect of the garment [115].

A drawback of using cellulases in stone washing to obtain fading look is the back staining or redeposition of indigo dye on cotton fibers. It diminishes the ideal white/blue contrast. This is due to the high affinity between indigo, cellulose, and cellulase proteins affecting the enzyme adsorption reversibility on cellulose. It depends on the enzyme origin, size of the cellulase, and binding domain (CBD). Preferred enzyme preparations for stone washing contain neutral and high endoglucanases to remove indigo from garments without damaging fibers, as back staining is more pronounced in acid pH ranges. Full-acid cellulases contain non-cellulase proteins that induce back staining. Treatment with neutral cellulases, such as those from *Humicola insolens* and *Penicillium echinulatum* with activity in both acid and neutral pH exhibited reduced indigo dye redeposition and fabric strength loss [115].

Many chemical and enzyme-based anti-back staining treatments, effectively reduce back staining. For example, proteases can reduce back staining and enhance contrast when used after cellulase washing, during rinsing, or post-washing steps. Additional washing with detergents or components like surfactants and dispersants, combined with vigorous mechanical stirring tends to enhance fading effects. The enzyme dosage depends on textile material type, density, porosity, hydrophilicity, treatment time (15–60 min), pH (4.5–7.0), temperature (45–65 °C), bath ratio (1:3–1:20), chemical auxiliaries (recommended non-ionic surfactants and dispersing agents), and equipment used. Combining endoglucanase and cellobiohydrolase also showed improved bio-stoning and reduced back staining. Bio-washing and biodesizing can now be done in one step utilizing laccases, cellulase, and amylases without affecting garment color. Combining stone washing with laccase-mediated bleaching can even improve the color of indigo-dyed denim. Laccase-based stoning is a promising alternative that degrades indigo dye without harming the fabric, resulting in less damage and a unique wash-down effect [4, 115].

4.5 Surface Modification of Synthetic Fibers

Synthetic fibers were introduced in the middle of the last century. Since then these fibers have dominated the textile industry, holding a substantial 54.4% market share compared to natural fibers. Notable synthetic fibers include polyethylene terephthalate (PET), polyamide (PA), polyacrylonitrile (PAN), and polypropylene (PP). These fibers are valued for their remarkable properties, such as high strength, stiffness, stretchability, resistance to wrinkles and abrasion, cost-effectiveness, ease of processing, customizable performance, and recyclability. Despite these advantages, these fibers exhibit drawbacks, including hydrophobicity, reduced comfort during wear, poor dye absorption, static buildup, pilling tendency, challenges in finishing, and inadequate washability due to their hydrophobic nature [116]. Their wetting-resistant surfaces make it challenging to apply finishing agents and dyes. The hydrophobic nature impedes water from permeating the fabric pores. Moreover, the lack of reactive groups on the backbone of these fibers make it difficult to modify their properties. Consequently, a variety of physical and chemical techniques are employed to enhance their characteristics. Surface modifications primarily aim to improve hydrophilicity while maintaining the inherent properties of synthetic fibers without altering their fundamental characteristics. .

Traditional methods for functionalizing synthetic materials involve harsh chemicals or plasma processes. Controlling the chemical hydrolysis of polymers with concentrated alkali or acids at high temperatures is challenging. Chemical treatments can cause up to 15% material loss, significantly reducing characteristics such as tensile strength. Plasma processes are more sophisticated and demand significant energy [115]. In recent years, enzyme-based methods have emerged as sustainable alternative for treating diverse synthetic materials. As PET, PA, and PAN collectively make up almost 90% of synthetic fibers, they are perhaps the most investigated materials for enzymatic functionalization [111].

4.5.1 Biofunctionalization of Polyester Fibers

PET is a widely used and extensively studied polymer due to its high tenacity, chemical and environmental resistance, and high glass transition temperature. Enzymes such as cutinases, esterases, and lipases have been shown to hydrolyze the ester link in PET. Proper usage of these enzymes results in the partial hydrolysis of ester bonds in PET chains on the fiber surface, producing defined carboxylic and hydroxyl groups on the polymer surface without affecting the bulk properties of the polymer. Cutinases are the most studied and active enzymes in this regard. In nature, cutinases hydrolyze cutin, a polyester found in the outermost parts of most plants. The chemical (ester bonds) and physical (hydrophobicity) similarities between cutin and PET contribute to the high activity of cutinases on this synthetic polymer [115, 129]. The most frequently reported fungal and bacterial sources for cutinase production include *Humicola insolens*, *Fusarium oxysporum*, *F. solani*, *Aspergillus niger*, *A. oryzae*, *A. nidulans*, *Colletotrichum gloeosporioides*, *Thermobifida fusca*, and *Pseudomonas mendocina* [8, 31, 90, 97, 133].

Cutinase can perform surface modification of PET and displays notable stability. Cutinase from *Humicola insolens* has shown 97 ± 3% degradation of low crystallinity PET films [133]. Similarly, the surface hydrolysis of PET polyester by recombinant cutinase from *F. oxysporum* resulted in successful surface modification without quality degradation. The tensile strength and dyeability increased up to 150% K/S . [71]. *T. fusca* cutinase treatment of PET fabrics enhanced the water contact and a 33% dye uptake for Turquoise Blue and Brilliant Red dyes due to improved hydrophilicity of modified PET textiles [168]. PET fabric treatment with recombinant thermophilic cutinase from *Saccharomonospora viridis* increased carboxyl end groups on the fiber surface resulting in a 25% K/S increase in the color strength [75].

Direct evolution and rational enzyme engineering by site-directed mutagenesis are also in use to enhance the hydrolytic rate of cutinases. For example, increased hydrolytic activity of cutinase from *Fusarium solani pisi* was obtained by expanding the active site of the enzyme by site-directed mutagenesis to accommodate a better fit of a larger polymer chain [11]. In another approach, PET hydrolyzing cutinase from *Thermomyces cellullosylitica* was combined with two hydrophobic binding modules, i.e., CBM (cellobiohydrolase I) from *Hypocrea jecorina* and PBM (polyhydroxyalkanoate depolymerase) from *Alcaligenes faecalis* and expressed in *E. coli*. This boosted the enzyme binding to polymer surface and the hydrolytic activity of the enzyme by 3.8 times [129]. The binding modules interact with polymer surfaces through hydrogen bonds and hydrophobic interactions. Increasing the hydrolytic rate by fusing hydrophobins to active cutinase makes this technique potent and applicable to other polymers where enzyme adsorption may be the rate-limiting step [115]. Another study of mutants from two native cutinases from *Thermobifida cellulolytica* revealed that replacing amino acids only on the enzyme surface led to a three-fold increase in activity compared to the native enzyme, resulting in a neutral area on the enzyme polymer interaction surface [61].

Various lipases from A*spergillus sp.*, *Beauveria brongniartii*, *Burkholderia cepacia*, *Candida antarctica*, *Pseudomonas mendocina*, *Thermomyces lanuginosus*, and *Triticum aestivum* have also been used to functionalize PET. The functionalization was assessed using indirect metrics such as increased hydrophilicity, moisture recovery, and dye uptake. However, no hydrolytic confirmation such as identification of hydrolysis products or increase in carboxylic and hydroxyl surface groups from protein adsorption, was provided [115]. In one study, researchers conducted a comparative analysis to evaluate the effects of lipase from *Thermomyces lanuginosus* and cutinase from *Thermobifida fusca* obtained from Novozyme. While the cutinase showed the highest activity, the lipase also demonstrated the ability to hydrolyze PET into MHET (mono 2-hydroxyethyl terephthalate), TA (terephthalic acid), and BHET (bis 2-hydroxyethyl terephthalate). The study reported approximately 100% and 200% increase in K/S values of Astrazone Blue dyeing of PET fabrics hydrolyzed with *T. lanuginosus* lipase and *T. fusca* cutinase, respectively, compared to the control [61]. In addition to adding reactive groups, *T. fusca* enzymes are also reported to cause nanoscale surface modifications, potentially enhancing wicking behavior. Both lipases and cutinases can remove hydrolytic oligoesters such as cyclic trimers from PET fibers, which can cause deposits on the fiber or machinery during hydrothermal treatments like dyeing [115].

4.5.2 Biofunctionalization of Acrylonitriles

The term acrylonitrile encompasses a wide range of polymers that are formed through the polymerization of acrylonitrile. Due to its high reactivity, acrylonitrile can polymerize with various monomers containing unsaturated ethylene groups [145]. Despite having low moisture regain (lower than polyamide but higher than polyester), acrylonitriles exhibit better water transport properties than PET and nylon, though lower toughness. Acrylic fibers in blended fabrics provide a natural feel, unlike PET and polyamides. Despite their presence in the synthetic fiber market, there are few enzymatic methods for functionalizing acrylonitriles. Polyacrylonitriles differ from polyesters and polyamides in having a backbone with overhanging nitrile groups, which are the target of most enzymatic changes. These nitrile groups can be converted into carboxylic acid, enhancing the dye adsorption. For example, nitrilase from *Micrococcus luteus* is reported to convert nitrile groups into carboxylic acid, enhancing Methylene Blue adsorption, a cationic dye. Polyacrylonitriles (PAN) differ from polyesters and polyamides by having a backbone with overhanging nitrile groups, which are the target for enzymatic modifications [115]. The enzymes nitrilase can transform these groups into carboxylic groups increasing the dye absorption. For instance, nitrile hydratase from *Micrococcus luteus* has been reported to convert nitrile groups into carboxylic acid, improving the absorption of Methylene Blue [46]. Similarly, *Nitrile hydratases* can convert nitrile groups into amides which can be further converted to carboxylic acid by amidase. This enzyme system enhances polyacrylonitrile hydrophilicity. Crude enzyme extract from *Rhodococcus rhodochrous* and *Agrobacterium tumefaciens*, containing both with nitrile hydratase and amidase activities, create amides or carboxylic groups on PAN surfaces. The extent of these modifications depends on substrate type, composition, and reaction conditions such as time and temperature [115].

4.5.3 Biofunctionalization of Polyamides

Polyamides belong to a broad polymer family with a repeating amide bond on their backbone. Their strength, flexibility, and resilience make them ideal for various applications, including textiles, automotive parts, and ropes. Polyamide can be functionalized enzymatically either through hydrolysis or oxidation. Hydrolysis of the amide bond between monomers of the polymer backbone creates a hydroxyl end group and a primary amine. The oxidative strategy, which involves a radical process, produces a wider range of products. Various hydrolases such as proteases and esterases, can hydrolyze polyamides. Among these, proteases are most commonly used because the structure of polyamide resembles their natural substrate [42, 115, 117]. Amidase from *Nocardia farcinica* and aminoacylase from *Aspergillus niger* are reported to increase fiber hydrophilicity. Lipase from *T. lanuginosus* and cutinase from *F. solani* have been reported to enhance direct chemical modification and hydrophilicity of polyamides. Among the oxidoreductases, peroxidase from *Bjerkandera adusta* and laccase from *Trametes versicolor* have been reported for oxidative degradation of polyamide fiber, respectively. Enzymes exhibit promising potential for surface polymer functionalization without affecting bulk characteristics, making them more ecologically friendly than current methods [115].

4.6 Enzyme-Assisted Dye Manufacturing

Enzyme-assisted dye manufacture is a process that utilizes enzymes to aid in the production or modification of dyes used in various industries, including textiles, paper, cosmetics, and food. Enzymes, being highly specific biological catalysts, can facilitate and enhance several stages of the dye manufacturing process, ranging from the extraction of natural dyes to the synthesis and modification of synthetic dyes.

Indigo holds significant economic importance, particularly in the textile sector, where it is extensively used for dyeing denim fabrics. Annually, approximately 80,000 tones of indigo are chemically manufactured, relying on non-renewable petrochemicals, resulting in the production of toxic compounds. The indole, precursor of indigo dye, is commonly present in the natural environment. Some microbes can also hydrolyze or transform it. Indio biosynthesis by bacteria was first discovered in 1983 by Ensley and his colleagues when they were working with heterologous expressing of naphthalene dioxygenase (NDO), an oxidoreductase from *Pseudomonas putida* G7. NDO is the most studied enzyme for the indigo biosynthesis. The initial substrate of the reaction is tryptophane which is converted into indole by tryptophanase. NDO catalyzes the transformation of indole into cis-indole-2,3-dihydroindole. This compound undergoes spontaneous dehydration, resulting in the formation of indoxyl. Indoxyl is then oxidized and undergoes dimerization to produce indigo dye (Fig. 13) [91]. Another enzyme, biphenyl dioxygenase, is also capable of synthesizing indigo from indole [100]. Recombinant cytochrome P450 monooxygenases, a specific CYP102A enzyme from *Streptomyces cattleya* (CYP102A_scat),

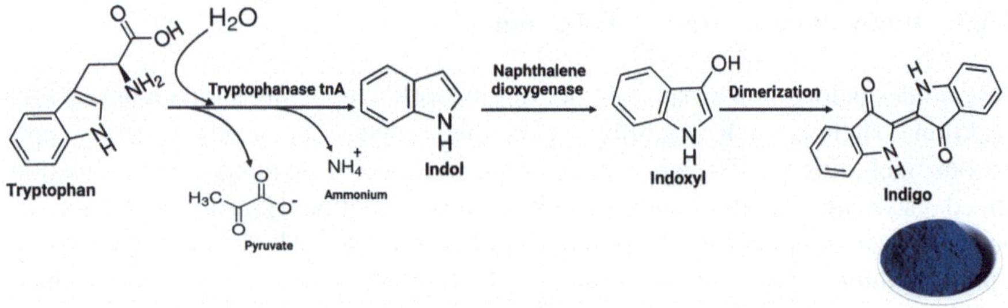

Fig. 13 Pathway for the biosynthesis of indigo dye

expressed in *E.coli* strain BL21 (DE3) have also been reported for significantly enhanced production of indigo. The enzyme achieved an impressive yield of up to 3.8 ± 0.1 g/L, using precursors such as glucose, L-tryptophan and indol [77].

Enzymes are also used for the extraction or modification of natural compounds used as dyes. For instance, enzymes like cellulases or pectinases assist in breaking down plant cell walls to release pigments from natural sources such as turmeric (curcumin), madder root (alizarin), and other plant-based colorants. This process aids in extracting and utilizing natural dyes for textiles [58]. Enzymes are also involved in dye enhancement and have the potential to serve as beneficial additives during the dyeing process of wool, particularly at reduced temperatures [72]. Proteolytic enzymes along with transglutaminase have been reported to serve as additives during the dyeing of wool fabrics with acid dyes [72]. Some fungal pigments, anthraquinone derivatives, resemble vat dyes, making fungi a promising source for the prouction of textile dyes or dye intermediates. Fungi-derived pigments include the vibrant red prodigiosin produced by *Vibrio* spp. and *Serratia* marcescens, the bluish-purple pigment from *Janthinobacterium lividum*, and the violet pigment violacein from *Chromobacterium violaceum*. The current research focus is the development of microbial pigments and their evaluation as textile coloring agents [122].

4.7 Textile Effluent Treatment and Degradation of Dyes

The textile industry is one of the oldest industries that satisfy the basic needs of human civilization i.e., clothing. This industry relies on production of natural, synthetic, and chemical fibers and yarns. As textile demand rises, textile mills and their effluent have also proliferated global pollution. Textile effluents released at different stages are loaded with dyes and auxiliary chemicals (Table 6). Textile industrial waste is a significant concern in many parts of the world due to its intense impact on aquatic ecosystems and human health. Thus, textile effluent must be extensively treated for reuse and environmental protection [3, 17, 66]. Successful textile process pollution reduction requires sustainable dyeing practices, the adaptation of less polluting technologies, effluent treatment, and waste recycling. Physicochemical, chemical, biological, and/or combined treatments are in use for the management of industrial effluents and dyes (Fig. 14). The biological method involves bioremediation of dyes and other chemicals using algae, fungi, and bacteria. Recently, these methods have been used in combination with other methods [3, 17, 66]. Most of these methods are currently based on adsorption or biotransformation of harmful chemicals. Direct use of enzymes for the biotransformation or degradation of dyes and other chemicals is a promising and emerging area of research.

Even with the advancements in dye structure, fiber chemistry, and dyeing equipment, less than 100% of the applied dye binds to fibers. Some dye molecules remain loosely bound to fiber by weak hydrophobic interactions or hydrogen bonding. In addition to excess dye, these unbound dyes must be washed off to maintain quality and enhance color fastness. Thus, significant amounts of dyes are present in the textile effluent. Consequently, proper processing of dye effluent is mandatory to comply with environmental legislation. Investigations are underway to develop effective methods for treating textile effluents containing dyes. The methods using aerobic or anaerobic microbial systems including bacteria, filamentous fungi, yeast, algae, genetically modified microorganisms, and microbial consortia, often combined with advanced oxidation processes are the subject of intense research [115]. Fungal laccases, with their higher redox potential compared to plant or bacterial counterparts, are effective in degrading diverse xenobiotics, including textile dyes found in water and soil [48].

White rot fungi possess a powerful lignin-degrading enzyme system that includes laccases, lignin peroxidases, manganese peroxidases, and multifunctional peroxidases. Due to structural similarities between dyes and lignin, these enzymes are the

Table 6 Textile effluent released during wet processing

Stage	Sizing	Desizing	Scouring	Bleaching	Mercerizing	Dyeing	Printing	Finishing
Effluent Component	• Starch • Waxes • PVA • Polyacrylate • polycyclic acid • carboxymethyl cellulose (CMC)	• Sizes • Enzyme • Starch • alkali • acid • CMC • PVA • fats • waxes • pectins	• NaOH • Soaps • ethers • fats • oils • Pectin • waxes • Surfactants • Defoamers • Lubricants • desiccants • sodium carbonate • Silica • cloth fragments • disinfectant • insecticide residue	• Sodium Hypochlorite • chlorine • sodium hydroxide • hydrogen peroxide • optical brighteners • Acids • Surfactants • sodium silicate • sodium phosphate • organic stabilizer	• Sodium hydroxide • cotton wax	• Dyestuffs • urea • heavy metals • salts • sulfides • surfactant • reducing agents • soda ash • oxidizing agents • acetic acid • detergents • wetting agents	• Pastes • Urea • formaldehyde • gums • oils • binders • soda ash • acids • solvents • thickeners • dyes • reducing agent • alkali	• Inorganic salts • Formaldehyde • resins • waxes • cross-linkers • softeners • Solvents • PVA • hydrocarbon
Nature	High in BOD, COD, SS/ Lower BOD for synthetic sizes	High in BOD, COD,SS, Dissolved salts (DS)	High pH, BOD,COD,SS	High alkalinity, High SS/ Low BOD	High pH, Low BOD, High DS	Strong colored, High BOD, DS, Low SS, Alkaline/Acidic	Highly colored, High BOD, Oily appearance, SS	Slightly alkaline, Low BOD, High pH

COD = chemical oxygen demand, BOD = biological oxygen demand, DS = dissolved solids, and SS = suspended solid.

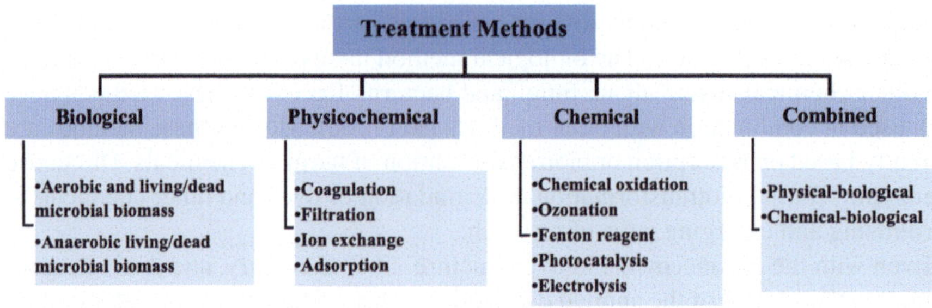

Fig. 14 Methods for the treatment of textile effluent

focus of interest as they can degrade *recalcitrant* substances with lignin-related structures, organophosphorus compounds, aromatic compounds, phenols, and chlorophenols. Numerous fungal species, including *Aspergillus, Bjerkandera, Chrysosporium, Daedalea, Dichomitus, Irpex, Neurospora, Pycnoporus, Pleurotus, Sclerotium, Trametes, and Polyporus*, effectively decolorize azo, triarylmethane, anthraquinonic, and indigo textile dyes. Studies have used various reactor designs, such as stirred tanks, packed beds, airlifts, bubble columns, and revolving discs, with differing decolorization efficacy [115].

Though there has been much research on the biodegradation of textile dyes as a potential solution for treating effluents, relatively less is known about the use of oxidoreductases after washes to eliminate excessive or unbound dye [115]. Laccase, multicopper oxidase, is a very efficient for the elimination of dyes from wastewater before release. Laccases, alone or in conjunction with redox mediators, are promising candidates for degrading various dyes. These enzymes decolorize textile dyes through polymerization or oxidation. Laccases degrade aromatic phenolic compounds, azo dyes, and insecticides by a redox process. Currently, researchers are focussed on isolating laccases that are both pH-stable and thermo-stable, and also demonstrate resistance to organic solvents [138]. Immobilized fungal laccase, in conjunction with SiO_2 support on magnetic separators, has been successfully employed for the efficient removal of azo dyes from wastewater. Additionally, researchers are now exploring nanocomposites of enzymes to remediate wastewater that contains dyes [138].

The limited degradation of stable dyes is attributed to the low redox potential of lignolytic enzymes. Most dye-degrading enzymes work best under acidic conditions, limiting their use on textile dyes. To overcome this limitation, researchers are exploring lignolytic enzymes of halo-alkaliphiles. In one such effort, lignolytic enzymes (lignin peroxidase, manganese peroxidase, and laccase) from *Halopiger aswanensis* produced by fermentation, were used for decolorization of malachite green (MG) and methyl orange. These enzymes efficiently decolorized the dyes like MG, achieving 82–93% decolorization within 2 h using lignin peroxidase and manganese peroxidase and 53.87% degradation of methyl orange with laccase [30]. Furthermore, efforts in genetic engineering and screening are focused on developing or discovering thermostable high redox potential laccases or those that work in salt-alkaline environments [115]. Polyester fibers are usually dyed with dispersion dyes (hydrophobic organic molecules with low water solubility). The most common among these are azo and anthraquinone dyes. Azoreductases are primarily synthesized by bacteria as intracellular or membrane-bounded enzymes. These enzymes are typically thermostable, hydrophilic and demonstrate superior efficiency in breaking down azo dyes (Gabriele Vaz [48]). The core challenge lies in the differential degradation of dyes by enzymes. Enzymes can be employed to eliminate thickeners and surplus dyes following the printing process [115].

5 Enzyme Immobilization for Textile Applications

5.1 Enzyme Immobilizaiton for textile processes

Enzyme immobilization (EI) involves the transformation of biocatalysts from a soluble to an insoluble state by attaching them to a solid support material. This technology has evolved significantly since the first introduction of catalysts by Elizabeth Fulhame in 1794 [67]. The primary objective of EI is to enhance the enzyme stability and functionality while reducing its mobility [49]. EI maintains the catalytic activity of the enzyme while limiting its mobility. Immobilization can be categorized into entrapment, adsorption, covalent ligation, and non-covalent immobilization. EI has diverse applications in various fields, including industrial, medical, bio-research, and food science. The solid support or carrier can be either natural polymers (e.g., cellulose, dextran, agar, and alginate), synthetic polymers (e.g., polystyrene and polyacrylamide), or inorganic materials (e.g., bentonite and silica). Additionally, processed materials such as porous glass and metal oxides [172] can also be employed. Nanomaterials including magnetic, copper, zinc oxide, gold, silica, chitosan-based nanoparticles, and carbon nanotubes are also being used as supports in EI techniques. An adaptable strategy for enhancing EI entails the utilization of three-dimensional (3D) printing scaffolds, which presents a straightforward, pliable, and economically viable method [94, 107, 167]. However, it is notable that each method and carrier for EI has its own merits and demerits. The choice of the method and carrier varies depending upon the type of the enzyme and its intended application.

Enzyme immobilization offers a sustainable approach in the textile industry, ensuring the reuse of enzymes, quick start and stop of reactions, lowering the reaction cost, and preventing deep penetration of enzymes in the fabric (Fig. 15) [32]. This eco-friendly method is particularly beneficial in the pre-treatment of cotton fabrics. Immobilized enzymes like amylase, pectinase, and glucose-oxidase, on carriers like chitosan and Eudragit S-100, have been successfully employed for cotton fabric pre-treatment, for the removal of impurities and maintaining the quality of the fabric over multiple cycles. Furthermore, immobilized enzymes showed increased stability and reusability, overcoming previous limitations of high costs and instability. The cotton fabric pre-treatment with immobilized enzyme showed fabric properties, uniform distribution of dye and fastness comparable to those treated conventionally. Thus, immobilized enzymes are a viable, eco-friendly alternative for textile processing, with performance matching traditional methods in fabric properties and color aspects [94].

Immobilization of cellulases by binding to Eudragit S-100 demonstrated a significant shift towards environmentally sustainable textile processing. The immobilized cellulases, in contrast to their native counterparts, minimize fabric damage by restricting their activity to the surface of fibers, thereby reducing weight loss and preserving tensile strength. This targeted action leads to enhanced fabric qualities such as improved softness, whiteness, and wrinkle recovery. The reusability of immobilized cellulases, retaining substantial activity over multiple cycles, presents an economically efficient approach, aligning with the industry's growing focus on environmental sustainability. Furthermore, the technical advancements in the immobilization process, particularly covalent coupling methods, ensure effective binding of cellulases to polymers, maintaining enzyme activity and enabling controlled interaction with textile fibers. Thus, EI can play a crucial role in developing eco-friendly, efficient, and sustainable textile processing techniques [169, 170].

The commercial protease, Esperase, was immobilized on Eudragit S-100 (a reversible soluble-insoluble polymer) for wool treatment in an attempt to control fiber damage. The enzyme exhibited higher thermal stability compared to the native form. It also showed a shift in optimum pH towards the alkaline side, maintained good storage stability, reusability, and caused less wool fiber damage due to limited enzymatic action on the cuticle surfaces of wool fibers. This approach promises an environmentally friendly alternative for wool shrink-resist finishing, potentially replacing conventional chlorine treatments [146].

Colorants and dyes are prevalently utilized across various sectors, including textiles, plastics, tanning, paper and pulp manufacturing, electroplating, petroleum-based products, pharmaceutical and cosmetic industries. These dyes pose severe,

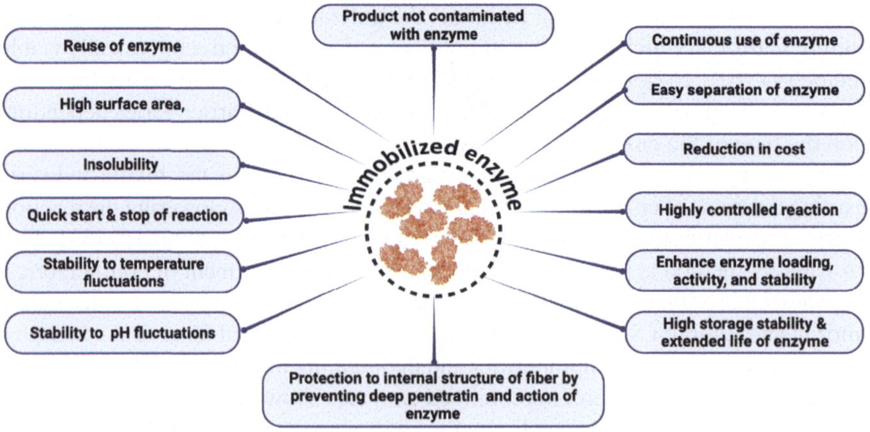

Fig. 15 Advantages of immobilized enzyme in the textile industry

potentially lethal risks to living organisms upon exposure. Their complex nature renders them resistant to traditional wastewater treatment techniques, including coagulation, flocculation, adsorption, filtration, and aerobic/anaerobic biological processes, even when these methods are combined. This resistance has necessitated focused research into innovative treatment technologies. Numerous studies have highlighted the effectiveness of heterogeneous enzymatic systems employing immobilized enzymes as an environmentally friendly and resource-efficient method for eliminating harmful dyes and other pollutants from water (Table 7) [2, 43, 73, 125].

Commercial laccase, immobilized on coconut fiber, has been reported to be economical and effective for repeated textile dye removal. Immobilized glucose oxidase on carbon felt can be employed as a heterogeneous catalyst for bio-Fenton and bio-electro-Fenton reactions to remove water contaminants. In one study, organophosphate degrading enzyme, immobilized on nonwoven polyester textiles, was employed for pesticide-contaminated water cleanup in a continuous system. The enzyme exhibited improved reaction control and reusability for over 60 days [107]

In another study, immobilized horseradish peroxidase (HRP) demonstrated a sustainable capacity to eliminate up to 98% of azo dyes from wastewater in 45 min. It retained 69.6% efficacy after 5 cycles and remained stable for additional 5 cycles [16]. Furthermore, the employment of immobilized microbial cells and enzymes in textile wastewater treatment has been established as an environmentally benign, user-friendly, and economically viable approach. The immobilization technique significantly augments the adaptability, durability, and effectiveness of biological treatment processes, particularly in extracting textile dyes from wastewater. This advancement marks a pivotal step in mitigating environmental issues related to textile manufacturing processes.

5.2 *Enzyme Immobilization for Textile Functionalization*

Fabrics are increasingly being used as support for EI. Industrial enzymes can be immobilized on fabrics through adsorption, covalent bonding, and entrapment. It enhances their activity and stability in numerous applications and creates new functionalized textile products. EI on textiles creates smart materials with unique capabilities including self-detoxification and antibacterial activity. Military applications involve covalently immobilizing organophosphorus hydrolases on cotton to detoxify warfare chemicals. Antibacterial functionalization in textiles potentially prevents disease transmission among consumers and the military [32]. Ester-crosslinked and Cu-chelated cotton fabrics coated with immobilized α-amylase, laccase, and alkaline pectinase showed antibacterial efficacy against gram-negative and gram-positive bacteria and fungi. The best activity was obtained for alkaline pectinase, followed by α-amylase, and laccase, regardless of the microorganism. The treated fabric maintained antimicrobial action for over 10 wash cycles [64].

Table 7 Role of immobilized enzymes in wastewater treatment

Pollutants name	Enzyme used	Pollutants category	Pollutant removal (%)	Reusability cycles
Triclosan	Laccase	PEDCs	65.0%	NR
2,4,6-trichlorophenol	Laccase	PEDCs	85.0%	10
Bisphenol A	Laccase	PEDCs	90.0%	NR
Bisphenol A, Triclosan and 2,4-dichlorophenol	Laccase	PEDCs	92.6–99.7%	10
Chlortetracycline, Carbamazepine, and Diclofenac	Laccase	Pharmaceuticals	48.6–72.7%	10
Catechol	Laccase	PEDCs	39.0% (COD)	NR
2,4,6-trichlorophenol	Laccase	PEDCs	95.4%	7
Naproxen, diclofenac	Laccase	Pharmaceuticals	90%	5
Chlortetracycline	Laccase	Pharmaceuticals	58.3%	7
Reactive Black 5, Reactive Yellow 15, Reactive Blue 114, Reactive Yellow 176, Reactive Red 180, Reactive Red 239, an	Laccase	Dyes	90–96% (individual) −45–63% (mixed)	2
Methyl parathion, Coumaphos	Organophosphate hydrolase	Pesticides	~ 100%	NR
Remazol blue RR	Glucose oxidase	Dyes	93.0%	6
Staphylococcus epidermidis, Escherichia coli	Glucose oxidase	Pathogenic contaminants	99.9%	NR
Methyl orange	Horseradish peroxidase	Dyes	98.0%	10
Pentachlorophenol	Horseradish peroxidase	PEDCs	83.0%	8
Phenol	Horseradish peroxidase	PEDCs	29.68%	25
Paracetamol	Horseradish peroxidase	Pharmaceuticals	83.5%	7
Bisphenol A	Tyrosinase	PEDCs	80.0%	10

Adapted from Morshad et al. [107].

In another study, peroxidase (HRP) immobilized on nonwoven polyester textiles with glutaraldehyde as a crosslinking agent retained 85% of its activity after 4 weeks of storage at 4°C. In contrast, free enzyme lost 90% of its activity. Catalase is commonly used to bleach textiles and sterilize liquid food by converting leftover peroxide to oxygen and water. Textiles have also been utilized for catalase immobilization. Photochemically immobilizing catalase on polyester and polyamide 6 with diallylphalate or cyclohexane-1, 4-dimethanoldivinylether crosslinking was stable and had a 3.5 times higher activity after 20 cycles than the free enzyme [32]. Several textiles are being used to immobilize enzymes as shown in Table 8 [107].

EI presents several advantages to the textile industry (Fig. 15). There are still challenges such as limited industrial acceptance due to high costs, the need for specialized equipment, and a reduction in enzyme activity and efficiency over multiple reuse cycles. Furthermore, it is also challenging to ensure uniform treatment and maintain consistent quality of textiles while using immobilized enzymes. Therefore, researchers are continuously trying to solve these issues. Despite these challenges, enzyme immobilization in the textile industry signifies a shift towards more sustainable and eco-friendly practices. While the technology faces these challenges, its potential benefits in terms of environmental impact, economic efficiency, and product quality make it a promising area for future development. Continuous advancements and innovations in this field are expected to overcome current limitations and broaden the scope of applications.

6 Enzymes as a Sustainable Approach in the Textile Industry

In textile processing, the utilization of enzymes offers several distinct advantages (Table 9). Firstly, enzymes are biodegradable, thus reducing the environmental impact of textile procedures. Their use results in lower pollution levels and a decline in hazardous waste generation, positioning them as a more sustainable solution compared to the traditional chemical processess [147]. Furthermore, enzymes not only maintain the natural quality of textiles but also enhance specific attributes. An illustrative example of this is the use of cellulases in denim finishing, which confers a premium quality fade and a softer texture, superior to conventional methods [24]. Additionally, enzymatic procedures frequently function at comparatively milder conditions ,characterized by lower temperatures and neutral pH. This results in significant energy conservation and a reduction in water consumption, an attribute especially valuable in areas grappling with water shortages. Cutting-edge developments in the field have underscored the efficacy of laccases in decolorizing textile waste, emphasizing the promise enzymes hold in waste treatment and recycling endeavors. Another salient advantage is the inherent specificity of enzymes, which guarantees precise action, safeguarding fibers from potential damage and ensuring the uniform quality of textile end-products [83].

The utilization of enzymes in textile manufacturing processes has introduced various innovative solutions and eco-friendly alternatives. However, the integration of enzymes in textile operations is not devoid of challenges. One of the primary concerns is the potential variatons in enzymes activities under different process conditions, which may lead to inconsistent treatment outcomes in textiles [102]. The specificity of enzymes necessitates rigorous parametric optimization such as pH, temperature, and

Table 8 Performance and stability of immobilized enzymes on textile matrix

Enzyme	Supporting chemicals/ treatments	Textile carrier matrix	Stability of biocatalysts immobilized textile	Performance
α-amylase	Cyanuric chloride	Acrylic textile	Retained 65% of the initial activity after storage at 4 °C for 8 weeks providing retained 53% activity after 15 reuses, improved thermal stability, pH, and heavy metal	81% of immobilized enzymes maintained their activity
β-galactosidase	Plasma treatment and $CaCl_2$	Polyester nonwoven	Exhibited potential for over 15 reuses	90% of immobilized absorbed remained active
Glucose oxidase	Plasma treatment	Carbon felt	Maintained 60% of their activity after 6 reuses	Resultant biocatalysts immobilized textiles possess good electrical and electrochemical characteristics
Laccase	No supporting chemical was used	Coconut fiber-based textile	Improved thermal and operational stabilities, maintaining 45% of activity after 13 reuses	High enzyme loading and excellent decolorization of all reactive colors tested in the study
Horseradish peroxidase	Cyanuric chloride	Acrylic textile	Retained 78% of original activity after 10 reuses, improved stability toward metal ions, temperature, pH, and some organic solvents	Removal of phenol from wastewater
Lipase	Glutaraldehyde	Polyvinyl alcohol/ alginate, polyethylene oxide/alginate nanofibrous textile	Maintained up to 60% of their activities after 14 reuses with improved pH and thermal stability	Effective in biodiesel production, transesterification, aminolysis, and alcoholysis

Table 9 Merits and demerits of enzymes and traditional textile methods

Enzymatic methods	
Advantages	Disadvantages
Eco-Friendly	Specificity
Energy and Water Savings	Cost
Improved Fabric Properties	Processing Time
Reduced Chemical Usage	Temperature and pH Sensitivity
Traditional Methods	
Advantages	Disadvantages
Proven Effectiveness	Environmental Impact
Familiarity and Expertise	Limited Color Range
Versatility	Health and Safety Concerns
Cost-Effective	Waste Generation

substrate concentration etc., often making the processes intricate and necessitating a deeper understanding of the underlying biochemistry [68]. Scaling up enzyme-based processes from laboratory to industrial levels also poses challenges, especially in ensuring uniform treatment across vast quantities of textile materials. Moreover, there are concerns about the potential for microbial contamination, especially when utilizing enzymes derived from microbial sources. Such challenges highlight the need for rigorous research and process controls to harness the full potential of enzymes in the textile industry (ul Aleem 2013).

In addressing the Sustainable Development Goals (SDGs) within the context of the textile and fashion industry, a holistic and multifaceted approach is imperative. The textile industry, accounting for a significant proportion of global greenhouse gas emissions and substantial environmental degradation, necessitates a transition towards more sustainable practices. This encompasses not only environmental considerations but also socio-economic factors. The use of enzyme-based processes will lead the textile industry to move toward sustainable industrialization (SDG 9) and responsible consumption and production (SDG 12) as evidenced by the adoption of plant-based materials and digital innovations [156]. Implementing enzyme-based practices will significantly reduce carbon footprints through energy-efficient processes, which is integral to mitigating climate change (SDG 13). The detrimental impact of dyes, microfibers, and other harmful pollutants on aquatic and terrestrial life will be mitigated by using biological and enzyme-deriven solutions in textile manufacturing and effluent treatment, aligning with the goals related to life below water (SDG 14) and life on land (SDG 15).

7 Future Prospects

The transformation of the textile landscape is primarily motivated by the imperative for sustainable and eco-friendly practices. Enzymes have become critical components in textile manufacturing. Enzymes such as lipases, laccases, pectinases, cellulases, catalases, amylases, and proteases are at the forefront. These enzymes are playing a pivotal role in the processes that once relied heavily on hazardous chemicals, as well as being intensive in term of water, energy, and time [82, 141]. Projections for the future role of enzymes in textiles indicate a groundbreaking shift. Enhancing productivity and efficiency is essential to establish the economic viability of the enzyme-based process compared to traditional methods.

Future research is imperative for the biomodification of both synthetic and natural fibers to enhance their properties. Other emerging trends include green synthesis of dyes, polymer grafting, and biomodification of natural and synthetic fibers [41, 103, 142]. With the advent of the biocatalytic processes for synthetic polymers, circular textile stands out as a significant development. This innovative approach involves a synergistic blend of multiple catalysts, and its potential extends beyond traditional textile materials, holding promise for broader applications.

The textile industry, a major contributor to pollution, is a critical target for green technology. Notably, enzyme-based bioremediation of textile effluent adulterated with a range of dyes and chemicals is a promising area of research. Enzymes offer sustainable solutions to mitigate the environmental impact of textile waste.

Enzyme immobilization stands at the forefront of cutting-edge textile processes. Immobilized enzymes are desirable due to several advantages such as reusability, longer shelf-life, and cost-effectiveness. However, despite significant progress in the field and ongoing research efforts, it remains an active and evolving area of research. Researchers are continuously exploring novel methods, carriers, and enzymes that can address the current limitations and challenges of immobilized enzymes in textiles to develop more economical and efficient processes [107, 132]. Furthermore, the ongoing focus of research is to optimize the immobilization process for industrial applications, etend shelf-life, and maximize reuse cycles [39].

Novel and integrated approaches such as DNA technologies, bioinformatics, metabolic engineering, and enzyme engineering are being employed to explore and gain a deeper understanding of novel metabolic pathways, growth patterns of host organisms, and biocatalysts that can bring transformative and sustainable shifts in the textile industry. The realm of genetic engineering combined with these approaches presents unprecedented opportunities for developing enzymes with superior characteristics and enhanced performance. Such enzymes are anticipated to find applications in various processes within the textile industry, serving as sustainable and eco-friendly alternatives in the foreseeable future [32, 122]. With the promising prospects of enzymes in the textile industry, they are expected to expand their role across many areas of the textile industry, gradually replacing the traditional methods still in practice. However, challenges such as scalability, economic feasibility, and the development of appropriate regulatory frameworks require further exploration to fully unlock their potential.

The prospects of enzyme application in textiles extend beyond mere processing enhancements to the potential biotechnological customization of microorganisms, enabling the production of specialized enzymes tailored to specific industrial requirements. This capability underscores the transformative potential of enzyme technology for the future of textile manufacturing. The convergence of scientific innovation, sustainability, and industry promises to redefine and revolutionize the very fabric of this age-old trade [58].

References

1. Industry Insights (2023) Industrial enzymes market size, challenges, scopes, market share, revenue, and forecasts 2023 to 2030
2. Aber S, Mahmoudikia E, Karimi A, Mahdizadeh F (2016) Immobilization of glucose oxidase on Fe 3 O 4 magnetic nanoparticles and its application in the removal of acid yellow 12. Water, Air, Soil Pollut 227:1–11
3. Adane T, Adugna AT, Alemayehu E (2021) Textile Industry Effluent Treatment Techniques. J Chem 2021:1–14. https://doi.org/10.1155/2021/5314404
4. Adeel S, Kamal S, Ahmad T, Bibi I, Rehman S, Kamal A, Saleem A (2020) Biotechnology: An Eco-friendly Tool of Nature for Textile Industries. In: Shahid M, Adivarekar R (eds) Adv Funct Finish Textes. pp 85–114
5. Ahmed HE, Kolisis FN (2011) An investigation into the removal of starch paste adhesives from historical textiles by using the enzyme α-amylase. J Cult Herit 12:169–179
6. Ahmed T, Mia R, Toki GFI, Jahan J, Hasan MM, Tasin MAS, Farsee MS, Ahmed S (2021) Evaluation of sizing parameters on cotton using the modified sizing agent. Clean Eng Technol 5:100320
7. Akkaya A, Ozseker EE (2019) Modification of polyacrylonitrile fabric for antibacterial application by tetracycline immobilization. Polym Test 78:105959. https://doi.org/10.1016/j.polymertesting.2019.105959
8. Alisch-Mark M, Herrmann A, Zimmermann W (2006) Increase of the Hydrophilicity of Polyethylene Terephthalate Fibres by Hydrolases from Thermomonospora fusca and Fusarium solani f. sp. pisi. Biotechnol Lett 28:681–685. https://doi.org/10.1007/s10529-006-9041-7
9. Anand P, Pandey JP, Pandey DM (2021) Study on cocoonase, sericin, and degumming of silk cocoon: computational and experimental. J Genet Eng Biotechnol 19:32. https://doi.org/10.1186/s43141-021-00125-2
10. Anastas PT, Warner JC (2000) Green chemistry: theory and practice. Oxford University Press
11. Araújo R, Silva C, O'Neill A, Micaelo N, Guebitz G, Soares CM, Casal M, Cavaco-Paulo A (2007) Tailoring cutinase activity towards polyethylene terephthalate and polyamide 6,6 fibers. J Biotechnol 128:849–857. https://doi.org/10.1016/j.jbiotec.2006.12.028
12. Arshad M (2010) Bioethanol: A sustainable and environment friendly solution for Pakistan. A Scientific J COMSATS–Science Vision 2011:16–7
13. Arshad M, Adil M, Sikandar A (2014) Hussain T (2014) Exploitation of meat industry by-products for biodiesel production: Pakistan's perspective. Pakistan J Life Soc Sci. 12:120–125
14. Arshad M (2017) Clean and sustainable energy technologies. InClean Energy Sustain Dev 1 (pp. 73–89). Academic Press

15. Arshad M, Ansari AR, Qadir R, Tahir MH, Nadeem A, Mehmood T, Alhumade H, Khan N (2022) Green electricity generation from biogas of cattle manure: an assessment of potential and feasibility in Pakistan. Front Energy Res 25(10):911485
16. Arslan M (2011) Immobilization horseradish peroxidase on amine-functionalized glycidyl methacrylate-g-poly (ethylene terephthalate) fibers for use in azo dye decolorization. Polym Bull 66:865–879
17. Azanaw A, Birlie B, Teshome B, Jemberie M (2022) Textile effluent treatment methods and eco-friendly resolution of textile wastewater. Case Stud Chem Environ Eng 6:100230. https://doi.org/10.1016/j.cscee.2022.100230
18. Babu KM (2019) 5 - The dyeing of silk. In: Babu KM (ed) Silk (Second Edition). Woodhead Publishing, pp 109–128
19. Banik S, Biswas S, Karmakar S (2018) Extraction, purification, and activity of protease from the leaves of Moringa oleifera. F1000Research 7
20. Bano I, Ghauri MA, Arshad M, Yasin T, Younus M (2017) Bioactivity of variant molecular weight chitosan against drug-resistant bacteria isolated from human wounds. Microb Drug Resist 1;23(8):958–65.
21. Bano I, Arshad M (2018) Climatic changes impact on water availability. Perspect Water Usage Biofuels Prod: Aquat Contam Clim Chang 2018:39–54
22. Basu S, Saha MN, Chattopadhyay D, Chakrabarti K (2009) Degumming and characterization of ramie fibre using pectate lyase from immobilized Bacillus pumilus DKS1. Lett Appl Microbiol 48:593–597. https://doi.org/10.1111/j.1472-765X.2009.02584.x
23. Bhala R, Dhandhania V, Periyasamy AP (2012) Bio-finishing of fabrics. Asian Dye 9:45–49
24. Bhat MMKM (2000) Cellulases and related enzymes in biotechnology. Biotechnol Adv 18:355–383. https://doi.org/10.1016/S0734-9750(00)00041-0
25. Breslauer DN, Kaplan DL (2012) Silks. In: Matyjaszewski K, Möller M (eds) Polymer Science: A Comprehensive Reference. Elsevier, Amsterdam, pp 57–69
26. Canal JM, Navarro A, Calafell M, Rodriguez C, Caballero G, Vega B, Canal C, Paul R (2004) Effect of various bioscouring systems on the accessibility of dyes into cotton. Color Technol 120:311–315
27. Cao Y, Li X, Ge J (2021) Enzyme catalyst engineering toward the integration of biocatalysis and chemocatalysis. Trends Biotechnol 39:1173–1183
28. Cassidy T, Goswami P (2017) Textile and clothing design technology. CRC Press
29. Çelem EB, Önal S (2022) Removal of Raffinose Family Oligosaccharides from Soymilk by α-Galactosidase Immobilized on Sepabeads EC-EA and Sepabeads EC-HA. ACS Food Sci Technol 2:1266–1275. https://doi.org/10.1021/acsfoodscitech.2c00115
30. Chauhan AK, Choudhury B (2021) Synthetic dyes degradation using lignolytic enzymes produced from Halopiger aswanensis strain ABC_IITR by Solid State Fermentation. Chemosphere 273:129671. https://doi.org/10.1016/j.chemosphere.2021.129671
31. Chen S, Tong X, Woodard RW, Du G, Wu J, Chen J (2008) Identification and Characterization of Bacterial Cutinase. J Biol Chem 283:25854–25862. https://doi.org/10.1074/jbc.M800848200
32. Choudhury AKR (2014) Sustainable Textile Wet Processing: Applications of Enzymes. In: Subramanian Senthilkannan Muthu (ed) Roadmap Sustain TextEs Cloth Environ Soc Asp Textes Cloth Supply Chain. pp 203–238
33. Chouhan D, Mandal BB (2020) Silk biomaterials in wound healing and skin regeneration therapeutics: From bench to bedside. Acta Biomater 103:24–51. https://doi.org/10.1016/j.actbio.2019.11.050
34. Çiğdem Gürsoy N, Lim S-H, Hinks D, Hauser P (2004) Evaluating hydrogen peroxide bleaching with cationic bleach activators in a cold pad-batch process. Text Res J 74:970–976
35. Cortez J, Bonner P, Griffin M (2012) Application of transglutaminase in the modification of wool textile. Enz Microb Technol 34:64–72. https://doi.org/10.1016/j.enzmictec.2003.08.004
36. Couto SR, Herrera JLT (2006) Industrial and biotechnological applications of laccases: a review. Biotechnol Adv 24:500–513

37. David Troncoso F, Alberto Sánchez D, Luján Ferreira M (2022) Production of plant proteases and new biotechnological applications: an updated review. ChemistryOpen 11. https://doi.org/10.1002/open.202200017
38. Dursun S, Yıldız SZ (2023) Eco-Friendly bleaching of cotton fabrics without heating using direct process water in the presence of sodium chlorite and phosphonate. J Nat Fibers 20. https://doi.org/10.1080/15440478.2022.2146248
39. Edwards JV, Prevost NT, Condon B, French A (2011) Covalent attachment of lysozyme to cotton/cellulose materials: protein verses solid support activation. Cellulose 18:1239–1249. https://doi.org/10.1007/s10570-011-9563-6
40. Eid BM, Ibrahim NA (2021) Recent developments in sustainable finishing of cellulosic textiles employing biotechnology. J Clean Prod 284:124701
41. El-hennawi HME (2021) New Approaches of Biotechnology in Textile Coloration. Egypt J Chem 0–0. https://doi.org/10.21608/ejchem.2020.34098.2714
42. El-Ola SMA, Moharam ME, Eladwi MM, El-Bendary A (2014) Optimum conditions for polyamide fabric modification by protease enzyme produced by Bacillus sp. Indian J Fibre Text Res 39:65–71
43. Eskandarian M, Mahdizadeh F, Ghalamchi L, Naghavi S (2014) Bio-Fenton process for Acid Blue 113 textile azo dye decolorization: characteristics and neural network modeling. Desalin Water Treat 52:4990–4998
44. Farooq A, Ali S, Abbas N, Fatima GA, Ashraf MA (2013) Comparative performance evaluation of conventional bleaching and enzymatic bleaching with glucose oxidase on knitted cotton fabric. J Clean Prod 42:167–171
45. Farooq MA, Ali S, Hassan A, Tahir HM, Mumtaz S, Mumtaz S (2021) Biosynthesis and industrial applications of α-amylase: a review. Arch Microbiol 203:1281–1292. https://doi.org/10.1007/s00203-020-02128-y
46. Fischer-Colbrie G, Matama T, Heumann S, Martinkova L, Cavaco Paulo A, Guebitz G (2007) Surface hydrolysis of polyacrylonitrile with nitrile hydrolysing enzymes from Micrococcus luteus BST20. J Biotechnol 129:62–68. https://doi.org/10.1016/j.jbiotec.2006.11.018
47. Fu K, Liu J, Lu D (2014) The pectin substance used to detect the activity of pectinase suitable for bioscouring. J Text Inst 105:12–19
48. Gabriele Vaz Souza A, Carolini Maria T, Maria Saran L, Maria Carareto Alves L (2022) Enzymatic Bioremediation of Dyes from Textile Industry Effluents. In: Toxic Environ Pollut. IntechOpen
49. Garcia-Galan C, Berenguer-Murcia Á, Fernandez-Lafuente R, Rodrigues RC (2011) Potential of different enzyme immobilization strategies to improve enzyme performance. Adv Synth Catal 353:2885–2904
50. Ghaffari-Moghaddam M, Eslahi H, Omay D, Zakipour-Rahimabadi E (2014) Industrial applications of enzymes. Rev J Chem 4:341–361
51. GlobeNewswire (2022) Global Enzymes Market Size, Segments, Outlook, and Revenue Forecast 2022–2028 by Product Type, Source, Type, Applications, and Major Regions. Res. Mark
52. GlobeNewswire (2023) Global Textile Enzymes Market, By Type (Amylases, Cellulase, Lipases, Proteases, Pectinases, Laccases, and Other Types), By Application (De-Sizing, Bio-Polishing, Bleaching, Bio-Sourcing, and Other Applicat
53. Gomaa SK, Zaki RA, Wahba MI, Taleb MA, El-Refai HA, El-Fiky AF, El-Sayed H (2022) Green method for improving performance attributes of wool fibres using immobilized proteolytic thermozyme. 3 Biotech 12:254. https://doi.org/10.1007/s13205-022-03323-y
54. Goodsell DS (2016) Seven Wonders of the World of Enzymes. Atomic Evidence. Springer International Publishing, Cham, pp 129–138
55. Goswami BC, Anandjiwala RD, Hall D (2004) Textile sizing. CRC Press
56. Gübitz GM, Paulo AC (2001) Biotechnology in the textile industry-perspectives for the new millennium
57. Gurung N, Ray S, Bose S, Rai V (2013) A broader view: microbial enzymes and their relevance in industries, medicine, and beyond. Biomed Res Int 2013:1–18. https://doi.org/10.1155/2013/329121

58. Haddar W, Adeel S, Salman M, Ghaffar A, Naseer M, Usama M, Ben Ticha M (2023) Ecological applications of enzymes in plants based textile dyeing. In: Dye Pigment—Insights Appl. IntechOpen
59. Hassan MM, Carr CM (2019) A review of the sustainable methods in imparting shrink resistance to wool fabrics. J Adv Res 18:39–60. https://doi.org/10.1016/j.jare.2019.01.014
60. Heikinheimo L, Buchert J, Miettinen-Oinonen A, Suominen P (2000) Treating denim fabrics with trichoderma reesei cellulases. Text Res J 70:969–973. https://doi.org/10.1177/004051750007001106
61. Herrero Acero E, Ribitsch D, Dellacher A, Zitzenbacher S, Marold A, Steinkellner G, Gruber K, Schwab H, Guebitz GM (2013) Surface engineering of a cutinase from Thermobifida cellulosilytica for improved polyester hydrolysis. Biotechnol Bioeng 110:2581–2590. https://doi.org/10.1002/bit.24930
62. Herrmann H, Tootle ML (1964) Specific and general aspects of the development of enzymes and metabolic pathways. Physiol Rev 44:289–371
63. Hossain KMG, González MD, Lozano GR, Tzanov T (2009) Multifunctional modification of wool using an enzymatic process in aqueous–organic media. J Biotechnol 141:58–63. https://doi.org/10.1016/j.jbiotec.2009.02.011
64. Ibrahim NA, Gouda M, El-shafei AM, Abdel-Fatah OM (2007) Antimicrobial activity of cotton fabrics containing immobilized enzymes. J Appl Polym Sci 104:1754–1761. https://doi.org/10.1002/app.25821
65. Jacomini D, Bussler L, da-Conceição Silva JL, Maller A, Kadowaki MK, Simão R de CG, (2023) Biopolishing of denim by the recombinant xylanase II of Caulobacter crescentus. Brazilian J Microbiol 54:1559–1564. https://doi.org/10.1007/s42770-023-01056-5
66. Jahan N, Tahmid M, Shoronika AZ, Fariha A, Roy H, Pervez MN, Cai Y, Naddeo V, Islam MS (2022) A comprehensive review on the sustainable treatment of textile wastewater: zero liquid discharge and resource recovery perspectives. Sustainability 14:15398. https://doi.org/10.3390/su142215398
67. Jarvis C (2020) Elizabeth Fulhame, a forgotten chemistry pioneer. Phys Today 2020:0617a. https://doi.org/10.1063/PT.6.4.20200617a
68. Kabir SMM, Koh J (2021) Sustainable textile processing by enzyme applications. Biodegrad Technol Org Inorg Pollut
69. Kadam V, Rani S, Jose S, Shakyawar DB, Shanmugam N (2021) Biomaterial based shrink resist treatment of wool fabric: A sustainable technology. Sustain Mater Technol 29:e00298. https://doi.org/10.1016/j.susmat.2021.e00298
70. Kalia S, Bhattacharya A, Prajapati SK, Malik A (2021) Utilization of starch effluent from a textile industry as a fungal growth supplement for enhanced α-amylase production for industrial application. Chemosphere 279:130554. https://doi.org/10.1016/j.chemosphere.2021.130554
71. Kanelli M, Vasilakos S, Nikolaivits E, Ladas S, Christakopoulos P, Topakas E (2015) Surface modification of poly(ethylene terephthalate) (PET) fibers by a cutinase from Fusarium oxysporum. Process Biochem 50:1885–1892. https://doi.org/10.1016/j.procbio.2015.08.013
72. Karanikas EK, Kosolia CT, Zarkogianni MC, Nikolaidis NF, Tsatsaroni EG (2013) Effect of enzymatic treatment on the dyeing properties of protein wool fibers. Fibers Polym 14:223–229. https://doi.org/10.1007/s12221-013-0223-3
73. Karimi A, Mahdizadeh F, Eskandarian M (2012) Enzymatic in-situ generation of H_2O_2 for decolorization of Acid Blue 113 by fenton process. Chem Ind Chem Eng Quarterly/CICEQ 18:89–94
74. Kaushal J, Mehandia S, Singh G, Raina A, Arya SK (2018) Catalase enzyme: Application in bioremediation and food industry. Biocatal Agric Biotechnol 16:192–199. https://doi.org/10.1016/j.bcab.2018.07.035
75. Kawai F, Kawase T, Shiono T, Urakawa H, Sukigara S, Tu C, Yamamoto M (2017) Enzymatic hydrophilization of polyester fabrics using a recombinant cutinase Cut 190 and their surface characterization. J Fiber Sci Technol 73:8–18. https://doi.org/10.2115/fiberst.fiberst.2017-0002

76. Kettlewell R, De Boos A, Jackson J (2015) Commercial shrink-resist finishes for wool. In: Paul R (ed) Functional Finishes for Textiles. Elsevier, pp 193–226
77. Kim H-J, Jang S, Kim J, Yang Y-H, Kim Y-G, Kim B-G, Choi K-Y (2017) Biosynthesis of indigo in Escherichia coli expressing self-sufficient CYP102A from Streptomyces cattleya. Dye Pigment 140:29–35. https://doi.org/10.1016/j.dyepig.2017.01.029
78. Kim S, Moldes D, Cavaco-Paulo A (2007) Laccases for enzymatic colouration of unbleached cotton. Enzyme Microb Technol 40:1788–1793. https://doi.org/10.1016/j.enzmictec.2007.01.002
79. Koronis G, Silva A (2019) Chapter 5 - Eco-impact assessment of a hood made of a ramie reinforced composite. In: Koronis G, Silva A (eds) Green Composites for Automotive Applications. Woodhead Publishing, pp 99–114
80. Kovačević S, Penava Ž (2004) Impact of sizing on physico-mechanical properties of yarn. Fibres Text East Eur 48:32–36
81. Kuffner H, Popescu C (2012) 8 - Wool fibres. In: Kozłowski RM (ed) Handbook of Natural Fibres. Woodhead Publishing, pp 171–195
82. Kumar B (2023) Dyes and Pigments-Insights and applications
83. Kumar D, Bhardwaj R, Jassal S, Goyal T, Khullar A, Gupta N (2023) Application of enzymes for an eco-friendly approach to textile processing. Environ Sci Pollut Res 30:71838–71848. https://doi.org/10.1007/s11356-021-16764-4
84. Kumar RS, Sundaresan S (2013) Mechanical finishing techniques for technical textiles. Adv Dye Finish Tech Text 135–153
85. Kumar V, Kumar A, Chhabra D, Shukla P (2019) Improved biobleaching of mixed hardwood pulp and process optimization using novel GA-ANN and GA-ANFIS hybrid statistical tools. Bioresour Technol 271:274–282. https://doi.org/10.1016/j.biortech.2018.09.115
86. Landage SM (2022) Biotechnological approaches in desizing of textile materials. In: Appl Biotechnol Sustain TextE Prod. Elsevier, pp 47–73
87. Lau PCK, Grosse S (2013) The fundamentals and fun of biocatalysis. In: Mar Enzym Biocatal. Elsevier, pp 17–69
88. Lenting HBM, Schroeder M, Guebitz GM, Cavaco-Paulo A, Shen J (2006) New Enzyme-based process direction to prevent wool shrinking without substantial tensile strength loss. Biotechnol Lett 28:711–716. https://doi.org/10.1007/s10529-006-9048-0
89. Li Y, Yang Z, Yang K, Wei J, Li Z, Ma C, Yang X, Wang T, Zeng G, Yu G, Yu Z, Zhang C (2022) Removal of chloride from water and wastewater: Removal mechanisms and recent trends. Sci Total Environ 821:153174. https://doi.org/10.1016/j.scitotenv.2022.153174
90. Liang X, Zou H (2022) Biotechnological application of cutinase: a powerful tool in synthetic biology. SynBio 1:54–64. https://doi.org/10.3390/synbio1010004
91. Linke JA, Rayat A, Ward JM (2023) Production of indigo by recombinant bacteria. Bioresour Bioprocess 10:20. https://doi.org/10.1186/s40643-023-00626-7
92. Liu X, Kokare C (2023) Microbial enzymes of use in industry. In: Biotechnol Microb Enzym. Elsevier, pp 405–444
93. Liu Z, Duan S, Sun Q, Peng Y, Feng X, Zheng K, Hu Z, Zhang Y (2012) A rapid process of ramie bio-degumming by Pectobacterium sp. CXJZU-120. Text Res J 82:1553–1559. https://doi.org/10.1177/0040517512444338
94. Madhu A, Chakraborty J (2022) Sustainable Approach for Cotton Fabric Pretreatment with Immobilized Enzymes. Fibers Polym 23:993–1007. https://doi.org/10.1007/s12221-022-4193-1
95. Madhu A, Chakraborty JN (2017) Developments in application of enzymes for textile processing. J Clean Prod 145:114–133. https://doi.org/10.1016/j.jclepro.2017.01.013
96. Madhu A, Chakraborty JN (2018) Recovery and reuse of immobilized α-amylase during desizing of cotton fabric. Res J Text Appar 22:271–290
97. Martínez A, Maicas S (2021) Cutinases: characteristics and insights in industrial production. Catalysts 11:1194. https://doi.org/10.3390/catal11101194
98. McLoughlin J, Hayes SG (2008) 13 - Automated fabric inspection. In: Fairhurst C (ed) Advances in Apparel Production. Woodhead Publishing, pp 266–303

99. Mei J, Zhang N, Yu Y, Wang Q, Yuan J, Wang P, Cui L, Fan X (2018) A novel "trifunctional protease" with reducibility, hydrolysis, and localization used for wool anti-felting treatment. Appl Microbiol Biotechnol 102:9159–9170. https://doi.org/10.1007/s00253-018-9276-y
100. Mendoza-Avila J, Chauhan K, Vazquez-Duhalt R (2020) Enzymatic synthesis of indigo-derivative industrial dyes. Dye Pigment 178:108384. https://doi.org/10.1016/j.dyepig.2020.108384
101. Mojsov K (2018) Enzymatic scouring and bleaching of cotton terry fabrics–opportunity of the improvement on some physicochemical and mechanical properties of the fabrics. J Nat fibers 15:740–751
102. Mojsov K (2012) Enzyme applications in textile preparatory process: A review. Int J Manag IT Eng 2:272–295
103. Mojsov K (2014) Biotechnological applications of laccases in the textile industry. Savrem Tehnol 3:76–79. https://doi.org/10.5937/savteh1401076M
104. Mondal M, Trivedy K, Kumar N (2006) The silk proteins, sericin and fibroin in silkworm, Bombyx mori Linn.—A review. 5
105. Montazer M, Dadashian F, Hemmatinejad N, Farhoudi K (2009) Treatment of wool with laccase and dyeing with madder. Appl Biochem Biotechnol 158:685–693. https://doi.org/10.1007/s12010-008-8403-0
106. Moreno AD, Ibarra D, Eugenio ME, Tomás-Pejó E (2020) Laccases as versatile enzymes: from industrial uses to novel applications. J Chem Technol Biotechnol 95:481–494. https://doi.org/10.1002/jctb.6224
107. Morshed MN, Behary N, Bouazizi N, Guan J, Nierstrasz VA (2021) An overview on biocatalysts immobilization on textiles: Preparation, progress and application in wastewater treatment. Chemosphere 279:130481. https://doi.org/10.1016/j.chemosphere.2021.130481
108. Mueller S, Riedel H-D, Stremmel W (1997) Determination of catalase activity at physiological hydrogen peroxide concentrations. Anal Biochem 245:55–60. https://doi.org/10.1006/abio.1996.9939
109. Mukhopadhyay A, Dutta N, Chattopadhyay D, Chakrabarti K (2013) Degumming of ramie fiber and the production of reducing sugars from waste peels using nanoparticle supplemented pectate lyase. Bioresour Technol 137:202–208. https://doi.org/10.1016/j.biortech.2013.03.139
110. Niaz A, Malik QJ, Muhammad S, Shamim T, Asghar S (2011) Bioscouring of cellulosic textiles. Color Technol 127:211–216
111. Nielsen PH, Kuilderd H, Zhou W, Lu X (2009) Enzyme biotechnology for sustainable textiles. In: Sustainable Textiles. Elsevier, pp 113–138
112. Ninpetch U, Tsukada M, Promboon A (2015) Mechanical properties of silk fabric degummed with bromelain. J Eng Fiber Fabr 10:69–78. https://doi.org/10.1177/155892501501000319
113. Niyonzima FN, More VS, Nsanganwimana F, Rao AS, Nair A, Anantharaju KS, More SS (2023) Chapter 23 - Microbial enzymes used in textile industry. In: Brahmachari G (ed) Biotechnology of Microbial Enzymes (Second Edition). Academic Press, pp 649–684
114. Nupur R, Kaul A, Sharma P, Kumar S, Barrueta M, Jha PC (2020) Block printing process performance evaluation and improvement using FMEA and Taguchi loss function for selecting print paste suppliers in apparel industry. Int J Adv Oper Manag 12:237–272
115. Nyanhongo G, Herrero Acero E, Matuchaki MDDJ, Rau M, Guebitz G, Andreaus J (2016) 2. Microbial applications for fabric and textile industries. In: Microb Appl. De Gruyter, pp 33–78
116. Parvinzadeh Gashti M (2012) Surface modification of synthetic fibers to improve performance: Recent approaches. Glob J Phys Chem 3:1–10
117. Parvinzadeh M, Assefipour R, Kiumarsi A (2009) Biohydrolysis of nylon 6,6 fiber with different proteolytic enzymes. Polym Degrad Stab 94:1197–1205. https://doi.org/10.1016/j.polymdegradstab.2009.04.017
118. Paul R (2015) Functional finishes for textiles: an overview. In: Paul R (ed) Functional Finishes for Textiles. Elsevier, pp 1–14

119. Prajapati CD, Smith E, Kane F, Shen J (2018) Laccase-catalysed coloration of wool and nylon. Color Technol 134:423–439. https://doi.org/10.1111/cote.12350
120. Puranen T, Alapuranen M, Vehmaanperä J (2014) Chapter 26 - Trichoderma Enzymes for Textile Industries. In: Gupta VK, Schmoll M, Herrera-Estrella A, Upadhyay RS, Druzhinina I, Tuohy MG (eds) Biotechnology and Biology of Trichoderma. Elsevier, Amsterdam, pp 351–362
121. Purushothama B (2019) Handbook of value addition processes for fabrics. Woodhead Publishing India PVT, Limited
122. Radhakrishnan S (2014) Application of biotechnology in the processing of textile fabrics. In: Subramanian Senthilkannan Muthu (ed) Roadmap to Sustainable Textiles and Clothing Environmental and Social Aspects of Textiles and Clothing Supply Chain. pp 277–325
123. Ragab MM, Othman H, Hassabo A (2022) An overview of printing textile techniques. Egypt J Chem 65:749–761
124. Rahman M, Bhowmik A, Das S, Chowhan K, Biswas T (2020) Green degumming of silk by enzyme extracted from natural sources. J Mater Sci Chem Eng 08:30–40. https://doi.org/10.4236/msce.2020.88003
125. Ravi S, Lonappan L, Touahar I, Fonteneau É, Vaidyanathan VK, Cabana H (2020) Evaluation of bio-fenton oxidation approach for the remediation of trichloroethylene from aqueous solutions. J Environ Manage 270:110899
126. Ray Chaudhuri S, Gogoi M, Biswas T, Chatterjee S, Chanda C, Jamatia R, Modak A, Sett SK, Mukherjee I (2020) Optimization of bio-chemical degumming of Ramie fiber for improved strength & luster. Biotechnol Reports (Amsterdam, Netherlands) 28:e00532. https://doi.org/10.1016/j.btre.2020.e00532
127. Rehman A, Saeed A, Asad W, Khan I, Hayat A, Rehman MU, Shah TA, Sitotaw B, Dawoud TM, Bourhia M (2023) Eco-friendly textile desizing with indigenously produced amylase from Bacillus cereus AS2. Sci Rep 13:1–9. https://doi.org/10.1038/s41598-023-38956-3
128. Riaz R, Ashraf M, Hussain N, Baqar Z, Bilal M, Iqbal HMN (2023) Redesigning robust biocatalysts by engineering enzyme microenvironment and enzyme immobilization. Catal Letters 153:1587–1601
129. Ribitsch D, Yebra AO, Zitzenbacher S, Wu J, Nowitsch S, Steinkellner G, Greimel K, Doliska A, Oberdorfer G, Gruber CC, Gruber K, Schwab H, Stana-Kleinschek K, Acero EH, Guebitz GM (2013) Fusion of binding domains to Thermobifida cellulosilytica Cutinase to tune sorption characteristics and enhancing PET hydrolysis. Biomacromol 14:1769–1776. https://doi.org/10.1021/bm400140u
130. Rippon JA (2008) 7 - Friction, felting and shrink-proofing of wool. In: Gupta BS (ed) Friction in Textile Materials. Woodhead Publishing, pp 253–291
131. Rippon JA (2013) The structure of wool. In: The Coloration of Wool and other Keratin Fibres. John Wiley & Sons, Ltd, pp 1–42
132. Rodrigues RC, Ortiz C, Berenguer-Murcia Á, Torres R, Fernández-Lafuente R (2013) Modifying enzyme activity and selectivity by immobilization. Chem Soc Rev 42:6290–6307. https://doi.org/10.1039/C2CS35231A
133. Ronkvist ÅM, Xie W, Lu W, Gross RA (2009) Cutinase-Catalyzed Hydrolysis of Poly(ethylene terephthalate). Macromolecules 42:5128–5138. https://doi.org/10.1021/ma9005318
134. Rony SY (2021) An Overview of Biopolishing in Textile. In: Text. Learn
135. Sharif N, Munir N, Hasnain M, Naz S, Arshad M (2021) Environmental impacts of ethanol production system. Sustain Ethanol Clim Chang: Sustain Assess Ethanol Distill. 2021:205–223
136. Salwan R, Sharma V, Das S (2022) Phylogenetic and structural analysis of bacterial nitrilases for the biodegradation of nitrile compounds. Curr Protein Pept Sci 23:874–882. https://doi.org/10.2174/1389203723666220921154409
137. Samant L, Jose S, Rose NM, Shakyawar DB (2022) Antimicrobial and UV protection properties of cotton fabric using enzymatic pretreatment and dyeing with Acacia catechu. J Nat Fibers 19:2243–2253

138. Sarkar S, Soren K, Chakraborty P, Bandopadhyay R (2020) Application of enzymes in textile functional finishing. In: Shahid M, Adivarekar R (eds) Adv Funct Finish TextEs. pp 115–127
139. Schumacher KA, Forster AL (2022) Textiles in a circular economy: An assessment of the current landscape, challenges, and opportunities in the United States. Front Sustain 3:146
140. Scott A (2015) Cutting out textile pollution. Chem Eng News 93:18–19
141. Shaarawy S (2020) A review on the development of innovative capabilities in the textile finishing of natural fibers. Egypt J Chem 0–0. https://doi.org/10.21608/ejchem.2019.19009.2169
142. Shahid M, Mohammad F, Chen G, Tang R-C, Xing T (2016) Enzymatic processing of natural fibres: white biotechnology for sustainable development. Green Chem 18:2256–2281. https://doi.org/10.1039/C6GC00201C
143. Sharma H, Upadhyay SK (2020) Enzymes and their production strategies. In: Biomass, biofuels, biochemicals. Elsevier, pp 31–48
144. Silva AS, Costa EC, Reis S, Spencer C, Calhelha RC, Miguel SP, Ribeiro MP, Barros L, Vaz JA, Coutinho P (2022) Silk sericin: a promising sustainable biomaterial for biomedical and pharmaceutical applications. Polymers (Basel) 14:4931. https://doi.org/10.3390/polym14224931
145. Silva C, Cavaco-Paulo AM, Fu JJ (2015) Enzymatic biofinishes for synthetic textiles. In: Funct Finish TextEs. Elsevier, pp 153–191
146. Silva CJSM, Zhang Q, Shen J, Cavaco-Paulo A (2006) Immobilization of proteases with a water soluble–insoluble reversible polymer for treatment of wool. Enzyme Microb Technol 39:634–640. https://doi.org/10.1016/j.enzmictec.2005.11.016
147. Singh A, Varghese LM, Battan B, Patra AK, Mandhan RP, Mahajan R (2020) Eco-friendly scouring of ramie fibers using crude xylano-pectinolytic enzymes for textile purpose. Environ Sci Pollut Res 27:6701–6710
148. Singh R, Kumar M, Mittal A, Mehta PK (2016) Microbial enzymes: industrial progress in 21st century. 3 Biotech 6:174. https://doi.org/10.1007/s13205-016-0485-8
149. Smith E, Shen J (2011) Surface modification of wool with protease extracted polypeptides. J Biotechnol 156:134–140. https://doi.org/10.1016/j.jbiotec.2011.08.012
150. Sneha PJP, Pandey DM (2022) Evaluating the role of trypsin in silk degumming: An in silico approach. J Biotechnol 359:35–47. https://doi.org/10.1016/j.jbiotec.2022.09.007
151. Solbak AI, Richardson TH, McCann RT, Kline KA, Bartnek F, Tomlinson G, Tan X, Parra-Gessert L, Frey GJ, Podar M (2005) Discovery of pectin-degrading enzymes and directed evolution of a novel pectate lyase for processing cotton fabric. J Biol Chem 280:9431–9438
152. Sooch BS, Kauldhar BS, Puri M (2014) Recent insights into microbial catalases: Isolation, production and purification. Biotechnol Adv 32:1429–1447. https://doi.org/10.1016/j.biotechadv.2014.09.003
153. Sreelakshmi SN, Vasanthi NS, Saravanan D (2013) Acidic bacterial proteases from Bacillus species as an alternative agent for scouring of cotton fabrics. J Text Inst 104:1118–1124
154. Takebira UM, Mohibullah ATM, Hossain MZ, Tanim SI, Redoy MSA, Rahman ML, Rejwan MS (2020) A regular study on yarn count, size box temperature and machine speed to increase weaving efficiency of cotton fabric. J Text Sci Technol 6:168
155. Tesfaw A (2014) Applications of transglutaminase in textile, wool, and leather processing. Int J Text Sci 3:64–69. https://doi.org/10.5923/j.textile.20140304.02
156. Thakker AM, Sun D (2023) Sustainable development goals for textiles and fashion. Environ Sci Pollut Res 30:101989–102009. https://doi.org/10.1007/s11356-023-29453-1
157. Ul-Haq N, Nasir H (2012) Cleaner production technologies in desizing of cotton fabric. J Text Inst 103:304–310
158. ul Aleem A (2013) An investigation of alternatives to reductive clearing in the dyeing of polyester
159. Verma P, Madamwar D (2002) Decolorization of synthetic textile dyes by lignin peroxidase of Phanerochaete chrysosporium. Folia Microbiol (Praha) 47:283–286. https://doi.org/10.1007/BF02817653

160. Vidya M, Rajagopal S (2021) Silk Fibroin: A Promising Tool for Wound Healing and Skin Regeneration. Int J Polym Sci 2021:e9069924. https://doi.org/10.1155/2021/9069924
161. Vigneswaran C, Ananthasubramanian M, Kandhavadivu P (2014) Bioprocessing of textiles. Elsevier
162. Wang N, Tang P, Zhao C, Zhang Z, Sun G (2020) An environmentally friendly bleaching process for cotton fabrics: mechanism and application of UV/H_2O_2 system. Cellulose 27:1071–1083. https://doi.org/10.1007/s10570-019-02812-3
163. Wilson TJ, Lilley DMJ (2015) RNA catalysis—is that it? RNA 21:534–537. https://doi.org/10.1261/rna.049874.115
164. Wu XT, Ansari AR, Pang XX, Li HZ, Zhang ZW, Luo Y, Arshad M, Song H (2019) Visfatin plays a significant role in alleviating lipopolysaccharide-induced apoptosis and autophagy through PI3K/AKT signaling pathway during acute lung injury in mice. Arch Immunol Ther Exp 1(67):249–261
165. Wulandari AP, Awis VPD, Budiono R, Kusmoro J, Hidayat SS, Masruchin N, Lubis MAR, Fatriasari W, Rachmawati U (2023) Tensile Strength Improvements of Ramie Fiber Threads through Combination of Citric Acid and Sodium Hypophosphite Cross-Linking. Mater (Basel, Switzerland) 16:4758. https://doi.org/10.3390/ma16134758
166. Xu F, Salmon S (2008) Potential applications of Oxidoreductases for the Re-oxidation of Leuco Vat or Sulfur Dyes in textile Dyeing. Eng Life Sci 8:331–337. https://doi.org/10.1002/elsc.200700070
167. Ye J, Chu T, Chu J, Gao B, He B (2019) A versatile approach for enzyme immobilization using chemically Modified 3D-Printed scaffolds. ACS Sustain Chem Eng 7:18048–18054. https://doi.org/10.1021/acssuschemeng.9b04980
168. Ye R, Wang Q, Fan X, Yu Y, Yuan J, Wang P (2013) A new model substrate for cutinase hydrolyzing polyethylene terephthalate. Fibers Polym 14:1128–1133. https://doi.org/10.1007/s12221-013-1128-x
169. Yu Y, Yuan J, Wang Q, Fan X, Ni X, Wang P, Cui L (2013) Cellulase immobilization onto the reversibly soluble methacrylate copolymer for denim washing. Carbohydr Polym 95:675–680. https://doi.org/10.1016/j.carbpol.2013.03.043
170. Yu Y, Yuan J, Wang Q, Fan X, Wang P, Cui L (2015) Noncovalent immobilization of cellulases using the reversibly soluble polymers for biopolishing of cotton fabric. Biotechnol Appl Biochem 62:494–501. https://doi.org/10.1002/bab.1289
171. Zaini MS, Arshad M, Syed-Hassan SS (2023) Adsorption isotherm and kinetic study of methane on palm kernel shell-derived activated carbon. J Bioresour Bioprod 1; 8(1):66–77.
172. Zdarta J, Meyer AS, Jesionowski T, Pinelo M (2018) A general overview of support materials for enzyme immobilization: Characteristics, properties, practical utility. Catalysts 8. https://doi.org/10.3390/catal8020092
173. Zhang N, Huang P, Wang P, Yu Y, Zhou M, Wang Q (2022) Combined Cutinase and Keratinolytic Enzyme to Endow Improved Shrink-resistance to Wool Fabric. Fibers Polym 23:985–992. https://doi.org/10.1007/s12221-022-4445-0
174. Zhang T, Bai R, Wang Q, Fan X, Wang P, Yuan J, Yu Y (2017) A novel strategy to improve the dyeing properties in laccase-mediated coloration of wool fabric. Color Technol 133:65–72. https://doi.org/10.1111/cote.12252
175. Zheng L, Du Y, Zhang J (2001) Degumming of ramie fibers by alkalophilic bacteria and their polysaccharide-degrading enzymes. Bioresour Technol 78:89–94. https://doi.org/10.1016/s0960-8524(00)00154-1
176. Zheng X, Zhang Y, Liu X, Li C, Lin Y, Liang S (2020) High-Level expression and biochemical properties of A Thermo-Alkaline pectate lyase from bacillus sp. RN1 in Pichia pastoris With Potential in Ramie Degumming. Front Bioeng Biotechnol 8:850. https://doi.org/10.3389/fbioe.2020.00850
177. Zhu L, Lin J, Pei L, Luo Y, Li D, Huang Z (2022) Recent advances in environmentally friendly and green degumming processes of silk for textile and Non-Textile applications. Polymers (Basel) 14:659. https://doi.org/10.3390/polym14040659

Bio- and Organo-Immobilized Textile Catalysis: Synthesis and Applications

Asma Khurshid and Aamer Saeed

Abstract Catalysts make use of materials that speed up a chemical process without being used itself. The efficiency of any chemical process could be significantly improved by making use of suitable in-expensive catalysts with desirable properties and reusability factors. Efforts to achieve the aforementioned goals have led to the design and development of more sustainable approaches thereby paving the way for the development of textile catalysis in the last decade. Textile catalysis makes use of an immobilization approach, where the biobased compounds (enzymes) and organic compounds/organometallics are anchored using natural and synthetic textiles as catalyst support for the development of a number of highly relevant new catalysts with enhanced catalytic efficiency and broad-spectrum applications. This led to the inclusion of a new concept in the literature named "immobilization of catalysts on textiles" or "textile catalysts/catalysis" in the field of heterogeneous catalysis systems (industrial biotechnology) as well as in technical textiles. This chapter highlights/explains two main categories of textile catalysts, i.e., organotextile catalysts and biobased textile catalysts taking into account the synthetic approaches, their types, and applications besides a highlight on SDG prospect. In addition, a brief account of limitations/challenges and future perspectives of the textile catalysts has been presented as a part of this chapter.

Graphical Abstract

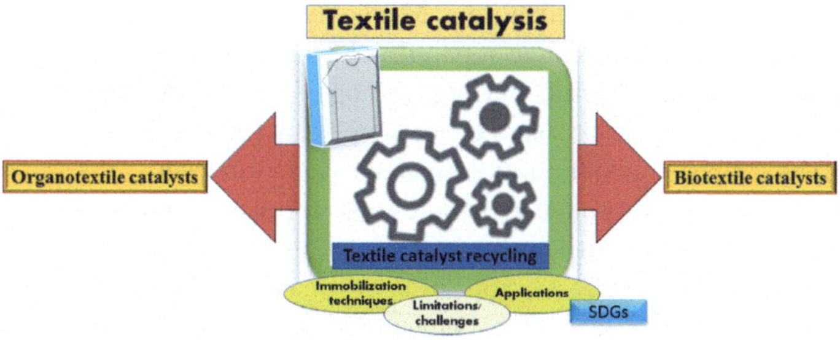

A. Khurshid · A. Saeed (✉)
Department of Chemistry, Quaid-I-Azam University, Islamabad 45320, Pakistan
e-mail: asaeed@qau.edu.pk

M. Arshad (ed.), *Enzymes in Textile Processing: A Climate Changes Mitigation Approach*, SDGs and Textiles, https://doi.org/10.1007/978-981-97-8058-7_4

Keywords Textile catalysts · Immobilization technology · Organocatalysts · Biocatalysts · Synthetic approaches · Wastewater treatment

1 Introduction

The efficiency of any chemical process can be enhanced significantly by employing suitable catalysts. Since catalysts are the substances that speed up the chemical process without being consumed in the process. They are broadly categorized into four categories namely: organometallic catalysts (metal–organic ligand complex), biocatalysts (enzymes), organic catalysts, and inorganic catalysts (metals and metal oxides) [10, 51]. Catalyst recycling/reusability is an important aspect linked to catalyst sustainability besides economic factors. Various strategies have been devised in past decades to achieve/address catalyst sustainability and catalyst immobilization or catalyst binding to base material (like polymers, resins, or metals particles) has achieved significant attention in material/chemical, science and biotechnology fields, etc., [18, 28, 29]. However, the use of exclusive binders, and resource-intensive process makes catalyst immobilization a complex chemical process with a resultant expensive catalytic system with no reusability factor thereby resembling the traditional catalytic system [23]. This is the case for either the biocatalyst or enantioselective organ catalysts or inorganic catalysts [15, 41]. Considering the aforementioned limiting aspects, the focus of researchers has diverted in recent years to the exploration of more robust, cost-effective, and readily available support materials for catalyst immobilization. In this context, the use of textiles has emerged as a promising catalyst carrier/support material that overcomes the mentioned limitation factors associated with other support materials. Hence, the use of textiles has led to the birth of an entirely new idea/concept in the field of heterogenous catalysis-organocatalysis, enzymes (bio-catalysts) industrial biotechnology, and in technical textiles termed as "textile catalysts" or "immobilization of catalysts on textiles". The benefits associated with textiles like their shape, size, and orientation help in random reactor design for specified applications. Moreover, the added benefit of the use of textiles is the efficient and complete separation of the immobilized catalysts without any residue traces in the reactor. Further, the open porous structure feature of the textile surface results in low-pressure drop, thereby facilitating rapid diffusion of the reaction mixture, mass transfer, and optimal substrate turnover.

Several relevant catalysts have evolved to date for use in fine synthesis and pharmaceutical products since textile catalysts offer high catalyst stability due to strong covalent bonding [2–4, 13, 37, 43, 56].

Considering the significance of this new emerging/progressing field of catalysts, this chapter provides an insight into the main types of textile catalysts, their synthetic methodologies, and catalyst activity and reusability.

2 Types of Textile Catalysts: Bio- and Organo-Immobilized Textile Catalysts

Based on the nature of catalysts textile catalysts are classified into two main types as follows.

2.1 *Organotextile Catalysts*

Small organic molecules that can catalyze a chemical reaction in the absence of metal/metal ions are termed organocatalysts. Hence, they are metal-free species that display Lewis acid/basic, Bronsted acidic, or basic reactivity [16, 69]. Besides the advancements in the design of organocatalysts, the further application of organocatalysts in the industry has been largely reduced/slowed down mainly because of the limitations associated with their reusability. Efforts by scientists have been documented to develop inexpensive and abundant polymeric solid supports that would support organocatalysts with aided practical benefits like high durability, adaptability, flexibility, and broad accessibility. In this regard, textiles have been initially used as solid support for nanocarbanaceous materials like graphene and carbon nanotubes via adhesion or adsorption methodologies. Later, Bredig et al. [17] reported the application of amine-functionalized organic fibers like cellulose and cotton as catalysts for cyanohydrin synthesis [17, 36]. This was followed using heterobimetallic complex (Pd/Co) synthesized by using wool as a ligand with application in asymmetric hydration of unsaturated carboxylic acids [39]. For the very first time, the application of textile as a carrier material of choice in selective heterogenous organocatalysis was highlighted by List et al., with demonstrated excellent activity, selectivity, and recyclability [17].

2.1.1 Organotextile Catalyst Immobilization Methodologies

The immobilization of the organocatalyst on textiles is achieved by various methodologies listed in the following section along with details on the catalyst recycling and reusability.

A. **Photochemical immobilization**

The general protocol for photochemical immobilization involves dissolving the catalyst in the appropriate solvent followed by textile wetting in the solution with ultimate irradiation of the wetted textile for a specific time period. Removal of unbound catalyst is achieved via extraction technique.

Textiles irradiation results in the generation of active surfaces via radical formation that serves as the functionalization points for anchoring the organic molecules to the textiles [63].

In 2013, List and Opwis et al. demonstrated a simple and permanent UV-based immobilization of organocatalysts on nylon textiles successfully by avoiding chemical modifications. All the prepared organocatalysts showed excellent reusability, stability, and catalytic activity for a number of organic transformations. Three organocatalysts used belong to Lewis basic, a Bronsted acidic, and a chiral organocatalyst class, namely dimethylaminopyridine (DMAP) derivative (1), sulfonic acid (2), and a bifunctional chiral organocatalyst (3) (Fig. 1).

Irradiation time periods are significant in determining the catalyst loading as well as catalyst activity. In addition, the ratio of catalyst to support material also determines the catalyst loading. In the case of enhanced irradiation time periods, loading versus irradiation time curves often show a maximum with a resultant decrease in catalyst loading indicating photodegeneration of either the catalyst or the polymer. Variation in catalyst to support ratio results in a saturation curve (such as a Langmuir isotherm) for the catalyst loading, however, after a certain ratio the loading stays constant. Two important factors relating to textile support are decisive and influence the catalyst loading. The first one is the fabric diameter which indicates the surface area that directly shows a correlation for catalyst amount bound to the surface. Another factor is the amount of ultraviolet radiation absorbed by polymer, the more the absorption the more the active surface generation with resultant high catalyst loading and vice

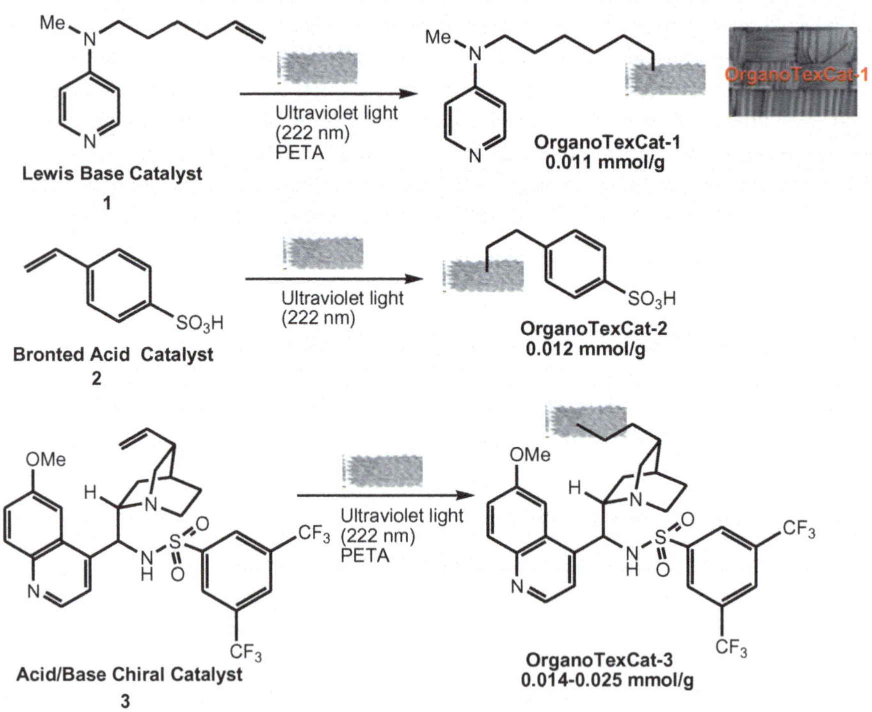

Fig. 1 Schematic illustration of various types of organo-textile catalysts

versa. In the designed experimental setup by List et al., it is found that shorter irradiation time periods of approximately 5 min for every side of the textile greatly ameliorate catalyst loading without due damage to the textile and organocatalysts. A bifunctional catalyst containing both acidic and basic groups was also found stable under photochemical reaction conditions.

The catalytic activity and recyclability were tested by performing an acylation reaction using sterically hindered phenol and isobutyric anhydride. A number of recycling experiments demonstrated the robustness of the designed organocatalysts without any observable catalyst erosion during the recycling experiments. Several washings with triethyl amine recovered the catalyst with good reactivity [9, 34].

Another means of photochemical immobilization is achieved by the use of UV excimer lamps that emit monochromatic light of high energy density with a wavelength range between 2 and 15 nm. The absence of IR irradiation in the UV excimer lamps makes their use beneficial as it avoids substrate heating. The wavelength range of common excimer UV lamps is between 100 and 350 nm (12.4–3.5 eV). Depending on the fiber type, it can be estimated whether the UV-induced cleavage has taken place because of UV irradiation. In literature, PET, PA, Cotton (CO), and polyethylene-based fabrics have been irradiated by UV excimer lamps and their remission spectra have been observed to find the generation of active surfaces and further reaction. In case of the absence of radical scavengers, the generated radicals recombine or undergo migration, however, in the presence of reactive molecules, the reaction occurs on the polymer surface to generate the modified form of fiber [50].

B. **Surface chemical reactivity for immobilization**

In case of catalyst degradation by UV light or catalyst reaction with radical scavengers to form inactive catalysts after immobilization, conventional chemical methodologies serve as a good alternative to produce immobilized organotextile catalysts. In such cases, the possible anchoring reaction is determined by the nature of terminal groups present at the polymer chain surface. This is exemplified by the presence of PA amino, carboxyl groups while in the case of PET hydroxyl and carboxyl groups. Further examples include cyanuric chloride that serves as the substrate for the binding of amino- or hydroxyl nucleophiles to the surfaces. This methodology offers modification in cellulose besides its application to PA. However, for PET, amide formation takes place via a reaction between the amino-based molecules and carboxylic or ester groups.

2.2 *Bio-based Textile Catalysts*

Due to their exceptional substrate specificity and moderate operating conditions (pH, temperature, and solvent), biocatalysts/enzymes responsible for catalyzing the biochemical reaction hold a unique place among the available catalysts [11, 26]. The extravagant properties of enzymes like lipases, amylases, and peroxidases, etc., widen their use in multiple areas like textile, food, beverage, and pharmaceuticals in addition to their application in green energy [34, 49].

First ever industrial application of immobilized enzyme named "invertase" was reported by Chibata [20], in which charcoal and aluminum hydroxide matrices were employed for enzyme immobilization with resultant no loss in enzyme activity [5, 25]. This initial success in biotechnology opened doors for the development of a number of methodologies for enzyme immobilization on textiles. A notable example in this regard includes lipase-immobilized silk woven fabric with notable improvement in operational stability while using standard esterification conditions. In addition, the biocatalyst showed a high turnover rate in 27 batches of use with 97% biocatalyst efficiency in comparison to the free lipase enzyme [24].

Various factors such as inadequate loading of catalysts, leaching of catalysts, or frictional instability affect the catalyst immobilization thus ultimately hampering the broad-spectrum application of bio-textile catalysis. Poor productivity and loss of catalyst during the immobilization process occur due to poor loading. The extra use of energy besides chemical, and system interruption during enzyme leaching leads to additional operational costs. Frictional instability of catalysts is directly linked to the application of the catalysts like in the case of wastewater treatment where frictional interaction of the catalysts is minimal [6]. Various protocols have been adapted to overcome the mentioned drawbacks among which surface modification of base material is of immense significance since it provides appreciable bonding between enzyme and support material to avoid leaching. Modification in surface functionalization of textile fiber to various functionalities via adhesion phenomenon has been adopted by researchers as a promising solution [42, 53] by adapting the principle of chemical finishing by padding. As an example, Mohamed et al. [44], Van Beilen et al. [59] showed the use of glutaraldehyde as an efficient crosslinking agent for the immobilization of horseradish peroxidase on nonwoven polyester fabric. A further notable example in this regard was demonstrated by Kahoush et al., Wu et al. [64] via modification of nonwoven textile-the carbon felt by cold remote plasma either the O_2 or N_2 gas that creates amine/amide groups thus favoring glucose oxidase enzyme immobilization. Amongst various functional materials, explored for surface modification of support material include hyperbranched dendrimers [20] metal–organic framework (MOFs) [62], and so on [27].

Subcategorization of biocatalysts immobilized textiles has been made based on the type of textile used. The subcategories are as follows:

(i) biocatalysts immobilized nonwoven textile,
(ii) biocatalysts immobilized woven textile,
(iii) biocatalysts immobilized knitted textile, and
(iv) biocatalysts immobilized nanofibrous textile

The introduction of textile support materials for immobilizing catalysts has recently been very well perceived by the scientific community due to inherent advantages associated with the textile like lightweight construction, flexibility, and arbitrary geometry which expands its usage in any reactor with aided quick separation from the reaction medium, and residue-free product generation [44].

Biocatalyst immobilized textiles is a pioneering technology for the design and development of enzyme-based robust catalytic systems to address the durability, cost factor (via cheap carrier material), loading efficiency (by using tailored textile surface), strength, catalyst leaching and reduction in enzymes denaturation. The catalytic activity of enzyme-immobilized textiles is governed by the enzyme nature and strength of interaction between the functional group of enzymes and the polymer used in the textile.

2.2.1 Factors Governing Biocatalysts Immobilization on Textiles

Immobilization of biocatalysts is governed by various parameters such as:

a. Lifetime of immobilized enzyme since it's important as the used textile can be recovered after the useful lifetime of immobilized enzyme
b. Binding interaction between textile and enzyme
c. Textile properties like shape, form, surface area of textile, porosity, density, and operational stability of textile directly affect the reactor configuration which utilizes immobilized biocatalyst textile
d. Retention and stability of enzyme performance are mainly governed by functional groups anchored on textile surface and microenvironmental conditions.

2.2.2 Biocatalysts Immobilization on Textile Types: Their Performance and Progression Update

The development of biocatalysts immobilization on textiles has opened extraordinary opportunities for the design and development of robust fiber-based catalytic systems. The advancements include the usage of microfibrous to nanofibrous textiles depending on the desired application. The progress in the technology utilizing various forms of fabrics is addressed as follows.

A. **For woven textiles**

Various enzymes have been immobilized on woven textiles like cotton. Earlier work includes the immobilization of Yeast invertase [30] followed by Jack bean urease [60] on cotton cloth for the preparation of biocatalyst immobilized textiles. The application of the latter was found in urea hydrolysis as a heterogenous catalyst with appreciable recyclability and catalytic activity in comparison to free urease. Similar high catalytic activity was then reported for papaya protinease immobilization on cotton textiles with 30% retention in activity after multiple usages. Gaining success in the performance of catalytic activity, in 2004, catalase enzyme was immobilized on natural fiber-based and synthetic fiber based, i.e., cotton and nylon fabric with resultant 10 times higher catalytic activity. The results of the study revealed crosslinking between catalase and glutaraldehyde by creating a three-dimensional structure on the whole fiber thus increasing catalyst loading. Gaining success on the cotton fiber, silk woven fabric was then tested by the researchers to expand the design methodology. Various enzymes like lipase [38] with 97% effectiveness even after 27 cycle application and enzyme named trypsin immobilization on N, N′-dicyclohexylcarbodiimide, and N-hydroxysuccinimide activated polyester woven textile indicated 18% enhanced initial activity even after reusing the immobilized enzyme 15 times [22]. Further advancements in the use of woven textile immobilized textile catalysis were reported by Yin et al. [68], Sheldon et al. [52] by immobilizing glucose dehydrogenase and bilirubin oxidase enzyme on cotton-based textile with demonstrated application in wearable biofuel cells. Further studies involved the immobilization of alpha-amylase amidrazone acrylic woven textiles with 53% retention in catalytic activity over 15 turnovers [7, 28, 29]. It is further demonstrated that microfibrous textiles formed by weaving technology result in dual porosity of woven textiles like in the case of yarn. For continuous and long-lasting catalytic activity during the enzyme immobilization process, it is important to maintain easy migration of enzyme molecules via woven textiles pores.

B. **For non-woven textiles**

Nonwoven textiles offer various advantages like breathability, strength, and barrier properties. Various factors determine the performance or activity of non-woven textile-based immobilized biocatalysts like air permeability, porosity, fiber arrangements, fiber type, and the overall chemical property of nonwoven fiber surface.

The use of urease [67] and various redox enzymes [31] on non-woven textiles like nylon silk fibroin, viscose rayon, polyester-based fabric materials have been documented in earlier studies as effective carrier materials for biocatalysts immobilization. Later in 1997, the immobilization of mycelia over nonwoven textile depicted increased turnover catalytic activity to approximately 10 times higher and without any added adverse effect on the biocatalyst activity than the free enzyme [19]. Thermolysin enzyme was covalently immobilized to polyamide nonwoven textile which demonstrated significant enzyme activity besides enhanced stability under elevated reaction conditions like pH, temperature, polarity, etc. Various other enzymes with improved operational stability and enzyme activity include inulinase enzyme and xylanase enzyme immobilization on nonwoven textile and polypropylene nonwoven, respectively. The challenges encountered so far in enzyme immobilization in terms of loading and stability of enzyme are [33] then addressed by modifying the fiber surface by either chemical or physical treatments or by making use of binders or cross-linkers. In this regard, Mohamed [44] and Shim et al. [55] provided an example of crosslinker-based biocatalyst immobilized textiles. They used glutaraldehyde and chitosan glutamate as cross-linking agent for immobilizing horseradish peroxidase enzyme on nonwoven polyester fabric. Further efforts by the same research group in 2016 [68] showed the robust immobilization of beta-galactosidase enzyme under the influence of plasma on polyethylene terephthalate nonwoven textile. Relatable work was then documented by Kahoush et al. who demonstrated the successful glucose oxidase immobilization on plasma-triggered fibrous carbon nonwoven i.e., felt textile. The advantages associated with plasma treatment-based biocatalysts immobilization on textiles include better reaction mixture flow and enhanced interaction between the enzyme and nonwoven fabric.

Further advancements in the synthetic design of biocatalysts include surface modification of nonwoven polyester fabric via grafting of amine functional groups using polyvinyl amine for better adhesion and immobilization properties of peroxidase enzyme as reported by Wunschik et al. [65], Wunschik et al. (2019). It is worth notable that immobilized enzyme possesses better operational stability and immobilization efficiency as a result of functionalization on the tailor-made surface. The useful properties of non-woven fabric like light weight, structure freedom make it a matrix of excellent choice for the preparation of robust immobilized biocatalyst-based enzyme catalytic assemblies.

C. **For nanofibrous textiles**

Nanofibrous textiles are associated with various advantages that make them a suitable matrix of choice for immobilizing biocatalysts. This includes large specific surface area, controllable pore sizes, small fiber diameters specifically on the nanometer scale, and facile surface modification thereby making them perfect fit for applications in multiple domains like functional materials, tissue engineering, and energy storage devices. Varying technologies like electrospinning, melt blown, and natural bacterial cellulose are used for synthesizing nanofibrous textiles. To immobilize biocatalysts on nanofibers, different methodologies have been reported for electron -spin nanofibrous textiles. The very first example of such an immobilized biocatalyst includes the covalent immobilization of alpha chymotrypsin on polystyrene with 65% retention of enzyme activity and enhanced resistance against organic solvents than in comparison to non-mobilized ones. Gaining initial success in use of nanofibrous matrix material for immobilization, there has been a growing concern in use of variety of enzymes like lysozyme and lipase on a mixture of biocompatible and biodegradable poly-e-caprolactone and poly-dl-lactic-co-glycolic acid, polyethylene glycol-NH_2 (PLGA-b-PEG-NH_2) block copolymer based nanofibrous textile. Such textile materials contain a primary amine group on the surface for facile immobilization of enzymes. With the help of coupling agent glutaraldehyde (GA), improvement in the relative loading and catalytic activity of the mentioned enzymes was then achieved by immobilizing them on chitosan/ polyvinyl alcohol (PVA) nanofibrous textiles. Later many reports have been documented mentioning the effective use of glutaraldehyde-GA-coupling agent. Later work with variations in the enzyme immobilization techniques on nanofibrous textiles was carried out. A notable example of such includes covalent immobilization of lipase enzyme via amidination reaction on polyacrylonitrile (PAN) nanofibrous textile surface with resultant retention of 81% of enzyme activity (It was stored for 20 days at ambient conditions). In later years immobilization experiments on nanofibrous-based textiles with and without multi-walled carbon nanotubes were performed with the resultant achievement of better loading efficiency and enhanced operational stability [8]. In continuation of advancement in the field various modified forms of multi-walled carbon nanotubes-based nanofibrous textiles are synthesized using catalyse for immobilization are reported. The results of modified immobilized biocatalyst nanofibrous textiles high relative activity, pH, and temperature stability and better loading, optimal as compared to free catalases.

The enzyme immobilization on nanofibrous textile is limited by the generation of a monolayer of enzyme that decreases the enzyme loading. This issue is sorted out by using enzyme aggregates where in the first attempt a group of seed enzymes on the nanofibrous textile surface followed by further covalent attachment of enzyme on seed enzyme layer was done thus developing a multilayer enzyme biocatalyst for enhanced operational performance.

In a nutshell, taking into account the applications associated with the nanofibrous textiles, enzyme immobilization could offer enhanced catalytic activity, and better mass-transfer during both enzyme immobilization i.e., between support matrix and enzyme molecules in addition to a high turn-over rate [54].

D. **For knitted textiles**

Knitting gives rise to knitted textiles with high flexibility and penetrability properties useful for their broad-spectrum applications. The mentioned properties of knitted textiles make them distinguishable from other textile types addressed before (nanofibrous, woven, and nonwoven). This class of biocatalysts was developed after nonwoven and woven textiles. The first example from this class of biocatalysts included immobilization lysozyme via covalent interaction on dialdehyde functionalized cellulose and polycaproamide knitted textiles through covalent binding with demonstrated bacteriolytic activity and effectiveness in wound healing. Achieving success for immobilized biocatalysts on knitted textiles opened the ways for the immobilization of a variety of other enzymes like trypsin and beta-galactosidase on polyester knitted (fleece) textiles and on knitted cotton textiles named terry-activated with p-toluenesulfonyl chloride. The results of both biocatalysts demonstrated high thermal, pH, operational stability, and enhanced turnover activity of biocatalysts in comparison to their free enzymes. Further progress with this type of textile involved advancement in biocatalyst generation by making use of cross-linkers as exemplified by the use of glutaraldehyde cross-linker assisted immobilization of marine pectinase on nylon 6,6 [43].

To sum up, knitted textiles are of less use than nonwoven/woven textiles. The covalent binding or crosslinking immobilization methodology in addition to physical adsorption methods is considered a method of choice in the case of knitted fabrics due to their loose structure. Knitted textiles with immobilized biocatalysts have been documented for high-end applications such as bioelectronics, i.e., actuators, sensors, and medical (wound dressing) besides their function in chemical synthesis, environmental remediation, etc.

2.2.3 Methods for Design of Biocatalysts Immobilized Textiles

Immobilization méthodologies have a significant rôle in determining the properties of the biocatalysts (enzymes) immobilized textile. The literature reported methodologies for immobilization of biocatalysts on textiles include covalent binding, cross-linking process, physical adsorption besides encapsulation. Two protocols are considered important for immobilization that is: (a) dipping/solution-based immobilization and (b) printing-based immobilization. In the former case, under a standard environment both enzymes and textiles are brought in contact with each other in a solution, while in the latter case, direct mixing of an enzyme with solvent is accomplished to prepare inkjet. Higher enzyme loading is encountered in the case of dipping protocol due to available open areas of the fibrous textile at once in the solution. While, additional chemicals, time, energy, and enzyme concentrations are necessary for the dipping method in comparison to the printing methodology.

A. **Immobilization via physical adsorption**

The simplest form of immobilization methodology is physical adsorption (Fig. 2) since it involves binding of the enzyme to the textile surface via weak interactions like van der Waals forces, ionic, hydrogen bonds, and affinity binding besides hydrophilic and hydrophobic interactions.

The exact process of physical adsorption involves textile surface activation before exposure to enzyme solution. The advantage of this method includes the high loading efficiency of the enzyme. However, this protocol suffers drawbacks due to the desorption of enzymes due to weak physical forces that require additional attention to interactions involved in carrier enzyme. Oxidoreductases such as laccase, glucose oxidase, and horseradish peroxidase and related enzymes used in biodegradation, bioremediation and oxyfunctionalization of organic substrates are the most common class of enzymes studied for immobilization on textile surfaces via physical adsorption means.

A summary of biocatalysts immobilized textiles via physical adsorption in view of various parameters like enzymes, carrier matrix, supporting chemicals performance, and stability is listed in Table 1.

B. **Imobilization via covalent binding**

The most common method of immobilization is covalent immobilization where the enzyme is covalently bound to the textile surface offering an advantage in terms of enzyme leaching prevention. This basic requirement for immobilization methodology is the presence of suitable functional groups on textile surfaces to be used in covalent functionalization. A general approach in this regard is to use a strong electrophilic functional group based textile to bind nucleophile-based enzymes on

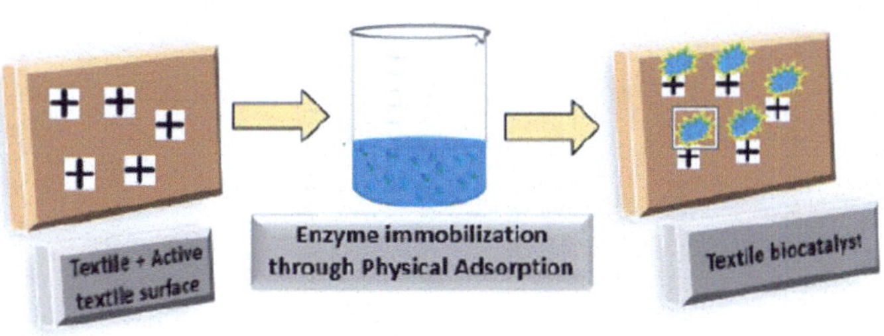

Fig. 2 Physical adsorption methodology illustration for biocatalyst immobilized textile design

Table 1 Summary of various parameters affecting immobilization methodologies

Sr. no.	Enzyme	Textile carrier matrix	Supporting chemicals/ treatments	Performance and stability	References
(i)	b-galactosidase	Polyester nonwoven	Plasma treatment and $CaCl_2$	90% enzyme activity retention; 15 reuses	[45]
(ii)	Glucose oxidase	Carbon felt	Hyperbranched dendrimers	31% enzyme loading, and 81% active immobilized enzymes; 50% enzyme retention6 reuses	[2–4]
(iii)	Horseradish	Acrylic textile Peroxidase	Cyanuric chloride	Phenol removal for wastewater treatment; 10 reuses	[1]
(iv)	a-amylase	Acrylic textile	Cyanuric chloride	81% enzyme activity;15 reuses	[27, 56]
(v)	Organophosphate Hydrolase	Polyester nonwoven	Ethylenediamine, glutaraldehyde, and methyl parathion	Used for remediation of pesticides; activity up to 4 weeks	[57]
(vi)	Alcohol textile dehydrogenase	Polyvinyl alcohol	Chloropropinoyl chloride, ethylenediamine, and glutaraldehyde	Better operational, pH, and thermal stability; 08 reuses	[40, 58]
(vii)	Horseradish.peroxidase	Poly (vinyl alcohol)/ polyacrylamide nanofibrous textile	Glutaraldehyde	High loading and stability; 25 reuses	[21, 48]
(viii)	Laccase	Polyacrylonitrile-biochar compositenanofibrous textile	Hydroxylamine hydrochloride	Improved storage, temperature, and pH stability	[12]
(ix)	Laccase	Polyamide 6/chitosan nanofibrous textile	Bovine serum albumin and hexamethylene diamine	Removal efficiency for a mixture of endocrine-disrupting chemicals in three treatment cycles	[66, 70]

(continued)

Table 1 (continued)

Sr. no.	Enzyme	Textile carrier matrix	Supporting chemicals/ treatments	Performance and stability	References
(x)	Laccase	Laccase-carrying electrospun fibrous textile	No supporting chemical was used	85% catalytic activity; better storage and operational stability	[48, 61]
(xi)	Horseradish peroxidase	Poly (d,l-lactide-co-glycolide)	Without support chemical	Efficient removal of phenolic contaminants and 50% retention in enzyme activity	[71]

Immobilization methodology types: Physical adsorption (i–iv); covalent binding (v–vii); cross-linking (viii–ix); encapsulation (x–xi)

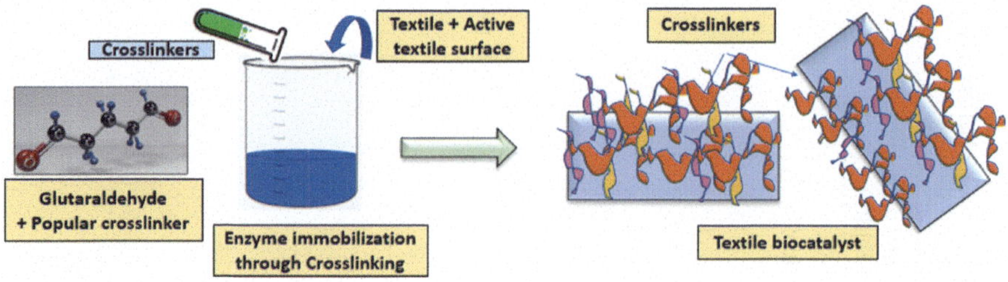

Fig. 3 Biocatalyst immobilized textile formation via cross-linking

its surface. Examples of notable functional groups used for covalent immobilization include free amino, hydroxyl, carboxyl, and sulfhydryl groups. Covalent binding occurs in three ways. In the first case, binding occurs through enzyme side chains, i.e., lysine-e-amino group while in the second case, binding to the thiol group in cysteine occurs. Whereas in third case which is the most common case involves the binding between the carboxylic acid with either the aspartic or glutamic acid. The summary of covalent functionalization-based biocatalysts immobilized textiles in view of various parameters like enzymes, carrier matrix, supporting chemicals performance, and stability is listed in Table 1.

C. **Immobilization via cross-linking**

A well-known methodology for immobilization of enzymes on textile surfaces is via cross-linking. This is an irréversible process and involves the formation of intermolecular cross-linkages among enzyme molecules through covalent bonding as shown in Fig. 3.

The non-functional part of the enzyme like the amino substituent of lysine residue is bound to one terminal of cross-linker while the other end/terminus is bound to functional groups on the textile surface. Using a suitable cross-linker is important for maintaining high catalytic activity. glutaraldehyde, hydroxylamine hydrochloride, hexamethylenediamine, are the most common types of cross-linkers used for immobilization. However, these cross-linkers have a toxic nature that results in conformational changes in enzyme surface thereby leading to reduced catalytic activity. Advancements in modification of crosslinkers were made keeping in view the green technology to subside the issues due to toxic cross-linkers. Examples of green cross-linkers include the use of bio-based naturally occurring crosslinkers like genipin, bio-compatible polymers such as chitosan, polyvinyl alcohol (PVA), polyethyleneimine, etc. Another disadvantage that comes with the crosslinking mode of enzyme-immobilized textile biocatalysts is their hindered performance mainly due to the poor availability of enzyme active sites by the substrate. Table 1 gives an illustration of examples for cross-linked based biocatalysts immobilized textiles with details on various performing parameters.

D. **Immobilization via encapsulation**

This type of immobilization occurs under mild operating conditions in comparison to the immobilization types already mentioned. In this technique, enzymes are either chemically or physically encapsulated inside textile material or encapsulated via a semipermeable membrane of 100 mm or less diameter. Textile or membrane pore size has a significant role in the design of biocatalyst-immobilized textiles via encapsulation. Therefore, it is necessary to prevent the passage of enzymes through the pores of either of before mentioned, and for this reason smaller pore size of the membrane is considered essential. This important aspect basically limits the use of this encapsulation methodology for many textile substrates as carrier material. Amongst the efforts made for this methodology involves the successful design of biocatalyst immobilized textile catalyst via encapsulation by making use of nanofibrous membranes where the pore size and properties can easily be controlled. Another disadvantage that makes the use of this methodology less likely is the limitations encountered in diffusion between the used enzyme and substrate in the process. Examples of encapsulation with varying affective parameters are listed in Table 1 as follows.

E. **Miscellanées methodologies for biocatalysts immobilization on textiles**

Digital printing such as inkjet and valve-jet, electrostatic binding like layer-by-layer deposition, electrochemical doping/polymerization, and other related methodologies have so far been reported for immobilization of enzymes on textiles. Some notable examples include immobilizing oxidoreductase enzymes as reported by Bal et al. (1997) on flexible fabric in predefined patterns via a new generation resource-efficient process named digital inkjet printing. Biswas and coworkers actively reported efforts in this regard by documenting the effectiveness of various parameters for enzyme immobilization on synthetic polyethylene terephthalate textile surfaces through inkjet printing technology. Further innovation in the methodology includes enzyme binding via electrostatic forces on the textile surface with enhanced enzyme loading and stability. Layer-by-layer assembly innovative methodology in line with this effort resulted in high enzyme loading to the textile substrate [43].

2.2.4 Reusability of Biocatalyst Immobilized Textiles

Catalyst recycling is an important aspect associated with enhanced catalyst efficiency. The high reusability of enzyme-immobilized textiles is directly linked to modification on the surface of textiles and immobilization strategies. It is considered important to have high catalyst turnovers as à good compensation for the cost of enzyme immobilization. Various findings have been documented for enzyme-immobilized textiles demonstrating high recyclability and reusability. Work by [35, 65], showed lipase immobilization on silk woven fabric with 97% effectiveness beyond 27 cycle applications. Following this, Chen and coworkers reported enhanced catalytic activity for trypsin immobilized textiles after 15 turnovers. Similarly, Shim et al. and Song in 2017 reported 05-to 35-cycle applications of lipase-immobilized nanofibrous textile [55].

3 Applications of Bio- and Organo-Immobilized Textile Catalysts

Enzyme immobilization is an important technology in terms of reusability, ease of separation, and associated higher stabilities (pH, thermal, storage) as compared to free enzyme usage. Biocatalysts immobilized textiles have broad spectrum applications from pharmaceuticals and medical devices, bioremediation, food industry involving (preparation and preservation), fine chemical synthesis, wastewater treatment, etc., [54]. Among the mentioned applications, the use of biocatalyst-immobilized textiles has been increasingly reported in wastewater treatment as a sustainable approach for climate change mitigation.

Multiple industries like textile, paper, electroplating, pulp, plastic, tannery, petroleum-based products, pharmaceuticals, and cosmetic industries are releasing dyes/colorants as industrial waste into the surface water thereby contaminating the freshwater with toxic aromatic compounds that are part of the waste and are associated with serious life-threatening problems when encountered with the human body. Conventional waste-water treatments like physical, chemical, and biological like coagulation, flocculation, adsorption, filtration, aerobic/anaerobic biological systems, or an amalgam of physical and biological systems seem less effective. Thus, they are replaced by heterogeneous enzymatic systems based on immobilized enzymes as a green alternative for toxic dyes and other contaminants removal from water effectively. Notable work in this regard includes a report by Ravi et al., who showed 98% removal of azo dyes from wastewater in a span of 45 min by using their immobilized Horseradish peroxidase enzyme. Later, Zdarta et al. reported the efficient removal (up to 90%) of pharmaceuticals (anti-inflammatories like naproxen, and diclofenac, etc.) from wastewater by using electrospun poly-L-lactic acid-co-poly-ε-caprolactone nanofiber textiles with immobilized laccase enzyme [14].

Further, the use of biocatalyst-immobilized textiles has been widened via their application in the depletion of phenolic endocrine-disrupting chemicals (PEDCs) from contaminated water and soil [47] threatening to life. In this regard, successive efforts by Xe et al. (2013, 2017), for the removal of contaminants like 2,4,6-trichlorophenol from wastewater by using immobilized laccase on polyacrylonitrile/polyvinylidene fluoride nanofibrous textile showed successful removal >85% with noticeable reusability of 65.9% up to 7-cycles [32]. In continuation of the efforts, Wang and his co-workers reported the successful withdrawal of catechol from contaminated water via mutual application of adsorption and biocatalysis processes. For this purpose, they used commercial laccase immobilized on polyacrylonitrile/montmorillonite/graphene oxide composite nanofibers [35]. Later reports also document the use of MWCNT's modified laccase-carrying electrospun fibrous membranes for phenolic contaminants removal from wastewater successfully. Niu et al. demonstrated the removal of 83% pentachlorophenol by using encapsulated horseradish peroxidase into nanofibrous textile under ambient reaction conditions [46]. Progress in the field of application was reported by Zdarta et al. [71] for the removal of bisphenol by immobilizing tyrosinase via response surface methodology with a report on retention of 90% biocatalyst activity even after storage of 30 days [70].

In conclusion, biocatalyst immobilized textiles have shown demonstrated potential for catalytic removal of pollutants taking into account as one of the sustainable development goals targeting environmental pollution remediation. However, the technology still suffers limitations like vulnerability to the surrounding environment, the necessity of pretreatment of wastewater, etc., that need to be addressed for exploiting the enhanced utility of the textile biocatalysts in the field on both the pilot and industrial scale. Introducing optimization in enzyme, reactor, mass-transfer, and support matrix are the key parameters to overcome or reduce the associated drawbacks.

4 An Outlook to Challenges/Limitations, Future Perspective and SDGs

Bio or organocatalysts immobilized textiles is an innovative approach with more room for progression in the field for further exploration of applications of aforementioned. The immobilization technology offers benefits in terms of recycling and reusability to both organo- and bio-textile catalysts.

For organo textile catalysts, the feasibility of an anchoring variety of commercial and synthetic organic molecules/catalysts though offers unexpected enhanced catalytic activity but at the same time poses limitations. A specific case is of the photoreactive catalytic centers which are responsible for loss in catalytic activity by destroying the chemical structure. However, on the positive aspect, the immobilization methodology for heterogenous organo textile catalysts offers new avenues for industrial scale flow chemical reactions which conventional catalysts fail to do.

Biocatalysts which catalyze the reaction under mild conditions show enhanced catalytic activity, recyclability besides stability, etc., added features under employed conditions as a result of immobilization technology. The biocatalysts immobilized textile is a pretty new concept with scope for further advancement in the design and development of novel biocatalyst immobilized textiles that surmount the present associated usage limitations. The limiting factors like enzyme leaching, enzyme loss during immobilization, textile fiber modification under harsh reaction conditions and storage (biocatalysts reduce relative activity thus limiting the potentiality of large-scale production and long-lasting application of "biocatalysts immobilized textile"), etc., could be overcome by using resource-efficient processes. This includes printing and the introduction of eco-friendly alternatives like bio cross-linkers and plasma treatment that can significantly improve the function of biocatalysts immobilized textiles.

Achievement of sustainable development goals is of utmost importance for a healthy planet and well-being toward which the research world is actively progressing. In this regard, the introduction of immobilization technology in textile catalysis (organo or bio-based) serves as an innovative technology as shown by its demonstrated potential to address the wastewater issue (one of the significant SDGs) and ultimately the human health—the underlying sustainable development goal. However, research in the field of textile catalysis is still in its initial phase with many open questions and avenues for further advancement in the field to target broad-spectrum sustainable development goals besides the aforementioned wastewater treatment.

References

1. Almulaiky YQ, El-Shishtawy RM, Aldhahri M et al (2019) Amidrazone modified acrylic fabric activated with cyanuric chloride: a novel and efficient support for horseradish peroxidase immobilization and phenol removal. Int J Biol Macromol 140:949–958
2. Al-Najada AR, Almulaiky YQ, Aldhahri M et al (2019) Immobilisation of a-amylase on activated amidrazone acrylic fabric: a new approach for the enhancement of enzyme stability and reusability Sci Rep 9(1):1–9
3. Al-Najada AR, Almulaiky YQ, Aldhahri M et al (2019) Immobilisation of a-amylase on activated amidrazone acrylic fabric: a new approach for the enhancement of enzyme stability and reusability. Sci Rep 9(1):19
4. Al-Najada AR, Almulaiky YQ, Aldhahri M et al (2019) Immobilization of a-amylase on activated amidrazone acrylic fabric: a new approach for the enhancement of enzyme stability and reusability. Sci Rep 9(1):1–9
5. Arshad M, Adil M, Sikandar A (2014) Hussain T (2014) Exploitation of meat industry by-products for biodiesel production: Pakistan's perspective. Pak J Life Soc Sci 12:120–125
6. Arshad M (ed) (2017) Perspectives on water usage for biofuels production: aquatic contamination and climate change. Springer, 24 November 2017
7. Arshad M, Bano I, Younus M, Khan A, Rahman A (2018) Health concerns associated with biofuel production. In: Perspectives on water usage for biofuels production: aquatic contamination and climate change, vol 2018, pp 97–105
8. Arshad M, Hussain T, Chaudhry N, Sadia H, Aslam B, Tahir U, Abbas M, Qureshi N, Nazir A, Rajoka MI, Iqba M (2019) Enhancing Profitability of ethanol fermentation through gamma ray mutagenesis of saccharomyces cerevisiae. Pol J Environ Stud 28(1)
9. Arshad M (ed) (2020) Sustainable ethanol and climate change: sustainability assessment for ethanol distilleries. Springer Nature, 13 November 2020
10. Arshad M (ed) (2023) Climate changes mitigation and sustainable bioenergy harvest through animal waste: sustainable environmental implications of animal waste. Springer Nature, 28 April 2023
11. Arshad M, Ansari AR, Qadir R, Tahir MH, Nadeem A, Mehmood T, Alhumade H, Khan N (2022) Green electricity generation from biogas of cattle manure: an assessment of potential and feasibility in Pakistan. Front Energy Res 25(10):911485
12. Asakura T, Kitaguchi M, Demura M et al (1992) Immobilization of glucose oxidase on nonwoven fabrics with Bombyx mori silk fibroin gel. J Appl Polym Sci 46(1):49–53
13. Badgujar KC, Bhanage BM (2014) Application of lipase immobilized on the biocompatible ternary blend polymer matrix for synthesis of citronellyl acetate in non-aqueous media: kinetic modelling study. Enzym Microb Technol 57:16–25
14. Bal N, Houshyar S, Gao Y et al (2014) Digital printing of enzymes on textile substrates as functional materials. J Fiber Bioeng Inf 7(4):595–602
15. Bano I, Ghauri MA, Arshad M, Yasin T, Younus M (2017) Bioactivity of variant molecular weight chitosan against drug-resistant bacteria isolated from human wounds. Microbial Drug Resist 23(8):958–965

16. Bano I, Arshad M (2018) Climatic changes impact on water availability. In: Perspectives on water usage for biofuels production: aquatic contamination and climate change, pp 39–54
17. Bredig G, Gerstner F, Lang H (1935) Biochem Z 282:88
18. Brena BM, Batista-Viera F (2006) Immobilization of enzymes. In: Immobilization of enzymes and cells, 15th ed. Springer, p 30
19. Chen B, Yin C, Cheng Y et al (2012) Using silk woven fabric as support for lipase immobilization: the effect of surface hydrophilicity/hydrophobicity on enzymatic activity and stability. Biomass Bioenergy 39:59–66
20. Chibata I (1996) Industrial applications of immobilized biocatalysts and biomaterials. Advan Mol Cell Biol Elsevier 15:151–160
21. Dai Y, Yao J, Song Y et al (2016) Enhanced adsorption and degradation of phenolic pollutants in water by carbon nanotube modified laccase-carrying electrospun fibrous membranes. Environ Sci: Nano 3(4):857–868
22. Datta S, Veena, R., Samuel MS et al (2020) Immobilization of laccases and applications for the detection and remediation of pollutants: a review. Environ Chem Lett 1–18
23. Dwevedi (2016) 100 years of enzyme immobilization. In: Enzyme immobilization. Springer, pp 1–20
24. Dyer DJ, Feng J, Schmidt R et al (2004) Macromol 37:7072–7074
25. Gam-Derouich S, Lamouri A, Redeuilh C et al (2012) Langmuir 28:8035–8045
26. Gao SL (2005) Appl Phys B 81:681–690
27. Gao Y, Truong YB, Cacioli P et al (2014) Bioremediation of pesticide contaminated water using an organophosphate degrading enzyme immobilized on nonwoven polyester textiles. Enzym Microb Technol 54:38–44
28. Homaei AA, Sariri R, Vianello F et al (2013) Enzyme immobilization: an update. J Chem Biol 6(4):185–205
29. Homaei AA, Sariri R, Vianello F, Stevanato R (2013) Enzyme immobilization. J Chem Biol 6(4):185–205
30. Kahoush M, Behary N, Cayla A et al (2019) Surface modification of carbon felt by cold remote plasma for glucose oxidase enzyme immobilization. Appl Surf Sci 476:1016–1024
31. Kamath N, Melo JS, DSouza SF (1988) Grease immobilized on polyethyleneimine cotton cloth. Appl Biochem Biotechnol 19(3):251–258
32. Karimpil JJ, Melo JS, D'Souza SF (2012) Immobilization of lipase on cotton cloth using the layer-by-layer self-assembly technique. Int J Biol Macromol 50(1):300–302
33. Kokoda H, Urata T, Nakanishi K (1997) Hydrolysis of xylan by Aspergillus Niger immobilized on non-woven fabrics. Biosci Biotechnol Biochem 61(4):583–587
34. Lee JW, Gall TM, Opwis K et al (2013) Organotextile catalysis. Sci 341
35. Li W, Chen B, Tan T (2011) Comparative study of the properties of lipase immobilized on nonwoven fabric membranes by six methods. Process Biochem 46(6):1358–1365
36. List B, Yang JW (2006) Science 313:1584–1586
37. Luna C, Verdugo C, Sancho ED et al (2014) Production of a biodiesel-like biofuel without glycerol generation, by using Novozym 435, an immobilized Candida Antarctica lipase. Bioresour Bioprocess 1(1):11
38. Ma K, Islamoglu T, Chen Z et al (2019) Scalable and template-free aqueous synthesis of zirconium-based metaleorganic framework coating on textile fiber. J Am Chem Soc 141(39):15626–15633
39. MacMillan DWC (2008) Nature 455:304–308
40. Maryskov M, Ardao I, García-Gonzalez CA et al (2016) Polyamide 6/chitosan nanofibers as support for the immobilization of Trametes versicolor laccase for the elimination of endocrine disrupting chemicals. Enzym Microb Technol 89:31–38
41. Mayer-Gall T, Lee JW, Opwis K et al (2016) Textile Catalysts dan unconventional approach towards heterogeneous catalysis. Chem Cat Chem 8(8):1428–1436
42. Messing R (2012) Immobilized enzymes for industrial reactors. Elsevier
43. Mohamed A, Nemeshwaree B, Brigitte M et al (2016) Activity of enzymes immobilized on plasma treated polyester. J Mol Catal B Enzym 134:261–272

44. Mohamed SA, Aly AS, Mohamed TM et al (2008) Immobilization of horseradish peroxidase on nonwoven polyester fabric coated with chitosan. Appl Biochem Biotechnol 144(2):169–179
45. Morshed MN, Behary N, Bouazizi N et al (2019) Surface modification of polyester fabric using plasma-dendrimer for robust immobilization of glucose oxidase enzyme. Sci Rep 9(1):1–16
46. Morshed MN, Behary N, Bouazizi N et al (2020) Modification of fibrous membrane for organic and pathogenic contaminants removal: from design to application. RSC Adv 10(22):13155–13173
47. Neaz MM, Behary N, Bouazizi N et al (2021) An overview on biocatalysts immobilization on textiles: preparation, progress, and application in wastewater treatment. Chemosphere 279:130–481
48. Niu J, Xu J, Dai Y et al (2013) Immobilization of horseradish peroxidase by electrospun fibrous membranes for adsorption and degradation of pentachlorophenol in water. J Hazard Mater 246:119–125
49. Opwis K, Knittel D, Schollmeyer E (2007) Biotechnol J 2:347–352
50. Praschak D, Bahners T, Schollmeyer E (1998) Appl Phys A 66:69–75
51. Rase HF (2000) Handbook of commercial catalysts: heterogeneous catalysts. CRC Press
52. Sheldon RA (2007) Enzyme immobilization: the quest for optimum performance. Adv Synth Catal 349(8–9):1289–1307
53. Sharif N, Munir N, Hasnain M, Naz S, Arshad M (2021) Environmental impacts of ethanol production system. In: Sustainable ethanol and climate change: sustainability assessment for ethanol distilleries, pp 205–223
54. Shemer L, Granot R, Freeman A et al (1979) Multilayer immobilized-enzyme filter reactors: urease bound to nylon fabric filters. Biotechnol Bioeng 21(9):1607–1627
55. Shim EJ, Lee SH, Song WS et al (2017) Development of an enzyme immobilized support using a polyester woven fabric. Textil Res J 87(1):3–14
56. Shinde P, Musameh M, Gao Y et al (2018) Immobilization and stabilization of alcohol dehydrogenase on polyvinyl alcohol fibre. Biotechnol Rep 19:00260
57. Taheran M, Naghdi M, Brar SK et al (2017) Degradation of chlortetracycline using immobilized laccase on Polyacrylonitrile-biochar composite nanofibrous membrane. Sci Total Environ 605:315–321
58. Temoçin Z, Inal M, Geokgeoz M et al (2018) Immobilization of horseradish peroxidase on electrospun poly (vinyl alcohol)epolyacrylamide blend nanofiber membrane and its use in the conversion of phenol. Polym Bull 75(5):1843–1865
59. Van Beilen JB, Li Z (2002) Enzyme technology: an overview. Curr Opin Biotechnol 13(4):338–344
60. Virgen-Ortíz JJ, Dos Santos JCS, Berenguer-Murcia A et al (2017) Polyethylenimine: a very useful ionic polymer in the design of immobilized enzyme biocatalysts. J Mater Chem B 5(36):7461–7490
61. Wang Q, Cui J, Li G et al (2014) Laccase immobilized on a PAN/adsorbents composite nanofibrous membrane for catechol treatment by a biocatalysis/adsorption process. Molecules 19(3):337–388
62. Wang SG, Zhang WD, Li Z et al (2010) Lipase immobilized on the hydrophobic polytetrafluoroethene membrane with nonwoven fabric and its application in intensifying synthesis of butyl oleate. Appl Biochem Biotechnol 162(7):2015–2026
63. Wang SQ (2007) J Mol Catal Chem 264:60–65
64. Wu XT, Ansari AR, Pang XX, Li HZ, Zhang ZW, Luo Y, Arshad M, Song H (2019) Visfatin plays a significant role in alleviating lipopolysaccharide-induced apoptosis and autophagy through PI3K/AKT signaling pathway during acute lung injury in mice. Arch Immunol Ther Exp 1(67):249–261
65. Wunschik DS, Ingenbosch KN, Süss P et al (2020) Enzymatic epoxidation of cyclohexene by peroxidase immobilization on a textile and an adapted reactor design. Enzym Microb Technol 136:109–512
66. Xu R, Cui J, Tang R et al (2017) Removal of 2, 4, 6-trichlorophenol by laccase immobilized on nano-copper incorporated electrospun fibrous membrane-high efficiency, stability, and reusability. Chem Eng J 326:647–655

67. Yamazaki H, Cheok RKH, Fraser ADE (1984) Immobilization of invertase on polyethylenimine-coated cotton cloth. Biotechnol Lett 6(3):165–170
68. Yin S, Jin Z, Miyake T (2019) Wearable high-powered biofuel cells using enzyme/ carbon nanotube composite fibers on textile cloth. Biosens Bioelectron 141:111–471
69. Yoon TP, Jacobsen EN (2003) Science 299:1691–1693
70. Zdarta J, Jankowska K, Wyszowska M et al (2019) Robust biodegradation of naproxen and diclofenac by laccase immobilized using electrospun nanofibers with enhanced stability and reusability. Mater Sci Eng C 103:109–789
71. Zdarta J, Staszak M, Jankowska K et al (2020) The response surface methodology for optimization of tyrosinase immobilization onto electrospun polycaprolactone chitosan fibers for use in bisphenol A removal. Int J Biol Macromol 165:2049–2059

Sustaining Amylase Applications Through Digitalization in the Textile Economy

B. A. Hadi

Abstract Amylases are enzymes that have gained significant popularity in textile processing due to their eco-friendly and efficient properties. Digitalization can enhance enzyme applications by improving efficiency, reducing production costs, and addressing chemical compatibility issues. Various digital technologies such as sensor technology, artificial intelligence, machine learning, big data analytics, and cloud computing are explored for their potential to enhance amylase performance. Amylase applications in textile processing include desizing, biopolishing, bioscouring, and bleaching. Strategies for enhancing amylase performance and overcoming these challenges through digitalization have been explained in this chapter. Challenges and limitations associated with amylase stability, production costs, and chemical compatibility are addressed in the subsequent section. The importance of consumer education on enzyme applications is emphasized, highlighting its role in increasing awareness, understanding, and acceptance of enzyme-based textile processes. Methods for educating consumers and the potential impacts of consumer education on enzyme applications are discussed. Finally, the chapter concludes with future directions and recommendations for further research in amylase applications. Trends and prospects in amylase applications are explored, and suggestions are provided for future research to optimize their effectiveness in the textile industry. Overall, this chapter provides a comprehensive overview of the applications of amylases in textile processing and highlights the potential of digitalization in enhancing their performance. It serves as a valuable resource for researchers, engineers, and industry professionals interested in sustainable and efficient textile processing methods.

B. A. Hadi (✉)
Faculty of Medicine, University of Kerbala, Karbala, Iraq
e-mail: Bahaa.h@uokerbala.edu.iq

M. Arshad (ed.), *Enzymes in Textile Processing: A Climate Changes Mitigation Approach*, SDGs and Textiles, https://doi.org/10.1007/978-981-97-8058-7_5

Graphic Abstract

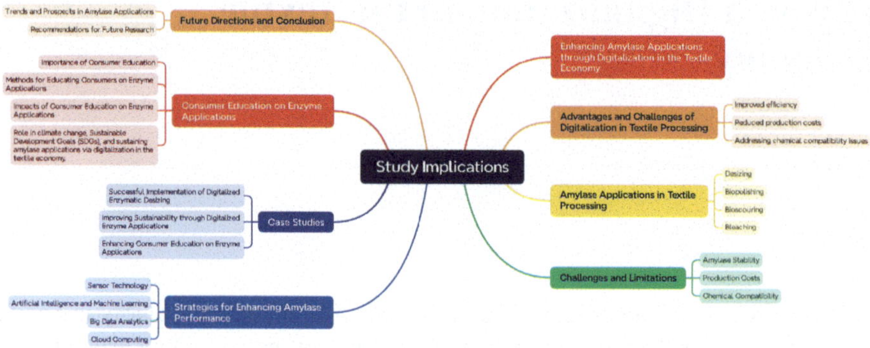

Keywords Amylases · Textile industry · Desizing · Challenges · Processing methods · Possible solutions

1 Introduction

The utilization of amylases in the textile industry has experienced a substantial increase since their discovery in the 1960s for desizing purposes. This surge can largely be attributed to the world's growing commitment to environmental sustainability and the implementation of continuous processes in textile wet processing, which revealed the technical advantages of amylases. Prior to the introduction of amylase enzymes, desizing textile fabric heavily relied on the use of acids. However, these acids were slow, polluting, and incompatible with modern high-speed processes. In contrast, amylases provide a much gentler, eco-friendly, and faster method for removing starch-based sizing from textile materials. This significant reduction in process time and environmental impact by substituting acids with amylases has brought about a revolution in the textile wet processing sector. The desizing process in textiles involves the elimination of starch-based sizing materials from the fabric. Sizing is performed to enhance the weight or stiffness of yarns, smoothen the fabric surface, improve abrasion resistance, and facilitate the handling of yarns during weaving. Starch is the most commonly used sizing material, followed by other natural polymers like cellulose ethers [80].

Amylases catalyze the hydrolysis of β-1, 4-glucosidic linkages in starch molecules, resulting in the production of dextrins, maltose, β-glucose, and ultimately glucose monomers. By changing starch molecules into smaller, water-soluble compounds, amylase enables easy removal from the fabric through washing. These amylase enzymes belong to the broad category of hydrolases, as they act on starch molecules and break them down into smaller units through the addition of water molecules, a process known as hydrolysis. Amylase enzymes can be classified as exo-amylases, which randomly hydrolyze β-1, 4-glucosidic linkages along the starch chain, generating shorter, water-soluble chain molecules [11]. This differs from endo-amylases, which attack the ends of long chains and release two β-glucose molecules with each hydrolytic step. The typical chemical equation representing the amylase enzyme-catalyzed reaction is as follows: $(C_6H_{10}O_5)_n + H_2O \rightarrow [(C_6H_{12}O_6)_x]_y + [(C_6H_{12}O_6)]_m + (C_6H_{10}O_5)_p$, where n represents the number of glucose molecules in the starch unit, x indicates the dextrin with varying chain length, y denotes the number of dextrin molecules, m represents the number of maltose molecules, and p signifies the number of unchanged starch molecules remaining [48, 107].

1.1 Overview of Amylase Applications

Amylose is an information carrier of heredity and catalyzes the formation of the so-called cyclodextrins. Work holds the potential of optimizing the yield and specificity of the predigesting enzyme in processing applications in textiles. Also, it indicates its commercial applications, for example, plasma substitute, which takes advantage of the ability of amylose to encapsulate liquids. Amylase finds diverse applications and many remain to be discovered, which means it is a very important material in the textile industry. Textile industry and clothes recycling has been a worldwide issue. The use of enzymes in breaking down the fibers in clothes is under development. Amylase is proved to be very effective in this area, making the process of biopolishing more efficient. Amylase has been tested successfully functioning in clothes recycling and has good potential in the future to replace some of the contemporary technologies in clothes recycling. Amylase appears to be a brown–red powder. It has a long chain structure and there is of uniform helix structure composed of 20 monomers, forming aliphatic groups. Amylase is hydrolyzed by exterior forces into glucose. Amylase is a material worth studying for engineering researchers. Amylase has good potential in the future in textiles as the technology develops. The researchers will keep learning the effect of material parameters on the action of amylose as they overcome the challenges of measuring and modeling the microscopic material behavior. Amylase, which has variations in actions on enzymes, will become an even more popular group of bioprocessing material through the technology on genetic modifications of cells [16]. Amylase has diverse applications in the world. New findings in this field have been discovered over time. With the technological advances and the development of new theories and hypothesis, many discoveries and inventions related to amylase will be found. This will indeed benefit the development of engineering and technology. Amylase, based on its applicability, is expected to be one of the most widely used materials in engineers' daily life. With the development of genetic modification of cells, new variations of amylase can be created to optimize its application in processing technologies. Cultivations of amylose separated from different sources are applied to researches. It is commercially used in industries such as detergent, pharmaceuticals, and clinical detections as a material in processing applications. Amylase is a highly effective catalyst with a large range of substrates and has extensive applications. Amylase is also a very stable and cost-effective material due to its good temperature and pH tolerance. Amylase can be a material selected for our better use by recycling old clothes. With the specific information established from the practices of clothes recycling, different materials and their applications can be assessed [44, 53, 59].

1.2 Historical Perspectives

The genesis of electric vehicles paralleled early enzymatic liquefier patents, with notable advancements in industrial enzymes, particularly diastatic ferments or amylases, beginning in the 1850s. The first patent for a starch sugar process using diastase was filed in 1898, coinciding with the American Enzyme Manufacturing Company's innovations in solid-state fermentation for enzyme production. The documentation of the first bacterial amylase in 1934 marked a significant scientific breakthrough, further accelerated by World War II's influence on enzyme production using bakers' yeast. These technological strides have remained relevant over decades. In 1964, American Amylum's patent for amyloglucosidase utilization heralded the advent of genetic recombinant techniques in enzyme manufacturing. The introduction of immobilized amylases, which stay fixed within mixtures, enhanced the efficiency and control of enzyme applications, continuing the legacy of innovation in the field [9, 57, 90].

The historical timeline of amylase development in the textile industry is marked by several key milestones, reflecting the evolution of this biotechnological tool to enhance textile processing. Here are some significant points in its developmental history:

1. Discovery of Amylases (Early twentieth century): Amylases were first discovered and isolated, marking the beginning of understanding their potential in industrial applications.
2. Initial Industrial Use (Mid-twentieth century): Amylases started being used in industries, including textiles, for processes like desizing, where they replaced traditional methods that were harsher on fabrics.
3. Advancements in Enzyme Technology (1970s–1980s): The development of more stable and specific amylase enzymes allowed for their broader application in textile processes under various conditions.
4. Introduction of Genetically Modified Enzymes (1990s): With the advent of genetic engineering, amylases were modified to be more efficient, cost-effective, and suitable for the specific needs of the textile industry.
5. Widespread Adoption in Textile Desizing (2000s): Amylases became the standard for desizing in the textile industry due to their environmental benefits and compatibility with a broad range of fabric types.
6. Digitalization of Enzyme Application (2010s-Present): The integration of digital technologies, like automation and real-time monitoring, has improved the precision and efficiency of amylase application in textiles.

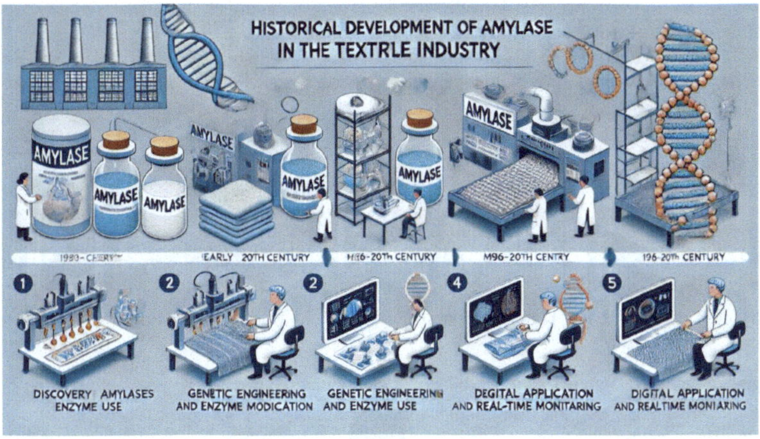

Fig. 1 Historical timeline of amylase development and applications in the textile industry

7. Sustainability and Circular Economy (2020s): Amylases are now a key component in sustainable textile processing, contributing to water and energy conservation, and aligning with the goals of the circular economy (see Fig. 1).

2 Digitalization in Textile Processing

The process of digitalization in the textile industry has been highly advanced for a number of years, focusing on new technological developments and to optimize lower production costs, improve textile product quality, and allow for increases in process engineering efficiency. Simply put, the process allows for digital data structures to be interlinked along the entire textile production and maintain this digitalization through textile material properties right through to product costings and simulation processes [21]. This is seen as a direct response to the increasing drive for higher levels of innovation in the textile industry and the integration of more sustainable and smart fibers, textiles, and processes. Towards sustainable and smart textiles. The importance of digitalization is evident in enabling textile products and processes towards sustainable and smart design work. It has seen the potential for new fibers and materials to be developed and contribute to multifunctional and intelligent textiles for a wide range of technical performance and healthcare applications. By enabling more efficient and detailed characterization of textile materials and the integration of 'digital materials' data sheets into the methods of the textile product development process, it is possible to take account and optimize through the life cycle analysis… this could lead to "smarter" materials selection and processing strategies, improved product quality and more innovative and sustainable… smart textile products and more integrated digitalization towards customized and responsive manufacturing and design methodologies. Overall, digitalization within the life cycle of textile products has the

Fig. 2 Digitalization in textile processing: A framework—An illustrative figure that outlines the framework of digitalization within the textile processing industry

potential to enable a more detailed understanding of the interconnected processes from material selection right through to recycling possibilities [37, 38, 45] (see Fig. 2).

2.1 Advantages and Challenges

Digitalization in textile processing has brought about a more environmentally friendly and efficient approach to textile production, but not without its own challenges. One of the most significant benefits is the reduction in the usage of water, energy, and chemicals. Digitalization techniques, for instance, the use of automation in the dyeing process, help to attain optimized consumption of water, energy, dyestuff, and chemicals. According to J. P. W. Carey in his report on the staff paper of the United States International Trade Commission, water and electricity take up the highest percentage of direct manufacturing costs [20]. By reducing water and energy consumption, it saves costs for the manufacturer. In addition, it also increases the

Table 1 Advantages and challenges of digitalization in textile processing

Advantages	Challenges
Enhanced efficiency	Implementation costs and infrastructure requirements
Reduced production costs	Data privacy and security concerns
Improved process control	Technical complexity and integration with existing systems
Real-time data analytics	Workforce training and adoption of digital technologies

lifespan of water resources and reduces water treatment costs as well as discharges of wastewater into the environment. On the other hand, limitations are present as newer and better technology is being introduced. For every different technology, it requires time, cost, and skill to adapt to a new technology. In other words, the investment of time and money in effectively having the technology live up to its promised benefits is a must [94] (see Table 1).

2.2 *Role in Enhancing Enzyme Applications*

When we talk about how digitalization enhances enzymes, it's really the accessibility that stands out. The capability of digitalization to collapse time and distance brings the enzyme designer closer to the end user, allowing us to tailor enzymes to their specific needs. Whether it's improving the speed, dye exhaustion or even working at lower temperatures, digitalization shortens the development times for new products and reduces the amount of experimental work required, maintaining our position at the cutting edge of enzyme technology. Because these new products can be developed for niche markets with digitalization, it may also be possible to further improve the sustainability offered by using enzymes in the first place, something which is becoming more and more important. By making it easier to identify the best possible enzymes and then scaling up their manufacture, digitalization will support the shift from traditional, chemically based processes to more sustainable enzyme-based ones. This has benefits for the environment, for the companies using these processes, and for the people who work in the industry [109]. There are also many possibilities offered by digitalization to further utilize enzymes in different parts of the textile process. For example, maintaining a constant bioreactor cycle that is managed remotely is a possibility and something I realized through this research, which discovery and improvement of existing enzymes. Also with the potential that enzymes can offer right from washing the raw cotton, to the final product leaving the production line, digitalization provides a gateway to the most sustainable and efficient methods possible. All of these aspects culminate in the belief that, should a sustainable industry be centered on enzymes, the digitalized platform will become the catalyst for further improvement and large uptake throughout the industry [22, 31, 40] (see Fig. 3).

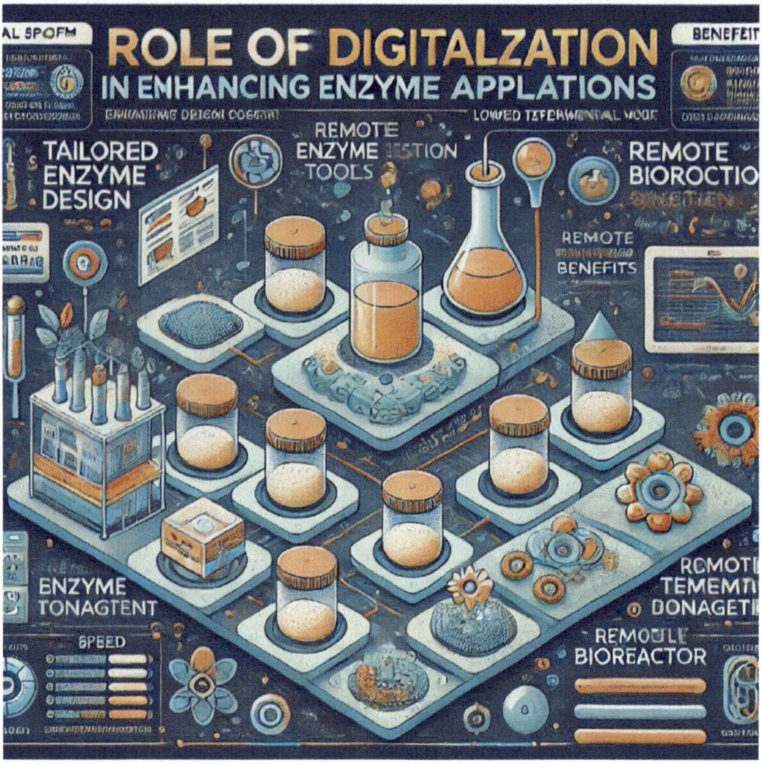

Fig. 3 Digitalization enhancing enzymes in the textile industry

3 Amylase Applications

Amylase is widely used in textile processing as it is environmentally friendly and an alternative to harsh chemicals. The objectives of using amylase for desizing are to remove starch from the fabric and retain the size of the fabric. There are two types of amylase used for desizing: amylase pancreatin or amylase from an animal source, and fungal amylase. Fungal amylase is widely used in the market and can be used in a wide range of pH and temperature, resulting in better fabric quality. Amylases give a moderate desizing rate, but the amylase of choice in the textile industry is high diastatic fungal amylase, as this enzyme provides a continuous push toward quicker and more efficient desizing, and it helps with modern processing in the continuous wet running system. For biopolishing, the common type of bacterial amylase used is Bacillus amyloliquefaciens sp. amylase, which works in alkaline pH. First of all, the absorbency of the fabric will increase due to the removal of the microfibrils and the surface of the fabric will be smoothed. Then, molecules will be removed from the fabric to an extent that the fabric will give brighter shades. On top of that, the fabric becomes softer and smoother because the fibers are more flexible after the fabric has been treated with amylase for biopolishing. For bioscouring, the most commonly used enzymes for the process are amylase and lipase. Through the enzymatic action of amylase on the surface of the cellulose, it is able to open the molecules of the cellulose by breaking the hydrogen and hydrophobic interactions. Then, it allows the enzyme to easily penetrate into the amorphous regions of the molecular polymers and expose water at the oxygen bonding sites. By breaking the molecule at the sides, it can enhance the hydrophilicity and swelling properties of the cellulose fiber. Alpha-amylase is widely used in textile processing as it is environmentally friendly and an alternative to harsh chemicals [3, 26, 59].

Amylase also can be used effectively to remove oil, starch, and carbohydrate size from the yarns and fabrics. When digested with alpha-amylases, these materials are effectively removed from the products. This makes these enzymes a versatile tool in preparing textiles for the various treatments and processes used to add designs, colors, and effects. Also, beta-amylase can be used to selectively remove 'fuzz' (microscopic surface fibers) from cellulosic materials such as cotton, flax, and jute to create a smooth, clean surface on the fabric. This is a physical effect; the enzyme acts by breaking the microscopic fibers to ease their removal in the washing and rinsing stages. Amylases also are used to 'biopolish' fabrics. The enzyme selectively attacks and releases the fiber ends and 'pills', leaving the fabric surface clean and even. This can be used either to improve fabric appearance or in the manufacture of 'peach skin' fabrics, where a soft, smooth handle is desired. Through biopolishing, less aggressive mechanical processes are needed, which reduces the damage to the surface of the fibers and hence increases the quality and lifespan of amylase-treated textiles. The application of enzymes like amylase in textile wet processing has been gaining popularity due to the increase in the awareness of sustainable and green processes in the textile industry. Also, advances in research are being made to develop patented amylase products specifically for substrates and conditions found in the textile industry, which will continue to expand the uses and benefits of amylases in textiles [53, 82, 95] (see Fig. 4 and Table 2).

3.1 Desizing

The primary enzyme used for starch desizing is amylases. Amylases are a group of enzymes that break down starch and glycogen into smaller carbohydrates [71, 82, 125]. Enzymatic desizing is a process that involves the breakdown of starch within a fabric. Starch is a natural substance that is a polymer made from repeating glucose units, meaning that it is made up of long chains of glucose molecules. These chains intertwine to make an organized structure, which is the main basis for binding the fibers collectively when making cloth [49, 60]. Desizing is a process within textile manufacture and plays a vital part in making sure that the final product is as desired. Desizing may be broken down into two additional categories: chemical desizing and

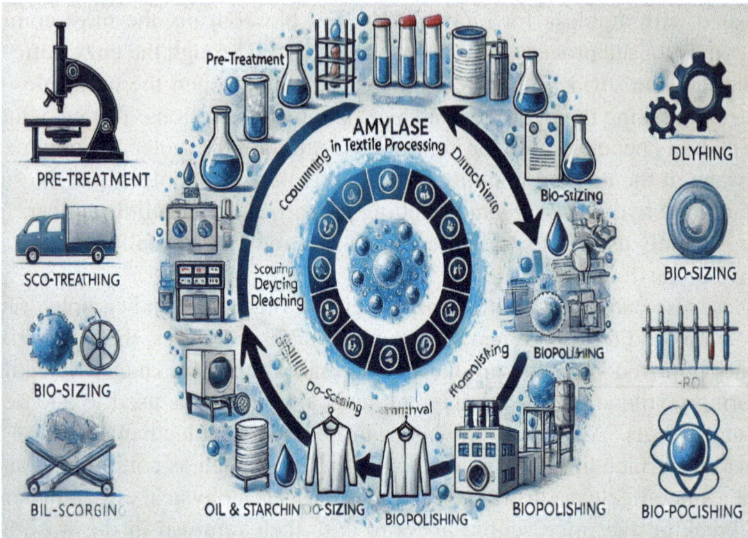

Fig. 4 Amylase applications in textile processing

Table 2 Overview of amylase applications in textile processing

Application	Description
Desizing	Removal of starch-based sizing materials from textiles, improving fabric quality and handle
Biopolishing	Enzymatic treatment to remove protruding fibers and create a smoother fabric surface
Bioscouring	Removal of non-cellulosic impurities from textiles, enhancing dye uptake and fabric quality
Bleaching	Enzymatic bleaching process to improve color brightness and reduce environmental impact

enzymatic desizing. Nonetheless, the latter has been proven to be more environmentally friendly and cost-effective in a variety of different textile manufacturing applications [74, 105, 128] (see Fig. 5).

3.2 *Biopolishing*

In biopolishing, the main target is to clean up one of the most refractory components of the cotton fabrics, that is, the cotton fiber long hairs, which are made up of cellulose, and not the short fiber furs, which are also made up of cellulose. Currently, the preparation starts with the treatment of the dry fabric which can have a certain

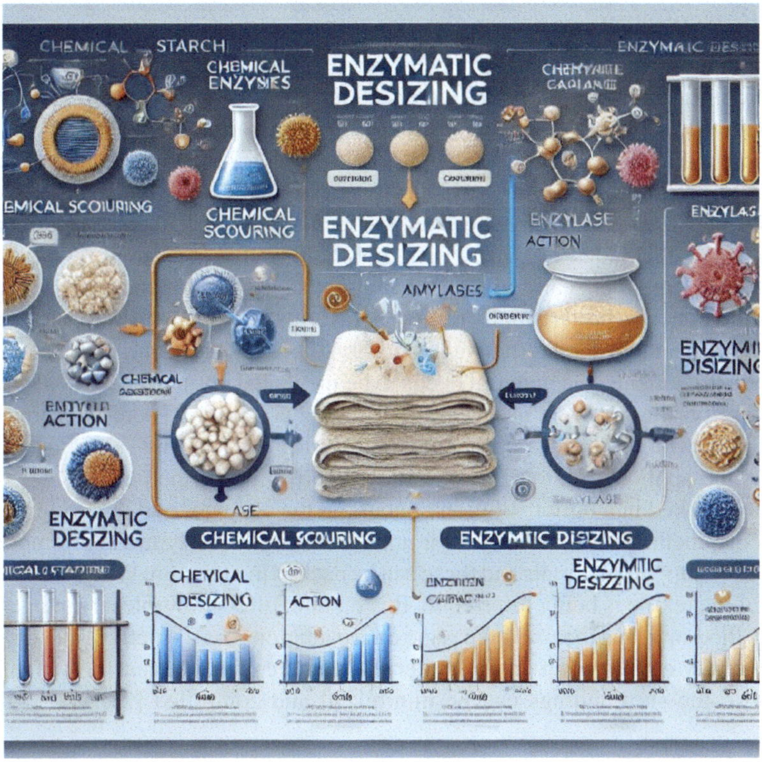

Fig. 5 Enzymatic desizing process in textile manufacturing with labeled enzymes, highlighting the breakdown of starch within fabric, the role of amylases, and comparison between chemical and enzymatic desizing methods

moisture content at the beginning of the process. Then the fabric will be passed through a bio-polishing machine, a rotor which comprises a lot of circular saw blades and also ball-like fiber impellers meant to keep the fabric in close contact with the rotary blades for better treatments. The bio-polishing machine has two rotary parts, with the fabric being treated passed and kept in close contact between them [121]. The treatment is done with cellulase enzyme so that the fine hairs can be selectively digested and orgalose in nature. The terminals of the cellulose polymer long chains will be progressively eroded in a hydrolyzing way as a result of the exocellular action of the cellulase enzymes which will diffuse into the inner of the cotton fiber walls due to the osmotic effect of the broken-down cell chains pulling water into the fiber interior to realize an even easier access to the enzymes. The fabric will change a lot of its physical and also chemical properties and appearance as a result of the bio-polishing treatment, as the surface hairs are being selectively ablated and removed from the fabric. The fabric will appear to be shinier and softer, and the treated fabric will have better vibrance on absorption of dyestuff in comparison with untreated materials [69, 127].

3.3 Bioscouring

Bioscouring is actually a type of wet treatment in textile processing. This is not only performed to remove impurities, but also to impart smoothness and luster to the textile. When this wet process is performed with the help of fermentative action, it is called bioscouring. So, in bioscouring or enzymatic scouring processes, plant cuticular materials can be removed more selectively. During the enzymatic action performed by the added enzymes, the amorphous and less ordered regions in the cuticles of cotton fibers can be attacked and the waxes, pectin, and other impurities in the cuticles can be preferentially removed. In terms of processing time, bioscouring is known to offer a faster alternative to traditional hot caustic scouring. However, for enzymatic treatments, the material needs to be incubated normally for about 24–48 h depending on the products used and the process condition. There are three main methods by which enzymes can be applied during bioscouring. The first method involves the application of enzymes in liquid form. In this method, the dry textile goods are soaked in a buffer solution containing the enzymes and other ingredients. The second method involves enzyme immobilization. The enzymes are immobilized onto related support materials and these can be packed in a column. When the compact health of the textile is being referred to, this method offers multiple advantages. The third method involves the use of enzyme encapsulation. In this method, the enzymes are physically entrapped in carrier materials, and therefore the leaching of enzyme activity to the surroundings can be minimized and controlled [23, 27, 116] (see Fig. 6).

3.4 Bleaching

Starch or size materials present in fabrics are removed in the desizing process to make the fabric ready for bleaching. The chemistry involved in the desizing process and the role of amylases have already been explained in the preceding section [15, 17]. The next process in line is known as bleaching. Bleaching is a process of removing natural color in the fibers and also any acquired color on the fiber. The process makes the fibers as white as possible. If necessary, hydrogen peroxide can be added to the amylase and other chemicals used in the bleaching process with the enzyme at a suitable temperature and pH in a jigger machine or beck. Amylase enzymes are used to remove any starch which might still be present. This is known as amylolytic desizing. This is particularly important in the production of textiles made from materials such as cotton or linen. Due to the raised environmental and cost concerns, the industry has moved from traditional batch processing to more modern and efficient continuous processes. New methods of digitalizing the production processes such as creating an inventory control system or linking the sales department to the factory's production schedule can further improve the sustainability of amylase use [24, 29, 32] (see Fig. 7).

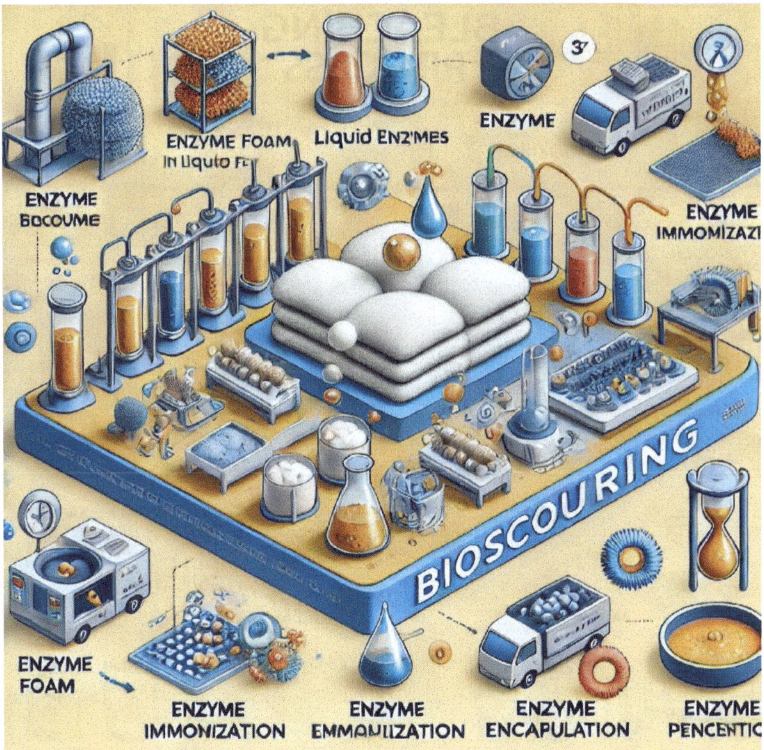

Fig. 6 Biosourcing process in textile manufacturing, illustrating the application of enzymes in liquid form, enzyme immobilization, and enzyme encapsulation, with emphasis on the removal of impurities and the imparting of smoothness and luster to the textile

4 Challenges and Limitations

Amylase is the modern solution to the decades of unsustainable chemical practices in the textile industry—but only if the enzymes used can survive the dyeing process. This is an unfortunately common problem when bio-based materials are put forward as replacements for traditional chemicals, and betrays a "general assumption" about the chemical stability of enzymes. If industry is to start a transition toward bio-based alternatives, then issues such as the stability of enzymes need to be better understood and industries need to support research into this. However, amylase cannot function effectively in alkaline conditions as required by some major dyes. This could severely limit its applicability in certain stages of the dyeing process, even though it is a great substitute for chemicals in the initial scouring of cotton. These types of problems are already seen with alternatives such as lipases. Lipases are a class of enzymes with great promise in bio-based alternatives, and research has been encouraged into areas such as better recovery of the products and immobilization

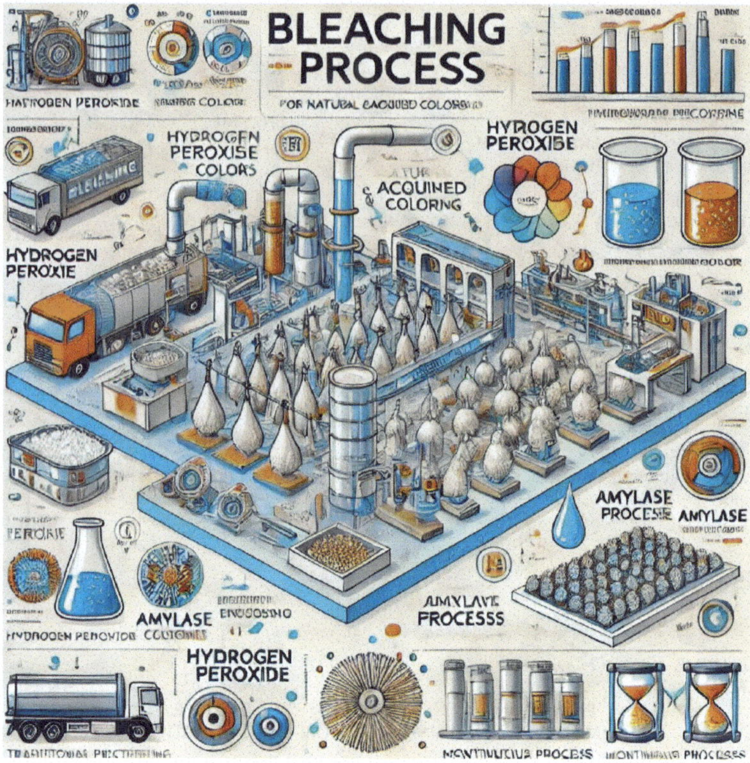

Fig. 7 Bleaching process in textile manufacturing, showing the removal of natural and acquired colors from fibers, the use of hydrogen peroxide and amylase enzymes, and the transition from traditional batch processing to modern continuous processes

techniques. However, one main reason for the lack of commercialization of lipases in the industry is their sensitivity to temperature fluctuations; if the temperature in the dyeing process fluctuates appreciably over the course of a shift, there is no guarantee the enzyme will remain stable, and effective. Conceivably in some large commercial processes, variations of several degrees could be experienced throughout the day. This again raises issues of time and expense involved in attempting to maintain the right conditions for effective use of the alternative. And ultimately the method itself, based on enzymatically catalyzing a surface-level reaction, is quite specific to pre-treatment. Every different type of pre-treatment would need a different enzyme and specific dye treatment would render ineffective these new "modern" bio-based alternatives [44, 70, 117].

4.1 Amylase Stability

Amy, for short, amylases are used in the textile industry to prepare the fabric for dyeing. Amylase enzymes break down the starch in the cotton fibers. Then, during the dyeing process, the amylase prevents the dye from adhering to the cotton, resulting in a pattern involving dyed sections and undyed sections. This method is called tie-dye. However, throughout the dyeing process, the dye is either in a solution or chemically bonded to the cotton. This means that the amylase is exposed to two types of environments: aqueous (the dye is dissolved in water) and non-aqueous (the cotton is a non-polar covalent compound) [14]. The amylase has to retain its 3D tertiary structure in both of these environments in order to work effectively and break down the starch in the cotton. The shape of an enzyme's active site is complementary to the structure of its substrate, meaning that if the amylase unfolds and loses its 3D shape, it can no longer carry out its intended function (in this case, catalyzing the breakdown of starch). This 3D structure is maintained by the bonds between the R groups of the amino acids that make up the amylase. These bonds are relatively weak and are easily broken by changes in temperature, pH, metal ions, and salinity. These factors are known as "denaturing agents" because they cause the denaturation of the protein (unfold), leading to a loss of protein function. As a result, maintaining the stability (its ability to remain in its native confirmation) of the amylase is crucial to ensuring that the dyeing process will run smoothly. Different denaturing conditions will aid in the production of cotton with varied sensory attributes. For example, using a higher temperature environment will break down the starch to a greater extent, resulting in a softer fabric. On the textile production line, cotton is first prepared (this is treated with amylase) and is then dyed. Digital control of such processes is now possible [4, 34, 51, 56, 59, 86, 106, 113].

4.2 Production Costs

In addition to the challenge of maintaining amylase stability throughout different phases of the textile production process, one of the more prevalent obstacles to a more widespread industrial implementation of amylase treatments is the production cost. Amylase treatments are generally relatively expensive, thereby hindering their use, especially for the treatment of textiles and garment pieces in developing countries, where labor costs are often the primary concern [13]. Also, because of its specificity, the extensive research and development costs incurred and required to further strengthen the treatment's robustness and efficiency are major drawbacks, especially for independent startups and businesses. These costs consist of the actual cost of the enzyme, which, due to the specialist development and production sought by global amylase suppliers such as Novozymes and DuPont, will always be pertinent, the energy costs of heating the amylase solutions and the costs of involving expert knowledge to both use and manage the WRC. Further high costs may be added for the application of strategies involving proprietary technology, such as that advocated by Bio-Cat, which may require additional payments or royalties upon initial setup or use. Such factors act in concert to create a high financial threshold for businesses or industrial practices seeking to adopt amylase treatments, thereby reducing the likelihood of seeing widespread industrial application and creating a cycle of underfunding with regard to further research and commercial diversification [28, 34, 53, 62, 82, 87, 99].

4.3 Chemical Compatibility

To analyze the impact of chemicals on enzymes and structure, ion chromatography was utilized. Before using the amylases for tests, both enzymes were pretreated with the same sodium chloride solution and stored in separate flasks. A sample of each was then heated to reduce the action of residual enzymes. These solutions were then used to determine specific ions or chemicals that are found in each type of amylase and their concentration. 0.01 M acetate buffer was prepared by weighing out 0.0248 g of acetic acid and 0.16 g of sodium acetate before dissolving them in 100 cm3 of deionized water. Leading on from that, a longer continuous investigation could be conducted to determine the optimum temperature of the amylase action or how the reaction would be affected by a change in temperature. In addition to diastase, another type of amylase known as Termamyl, obtained from Bacillus licheniformis, was tested for the action of dilute hydrochloric acid and the reducing sugars [12]. The method that was used was essentially the same as that for diastase, including the spectrophotometric analysis. This would allow a comprehensive understanding of amylase and the digestion process as a whole and the factors that can affect it such as the corrosive thyoilic acid. Amylase is widely used; hence it would be difficult to produce a solution that showed the optimal temperature for the action of amylase. However, by discerning the points at which the rate of reaction increases most rapidly, the possible temperature conditions inside the human body and those outside could be estimated. Also, the effects of temperature and chemical concentration could be studied and evaluated by considering the rate of reaction at different temperatures [25, 53, 100, 106, 126].

5 Enhancing Amylase Performance

The technology and strategies for enhanced amylase performance are investigated and compiled in this section to provide a broad and coherent review of the current capabilities and future research directions for intelligent amylase treatments. By assessing the environmental conditions, offering progressive knowledge on the development of intelligent amylase systems through background information on enzyme-based cleaning, and offering a case study on a current method of specific enzyme selection, the information in this section acts as a comprehensive literature review and research proposal for the project's first aim of creating new testing methods to allow step-wise developments in intelligent systems [1, 46, 50].

Concisely, sensors such as those in Fig. 1 provide the input for smart control systems to assess the environmental conditions and select the best treatment to ensure effective substrate digestion and cleaner surfaces. Such technology has been successfully implemented in automated irrigation systems that provide plants and crops with water when soil moisture is low. By assessing the real-time requirements of the plants against the environmental conditions, water usage is optimized and effectively reduced, directly leading to lower water costs and a lower burden on increasingly scarce water resources. This has been documented by two Austrian researchers who found that these systems resulted in an average water usage reduction of 66 percent compared to more traditional irrigation techniques [64, 75, 79].

Consequently, to improve the functional capabilities of amylases in a variety of conditions, smart digital tools are needed to provide accurate and adaptive methods that either assess the environmental conditions and select the appropriate option or continually adapt the treatment process. Currently, sensors, especially those linked to the emerging field of the Internet of Things, provide great potential for the future success of amylase treatments in building maintenance due to their ability to provide real-time demands and options based on the specific environmental conditions at a particular time [5, 63, 118].

Existing surfaces benefit from amylase enzyme products due to the inclusion of potent blends of specially selected enzymes that work synergistically to degrade complex organic matter, such as starches, carbohydrates, proteins, and fats. Amylases, like all enzyme-based cleaning products, rely on an internal chemical, physical, or biological process to remove unwanted matter from surfaces. These enzymes perform optimally under explicitly controlled conditions, usually within a finite range of temperature and pH. However, in reality, the environmental conditions—temperature, pH, humidity, and organic load—can fluctuate widely [44, 87, 91].

5.1 Sensor Technology

These include, for example, forward wave sensors for liquid biofilm monitoring, 3D optochemical dissolved oxygen sensors, and UV/Vis dissolved carbon dioxide sensors. Forward wave sensors have the capacity to transmit biosensor mass data online and can be used for monitoring the amylosic bacteria in continuous in situ flow mode while at the same time, the performance and optimization of the biofilm sensors are evaluated. The 3D optochemical and UV/Vis sensors have the potential to provide sustainable continuous monitoring of the nutrients and the toxic and limiting variables in the fermentation unit operations. In industries such as wastewater treatment, aeration control strategies and energy uses are large unknowns due to insufficient real-time and toxin in situ monitoring. But with the advent of these biofilm

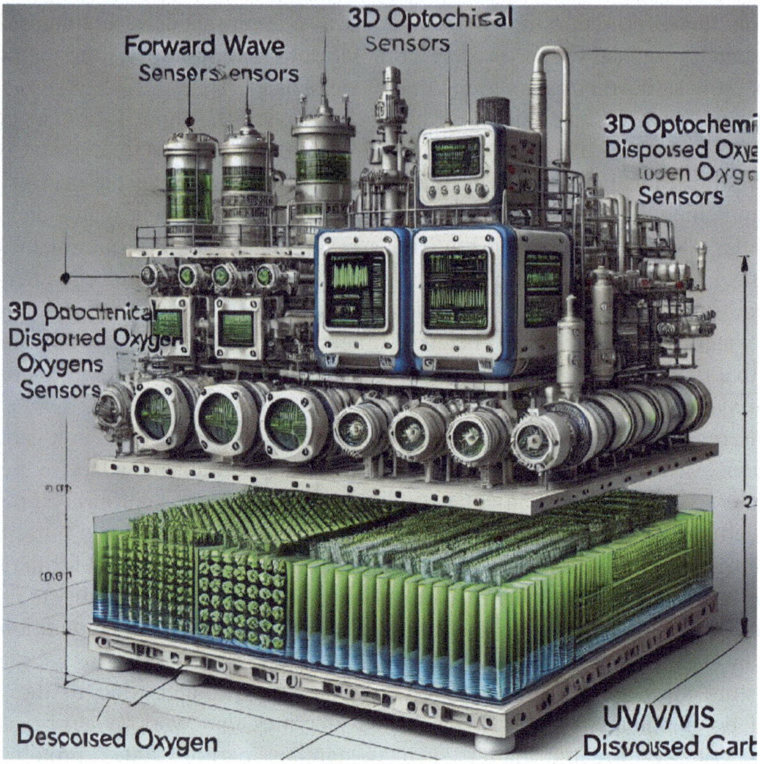

Fig. 8 3D labeled image of forward wave sensors, 3D optochemical dissolved oxygen sensors, and UV/Vis dissolved carbon dioxide sensors for biofilm monitoring

sensors, it is anticipated that a development and technology transfer to the industrial sector can be envisaged. These sensors in research and pilot full scale followed by sensor validity, sensitivity, and accuracy analysis would ultimately benefit sustainable microbial amylase sensors and act as a conduit in influencing policymakers and environmentalists in the eco-industrial revolution [18, 30, 65] (see Fig. 8).

5.2 *Artificial Intelligence and Machine Learning*

AI in the textile economy, including amylase application, is already showing its potential, especially in understanding fiber behavior and amylase reaction. These applications are evolving rapidly, and several companies are investing in AI and machine learning technologies to keep themselves updated in the race of digitalization and sustainability. Unlike traditional methods, the use of AI and machine learning will eventually replace human intervention in amylase processing. For instance, the amylase processes using AI can be run continuously, and it can self-diagnose problems as well. The fiber testing and amylase reaction report generation process is another area where AI will play a large role. The advanced machine learning-based systems can already identify fiber types and provide a clearer picture of fiber behavior over time. Moreover, the amylase reaction reports can be integrated with these systems, and the application can provide a complete analysis of the amylase process in a more intelligent and informative way. With the computational power of cloud computing, AI-based fiber analytics and amylase applications will see a massive improvement. For instance, the amylase process can be optimized by the AI machines and expert systems over time, and the industries will easily implement amylase digitalization with minimum capital investment [6, 76, 115].

5.3 Big Data Analytics

Firmicute amylase technologies and applications development are driven by new progress in large data analytics for production processes in all stages of textile economical operations. Beginning from regulatory compliance and product quality tracking to constant process optimizations and innovative product development, large data analytics has become a significant advancement in the sustainable use of amylase in the textile market. Today, textile industry leaders agree that the future of amylase usage in the textile economy is going to be dictated more and more by the viability of this emerging big data analytics technological instruction for sustainable production. The most important step in implementing a large data analytics service for a sustainable amylase application is defining the aim of the project in regard to the particular sustainable production process. By way of example, constant quality monitoring through real-time sensor data analytics could be a project goal in order to develop innovative products or to optimize production. Then, next-generation sequencing for enzyme screening in product development or in silico mechanistic modeling to help process optimization and innovation making may also be introduced in accordance with project requirements and budgetary limitations. On account of the computational complexity of the routine, it is extremely important to think about the advancement in the algorithm and computational power. Amylase itself has become the interest of researchers in the textile business, especially in the area of enzyme immobilization, large-scale application, and optimizing sustainable procedures. By incorporating amylase research and emerging discoveries, for example, the recent application of amylase from a bacterial source of Bacillus in the bioremediation of water suggesting a circular and integrated method of amylase from wastewater to water remediation products, it is anticipated that this large data analytics technology could be conveniently used in creating more sustainable strategies in the future for both product and processes. From the section of "cloud computing," the data communication monitoring to the Cloud and the Cloud analytics section were amended in respect of the recent progress in cloud big data analytics, in which Davenport's process in real big data which consists of the 'large data understanding,' 'large data solutions,' and 'large data inferences,' i.e., The "government of Singapore" research data on the usefulness of large data in public policy was incorporated [43, 53, 95].

5.4 Cloud Computing

Cloud computing offers a way to improve amylase performance in the textile market by providing advanced storage and computational solutions for amylase simulation technology. Storing information on a cloud means that amylase and other related databases can be accessed from anywhere in the world at any time. As well as making amylase applications more user-friendly, it also helps ensure that any data produced by the amylase system is securely and permanently stored. This is important in ensuring that amylase technology can continue to be developed, as new ways of improving the amylase system are based on the collection and analysis of data produced by the amylase application. Access to an immediate, high-powered computational network is another key advantage of amylase digitization through cloud storage. Amylase simulation tasks that require the pooling of many small computational processing units can be completed without the need for expensive computational hardware. Cloud technology makes this possible by offering the use of thousands of interconnected yet remotely stored computational units that can be accessed and used as required. At present, such a facility would be prohibitively difficult and expensive to establish using traditional computational hardware; even systems that use grid technologies and connect together multiple local computational systems cannot offer the same fluid, on-demand computational power as cloud networks. By running the amylase system on a cloud service provider, it is possible to allocate and use the most computational power available at any given time. Such flexibility when managing extremely computationally intensive amylase processes can make a significant and positive difference to both the development and practical usage of amylase systems in the textile economy. However, the move towards utilizing cloud computing for amylase technology should be tentative and considered. Any substantial changes to existing amylase platforms would first require extensive redevelopment of the system to create network applications and incorporate the use of digital storage methods. Nonetheless, it is clear that the potential to benefit from using cloud technology and allowing amylase to continue to advance within the scientific and industrial communities is there, so long as a careful and managed digitization strategy is pursued [58, 93, 131].

6 Case Studies

The application of digital technologies to the sustenance of the performance of enzymes in the textile economy. This chapter will critically provide an in-depth analysis of the successful stories on the sustenance of amylase applications through digitalization in the textile economy. The chapter will set off by outlining digital technologies and their potential advantages and challenges in the textile sector. The rest of the chapter will discuss various case studies on the successful adoption of digitalized amylase applications in the modern-day textile industry with a view to sustaining the effectiveness of the enzyme. The chapter will analyze each case study and outline the main findings as well as the impacts of digitalized enzyme applications as well as the rationale for choosing the particular case study. At the end of the chapter, a summary of the major themes from the case studies will be provided so as to give a digital mechanism of amylase applications. The first part of the case study outlines some fundamental elements. First, the location of the industry and the period of study are outlined through a brief introduction of the particular industry and its product lines. Secondly, the case study provides a detailed description of the amylase application process which was hitherto very analog before the digitalization. Key processes in the application of enzymes have been examined including the pre-treatment, the actual enzyme application, post-treatment, and the drying and curing processes. Machine and technologies involved in each process have been identified and the whole process has been critically explored with a view to presenting the improvements brought about by digitalization. Finally, the chapter outlines the research methodology and the methods used for data collection and analysis. These include literature review, observation, interviews, group discussions, and surveys as well as the systematic SWOT and PESTLE analysis of the industry [66, 114, 119].

6.1 Case Study 1: Successful Implementation of Digitalized Enzymatic Desizing

However, not the whole textile industry is prosperous in digitalization. The development of digitalized enzymatic desizing, which is the most broadly utilized technique in modern textile wet processing, is still in its infancy stage. The conventional desizing procedure requires a couple of steps, such as the application of soap and also the scouring step. The desizing step includes spraying the fabric with a diluted mineral acid to degrade the starch size and this really is followed by washing and drying. Nowadays, a single-step desizing process also known as enzymatic desizing primarily uses alpha-amylase enzyme to degrade the starch sizes. Many researches have been carried out to improve the efficiency of this process. It is proven that, with the correct formula of digitoxin, the alpha-amylase may be well maintained at its maximum productivity when the digitalized enzymatic desizing is successfully implemented. Every enzymatic and desizing process parameter can be digitally controlled through the efficient use of programmable logic control (PLC) and software such as supervisory control and data acquisition (SCADA) system. Amylase enzyme use is optimized at a really low dosage and the sizes are effectively removed by the very fast reaction between the enzyme and the substrates. Such a successful story of applying digitalized enzymatic desizing is present in a renowned international company, which specializes in woven fabrics, apparels, performance activewear and leisure wear production. All elements along the wet processing from fabrics to garments. Options resolver can be digitally controlled and monitored. Post washing, the fabric is sent to dry and it is ready for the subsequent stage of manufacturing. With the successful implementation of this newly developed enzymatic desizing, the business is able to improve productivity by 10% and also the very initial time [52, 68, 83] (see Fig. 9).

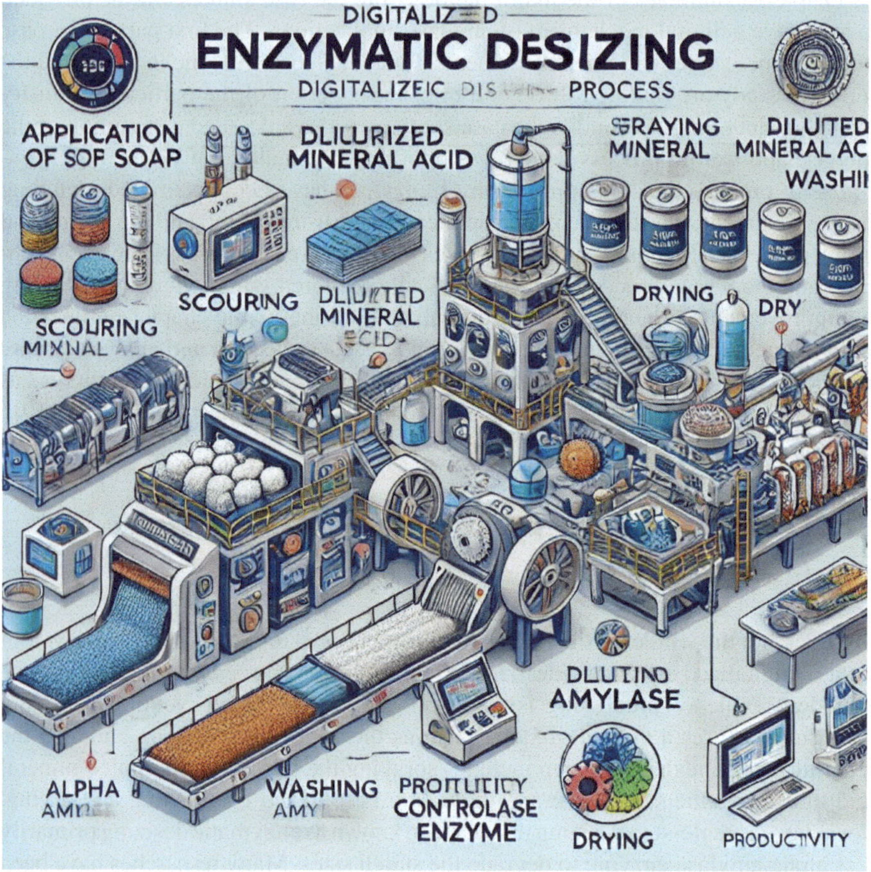

Fig. 9 Desizing process in textile manufacturing, precisely labeled, showing the steps from the application of soap, scouring, spraying with diluted mineral acid, washing, drying, use of alpha-amylase enzyme, control with PLC and SCADA system, and the impact on productivity

6.2 Case Study 2: Improving Sustainability Through Digitalized Enzyme Applications

This approach resulted in a full-scale trial in a mill in Pakistan over three seasons. The water, energy, and chemical use for Eco Applicable 3, a targeted enzyme used to prevent pilling, were monitored for a conventional process and a digitalized process. The conventional process monitored was the one originally used to set the benchmark for water and energy savings in R&D in our earlier project on digitalization. The trial results and project outputs demonstrated that significant sustainability benefits can be realized through cost-effective, practical approaches. The results of the trial showed that water, energy, and chemical usage reduced significantly with the use of digitalized enzyme application in comparison to the benchmark and the conventional process. Firstly, water consumption was reduced by between 20 and 60%, and carbon emissions from energy decreased by between 10 and 65%. Waste from excessive chemical application in the conventional process was also eliminated. Through the chemical, water, and energy savings, total environmental, economic, and social benefits were calculated. This suggested that for every ton of material produced and treated with the digitalized process, up to £2500 of social, economic, and environmental benefits could be realized, indicating the substantial potential of this method. Bespoke digitalization of amylase-based processes presents a solid opportunity for environmental compliance in the textile industry, as well as reducing costs and providing increased flexibility. Based on the trial results and the fact that innovation spawned a new service for customers, the company was successful in achieving funding for the development of digitalized, amylase-based research. This not only demonstrates the potential of digitalization in providing long-term, sustainable, and cost-effective solutions for clients and their products but additionally illustrates that a proactive approach towards sustainable processes can open new, innovative pathways to providing customer satisfaction through cutting-edge technology [31, 112, 129] (see Figs. 10 and 11).

6.3 Case Study 3: Enhancing Consumer Education on Enzyme Applications

The study revealed that consumers can find information regarding "what is in the products" and "what can be the possible effects," and basic information about "how the product works" and "how the products are made" could be found. By incorporating both theoretical and practical components in consumer education, it was concluded that there may be diverse and far-reaching opportunities to increase the representation of scientific. The flexibility provided on the online platform can be used to turn consumer education into a stone that can support a number of objectives related to market transformation. For instance, instead of simply providing the complete publication of the research study which can be more regulatory-focused,

Fig. 10 Sustainability improvements through digitalized enzyme applications in textile manufacturing, showing the full-scale trial results, water, energy, and chemical usage reductions, environmental, economic, and social benefits, and the digitalization of amylase-based processes

the information as suggested by the research can be provided [41, 67, 84]. In a project that aimed to find ways to support consumer engagement and market transformation through newer technologies, a web platform where consumers and experts can discuss products and newer technologies was established. By linking several pieces of information to explain the benefits and process of using newer enzyme technologies in a particular treatment, this online platform, namely "Dialogue in the (Enzyme) Technology and Society," is able to provide a material selection that captures the experts' view in promoting newer technologies that would augur environmental benefits for a particular treatment. This tool is expected not only to support the consumer but also as a means for the experts who can provide valuable knowledge in public engagement and also in the dissemination of relevant technology. The application of enzymes in industries may not be the most straightforward concept for the

Fig. 11 Consumer education on enzyme applications in textile manufacturing, showing the online platform 'Dialogue in the (Enzyme) Technology and Society', the benefits and process of using newer enzyme technologies, and the environmental advantages of these technologies

consumer to understand, especially in the textile industry. In view of the tremendous potential benefits that newer enzyme technology could offer, from energy reduction to enhancing textile properties, educating the consumer has become a priority. In recent years, consumers are demanding more eco-friendly and sustainable products. This demand is expected to drive brands and retailers to proactively seek ways to highlight the environmental and cost benefits of using newer technologies, such as enzymatic treatments. However, before consumers can appreciate why professionals are championing such technologies to be adopted, consumer education and a clear understanding of these technologies are paramount [89, 102, 124] (see Table 3).

Table 3 Case studies on digitalized enzyme applications in textile processing

Case study	Description
Case study 1	Successful implementation of digitalized enzymatic desizing, resulting in improved efficiency
Case study 2	Demonstrating the enhancement of sustainability through digitalized enzyme applications
Case study 3	Focus on enhancing consumer education and awareness regarding enzyme applications in textiles

7 Consumer Education on Enzyme Applications

In recent years, digitization and automation have immensely transformed the textile economy, particularly in the dyeing and finishing industries. At present, various novel and extremely sophisticated machinery and technology are being broadly applied with the view to ensuring sustainability and business growth in this sector. However, this research work focuses on the enzymes, specifically amylase, and how digitalization can sustain their application in the dynamic and ever-changing textile economy. In this section, we will appraise consumer education on amylase applications and how this can potentially affect their usage in the context of the textile economy. First and foremost, there is a wide range of enzymatic applications that are being used in various fields. However, states Goddard [39], many individuals, consumers, and some players in the textile industry are not well familiar with the numerous advantages of using amylase in the dyeing process and finishing of textiles. This is one crucial point for which digitalization may be required in order to simulate and analyze how amylase and other substances can be used in the dyeing and finishing process. When this is achieved, consumers can be shown how such kind of simulation works and how the end results, that is the quality of the textile after such applications, is going to look and benefit everybody right from the industry individuals to the end users such as consumers. This is essential because when knowledge is imparted by consumers, it will not only facilitate informed choices on the market shelves but also significantly reduce the likelihood of prescribing the usage of hazardous substances in textile finishing. Amylase is classified under enzymes and according to Ranjan et al. [92], enzymes are biological catalysts that typically speed up the rate of a reaction by reducing the energy of activation. He explained that amylase is used in the dyeing process to remove starch that may be present in the fiber, thus allowing for a much smoother chemical reaction between the dyestuff and the fiber. Such kind of information, which is basically the scientific process through which amylase operates, is very important to be impacted on the consumers. However, as noted by Sempere et al. [104], a common problem for scientists and chemical experts who have advanced these eco-friendly methods is that there has been very little knowledge and availability of this scientific information to the common people. His study found that with the new methods of education, such as ensuring that digital simulation suites are availed close to classes and to consumers, there has been increased knowledge and acceptance of new and eco-friendly scientific methods [81, 83, 85, 110].

7.1 Importance of Consumer Education

First and foremost, effective consumer education is paramount in ensuring that buyers of textile products such as clothes develop a clearer understanding of their interactions with manufacturers and producers of various products. According to Muscat et al. [78], consumer education enhances the ability of individuals to develop a deeper understanding and the confidence to engage in consumption activities that are mindful of their individual needs and preferences as well as those of society. Amylase is an enzyme whose applications in the textile industry cannot be overlooked. With increasing scrutiny on the suitability and environmental friendliness of the chemicals used in various industrial processes, it is undeniable that the textile industry will experience more reliance on eco-friendly techniques such as the use of enzymes in processing fibers and fabrics [72]. However, for the sustainable utilization of enzymes such as amylase, there is a fundamental need to ensure that consumer awareness and education run parallel to the implementation of strategies designed to enhance both the economic and environmental attractiveness of enzyme applications in the textile industry. This is necessary considering that as more consumer education goes on, different strategies and opinions can emerge and act as instrumental forces that may necessitate a shift from the usual norms in the competitive textile market [33]. By involving consumers in the generation of knowledge concerning the benefits of using enzyme-based technologies as alternatives to processes that are heavily reliant on chemicals in processing textile products, the efforts to create a sustainable textile industry will be made not only easy but also clear and transparent to the beneficiaries of such innovative ideas [47, 87]. Consumer education is very essential as far as enzyme applications like amylase production and its usage in the textile industry are concerned. Consumer education helps consumers gain knowledge of the positive impacts that may be brought about by application of the enzymes in various industrial processes. It is through education that the public and in this case the consumers get to understand the aims of improving the industrial processes through the adaptation of more advanced and environmentally friendly technologies. In addition, knowledge of the processes of amylase production and other enzymes makes the consumers appreciate the products in the market more. They gain confidence in such products as compared to conventional products and this in turn influences their purchasing decisions. What is even more important when it comes to consumer education is the fact that it has a big role to play in matters concerning environmental conservation. It is through education that people including the consumers get to understand the consequences of using products that are considered to be unfriendly to the environment. By highlighting the positive impacts of amylase application in the textile industry, education will definitely motivate more industries and research institutions to embrace further innovations and research of more new enzyme applications. Through consumer education, the operators in the textile industry and the entire industrial world will shift from the conventional processes and start appreciating better enzyme-based methods of improving the general quality of the products [111, 120, 130].

7.2 Methods for Educating Consumers on Enzyme Applications

The use of more recent technological developments such as the internet and social media is also provided as one of the ways in which consumers can be educated on enzyme applications. The various strategies and approaches that can be adopted when using such modern forms of media are analyzed. This includes the use of web "pop-ups" to draw attention to educational material and the use of social media sites to deliver messages and information. The section is also elaborated by providing recommendations. Two recommendations have been provided. First, the need to continue with current initiatives of different stakeholders involved in educating consumers is emphasized. Secondly, the recommendation states the need to consider the changing consumer demographics and to adapt education strategies in order to effectively reach a wider section of the market [8, 10, 95].

The second method outlined is the use of focus groups as a means of assessing the effectiveness of education initiatives and determining the impact such programs have on consumer behavior. When a member of a focus group takes part in a discussion, their opinions and views are quite revealing. After a particular campaign, the facilitating agent is able to tell the extent of attitude and behavior change based on what is observed in the group discussions. By incorporating the study of consumer behaviors and attitudes in a given social context, more knowledge will be gained about different consumer education strategies and how best to tailor them to different societal settings [54, 73, 103]. The first method for consumers is to ensure their awareness and ability to differentiate between different products is increased. This can be done by lobbying for the production of safer biological chemicals that are friendly to the environment and having these clearly labeled. Furthermore, research can be carried out on how to improve the current methods of educating consumers on enzyme applications. This will make it possible for education outreach programs to adopt new and more effective strategies. Also, new and more effective methods might be developed, leading to improved consumer education [35, 88, 101].

7.3 Impacts of Consumer Education on Enzyme Applications

Another significant argument of consumer education pertains to better judgment from the consumers in selecting products and understanding which processes are the most sustainable. Over time, with increased consumer awareness about sustainable products and production, market demand for sustainable products is likely to increase. This, in turn, provides a more financial incentive towards innovation for more sustainable methods and materials by the industry. As mentioned by Garcés-Perez et al., when consumer demands shift to 'greener' alternatives, industries will have no choice but to adopt more sustainable practices to maintain or acquire greater market shares, thereby allowing for more environmentally friendly products and processes to dominate the economy. This is a widely accepted argument about sustainable consumer education and it is also noted in the European Sustainable Development Strategy that encouraging sustainable consumption is a key component for achieving sustainability. The importance of public awareness and public participation in the wider picture of sustainable development highlights the critical importance of not only public awareness and education but also choosing education that will activate citizens in the process. Therefore, continuing education should be a focus of the developers and providers of educational programs to ensure that citizens are encouraged to partake in the wider sustainable developmental processes. This, however, detracts from the main argument that the propensity for sustainable practice concurrently provides an advantage for both industry practice and general sustainable developmental targets [7, 55, 98] (see Table 4).

7.3.1 The Climate Change, Sustainable Development Goals (SDGs), and Sustaining Amylase Applications via Digitalization in the Textile Economy

In the textile sector, the integration of digital technologies is reinforcing the use of enzymatic methods, such as amylases, to support the Sustainable Development Goals and mitigate climate change impacts. This digital shift is pivotal for sustainability,

Table 4 Methods for educating consumers on enzyme applications

Method	Description
Product labeling	Clear and informative labeling on textile products regarding the use of enzymes
Consumer education campaigns	Public awareness campaigns to educate consumers about the benefits and safety of enzyme processes
Online resources	Websites, blogs, and social media platforms providing information on enzyme applications
Collaborations with brands	Collaboration between textile brands and educational institutions to promote enzyme education

fostering seamless communication throughout the supply chain, thereby enhancing operational efficiency and responsiveness [36]. Nonetheless, the nexus between digitalization and sustainability presents a nuanced landscape, where environmental benefits may be offset by the entrenched economic systems that could pose sustainability challenges [36]. To navigate this complexity, the Digitainability Assessment Framework has been introduced, offering a comprehensive analysis tool to evaluate the influence of digital strategies on sustainability objectives, ensuring that digital advancements contribute positively to the SDGs [42].

Climate Change and Amylase Applications

Digitalization's impact on the textile industry is multifaceted, with significant implications for energy consumption and environmental sustainability. The reliance on digital infrastructure, such as data centers and cloud services, necessitates a considerable amount of energy, which can be exacerbated by climate change phenomena like heatwaves and droughts. These events pose risks to energy stability, potentially increasing operational costs for digital systems and enzymatic processes in textiles [61].

Furthermore, water scarcity, a growing concern due to climate change, poses a challenge for the textile industry, which depends heavily on water for fabric processing and for maintaining the operational efficiency of cooling systems in data centers. The scarcity of water resources could constrain the industry's capacity to utilize amylase-based treatments and digital solutions [61].

Additionally, climate change has a direct effect on the production of natural fibers such as cotton and wool, which are integral to the textile sector. Fluctuations in fiber availability can influence the demand for amylase applications, as these enzymes are commonly used in the desizing process of cotton. A decline in cotton production, driven by climate change, may lead to a reduced need for amylase enzymes, impacting the market for these biotechnological solutions [61] (see Fig. 12).

SDGs and Amylase Applications

Sustainable Development Goals (SDGs) are closely intertwined with the advancements in enzymatic treatments, such as the use of amylases in the textile industry, which are being supported by digitalization efforts:

1. **SDG 7: Affordable and Clean Energy**:
    - The textile sector is increasingly turning to renewable energy sources like solar and wind power to operate its digital infrastructures. This transition not only aligns with the goal of affordable and clean energy but also significantly lowers the industry's carbon footprint, contributing to a more sustainable future [61].

Fig. 12 Multifaceted impact of digitalization on the textile industry, highlighting energy consumption, environmental sustainability, water scarcity, and the production of natural fibers like cotton and wool

2. **SDG 9: Industry, Innovation, and Infrastructure**:
   - Digitalization is a catalyst for innovation within the textile industry, fostering the development of new business models and processes. By integrating digital technologies, the industry is able to boost both efficiency and sustainability. This includes the application of amylases, which are enzymes used to enhance textile processing, making them more eco-friendly and efficient [61].
3. **SDG 12: Responsible Consumption and Production**:
   - Through digitalization, the textile industry is optimizing the use of critical resources such as water and energy. This optimization leads to a reduction in waste generation and emissions, thereby promoting responsible consumption and production practices. Such measures are essential for minimizing the environmental impact of textile manufacturing and ensuring that it adheres to sustainable development principles [61].

8 Future Directions

When starting to write this section, the first thing was considering all of the possible ways in which amylase and sustainable digital design processes could be taken further using technologies and understanding. Considering the way in which amylase is used in the current market along with digital software and material design, it became apparent that the scope for more digital practices and newer material developments is very promising. One of the technological advancements that the chapter discusses was pioneering work into 3D digital body scanning tools and programs to create perfectly tailored garments specific to an individual's body shape and height—3D modeling for amylase printing processes can now begin to be integrated into these software packages. This meant that the chapter wasn't only about discussing in depth the theories and practices surrounding amylase and sustainable resources; it brought together personal experiences and the knowledge of recent technological advancements in the textile market. Ideas and practices associated with wearable technologies and attached digital practices were also evaluated—this guided me to understand that such research was important nowadays in not just shaping more futuristic clothing designs and practices, but that it was crucial to identify and work towards more sustainable methodologies and processes [97, 108, 122].

Future Directions as a part of the book Sustaining Amylase Applications Through Digitalization in the Textile Economy was a chapter that highlights the approaches and predictions in using the currently available technologies and resources in textiles and clothing design with the intention of becoming more sustainable in the long term. This was a very interesting section to write because it involved a lot of reflective practice and analyzing how different technologies, practices, and materials in textiles and fashion as a whole have developed over time and why. I think that one of the most vital elements of this chapter was to encourage readers to think about their work and practice in a wider context and not to become complacent with using the same techniques and resources—keeping updated with all such things could not only mean working in more modern, efficient ways but also to start making a bigger push towards more sustainable designs which is what is needed for the future.

8.1 Trends and Prospects in Amylase Applications

Trends and prospects in amylase applications indicate considerable and enduring growth in the coming years. This is especially so for industrial processes. Nowadays, amylase applications are shifting from the use of organic solvents to more eco-friendly alternatives. In this case, enzyme-based procedures are becoming more popular than traditional physiochemical ones. Enzyme immobilization is the future of amylase research. Amylase, which refers to a collective of α, β, and γ amylase, has its roots in ancient research, dating back to the 1950s. Its popularity has been growing through the years, being commercially exploited in the food and beverages industry, organic chemistry, molecular biology, and now in the energy industry too. Forecasts indicate that γ amylase, frequently applied in the malt industry and also known as γ amylase, will have the highest growth rate among all enzymes in the sector. The molecule has been studied as an efficient model for the safety assessment in genetically modified organisms. With regard to the food and beverages industry, there is an increasing demand for various digestive enzymes, such as amylase. This is primarily due to the growing millennial attitude change whereby they are constantly seeking healthy and convenient options of food. Modern amylase research is employing computational methods to study and even predict various modes of interactions and, most importantly, to find ways of ensuring that these molecules maintain their impressive catalytic performances even when subjected to new conditions such as immobilization and genetic modifications. Such in silico research is critical in providing qualitative changes into inquiry methods, hence creating efficiency. Amylase, especially the immobilized variety, proves to be a cheap and readily available biocatalyst for a long list of industrial processes. It is then not surprising to find that the demand for these molecules is on a trend of steady growth in the global market [19, 77, 82].

8.2 Recommendations for Future Research

The use of amylase in desizing processes is already well established, but the wide range of factors involved in optimizing its performance means that significant opportunities remain—this is reflected in the diversity of research topics several authors recommend. One such factor is the development of novel enzymes. For example, Du et al. suggest exploring genetic and protein engineering in order to modify and improve existing amylase strains, while Kaur and Singh advocate research into the isolation and characterization of new bacterial strains such as alkaliphilic or thermophilic varieties that are able to work effectively at modern high temperature, high pH desizing conditions. Newer technologies such as cold plasma treatment have been proposed as ways to improve the effectiveness of amylase by creating free radicals in the reaction solution, and this is reflected in the recommendations of several authors, such as Tyagi and Manas, who suggest further studies into the synergistic effects of cold plasma and biopolymers on the removal of starch as well as investigating the kinetics of the desizing process under plasma conditions. Furthermore, with international research into environmentally friendly processes gaining traction, a stream of papers suggests investigations into how amylase can be best utilized in low solvent or solvent-free protocols, including the development of methods such as ultrasonic and microwave energy strategies as outlined by Kaur and Singh, and the exploration of 'greener' and more sustainable additives such as deep eutectic solvents, recommended by Braun and co-workers. Raising microbe or enzyme stability is another area of interest; for example, Botha, Gorgens and Kock call for studies into molecular and cellular mechanisms to enhance amylase productivity. On the other hand, Xue, Tan and Xu advocate the improvement of biotechnology and microbiology processes to produce purer amylase preparations from natural cultures in genetic investigations—a sign of the increasingly wide ranging applications of modern biotechnology techniques in amylase research. Lastly, and increasingly importantly, researchers such as Kaur and Singh and Haque suggest that alongside technological research attention should also be focused on the development of predictive biological or computer models to optimize amylase application and reduce overall process times and costs. Such 'in silico' methods are a growing area of interest in the academic field and potentially offer a means to integrate amylase desizing with fast-developing technologies [2, 82, 123].

9 Conclusions

This chapter dive in sustaining amylase applications through digitalization in the textile economy provides a comprehensive understanding of the role of amylases in textile processing and the potential benefits of digital technologies in enhancing their effectiveness. Amylases have emerged as eco-friendly alternatives to traditional acid-based methods for desizing textiles, offering faster and more sustainable processes. The chapter highlights the advantages and challenges of digitalization in textile processing, emphasizing how technologies such as sensor technology, artificial intelligence, machine learning, big data analytics, and cloud computing can improve the performance of amylases. These digital tools have the potential to enhance efficiency, reduce production costs, and address chemical compatibility issues, thereby contributing to more sustainable and cost-effective textile processing.

Specific amylase applications in textile processing, including desizing, biopolishing, bioscouring, and bleaching, are discussed in detail. The benefits and limitations of using amylases in each application are examined, providing valuable insights for industry professionals seeking to optimize their processes. Challenges related to amylase stability, production costs, and chemical compatibility are acknowledged, and the chapter explores strategies for enhancing amylase performance through digitalization. The case studies presented demonstrate successful implementations of digitalized enzymatic desizing, sustainability improvements through digitalized enzyme applications, and the importance of consumer education on enzyme applications.

The chapter emphasizes the significance of consumer education in promoting awareness, understanding, and acceptance of enzyme-based textile processes. It discusses methods for educating consumers and highlights the potential impacts of consumer education on enzyme applications. Looking ahead, the chapter identifies future directions and provides recommendations for further research in amylase applications. It explores trends and prospects in amylase applications and suggests avenues for optimizing their effectiveness in the textile industry. Overall, this chapter serves as a valuable resource for researchers, engineers, and industry professionals interested in sustainable and efficient textile processing methods. It underscores the importance of leveraging digitalization to enhance the performance of amylases and highlights the potential of these enzymes in contributing to a more sustainable textile economy. By embracing digital technologies and advancing research in this field, the textile industry can continue to evolve towards more environmentally friendly and economically viable practices.

References

1. Abchir O, Yamari I, Shtaiwi AM, Nour H, Kouali ME, Talbi M, Errougui A, Chtita S (2024) Insights into the inhibitory potential of novel hydrazinyl thiazole-linked indenoquinoxaline against alpha-amylase: a comprehensive QSAR, pharmacokinetic, and molecular modeling study. J Biomol Struct Dyn, 1–18
2. Abootalebi SN, Saeed A, Gholami A, Mohkam M, Kazemi A, Nezafat N, Mousavi SM, Hashemi SA, Shorafa E (2020) Screening, characterization and production of thermostable alpha-amylase produced by a novel thermophilic Bacillus megaterium isolated from pediatric intensive care unit. J Environ Treat Tech 8(3):952–960
3. Abou Taleb M, Gomaa SK, Wahba MI, Zaki RA, El-Fiky AF, El-Refai HA, El-Sayed H (2022) Bioscouring of wool fibres using immobilized thermophilic lipase. Int J Biol Macromol 194:800–810
4. Adeel S, Kamal S, Ahmad T, Bibi I, Rehman S, Kamal A, Saleem A (2020) Biotechnology: an eco-friendly tool of nature for textile industries. Adv Funct Finish Text, 85–114
5. Adhikary RR, Banerjee R (2021) Development of smart core-shell nanoparticle-based sensors for the point-of-care detection of alpha amylase in diagnostics and forensics. Biosens Bioelectron 184:113244
6. Ahmad S, Miskon S, Alabdan R, Tlili I (2020) Towards sustainable textile and apparel industry: exploring the role of business intelligence systems in the era of industry 4.0. Sustainability 12(7):2632
7. Alamsyah D, Othman N, Mohammed H (2020) The awareness of environmentally friendly products: the impact of green advertising and green brand image. Manag Sci Lett 10(9):1961–1968
8. Almomani H, Patel N, Donyai P (2023) Reasons that lead people to end up buying fake medicines on the internet: qualitative interview study. JMIR Form Res 7(1):e42887
9. Ameer K, Das PR, Eun J-B (2020) Advances in enzymatic applications in food industry. In: Microbial fermentation and enzyme technology. CRC Press, pp 85–97
10. Anwar A, Malik M, Raees V, Anwar A (2020) Role of mass media and public health communications in the COVID-19 pandemic. Cureus 12(9)
11. Arshad M, Adil M, Sikandar A, Hussain T (2014) Exploitation of meat industry by-products for biodiesel production: Pakistan's perspective. Pakistan J Life Soc Sci 12:120–125
12. Arshad M (ed) (24 Nov 2017) Perspectives on water usage for biofuels production: aquatic contamination and climate change. Springer
13. Arshad M, Bano I, Younus M, Khan A, Rahman A (2018) Health concerns associated with biofuel production. Perspectives on water usage for biofuels production: aquatic contamination and climate change, vol 2018, pp 97–105
14. Arshad M, Hussain T, Chaudhry N, Sadia H, Aslam B, Tahir U, Abbas M, Qureshi N, Nazir A, Rajoka MI, Iqba M (2019) Enhancing profitability of ethanol fermentation through gamma ray mutagenesis of saccharomyces cerevisiae. Pol J Environ Stud 28(1)
15. Arshad M (ed) (13 Nov 2020) Sustainable ethanol and climate change: sustainability assessment for ethanol distilleries. Springer Nature
16. Arshad M, Ansari AR, Qadir R, Tahir MH, Nadeem A, Mehmood T, Alhumade H, Khan N (2022) Green electricity generation from biogas of cattle manure: an assessment of potential and feasibility in Pakistan. Front Energy Res 25(10):911485

17. Arshad M (ed) (28 Apr 2023) Climate changes mitigation and sustainable bioenergy harvest through animal waste: sustainable environmental implications of animal waste. Springer Nature
18. Arumugam S, Colburn DA, Sia SK (2020) Biosensors for personal mobile health: a system architecture perspective. Adv Mater Technol 5(3):1900720
19. Ban X, Guo Y, Kaustubh B, Li C, Gu Z, Hu K, Li Z (2023) The Global amylase research trend in food science technology: a data-driven analysis. Food Rev Intl 39(5):2492–2506
20. Bano I, Ghauri MA, Arshad M, Yasin T, Younus M (2017) Bioactivity of variant molecular weight chitosan against drug-resistant bacteria isolated from human wounds. Microbial Drug Resistance 1;23(8):958–65
21. Bano I, Arshad M (2018) Climatic changes impact on water availability. Perspectives on water usage for biofuels production: aquatic contamination and climate change, vol 2018, pp 39–54
22. Banovetz JT, Manimaran S, Schelske BT, Anand RK (2023) Parallel dielectrophoretic capture, isolation, and electrical lysis of individual breast cancer cells to assess variability in enzymatic activity. Anal Chem 95(20):7880–7887
23. Bashir T, Iftikhar T, Majeed H (2024). Bulk industrial textile production of bio scouring for cellulosic fabric utilizing indigenous hot springs triacylglycerol acylhydrolases from Bacillus toyonensis and Bacillus thuringiensis. Cellulose, 1–29
24. Baumann M, Moody TS, Smyth M, Wharry S (2020) A perspective on continuous flow chemistry in the pharmaceutical industry. Org Process Res Dev 24(10):1802–1813
25. Blaga AC, Caşcaval D, Galaction AI (2022) Improved production of α-amylase by Aspergillus terreus in presence of oxygen-vector. Fermentation 8(6):271
26. Choudhury AKR (2020) Enzyme applications in textile chemical processing. In: Sustainable technologies for fashion and textiles. Elsevier, pp 91–115
27. Colombi BL, Palozi MD, de Cássia Siqueira Curto Valle R, Andreaus J, Arias MJL, Valle JAB (2022) A sustainable approach for cotton bioscouring: reuse of the pectate lyase containing treatment bath. Bioprocess Biosyst Eng 45(8):1391–1405
28. Compart J, Singh A, Fettke J, Apriyanto A (2023) Customizing starch properties: a review of starch modifications and their applications. Polymers 15(16):3491
29. Dallinger D, Gutmann B, Kappe CO (2020) The concept of chemical generators: on-site on-demand production of hazardous reagents in continuous flow. Acc Chem Res 53(7):1330–1341
30. Das CM, Yang F, Yang Z, Liu X, Hoang QT, Xu Z, Neermunda S, Kong KV, Ho HP, Ju LA (2023) Computational modeling for intelligent surface plasmon resonance sensor design and experimental schemes for real-time plasmonic biosensing: a review. Adv Theory Simul 6(9):2200886
31. De Santis P, Meyer L-E, Kara S (2020) The rise of continuous flow biocatalysis–fundamentals, very recent developments and future perspectives. React Chem Eng 5(12):2155–2184
32. Domokos A, Nagy B, Szilagyi B, Marosi G, Nagy ZK (2021) Integrated continuous pharmaceutical technologies—a review. Org Process Res Dev 25(4):721–739
33. Elwood D (2017) The international politics of human rights: rallying to the R2P cause? Int Polit 54(2):182–200
34. Farooq MA, Ali S, Hassan A, Tahir HM, Mumtaz S, Mumtaz S (2021) Biosynthesis and industrial applications of α-amylase: a review. Arch Microbiol 203:1281–1292
35. Fowkes A, Foster R, Kane S, Thresher A, Werner A-L, de Oliveira AAF (2023) Enhancing global and local decision making for chemical safety assessments through increasing the availability of data. Toxicol Mech Methods 33(5):337–348
36. Fuster-Muñoz E, Roche E, Funes L, Martínez-Peinado P, Sempere JM, Reig JA (2016) Effects of pomegranate juice on circulating parameters, cytokines, and oxidative stress markers in endurance-based athletes: a randomized controlled trial. Nutrition 32(5):539–545
37. Ghoreishi M, Happonen A (2021) The case of fabric and textile industry: the emerging role of digitalization, internet-of-things and industry 4.0 for circularity. In: Proceedings of sixth international congress on information and communication technology: ICICT 2021, London, vol 3

38. Ghoreishi M, Happonen A, Pynnönen M (2020). Exploring industry 4.0 technologies to enhance circularity in textile industry: role of internet of things. In: Twenty-first international working seminar on production economics
39. Goddard W, Muscat A (2017) Towards genre as a game design research approach. In: Proceedings of the 10th digital games research association (DiGRA) conference
40. Goswami L, Kayalvizhi R, Dikshit PK, Sherpa KC, Roy S, Kushwaha A, Kim BS, Banerjee R, Jacob S, Rajak RC (2022) A critical review on prospects of bio-refinery products from second and third generation biomasses. Chem Eng J 448:137677
41. Grasso S, Asioli D (2020) Consumer preferences for upcycled ingredients: a case study with biscuits. Food Qual Prefer 84:103951
42. Gupta R, Rhyner PM (2022) Practice guidelines for the management of multiligamentous injuries of the knee. J Orthop Surg 30(1)
43. Hack-Polay D, Rahman M, Billah MM, Al-Sabbahy HZ (2020) Big data analytics and sustainable textile manufacturing: decision-making about the applications of biotechnologies in developing countries. Manag Decis 58(8):1699–1714
44. Haddar W, Adeel S, Salman M, Ghaffar A, Naseer M, Usama M, Ticha MB (2023) Ecological applications of enzymes in plants based textile dyeing. In: Dyes and pigments-insights and applications. IntechOpen
45. Happonen A, Ghoreishi M (2021) A mapping study of the current literature on digitalization and industry 4.0 technologies utilization for sustainability and circular economy in textile industries. In: Proceedings of sixth international congress on information and communication technology: ICICT 2021, London, vol 4
46. Haque Y, Zawad RS, Rony CSA, Al Banna H, Ghosh T, Kaiser MS, Mahmud M (2023) State-of-the-art of stress prediction from heart rate variability using artificial intelligence. Cogn Comput, 1–27
47. Harsanto B, Primiana I, Sarasi V, Satyakti Y (2023) Sustainability innovation in the textile industry: a systematic review. Sustainability 15(2):1549
48. Hieltjes T, Lahiri S (2009) Explorative research on the transport biofuel market
49. Hoque MT, Mazumder NUS, Islam MT (2021) Enzymatic wet processing. Sustainable practices in the textile industry, 87–110
50. Hu J-X, Zhao C-F, Wang S-L, Tu X-Y, Huang W-B, Chen J-N, Xie Y, Chen C-R (2023) Acute pancreatitis: a review of diagnosis, severity prediction and prognosis assessment from imaging technology, scoring system and artificial intelligence. World J Gastroenterol 29(37):5268
51. Iram N, Shakir HA, Irfan M, Khan M, Ali S, Anwar A, Saeed S, Qazi JI (2021) Statistical optimization of amylase production using grape fruit peels in submerged fermentation. Acta Scientiarum Technol 43:e50538–e50538
52. Kabir SMM, Koh J (2021) Sustainable textile processing by enzyme applications. Biodegradation technology of organic and inorganic pollutants, 339–366
53. Kalia S, Bhattacharya A, Prajapati SK, Malik A (2021) Utilization of starch effluent from a textile industry as a fungal growth supplement for enhanced α-amylase production for industrial application. Chemosphere 279:130554
54. Khalil R, Mansour AE, Fadda WA, Almisnid K, Aldamegh M, Al-Nafeesah A, Alkhalifah A, Al-Wutayd O (2020) The sudden transition to synchronized online learning during the COVID-19 pandemic in Saudi Arabia: a qualitative study exploring medical students' perspectives. BMC Med Educ 20:1–10
55. Khan MS, Saengon P, Alganad AMN, Chongcharoen D, Farrukh M (2020) Consumer green behaviour: an approach towards environmental sustainability. Sustain Dev 28(5):1168–1180
56. Kikani BA, Kourien S, Rathod U (2020) Stability and thermodynamic attributes of starch hydrolyzing α-amylase of Anoxybacillus rupiensis TS-4. Starch-Stärke 72(1–2):1900105
57. Kim J, Shim J, Chang S, Chung W (2022) Purification and characterization of strong simultaneous enzyme production of protease and α-amylase from an extremophile-bacillus sp. FW2 and its possibility in food waste degradation. Fermentation 8(1):12
58. Küfeoğlu S (2022) Emerging technologies: value creation for sustainable development. Springer Nature

59. Kumar D, Bhardwaj R, Jassal S, Goyal T, Khullar A, Gupta N (2021) Application of enzymes for an eco-friendly approach to textile processing. Environ Sci Pollut Res, 1–11
60. Landage SM (2022) Biotechnological approaches in desizing of textile materials. In: Applications of biotechnology for sustainable textile production. Elsevier, pp 47–73
61. Leal Filho W, Azeiteiro UM, Alves F (2016) Climate change and health: an overview of the interconnections. In: Climate change and health
62. Li X, Bai Y, Jin Z, Svensson B (2022) Food-derived non-phenolic α-amylase and α-glucosidase inhibitors for controlling starch digestion rate and guiding diabetes-friendly recipes. Lwt 153:112455
63. Liang Y, Wang S, Jia H, Yao Y, Song J, Yang W, Cao Y, Zhu F, Huo Z (2022) PH/redox/α-amylase triple responsive metal-organic framework composites for pest management and plant growth promotion. Microporous Mesoporous Mater 344:112230
64. Liao R, Zhang S, Zhang X, Wang M, Wu H, Zhangzhong L (2021) Development of smart irrigation systems based on real-time soil moisture data in a greenhouse: Proof of concept. Agric Water Manag 245:106632
65. Lin Y, Bariya M, Javey A (2021) Wearable biosensors for body computing. Adv Func Mater 31(39):2008087
66. Lokhorst C (2022) Practical experiences of IoT applications in the IoF2020 Dairy Trial. In: Practical precision livestock farming. Wageningen Academic, pp 269–292
67. Machová R, Ambrus R, Zsigmond T, Bakó F (2022) The impact of green marketing on consumer behavior in the market of palm oil products. Sustainability 14(3):1364
68. Madhu A (2022) Immobilization as sustainable solutions to textiles chemical processing. In: Sustainable approaches in textiles and fashion: manufacturing processes and chemicals. Springer, pp 21–67
69. Mafaesa MAM (2020) Extraction, variability, enzymatic biosoftening and evaluation of physico-mechanical properties of Agave americana L. Fibre. University of the Free State]
70. Mahdi MS, Ibrahim HK, Zbarr NS, Hadi SA (2022) Microbiological degradation of food dyes and amylase production, immobilization by Iraqi isolate Anoxybacillus rupiensis strain Ir3 (JQ912241). Int J Sci Res Arch 6(1):257–262
71. Mamo G, Mattiasson B (2020) Alkaliphiles in biotechnology. Springer
72. Manusadžianas R, Balkelytė A, Sadauskas J, Blinova A (2009) Ecotoxicological studies of landfill leachates using biotests and biomarkers. Environ Res 109(1):64–72
73. McNamara KE, Clissold R, Westoby R, Piggott-McKellar AE, Kumar R, Clarke T, Namoumou F, Areki F, Joseph E, Warrick O (2020) An assessment of community-based adaptation initiatives in the Pacific Islands. Nat Clim Chang 10(7):628–639
74. Menezes E, Singh N, Athalye A Application of enzymes for sustainable textile chemical processing. In: Sustainable textile chemical processing. CRC Press, pp 14–40
75. Mohammed M, Riad K, Alqahtani N (2021) Efficient iot-based control for a smart subsurface irrigation system to enhance irrigation management of date palm. Sensors 21(12):3942
76. Mohiuddin Babu M, Akter S, Rahman M, Billah MM, Hack-Polay D (2022) The role of artificial intelligence in shaping the future of Agile fashion industry. Prod Plan Control, 1–15
77. Mondal S, Mondal K, Halder SK, Thakur N, Mondal KC (2022) Microbial Amylase: Old but still at the forefront of all major industrial enzymes. Biocatal Agric Biotechnol 45:102509
78. Muscat A, de Olde EM, de Boer IJM, Ripoll-Bosch R (2020) The battle for biomass: a systematic review of food-feed-fuel competition. Glob Food Secur 25:100330
79. Obaideen K, Yousef BA, AlMallahi MN, Tan YC, Mahmoud M, Jaber H, Ramadan M (2022) An overview of smart irrigation systems using IoT. Energy Nexus 7:100124
80. Oyewole OA, Yakubu JG, Gana M, Adamu BB, Idris AD (2022) Applications of microbial enzymes in industries and medicine. In: Ecological interplays in microbial enzymology. Springer, pp 293–327
81. Panda SKBC, Sen K, Mukhopadhyay S (2021) Sustainable pretreatments in textile wet processing. J Clean Prod 329:129725
82. Paul JS, Gupta N, Beliya E, Tiwari S, Jadhav SK (2021) Aspects and recent trends in microbial α-amylase: a review. Appl Biochem Biotechnol 193:2649–2698

83. Periyasamy AP, Periyasami S (2023) Critical review on sustainability in denim: a step toward sustainable production and consumption of denim. ACS Omega 8(5):4472–4490
84. Peschel AO, Aschemann-Witzel J (2020) Sell more for less or less for more? The role of transparency in consumer response to upcycled food products. J Clean Prod 273:122884
85. Radhakrishnan S (2022) Biotechnological approaches for sustainable pretreatment of textile materials. In: Applications of biotechnology for sustainable textile production. Elsevier, pp 99–133
86. Rafikov AS, Fayzullaeva K, Shonakhunov TE, Soyibova DBQ, Yasinskaya NN (2023) Enzymatic treatment of cotton fabric for desizing. J Chem Eng Res Updat 10:31–41
87. Rahman M, Hack-Polay D, Billah MM, Un Nabi MN (2020) Bio-based textile processing through the application of enzymes for environmental sustainability. Int J Technol Manag Sustain Dev 19(1):87–106
88. Rahman MS, Adegoke EO, Pang M-G (2021) Drivers of owning more BPA. J Hazard Mater 417:126076
89. Rajesh R (2023) Study of biodegradable materials in consumer products. I-Manager's J Mater Sci 11(1)
90. Ramesh A, Harani Devi P, Chattopadhyay S, Kavitha M (2020) Commercial applications of microbial enzymes. Microbial enzymes: roles and applications in industries, 137–184
91. Rani H, Bhardwaj RD (2021) Quality attributes for barley malt: "The backbone of beer." J Food Sci 86(8):3322–3340
92. Ranjan R, Kumar D, Meena RS (2021) Conservation agriculture for carbon sequestration and mitigation of climate change. In: Conservation agriculture: a sustainable approach for soil health and food security (pp [pages]). Springer
93. Rashmi M, Girija D, Yogeesh N (2024) Exascale computing: the next frontier of high-performance computing. Hum Cancer Diagn Detect Using Exascale Comput 279:279
94. Rathore B (2022) Textile Industry 4.0 transformation for sustainable development: prediction in manufacturing & proposed hybrid sustainable practices. Eduzone: Int Peer Rev/Ref Multidiscip J 11(1):223–241
95. Rehman A, Saeed A, Asad W, Khan I, Hayat A, Rehman MU, Shah TA, Sitotaw B, Dawoud TM, Bourhia M (2023) Eco-friendly textile desizing with indigenously produced amylase from Bacillus cereus AS2. Sci Rep 13(1):11991
96. Rehman IU, Ganaie SA, Wani JA (2023). Research that sparked attention on the social Web in 2020: an Altmetric analysis of "top 100" articles. Glob Knowl Mem Commun 72(6/7):554–571
97. Rofeal M, Abdelmalek F, Pietrasik J (2023) Sustainable polyhydroxyalkanoate production from food waste via Bacillus mycoides ICRI89: enhanced 3D printing with poly (methyl methacrylate) blend. Polymers 15(20):4173
98. Rustam A, Wang Y, Zameer H (2020) Environmental awareness, firm sustainability exposure and green consumption behaviors. J Clean Prod 268:122016
99. Sahoo B, Roy A (2023) Structure–function relationship of resistant starch formation: enhancement technologies and need for more viable alternatives for whole rice grains. J Food Process Eng 46(2):e14214
100. Salem K, Jabalera Y, Puentes-Pardo JD, Vilchez-Garcia J, Sayari A, Hmida-Sayari A, Jimenez-Lopez C, Perduca M (2021) Enzyme storage and recycling: nanoassemblies of α-amylase and xylanase immobilized on biomimetic magnetic nanoparticles. ACS Sustain Chem Eng 9(11):4054–4063
101. Sarkar S, Gil JDB, Keeley J, Jansen K (2021) The use of pesticides in developing countries and their impact on health and the right to food. European Union.
102. Sasounian R, Martinez RM, Lopes AM, Giarolla J, Rosado C, Magalhães WV, Velasco MVR, Baby AR (2024) Innovative approaches to an eco-friendly cosmetic industry: a review of sustainable ingredients. Clean Technol 6(1):176–198
103. Sedrak MS, Freedman RA, Cohen HJ, Muss HB, Jatoi A, Klepin HD, Wildes TM, Le-Rademacher JG, Kimmick GG, Tew WP (2021) Older adult participation in cancer clinical trials: a systematic review of barriers and interventions. CA: Cancer J Clin 71(1):78–92

104. Sempere JM, Martínez-Peinado P, Arribas MI, Reig JA, De La Sen ML, Marco FM (2014) Single cell-derived clones from human adipose stem cells present different immunomodulatory properties. Clin Exp Immunol 176(3):457–465
105. Sen A, Kapila R, Chaudhary S, Nigam A (2021) Biotechnological applications of microbial enzymes to replace chemicals in the textile industry-a review. Text Assoc 82(2):68–73
106. Shad M, Hussain N, Usman M, Akhtar MW, Sajjad M (2023) Exploration of computational approaches to predict the structural features and recent trends in α-amylase production for industrial applications. Biotechnol Bioeng 120(8):2092–2116
107. Shakilanishi S, Shanthi C (2024) An overview on preparation of enzymes for industrial use. Biocatal Biotransform, 1–12
108. Shao Y, Liao Z, Gao B, He B (2022) Emerging 3D printing strategies for enzyme immobilization: materials, methods, and applications. ACS Omega 7(14):11530–11543
109. Sharif N, Munir N, Hasnain M, Naz S, Arshad M (2021) Environmental impacts of ethanol production system. Sustain Ethanol Clim Chang: Sustain Assess Ethanol Distill 2021:205–223
110. Sheikh JN, Teli M (2024) Sustainable textile chemical processing. CRC Press
111. Siddiqui SA, Asif Z, Murid M, Fernando I, Adli DN, Blinov AV, Golik AB, Nugraha WS, Ibrahim SA, Jafari SM (2022) Consumer social and psychological factors influencing the use of genetically modified foods—a review. Sustainability 14(23):15884
112. Soozanipour A, Ejeian F, Boroumand Y, Rezayat A, Moradi S (2023) Biotechnological advancements towards water, food and medical healthcare: a review. Chemosphere 312:137185
113. Thatoi H, Mohapatra S, Das SK (2021) Bioprospecting of enzymes in industry, healthcare and sustainable environment. Springer
114. Türkistanli TT (2024) Advanced learning methods in maritime education and training: a bibliometric analysis on the digitalization of education and modern trends. Comput Appl Eng Educ 32(1):e22690
115. Vannuccini S, Prytkova E (2023) Artificial intelligence's new clothes? A system technology perspective. J Inf Technol, 02683962231197824
116. Vashishtha A Comparative analysis of conventional scouring & bio-scouring on cotton fabrics
117. Visvanathan R, Qader M, Jayathilake C, Jayawardana BC, Liyanage R, Sivakanesan R (2020) Critical review on conventional spectroscopic α-amylase activity detection methods: merits, demerits, and future prospects. J Sci Food Agric 100(7):2836–2847
118. Wang B, Huang D, Cao C, Gong Y (2023) Insect α-amylases and their application in pest management. Molecules 28(23):7888
119. Wang M, Chen Y (2023) Electronic nose and its application in the food industry: a review. Eur Food Res Technol, 1–47

120. Wongsa P, Bhuyar P, Müller J (2023) Assessment of phenolic profile, allicin content, and inhibitory potential against α-amylase and α-glucosidase on conventional and organic garlic (Allium sativum L.). Eur Food Res Technol 249(11):2911–2923
121. Wu XT, Ansari AR, Pang XX, Li HZ, Zhang ZW, Luo Y, Arshad M, Song H (2019) Visfatin plays a significant role in alleviating lipopolysaccharide-induced apoptosis and autophagy through PI3K/AKT signaling pathway during acute lung injury in mice. Arch Immunol Ther Exp 1(67):249–261
122. Wu H, Sang S, Weng P, Pan D, Wu Z, Yang J, Liu L, Farag MA, Xiao J, Liu L (2023) Structural, rheological, and gelling characteristics of starch-based materials in context to 3D food printing applications in precision nutrition. Compr Rev Food Sci Food Saf 22(6):4217–4241
123. Xia W, Zhang K, Su L, Wu J (2021) Microbial starch debranching enzymes: developments and applications. Biotechnol Adv 50:107786
124. Xu H, Song H-Y (2024) Key factors influencing Chinese consumers' demand for naturally dyed garments: data analysis through KJ method and KANO model. Sustainability 16(3):1189
125. Yahya S, Sohail M, Khan SA (2022) Characterization, thermal stabilization and desizing potential of amylase from A. tubingensis SY 1. J Text Inst 113(6):993–1000
126. Yandri Y, Ropingi H, Suhartati T, Irawan B, Hadi S (2023) Immobilization of aspergillus fumigatus α-amylase via adsorption onto bentonite/chitosan for stability enhancement. Emerg Sci J 7(5):1811–1826
127. Yu H (2022) Secretory pathway of the filamentous fungus Trichoderma reesei. Macquarie University
128. Yusuf M (2021) Current and future perspectives of enzyme treatments for cellulosic fibers: a review. *Green Chem Sustain Text*, 69–76
129. Zafar A, Shaheen M, Tahir AB, da Silva APG, Manzoor HY, Zia S (2023) Unraveling the nutritional, biofunctional, and sustainable food application of edible crickets: a comprehensive review. Trends Food Sci Technol, 104254
130. Zhang T, Spence C (2023) Orthonasal olfactory influences on consumer food behaviour. Appetite, 107023
131. Zhang Y, Hu Y, Jiang N, Yetisen AK (2023) Wearable artificial intelligence biosensor networks. Biosens Bioelectron 219:114825

Practical Aspect of Handling the Use of Cellulases in the Textile Industry

Atiq Ur Rehman, Hamid Masood, Syed Hamza Abbas, Humaira Nawaz, Khurshid Ahmad, Abdullah Abdullah, and Shujaul Mulk Khan

Abstract The sources of fossil fuels are diminishing at a very high rate, developing a dire situation that has caught the interest of scientists. Biofuel generation from organic and biomass materials is an alternative source to fossil fuel sources as they are comparatively less expensive. According to research conducted worldwide, cellulase production reaches 8% required by the industry, but in the future, the market will reach US$ 400 million by the year 2020. The focus of investigations globally is on developing new technologies for the efficient use of novel biomass that could produce a variety of cheap and bulk biofuels. These enzymes have remarkable applications in many industries for efficient conversion of the polymer. The primary barrier to success, despite the massive amount of research being done, is the expensive price of cellulases, which makes it necessary to either find commercially viable methods including the repurposing of the enzymes or reduce their production costs.

Keywords Cellulase · Taxonomic diversity · Sources · Determination methods · Isolation · Applications

A. U. Rehman · H. Masood · S. H. Abbas · H. Nawaz
Department of Microbiology, Quaid-I-Azam University, Islamabad, Pakistan
e-mail: atiq_dvm@gu.edu.pk

A. U. Rehman
Institute of Microbiology, Faculty of Veterinary & Animal Sciences, Gomal University, Dera Ismail Khan, KPK, Pakistan

K. Ahmad
State Key Laboratory of Marine Food Processing & Safety Control, College of Food Science and Engineering, Ocean University of China, No. 1299, Sansha RoadShandong Province, Qingdao 266404, People's Republic of China

A. Abdullah · S. M. Khan (✉)
Department of Plant Sciences, Quaid-I-Azam University, Islamabad, Pakistan
e-mail: shujaqau@gmail.com

S. M. Khan
Pakistan Academy of Sciences, Islamabad, Pakistan

M. Arshad (ed.), *Enzymes in Textile Processing: A Climate Changes Mitigation Approach*, SDGs and Textiles, https://doi.org/10.1007/978-981-97-8058-7_6

1 Introduction

The products of textile industry fall into several categories: Sports textiles, agro-textiles, medical textiles for implants, industrial textiles, geotextiles (amazing for reinforcements), firefighting clothes, bulletproof clothes, welders' protective clothes, and aerospace textiles are just a few examples of the many types of textiles used in manufacturing [1]. Synthetic fiber production is responsible for around half of the world market [2]. The total volume of synthetic fiber produced worldwide was predicted to be 5.4 million tons in 2016. Chemicals and raw materials derived from petroleum are used to make synthetic fibers. These substances go through a protracted polymerization process to create various chemicals. Around half of all fibers are synthetic, and they are extensively employed in numerous applications related to textile technology [3]. The four main kinds of fiber based on synthetic polymers are acrylic, cotton, nylon, and polyethene. Cotton and polyester are the two most widely used textile fibers worldwide. Products used in clothes, cosmetics, and personal hygiene are key producers of primary microplastics. Furthermore, the breakdown of big plastics is the main source of secondary microplastics. There was 6.3% increase in wool production in textile industry between 2000 and 2018 worldwide. During this time, the production of cotton increased by 1.3% yearly, while the production of chemical textile fibers increased by 5.1% annually. Because of the increasing need, textile production has grown exponentially over time. Hence, the recycling of bioproducts from wastes not only produce valuable products but also minimize the detrimental effect on the environment [4]. Wool and silk are the two primary textile wastes that originate from animal sources. The wool fiber property depends on the type of animal source. The complex structure of the wool fiber's surface controls the dyes' ability to adsorb. Likewise, the dense and the layering of the wool fiber cuticle surface also regulate the changes in the fabric material including shrinkage and flexibility [5]. Silk materials have innate antiviral and antimicrobial characteristics in addition to their hydrophobic nature [6, 7]. Textile wastes derived from plants include cotton, flax, bamboo, jute, and yarn, textile wastes derived from minerals include asbestos and glass fiber. Moreover, wastes made of synthetic textiles include nylon, rayon, polyester, and acrylic. Cotton fibers are made mainly of crystalline in nature and cellulosic in [8]. Meanwhile, bamboo fiber is better suited for applying color than cotton as it has a large capacity to absorb moisture. Its strength and flexibility depend on chain molecules and axis. Compared to cotton, viscose rayon fiber is more hygroscopic, has a lower mass density, and is non-abrasive, which make it a better fit for liquid transfer applications like medical items. The cellulose microfibrils incorporated in the polysaccharide matrix of flax fibers make them suitable for use as reinforcement in composites. Similarly noted that jute fiber's interfacial compatibility, low density, and superior biodegradable qualities make it a promising filler for composite reinforcement. Glass fiber, when used as reinforcement in concrete, improves mechanical qualities such as internal resistance [3, 9]. The authors found that adding glass fiber to concrete increased its compressive strength. This was made clearer by emphasizing the importance of the matrix's interfacial strength, which provides crack resistance. Another study revealed that the capacity for cellulose fiber to reinforce in a cement matrix is less than that of asbestos fiber reinforcement. The scientists explained the more skillful pore structure by pointing to the increased fiber content in the microcracks as a consequence of the cement matrix's better responsiveness to pozzolanic and particle bonding. To adding nylon to composite material improves its cushioning properties as a possible packing material. Jute-epoxy composite outperforms jute polyester in terms of impact energy, but the former composite is stronger in terms of tensile and flexural strength [10]. Carbon fiber has a high tensile strength and modulus that is derived from acrylic fiber. The authors attributed the altered physical characteristics of carbon fiber to the impact of active oxygen functional groups and the modification of acrylic fiber. Conventional textile wastes made of cotton, acrylic, polyesters, nylon, polyvinyl chloride, polyolefin, and polyethene are among the most common forms. The most widely used thermoplastic polymer of polyesters is polyethylene terephthalate (PET). It is extensively used to make films, textiles, tapes, bottles, and food packaging. PET is one of the most extensively used plastics in the world. It was projected that around 27.8 million tons of PET-based bottles and packaging items were produced worldwide in 2015. PET is widely used in both residential and industrial settings because of its excellent tensile and impact strength, thermal stability, and economic viability [11]. According to estimations, the annual global production of nylon filaments is approximately 5.4 million tons. These filaments are a necessary raw material used to make paintbrushes, hairbrushes, and toothbrushes. During the production phase, 10% of the filaments are thrown away, which could lead to pollution if waste management is not well addressed. However, disposal has presented significant environmental issues because of the growth in PET production brought on by the rising demand. Every day, a significant quantity of PET-made bottles are thrown away after being used. Acrylic fiber is categorized as a synthetic fiber. Other subclasses, include modacrylic fibers, which contain roughly 85% acrylonitrile [12, 13]. Polymeric polymers based on acrylonitrile are widely used in production settings due to their strong mechanical strength and effective abrasion resistance. To increase qualities like dye improvement, acrylic fiber is typically treated during production using acrylic monomer, acrylic acids, or esters [14]. The wastewater produced during the production of acrylic fiber comprises highly hazardous chemical molecules such as nitriles, aromatic compounds, and alkanes that are complex and poorly biodegradable. But given the effects of CO2 emissions on global warming, as well as the pollution that comes from burning The synthesis of activated carbon, which is an essential requirement for CO2 adsorption, from PET wastes and wastewater dye leaks may offer a possibility to remove environmental pollutants [15].

2 Cellulases and Taxonomic Diversity

Different microorganisms produce cellulases, including actinomycetes, bacteria, and fungi. The different types of microorganisms that break down cellulose have been described as mentioned in Fig. 6.1. Although it has recently been shown that certain animals, such as termites and crayfish, also make cellulases, it is yet unknown how these enzymes work in the animal system probable cellulolytic microorganisms. The entire kingdom of fungi contains organisms that are capable of breaking down cell walls, ranging from basic groupings called Chytridiomycetes to more complex ones called Agaricomycetes. Although numerous cellulolytic species have been identified in the Zygomycetes genus, only a tiny percentage of these individuals are able to degrade cellulose for example Ascomycetes, Agaricomycetes, and Deuteromycetes subdivisions [16, 17]. Most of the cellulolytic species in the Eubacteria are found in the phyla *Fibrobacteres, Proteobacteria, Thermotogae, Actinobacteria, Firmicutes, Bacteroids* and *Spirochaetes*. The majority (about 80%) of the cellulolytic members are located in the mostly aerobic order Actinomycetes (phylum Actinobacteria) and aerobic gliding bacteria, fermentative anaerobic bacteria comprising of gram positive like *Ruminococcus, Caldicellulosiruptor* and *Clostridiales*, and very few number of gram negative (*Butyrivibrio, Acetivibrio* and *Fibrobacter*). These are the different physiological groups of bacteria that break down cellulose [18]. The way that anaerobic and aerobic bacterial groups hydrolyze cellulose varies. With a few exceptions, anaerobic bacteria break down cellulose mostly through a complexed cellulase system called a cellulosome. Extracellular free cellulases, which can be obtained via the liquid left over after fermentation, are a simple process used by aerobes to break down cellulose. Due to their rapid development and high enzyme yields, aerobic fungi and bacteria are receiving more attention in the field of biotechnology when it comes to the degradation of cellulose. To obtain novel cellulases with distinct properties, it is essential to screen and isolate cellulose-degrading bacteria from various habitats [19].

3 Bacterial-Producing Cellulase

Bacteria belonging to numerous genera within a certain domain are the source of cellulase-producing microorganisms found in soil [23, 24]. High cellulase activity microbial groups that are enriched in uncultivated or forest soil are important for studying novel species and their roles related to basic issues. Unique or new species include cellulolytic living things like Micromonospora, Acidothermus, Paenibacillus, Streptomyces, and Pseudomonas. These findings imply that the ecosystem may offer an appropriate environment for study as well as strategies for creating new enzymes that are capable of effectively breaking down polysaccharides and cellulose [25, 26]. It has been studied that a variety of bacteria can degrade synthetic textile colors, including anthraquinone, triphenylmethane, and azo dyes.

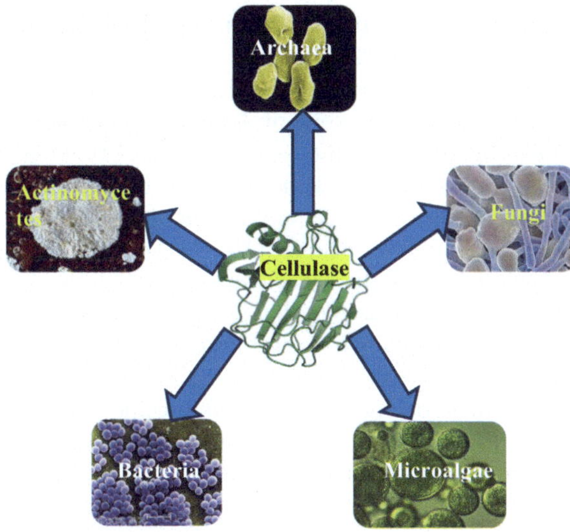

Fig. 6.1 The main microbes that produce cellulase [20]

Fig. 6.2 Applications of microbial cellulase in the textile industry [21, 22]

To achieve bacterial breakdown, a single bacterial isolate or a collection of microbes might be employed [27].

Bacterial cellulose serves as an inexpensive feedstock since microorganisms generate cellulose devoid of higher biopolymers. Since microbial cellulose is without lignin, hemicelluloses, and other contaminants, and pectin, bacterial cellulose synthesis is a more economical way to produce a lot [28, 29]. It is expensive and difficult to recover the cellulose from plants using nondegradable sources of these components' polysaccharides [30, 31]. The same is prepared from the coconut water by bacteria called gluconacetobacter (*Acetobacter*) xylinus for a variety of applications [24, 32]. One trait that sets bacterial cellulose apart is the extremely small collection of nanofibers of cellulose (3–8 nm) that comprise its three-dimensional structure. Their purity enables their practical use in biomedical goods such as animal feed, wound-covering scaffolds, and artificial cardiovascular tissues. Cellulase is produced using bacterial species strains to optimize conditions for numerous criteria required for commercial usage [33].

4 Fungus that Produces Cellulase

Fungi are considered dangerous bacteria even though they represent a vital raw material for industry for many different purposes today. The decomposition of the globe could potentially be attributed to dynamic cellulose decomposers. Additionally, because it was more resilient to environmental fluctuations, the framework used by fungal cellulosomes for the manufacture of cellulases was preferable. Of the Aspergillus and Humicola species, Trichoderma species have been found to be the most beneficial for cellulase production and industrial use [34, 35].

Moreover, Aspergillus species that have undergone genetic modification can produce cellulase in a disproportionately greater quantity. Numerous cellulolytic fungus spectrums have been gathered and studied over time, and the total of them has continued to yield important discoveries. A previously reported study included an outstanding collection of over 14,000 active fungus along with cellulose and other insoluble fibers. In favorable circumstances, such as solid and submerged fermentation, For example, Trichoderma viride and Trichoderma reesei aid in the synthesis of cellulase [36, 37]. It is advised to use fungal species for cellulase synthesis due to their large production of extracellular cellulase, which is extraordinarily adaptable [38, 39]. By hydrolyzing lignocellulosic biomass, fungal cellulases can produce a range of oxidative and hydrolytic catalysts. Fungi (white-rot) for example *P. chrysosporium* and fungi (soft-rot) like *Talaromyces emersonii, Penicillium funiculosum, Aspergillus solani* and *Trichoderma koningii* and *reesei* are the best-characterized cellulolytic organisms. The genera Fusarium nidulans, Fungus aculeatus, as well as Aspergillus oryzae, and Aspergillus niger, among others, are the most widely employed commercial microorganisms with a strong aptitude for cellulolysis. The following studies were conducted in 2022 [40–42] and demonstrated that 88 filamentous fungal strains were found after searching for fungi that make cellulase. Penicillium, Trichoderma,

Aspergillus and Talaromyces taxa have the ability to display the cellulolytic activity from about sixteen different samples. Following a search for fungi that produce cellulase, 88 filamentous fungal strains were discovered of these, 16 isolates from the taxa Penicillium, Trichoderma, an Aspergillus, and Talaromyces were documented with different intensity of cellulose activity [38, 43], which provides an industry-specific analysis of the optimization of cellulase production using strains of fungi. According to [33, 44], optimizing the enzymes synthesis parameters at the optimal settings of different parameters is essential for industrial applications.

5 Actinomycetes Producing Cellulase

Positive Grammatical—the majority of isolates show this ability. Mycelial actinomycetes, which are frequently found in soil, are essential for the absorption of organic substances and the dispersion of bioactive chemicals [45]. Actinomycetes from marine sediments can be isolated to produce unique actinomycetes that could result in a valuable new product, according to a few research studies. Conversely, actinomycetes are recognized for generating an array of extracellular enzymes, such as lipase, chitinase, and cellulase, which break down polymers [46, 47]. Based on the findings, it may be possible to isolate unique actinomycetes that have the capability to yield a valuable product by separating them from marine sediments. Conversely, a range of extracellular enzymes that break down polymers, such as cellulase, lipase, and chitinase, are produced by actinomycetes [46–48]. They are composed of the aerobics consisting of *Dermatophilus, Nocardia* and *Actinomadura* along with *Streptomyces* kinds as well as the anaerobes like *Arachnia, Bifidobacterium* and *Actinomyces*, as well as Rothia species. According to Subathra Devi et al. [49], *Actinomyces* is the most common pathogenic genus and is not acid fast, despite the fact that certain kinds of *Nacardia* are aerobic.

6 Cellulase-Producing Archaea

Maus et al. [50] have undertaken several researches on the morphology and history of evolution of the methanogenic archaeal the species that make up the communities of microbes in biogas producers that break down and consume biomass. Most Archaea cellulase is found in densely populated environments. Certain cellulase genes have been shown to be present in *Thermogladius cellulolyticus* and *Desulfurococcus fermentans, Sulfolobus solfataricus, Pyrococcus furiosus*, and *Pyrococcus horikoshii* are a few more hyperthermophilic Archaea [25, 51]. Seven of the methanogenic Archaea under investigation are predicted to produce CH4 through the hydrogenotrophic route, the last link in the anaerobic digestion (AD) chain. Methanoculleus bourgensis with Defuviito gatunisiensis have been shown to be the most influential microorganisms among those analyzed in the biogas microbial populations [52, 53]. Numerous studies focused on the characterisation of highly halophilic in archaeal isolated in Indian salt pans and the synthesis of hydrolytic enzymes. The ability of halophilic archaea to endure many stresses means that their catalysts are essential to the organisms' ability to be stable and functional even in the face of very high temperatures and very low water activity. Due to their improved performance under challenging conditions that are encountered in a variety of industries, researchers are constantly searching for new extremozymes. The six genera that were examined were *Haloarcula, Halorubrum, Halococcus, Haloferax, Halogeometricm*, as well as *Haloterrigena* [54, 55].

7 Microalgae Producing Cellulase

The chlorophyll-based microalgae which are microscopic in nature are mostly resent in marine and freshwater environments [56–58]. *Haematococcus pluvialis, Chlorococcum infusionum, Chlorella homosphaera, Rhizoclonium* sp, *Nannochloropsis* sp. and the *Scenedesmus* sp. are microalgae that generate cellulase. Microalgal biomass's high concentration of essential nutrients and minerals has piqued the interest of the industrial sector. The primary obstacles to algal biomass valorization include the minimum production and yield of the biomass, the carbon to nitrogen ratio (C/N) is uneven, and the resistant cells require huge expensive microalgal harvest [59, 60].

8 Fermentation-Based Microbial Cellulase Synthesis

There are two production methods used for the microbial cellulolase generation, which are commonly called submerged fermentation (SMF) and solid-state fermentation (SSF), which are the modern technologies to meet the industrial demand on large scale uses. Cellulolytic bacteria are known as true cellulolytic microorganism since they are able to degrade natural cellulose [40, 42]. Free cellulases can be produced by microbial enzymes. Free cellulases can be produced by microbial enzymes. These are widely used in commercial applications, together with the fact that their expression and secretion levels are high. The two primary methods utilized to manufacture cellulases are submerged digestion (SMD) and solid-status fermentation (SSF); their transmission mechanisms and stimuli from the outside differ [61–63]. In order to distinguish between these several cycles, a reliable investigation of the water content of the reaction is necessary. However, when paired with a solid matrix, water activity is not as effective as it may be in promoting cell growth and metabolism [64–66].

9 Methods for Determining Cellulase

All currently available techniques for assessing cellulase activity are categorized. The methods for determining cellulose are based on the following: (a) quantity of products formed following substrate hydrolysis, (b) reduction in substrate quantity during the test, and (c) modification in the substrate's physical characteristics seen following assay. The majority of test techniques quantify the resulting hydrolysis products, such as chromophores, reducing sugars, and total sugars. The Nelson-Somogyi method and the Dinitro salicylic acid (DNS) method are two widely used methods for determining cellulase activity due to their low protein interference and broad detection range for reducing sugars. Sugar assays can be made more sensitive by using one of two techniques: (1) color appearing after dilution or (2) adjusting the volume of sample added to each sample before the reaction. Nevertheless, there is a weak stoichiometric correlation in these test techniques between cellodextrins and the glucose standard. Nevertheless, a significant disadvantage of the extremely sensitive ferricyanide in 2, 2 bicinchoninic acids (BCA), the p-hydroxy benzoic acid and hydrazide (PAHBAH) testing techniques for the identification of reducing sugars is the non-specific interaction of protein. Assay methods that detect total sugars, including phenol-H2SO4 and anthrone-H2SO4, are only applicable to pure cellulases. These methods are less affected by proteins and significantly more sensitive [67, 68]. To measure cellulase activity, numerous additional assay techniques have also been documented. A decrease in substrate quantity also made it more important to assess cellulase activity using gravimetric and chemical techniques. However, because these techniques were so laborious to execute, they did not become extremely popular. The phenol-H2SO4 and anthrone-H2SO4 methods are two chemical techniques for calculating substrate loss. Cellulase activity has also been measured using activity assays that take into account changes in the substrate's physical characteristics [23, 69]. The rheological properties of carboxymethyl cellulose, or CMC, mixtures may be used to accurately quantify cellulase activity since viscosity and polymer content are related. For cellulase activity, numerous commercial kits that are based on these test techniques are available. The megazyme glucose oxidase/peroxidase (GOPOD) test kit, the NZYTech glucose GOPOD assay kit, and the sigma glucose assay are examples of assay kits based on the calorimetric approach, the Abcam cellulase assay kit and the MarkerGeneTM fluorescence cellulase assay kit are examples of assay kits based on the fluorometric estimating method. All of these test kits are extremely sensitive and function at the microplate levels [70].

9.1 Nanocellulose from Textile Wastes

As it is made from biomass such as vegetables, cotton, and wood, cellulose is the most abundant and continuous natural resource in the world. Fibrils of nanostructured cellulose range in size from 5 to 200 nm [25]. These materials are light, highly porous, have a high specific area, and are highly biodegradable. Nanocellulose has special qualities that make it useful in a wide range of sectors, including the biomedical, construction, packaging, and automotive [71]. In the field of nanotechnology, the synthesis of nanocellulose materials from textiles has created amazing prospects and applications. The present state of knowledge on the essential elements of isolation and pretreatment techniques for removing nanocellulose from textile wastes is reviewed critically in this paper. A recent development is the use of enzymatic hydrolysis to produce nanocellulose from textile wastes; cellulose extraction from basic materials, including cotton, is more widely documented. The separation of cellulose as a functional material from textile wastes has the potential for a variety of uses because of its high water holding capacity, excellent crystallinity, biocompatibility, thermal stability, and good mechanical characteristics. The idea of producing nanocellulose via textile wastes using an enzymatic-assisted isolation approach has not been covered in any review article, despite the fact that few have reported a nanocellulose from textile wastes using chemicals or physicochemical processes. It was investigated how to employ nanocellulose made from textile waste in a number of biotechnology fields to promote environmental sustainability and public health. Researchers will probably pay more attention to textile waste as a novel source of nanocellulose as a result of this review.

9.2 Production and Isolation of Cellulose Nanofibers by Cellulases

Two criteria are used to categorize cellulases: substrate specificity and structural similarity. All cellulolytic enzymes are hydrolases of glycosyl chains. The glycosidic bond that joins two or more carbs to a non-carbohydrate material is dissolved and broken down by this family of enzymes. Cellulases are often classified into two categories based on how specific they are to a certain substrate. Cellobiohydrolases called exo-1,4-Dglucanases have an affinity for the ends of the sugar chain, whereas endo-1,4-Dglucanases have an affinity for the middle of the glucose chain. Enzyme classification is difficult because of the frequent overlap of specificities [72]. Cellulases are produced by a wide variety of bacterium, plant, animal, fungal, and protozoa species. By regulating the depolymerization of cellulose, they are essential to many physiological processes occurring within cells [73]. There are several different types of bacteria that create cellulases, including mesophilic, anaerobic, aerobic, and thermophilic environments. The cellulose-producing bacteria Clostridium, Cellulomonas, Thermomonaspora, Trichoderma, and Aspergillus have been the most thoroughly researched [73]. Nowadays, cellulases make up a significant share of the worldwide enzyme market and are widely utilized for the modification of fiber. Cellulases are widely used in the paper and pulp, feed for animals, textiles farming, and food processing industries as well as in the production of biofuels like ethanol. Cellulases are reversible inducible enzymes that are tightly regulated by activating and repression mechanisms. Cellulase production can therefore be quickly inhibited in the presence of readily utilizable carbohydrates and boosted in the case of cellulosic substrates. In aerobic microbes and fungal organisms, when free enzymes predominate, the same mechanism breaks down cellulose. However, in order to enable the transfer of cellulosomal subunits necessary for the ongoing digestion of cellulose, anaerobic bacteria must go through a supramolecular rearrangement. Cellulases employ acid–base catalysis to selectively break the glycosidic linkages in cellulose. Since the residues of glucose are among the most persistent linkages in biopolymers—two times more resilient than DNA—and have a spontaneous half-life, or of around 5 million years, the hydrolysis of cellulose is a challenging procedure. As a poly(amidoamine)-coated gel made from carrageenan, cellulases were frozen and used [74]. The immobilization of the enzyme increased the related hydrolytic activity throughout a wide pH and temperature range. Moreover, enzymes can maintain as much as 85% of the initial efficiency even after six cycles. The fiber structure was preserved by immobilized cellulases, in contrast to traditional hydrolysis, producing nanofibers having a large ratio of aspect to diameter, 15–35 nm in diameter, with length of several micrometers. As an alternative to CNF separation, this method could be useful for the food and pharmaceutical sectors.

9.3 Application of Microbial Cellulase for Textile Industries

One kind of hydrolase that may break down lignocelluloses is called a cellulase. Because they originate from waste and renewable materials, they are widely used in many different fields. With a focus on the textile industry recently, they find extensive applications in detergents, textiles, and other biotechnology fields [75].

An eco-friendly technique that is widely used to give cotton textures and denim a pleasant look and non-abrasiveness is cellulase-based denim washing. But prior enzymatic denim cleaning methods employed neutral cellulase (Humicola isolens) and acid cellulase (Trichoderma reesei), which had the drawback of leaving an indigo backstain on the fabric. Although pH is said to be the most crucial element in preventing back stains, there is little information on the use of cellulase for denim cleaning in soluble situations. For denim finishing, thermomonospora sp. (T-EG), an alkalothermophilic soluble base stable endoglucanase, was used in rudimentary circumstances [76, 77].

In addition to providing for one of mankind's most basic needs, the textile sector boosts the economies of several nations. With the increase in both population and per capita textile requirements, the demand for textile related industry is also rising [78]. However, before producing a completed fabric, the traditional textile wet processing procedure necessitates a number of processes. Many resources, including water, salt, energy, and harsh chemicals, are used in these phases, all of which are known to have an adverse effect on the environment. Wet textile manufacturing uses enzymes, but their use is guided by environmental conscience [79, 80].

Biotechnology, which uses enzymes in the textile industries, seems to be a decent compromise between generating environmentally friendly products and satisfying commercial demands [32, 81]. Recent advances in bio-based processing technology have fundamentally altered the textile sector and elevated environmental issues to the top of the industry's worries list [82, 83].

In Fig. 5.2, wet processing of textiles is used primarily for five purposes: sizing, scouring, dyeing, coloring, and finishing. However, chemical-based process for wet textile preparation activities is becoming less and less efficient as industrial biotechnology develops in comparison to the most widely used textile and enzymes industrial procedures such biodesizing, bioscouring, and biobleaching [84].

9.4 Biostone Washing

One of the most important steps in achieving a faded look is denim washing. Giving stone the right appearance and velvety feel was once part of the stone cleaning process. After washing, the pumice stone takes color fragments from the denim's yarn surface. Ring-dying denim and severe abrasion during the stone-washing procedure are the methods used to get the faded look [85, 86]. Chemicals known as oxidative bleaching have also been utilized, either alone or in combination with stones, to produce the desired fading results. Washing your jeans is one of the most crucial procedures in getting a faded look. To get the ideal look and a supple feel, earlier stone washing was necessary. During washing, the pumice stone and any remaining yarn color particles come to the surface of the denim. To get the faded effect, denim is first ring-dyed and then severely abraded throughout the stone-washing procedure. To produce the fading effects, oxidative bleaching substances have also been utilized, either alone or in combination with stones [87, 88]. With its well-worn appearance and indigo-colored twill weave fabric, denim has become increasingly popular over time. There are a few disadvantages to using pumice stones for stone washing denim, whether or not an oxidizing agent like potassium permanganate is used [89]. Among these drawbacks include drainage system blockages and equipment damage, problems washing the pumice stones clean, the need for several stones for even tiny amounts, and the risk of severe abrasion damaging the cloth. Indigo colors and denim-heavy wool twill with a worn appearance have been incredibly successful throughout the years. When "stone washing" denim, pumice stones are usually used, either in combination or instead of an oxidizing agent such as potassium permanganate. Nevertheless, this process has several practical drawbacks, such as the potential for severe abrasion that could harm the fabric, damage to machine components, blockage of the drainage system, challenges in removing residue from the stones, and the requirement to use a large quantity of stones even for small quantities [90–92].

The textile industry has been interested in cellulases for a number of applications, such as "stone" finishing, washing, mercerization, scouring, and bio-polishing [87]. Periyasamy and Venkatesan [86] investigated the mechanism by which cellulase triggers "biostoning," or the indigo dye being released from the denim surface. A small amount of enzymes can be used in place of many pumice stones to facilitate handling. The biostoning technique reduces the wear and tear on processing gear, the formation of pumice dust, and the degradation of denim fabrics [84, 93, 94]. If a stone made of pumice losses as much as 50% of its total weight throughout the process and generates a significant amount of pumice grit, pumice sludge may result. It is more environmentally beneficial to utilize enzymes rather than pumice stones [92, 93, 95]. Acid cellulases, such those generated by Trichoderma, have not always shown to be appropriate for biostoning because of tissue stiffness and backstaining. To combat this, bleaching agents or anti-redeposition chemicals have been used during the washing phases. Conversely, neutral cellulases function in a less forceful manner. Biopolishing is another common way that enzymes are used to finish textile items made of cotton and other cellulose-based fibers. The white string is exposed as cellulases break down the surface fibers that contain the indigo dye [96, 97]. According to Rahman's and Rashid, (2020) research, acid washing denim jeans is growing in popularity because of its striking contrasts and eye-catching color scheme. If your clothing has an indigo or sulfur basis, you can wash it in acidic water. According to [98] findings, the most often used method for acid washing denim is to use pumice stones which have been presoaked within a solution including the following components: potassium permanganate or hypochlorite of sodium (5 to 10%) (3 to 6%).

9.5 Conclusion

The ecologically friendly process of biopolishing produces a smooth, glossy look with improved color intensity, hydrophilicity, and moisture absorption. It also produces consistently superior finishing. Because cellulases impact small fiber ends that protrude from the fabric surface, the fibers must be physically removed and polished [99, 100]. Endoglucanase II from Trichoderma reesei is said to be the greatest enzymes for biostoning denim garments and finishing cotton textiles. However, sometimes while finishing and biostoning, publicly accessible endoglucanase II is combined with various other cellulase components, such as endoglucanase I, resulting in hydrolysis and garment reduction in weight [94]. To rule out the existence of excess cellulose components, we extracted the endoglucanase II genes from Trichoderma reesei and generated it from Pichia pastoris by a methanol-inducible AOX1 promoter. It is far better than using commercially available Trichoderma reesei cellulose [101–103]. This is the result of heterologously generating the reesei Trichoderma endoglucanase II genes in Pichia pastoris. When using this enzyme in a jeans wash, the weights of the cellulosic fibers are maintained [104, 105].

References

1. Roy M, Sen P, Pal P (2020) An integrated green management model to improve environmental performance of textile industry towards sustainability. J Clean Prod 271:122656
2. Almroth BMC, Åstr¨om, L, Roslund, S, Petersson, H, Johansson, M, Persson, N.-K. (2018) Quantifying shedding of synthetic fibers from textiles; a source of microplastics released into the environment. Environ Sci Pollut Res 25(2):1191–1199
3. Deng H, Wei R, Luo W, Hu L, Li B, Shi H (2020) Microplastic pollution in water and sediment in a textile industrial area. Environ Pollut 258:113658
4. Kallio AMI (2020) Wood-based textile fibre market as part of the global forest-based bioeconomy. Forest Policy Econ 123:102364
5. Erdogan UH, Seki Y, Selli, F (2020). Wool fibres. In: Handbook of Natural Fibres. Elsevier, pp. 257–278
6. Parlin AF, Stratton SM, Culley TM, Guerra PA (2020) A laboratory-based study examining the properties of silk fabric to evaluate its potential as a protective barrier for personal protective equipment and as a functional material for face coverings during the COVID-19 pandemic. PLoS ONE 15(9):e0239531
7. Zaini MS, Arshad M, Syed-Hassan SS (2023) Adsorption isotherm and kinetic study of methane on palm kernel shell-derived activated carbon. J Bioresourc Bioproducts 1:8(1):66–77
8. Wang H, Farooq A, Memon H (2020) Influence of cotton fiber properties on the microstructural characteristics of mercerized fibers by regression analysis
9. Arshad M (2017) Clean and sustainable energy technologies. In: Clean energy for sustainable development, vol 1. Academic, pp 73–89
10. Kumar SS (2020) Effect of natural fiber loading on mechanical properties and thermal characteristics of hybrid polyester composites for industrial and construction fields. Fibers Polym 21(7):1508–1514
11. Zhang R, Ma X, Shen X, Zhai Y, Zhang T, Ji C, Hong J (2020) PET bottles recycling in China: an LCA coupled with LCC case study of blanket production made of waste PET bottles. J Environ Manag 260:110062
12. Chaudhari S, Shaikh T, Sharma D, Vadodaria K (2020) A review on influence of coagulation rate onto structural and physical properties of wet-spun acrylic fibres. Man-Made Text. India 48(9)
13. Wu XT, Ansari AR, Pang XX, Li HZ, Zhang ZW, Luo Y, Arshad M, Song H (2019) Visfatin plays a significant role in alleviating lipopolysaccharide-induced apoptosis and autophagy through PI3K/AKT signaling pathway during acute lung injury in mice. Arch Immunol Ther Exp 1(67):249–261
14. Yavuz G, Felgueiras HP, Ribeiro AI, Seventekin N, Zille A, Souto AP (2018) Dyed poly (styrene-methyl methacrylate-acrylic acid) photonic nanocrystals for enhanced structural color. ACS Appl Mater Interfaces 10(27):23285–23294
15. Zheng T, Wang Q, Shi Z, Fang Y, Shi S, Wang J, Wu C (2016) Advanced treatment of wet-spun acrylic fiber manufacturing wastewater using three-dimensional electrochemical oxidation. J Environ Sci 50:21–31
16. Arshad M, Adil M, Sikandar A (2014) Hussain T (2014) Exploitation of meat industry by-products for biodiesel production: Pakistan's perspective. Pakistan J Life Soc Sci 12:120–125
17. Nargotra P, Sharma V, Lee YC, Tsai YH, Liu YC, Shieh CJ, Kuo CH (2022) Microbial Lignocellulolytic enzymes for the effective valorization of lignocellulosic biomass: a review. Catalysts 13(1):83
18. Korsa G, Konwarh R, Masi C, Ayele A, Haile S (2023) Microbial cellulase production and its potential application for textile industries. Ann Microbiol 73(1):13
19. Balla A, Silini A, Cherif-Silini H, Bouket AC, Boudechicha A, Luptakova L, Belbahri L (2022) Screening of cellulolytic bacteria from various ecosystems and their cellulases production under multi-stress conditions. Catalysts 12(7):769

20. Leo VV, Ramesh N, Singh BP (2019) Microorganisms as an efficient tool for cellulaseproduction: availability, diversity, and efficiency. New Fut Devnt Microbial Biotechnol Bioeng 45–61. Elsevier. https://doi.org/10.1016/B978-0-444-64223-3.00004-7
21. Biernat K (2019) Elements of bioeconomy. Overv Proc Enzy Transform Biomass (Chapter 4):1–30. https://doi.org/10.5772/intec hopen.78099. https://doi.org/10.5772/intec.hopen.88966
22. Jayasekara S, Ratnayake R (2019) Microbial cellulases: an overview and applications. Cellulose 2:22. https://doi.org/10.5772/intechopen.84531
23. Sharif N, Munir N, Hasnain M, Naz S, Arshad M (2021) Environmental impacts of ethanol production system. In: Sustainable ethanol and climate change: sustainability assessment for ethanol distilleries. 2021, pp 205–223
24. Tureck BC, Hackbarth HG, Neves EZ, Garcia MCF, Apati GP, Recouvreux DD OS, Schneider ALDS (2022) Obtaining and characterization of bacterial cellulose synthesized by Komagataeibacter hansenii from alternative sources of nitrogen and carbon. Matéria (Rio de Janeiro) 26. https://doi.org/10.1590/S1517-707620210004.1392
25. Larson JA, Bagley DM (2022) Sessile and planktonic microbial taxonomy of a methanogenic cellulolytic enrichment reactor sourced from the organic fraction of municipal solid waste. J Environ Eng 148(4):04022004. https://doi.org/10.1061/(ASCE)EE.1943-7870.0001978
26. Poulsen JS, de Jonge N, Macêdo WV, Dalby FR, Feilberg A, Nielsen JL (2022) Characterisation of cellulose-degrading organisms in an anaerobic digester. Biores Technol 351:126933. https://doi.org/10.1016/j.biortech.2022.126933
27. Shukla A, Parmar P, Goswami D, Gehlot Y, Vala J, Parmar N, Saraf M (2021) Microbial technologies in textile industries: an elixir for the greener Korsa et al. Ann Microbiol (2023) 73:13 Page 20 of 21 environment. Green Chem Sustain Text 173–189. Woodhead Publishing. https://doi.org/10.1016/B978-0-323-85204-3.00028-2
28. Avcioglu NH (2022) Bacterial cellulose: recent progress in production and industrial applications. World J Microbiol Biotechnol 38:1–13. https://doi.org/10.1007/s11274-022-032 71-y
29. Gedarawatte ST, Ravensdale JT, Al-Salami H, Dykes GA, Coorey R (2021) Antimicrobial efficacy of nisin-loaded bacterial cellulose nanocrystals against selected meat spoilage lactic acid bacteria. Carbohydr Poly 251:117096. https://doi.org/10.1016/j.carbpol.2020.117096
30. Indumathi T, Jayaraj R, Kumar PS, Krishnaswamy VG, Ghfar AA, Govindaraju S (2022) Biological approach in deinking of waste paper using bacterial cellulose as an efective enzyme catalyst. Chemosphere 287:132088. https://doi.org/10.1016/j.chemosphere.2021.132088
31. Krishnaswamy VG, Sridharan R, Kumar PS, Fathima MJ (2022) Cellulase enzyme catalyst producing bacterial strains from vermicompost and its application in low-density polyethylene degradation. Chemosphere 288:132552. https://doi.org/10.1016/j.chemosphere.2021.132552
32. Singhania RR, Patel AK, Tseng YS, Kumar V, Chen CW, Haldar D, Saini JK, Dong CD (2022) Developments in a bioprocess for bacterial cellulose production. Biores Technol 344:126343. https://doi.org/10.1016/j.biortech.2021.126343
33. Gad AM, Suleiman WB, El-Sheikh HH, Elmezayen HA, Beltagy EA (2022) Characterization of cellulase from Geotrichum candidum strain Gad1approaching bioethanol production. Arab J Sci Eng 1–14. https://doi.org/10.1007/s13369-021-06391-z
34. Christopher M, Sreeja-Raju A, Kooloth-Valappil P, Gokhale DV, Sukumaran RK (2022) Cellulase hyper-producing fungus Penicillium janthinellum NCIM 1366 elaborates a wider array of proteins involved in transport and secretion, potentially enabling a diverse substrate range. Bioenerg Res 16:1–13. https://doi.org/10.1007/s12155-022-10407-3
35. Mattam AJ, Chaudhari YB, Velankar HR (2022) Factors regulating cellulolytic gene expression in flamentous fungi: an overview. Microbial Cell Fact 21(1):1–19. https://doi.org/10.1186/s12934-022-01764-x
36. Arshad M, Ansari AR, Qadir R, Tahir MH, Nadeem A, Mehmood T, Alhumade H, Khan N (2022) Green electricity generation from biogas of cattle manure: an assessment of potential and feasibility in Pakistan. Front Energy Res 25(10):911485

37. Zhao Q, Liu Q, Wang Q, Qin Y, Zhong Y, Gao L, Qu Y (2021) Disruption of the Trichoderma reesei gul1 gene stimulates hyphal branching and reduces broth viscosity in cellulase production. J Indust Microbiol Biotechnol 48(1–2):kuab012. https://doi.org/10.1093/jimb/kuab012
38. Lubeck M, Lubeck PS (2022) Fungal cell factories for efficient and sustainableproduction of proteins and peptides. Microorg 10(4):753. https://doi.org/10.3390/microisms1 0040753
39. Monclaro AV, Silva CDOG, Gomes HAR, de Souza Moreira LR, Ferreira Filho EX (2022) The enzyme interact come concept in filamentous fungi linked to biomass valorization. Biores Technol 344:126200. https://doi.org/10.1016/j.biortech.2021.126200
40. Faheina Junior GS, Sousa KA, Zilli JE, Vergara C, Pinto GA, Santiago-Aguiar,RS (2022) Enhanced cellulase production by Talaromyces amestolkiaeCMIAT055 using banana Pseudostem. Wast Biomas Valori 1–12. 10.21203/ rs.3.rs-726111/v1
41. Vasco-Correa J, Capouya R, Shah A, Mitchell TK (2022) Sequential fungal pretreatmentof unsterilized Miscanthus: changes in composition, cellulose digestibility and microbial communities. Applied Microbiol Biotechnol 106(5):2263–2279. https://doi.org/10.1007/s00253-022-11833-1
42. Santos GB, de Sousa Francisco Filho Á, da Silva Rodrigues JR, de Souza RR (2022) Cellulase production by Aspergillus niger using urban lignocellulosic waste as substrate: evaluation of different cultivation strategies. J Environ Manage 305:114431
43. Tomico-Cuenca I, Mach RL, Mach-Aigner AR, Derntl C (2021) An overviewon current molecular tools for heterologous gene expression inTrichoderma. Fungal Biol Biotechnol 8(1):1–17. https://doi.org/10.1186/s40694-021-00119-2
44. Helal GA, Khalil RR, Galal YG, Soliman SM, Abd Elkader RS (2022) Studies on cellulases of some cellulose-degrading soil fungi. Arch Microbiol 204(1):1–16. 10.1007/ s00203-021-02705-9
45. Al-Shaibani MM, Radin Mohamed RMS, Sidik NM, Enshasy HAE, Al-GheethiA, Noman E, Zin NM (2021) Biodiversity of secondary metabolitescompounds isolated from phylum actinobacteria and its therapeuticapplications. Molecules 26:4504. 10. 3390/ molec ules261545 04
46. Javed Z, Tripathi GD, Mishra M, Dashora K (2021) Actinomycetes–the microbialmachinery for the organic-cycling, plant growth, and sustainable soilhealth. Biocatal Agric Biotechnol 31:101893. https://doi.org/10.1016/j.bcab.2020.101893
47. Sudarshan A, Renuka S, Reshma S, Shilanjali B, Agsar D (2022) Upsurge productionof cellulase from maize stover under solid state conditions mediatedby Streptomyces enissocaesilis DQ026641. J Appl Biol Biotechnol (01):136–44. https://doi.org/10.7324/JABB.2021.100117
48. Phuoc NN, Le Hoang Minh Nhut NT, Hong X, Linh NT, Le B (2020) Isolation andcharacterization of actinomycetes from white leg shrimp (Litopenaeusvannamei) ponds antagonistic to Vibrio parahaemolyticus. Microbiol ResInt 8:35–42. https://doi.org/10.30918/MRI.82.20.015
49. Subathra Devi C, Merlyn Keziah S, Jemimah Naine S, MohanasrinivasanV (2022) Actinomycetes: microbiology to systems biology. Actinobact 1–35 Springer, Singapore. https://doi.org/10.1007/978-981-16-5835-8_1
50. Maus I, Rumming M, Bergmann I, Heeg K, Pohl M, Nettmann E, Jaenicke S,Blom J, Puhler A, Schluter A, Sczyrba A (2018) Characterization of Bathyarchaeota genomes assembled from metagenomes of biofilmsresiding in mesophilic and thermophilic biogas reactors. Biotechnol Biofuel 11(1):1–3. https://doi.org/10.1186/s13068-018-1162-4
51. Kabaivanova L, Hubenov V, Dimitrova L, Simeonov I, Wang H, Petrova P (2022)Archaeal and bacterial content in a two-stage anaerobic system for efficient energy production from agricultural wastes. Mol 27 (5):1512. https://doi.org/10.33448/rsd- v11i3.26368
52. Jo SU, Lee SJ, Kim HS, Eom JS, Choi Y, Oh DS, Bae D, Lee SS (2022) Effects of oriental medicinal plants on the reduction of methane production mediated by microbial population. Ital J Animal Sci 21(1):522–531. https://doi.org/10.1080/18280 51X.2022.2046192
53. Malik AD, Furtado IJ (2022) Isolation of Halomicroarcula pellucida strain GUMF5, an archaeon from the Dead Sea-Israel possessing cellulase. 3 Biotechnol 12(1):1–11. https://doi.org/10.1007/s13205-021-03090-2

54. Junior AF, Guedes EH, da Silva Soares CM, Cunha MG, dos Santos AL, AguiarAO, Martins AL (2022) Isolation and characterization of halophilic strains from beef jerky. Res Soci Dev 11(3):11(3):e21511326368. https://rsdjournal.org/index.php/rsd/article/view/26368
55. Leoni C, Manzari C, Tran H, Golyshin PN, Pesole G, Volpicella M, Ceci LR (2022) Identification of an amylomaltase from the halophilic archaeon Haloquadratum walsbyi by functional metagenomics: structural and functional insights. Life 12(1):85. https://doi.org/10.3390/life120100 85
56. Bano I, Arshad M (2018) Climatic changes impact on water availability. Perspectives on water usage for biofuels production: aquatic contamination and climate change 2018:39–54
57. Melendez JR, Matyas B, Hena S, Lowy DA, El Salous A (2022) Perspectives in theproduction of bioethanol: a review of sustainable methods, technologies,and bioprocesses. Renewab Sustain Energ Rev 160:112260. https://doi.org/10.1016/j.rser.2022. 112260
58. Shokrkar H, Keighobadi A (2022) Effect of fluid hydrodynamic situations on enzymatic hydrolysis of mixed microalgae: experimental study and simulation. Energ 241:122804. https://doi.org/10.1016/j.energy.2021.122804
59. Shah F, Mishra S (2020) In vitro optimization for enhanced cellulose-degradingenzyme from Bacillus licheniformis KY962963 associated with amicroalgae Chlorococcum sp. using OVAT and statistical modeling. SNApplied Sci 2(11):1–10. https://doi.org/10.1007/s42452-020-03697-9
60. Tawfik A, Ismail S, Elsayed M, Qyyum MA, Rehan M (2022) Sustainable microalgal biomass valorization to bioenergy: key challenges and future perspectives. Chemosphere 133812. https://doi.org/10.1016/j.Chemosphere.2022.133812
61. Chmelova D, Legerska B, Kunstova J, Ondrejovič M, Miertuš S (2022) The productionof laccases by white-rot fungi under solid-state fermentation conditions. World J Microbiol Biotechnol 38(2):1–20. 10.1007/ s11274- 021- 03207-y
62. El Sheikha AF, Ray RC (2022) Bioprocessing of horticultural wastes by solid state fermentation into value-added/innovative bioproducts: a review. Food Rev Int 1–56. https://doi.org/10.1080/87559 129.2021.20041.61
63. Nascimento FVD, Lemes AC, Castro AMD, Secchi AR, Zarur Coelho MA (2022)A temporal evolution perspective of lipase production by yarrow lipolytica in solid-state fermentation. Proc 10(2):381. https://doi.org/10.3390/pr10020381
64. Bano I, Ghauri MA, Arshad M, Yasin T, Younus M (2017) Bioactivity of variant molecular weight chitosan against drug-resistant bacteria isolated from human wounds. Microbial Drug Resist 1;23(8):958–965
65. Kalogeropoulou A, Plioni I, Dimitrellou D, Soupioni M, Nigam PS, Kanellaki M,Koutinas AA (2022) Biosynthesis of fuel-grade ethanol from cell obiose by a cell-factory of non-GMO Saccharomyces cerevisiae/starch-gel-cellulase. Fuel 313:122986. https://doi.org/10.1016/j.fuel.2021.122986
66. Nisar K, Abullah R, Kaleem A, Iqtedar M, Iftikar T (2022) Optimization and productionkinetics for cellulases by a wild and mutant strain of Thermomycesdupontii in stirred tank reactor. Acta Scientiarum Technol 44:e57664. https://orcid.org/0000-0003-3932-6297
67. Qasim M, Bukhari SA, Ghani MJ, Masoud MS, Huma T, Arshad M, Haque A, Ibrahim Z, Javed S, Rajoka MI (2016) Relationship of oxidative stress with elevated level of DNA damage and homocysteine in cardiovascular disease patients. Pak J Pharm Sci 2:29
68. Reis CER, Junior NL, Bento HB, de Carvalho AKF, de Souza Vandenberghe LP, Soccol CR, Chandel AK (2023) Process strategies to reduce cellulase enzyme loading for renewable sugar production in biorefineries. Chem Eng J 451:138690
69. Manyi-Loh CE, Lues R (2023) Anaerobic digestion of lignocellulosic biomass: substrate characteristics (challenge) and innovation. Fermentation 9(8):755
70. Wang B, Rutherfurd-Markwick K, Zhang XX, Mutukumira AN (2022) Kombucha: production and microbiological research. Foods 11(21):3456
71. Rizal S, Lai TK, Muksin U, Olaiya N, Abdullah C, Yahya EB, Chong E, Abdul Khalil H (2020) Properties of macroalgae biopolymer films reinforcement with polysaccharide microfibre. Polymers 12(11):2554

72. Chauve M, Barre L, Tapin-Lingua S, da Silva Perez D, Decottignies D, Perez S, Ferreira NL (2013) Evolution and impact of cellulose architecture during enzymatic hydrolysis by fungal cellulases. In: Advances in bioscience and biotechnology 2013.
73. Pirich CL, Picheth GF, Fontes AM, Delgado-Aguilar M, Ramos LP (2020) Disruptive enzyme-based strategies to isolate nanocelluloses: a review. Cellulose 27(10):5457–5475
74. Yassin MA, Gad AAM, Ghanem AF, Rehim MHA (2019) Green synthesis of cellulose nanofibers using immobilized cellulase. Carbohydr Polym 205:255–260
75. Ejaz U, Sohail M, Ghanemi A (2021) Cellulases: from bioactivity to a variety of industrial applications. Biomim 6(3):44. https://doi.org/10.3390/biomimetics6030044
76. Imran M, Bano S, Nazir S, Javid A, Asad MJ, Yaseen A (2019) Cellulases production and application of cellulases and accessory enzymes in pulp and paper industry: a review. PSM Biol Res 4(1):29–39. https://core.ac.uk/download/pdf/327166027
77. Sampathkumar K, Kumar V, Sivamani S, Sivakumar N (2019) An insight into fungal cellulases and their industrial applications. Approach Enhance Indust Prod Fung Cell 19–35. https://doi.org/10.1007/978-3-030-14726-6_2
78. Skiba EA, Gladysheva EK, Budaeva VV, Aleshina LA, Sakovich GV (2022) Yield and quality of bacterial cellulose from agricultural waste. Cell 1–13. https://doi.org/10.1007/s10570-021-04372-x
79. Ambaye TG, Vaccari M, Prasad S, van Hullebusch ED, Rtimi S (2022) Preparation and applications of chitosan and cellulose composite materials. J Environ Mgt 301:113850. https://doi.org/10.1016/j.jenvman.2021.113850
80. Son J, Lee KH, Lee T, Kim HS, Shin WH, Oh JM, Koo SM, Yu BJ, Yoo HY, Park C (2022) Enhanced production of bacterial cellulose from miscanthus as sustainable feedstock through statistical optimization of culture conditions. Int J Environ Res Publ Health 19(2):866. https://doi.org/10.3390/ijerph19020866
81. Aggarwal R, Dutta T, Sheikh J (2020) Extraction of pectinase from Candida isolated from textile mill efuent and its application in bio-scouring of cotton. Sust Sustain Chem Pharm 17:100291. https://doi.org/10.1016/j.scp.2020.100291
82. Kabir SMM, Koh J (2021) Sustainable textile processing by enzyme applications. Biodegrad. https://doi.org/10.5772/intechopen.97198.IntechOpen
83. Nursyirwani N, Feliatra F, Tanjung A, Harjuni F (2020) Isolation of cellulolytic bacteria from mangrove sediment in Dubai marine station Riau and the antibacterial activity against pathogens. IOP Conf Ser: Earth Environ Sci 430(1):012012. IOP Publishing. https://doi.org/10.1088/1755-1315/430/1/012012
84. Rahman M, Billah M, Hack-Polay D, Alam MA (2020) The use of biotechnologies in textile processing and environmental. Technol Forecast Soci Chang 159:120204. https://doi.org/10.1016/j.techfore.2020.120204
85. Arshad M (2010) Bioethanol: A sustainable and environment friendly solution for Pakistan. A Sci J. COMSATS–Science Vision 2011:16–7
86. Periyasamy AP, Tehrani-Bagha A (2022) A review of microplastic emission from textile materials and its reduction techniques. Polymer Degrad Stab 199:109901. https://doi.org/10.1016/j.polymdegradstab.2022.109901
87. Costa FN, de Souza LJ, Valério A, de Souza AA, de Oliveira D (2021) Utilization of montmorillonite in biostoning process as a strategy for effluent reuse. J Chem Technol Biotechnol 96(4):890–898. https://doi.org/10.1002/jctb.6597
88. Mustafa M, Ali L, Islam W, Noman A, Zhou C, Shen L, Zhu T, Can L, Nasif O, Gasparovic K, Gao J (2022) Heterologous expression and characterization of glycoside hydrolase with its potential applications in the hyperthermic environment. Saudi J Biol Sci 29(2):751–757. https://doi.org/10.1016/j.sjbs.2021.09.076
89. Korsa G, Masi C, Konwarh R, Tafesse M (2022) Harnessing the potential use of cellulolytic Klebsiella oxytoca (M21WG) and Klebsiella sp. (Z6WG) isolated from the guts of termites (Isoptera). Ann Microbiol 72(1):1–15. https://doi.org/10.1186/s13213-021-01662-4
90. Bağıran C, Körlü A, Yapar S (2021) Wastewater treatment of a denim washing plant by using waste pumice stones to recycle wastewater and reuse. Sustain Text Fash Value Chain 151–168. Springer. https://doi.org/10.1007/978-3-030-22018-1_9

91. Hasan MZ, Asif AAH, Razzaque A, Hasan MR, Sur S, Faruque MO (2021) An experimental investigation of diferent washing processes on various properties of stretch denim fabric. J Mater Sci Chem Eng 9(01):1. https://doi.org/10.4236/msce.2021.91001
92. Hoque MT, Mazumder NUS, Islam MT (2021) Enzymatic wet processing. In: Sustainable practices in the textile industry, pp 87–110
93. Eid BM, Ibrahim NA (2021) Recent developments in sustainable finishing of cellulosic textiles employing biotechnology. J Clean Prod 284:124701. https://doi.org/10.1016/j.jclepro.2020.124701
94. Pandit P, Maiti S, Singha K, Maity S, Shanmugasundaram OL (2022) Biotechnological and nano-biotechnological approaches in finishing of textile materials. Appl Biotechnol Sustain Text Prod 173–186 Woodhead. Publishing. https://doi.org/10.1016/B978-0-323-85651-5.00011-6
95. Mevada VA, Beladiya UH, Gandhi HR, Mangrola AV, Patel RK (2022) Alkalophiles: environmental distribution, taxonomy, physiology, bioenergetics, a survival mechanism, and enzymes. Phys Genom Biotechnol Appl Extrem 35–64. IGI Global. https://doi.org/10.4018/978-1-7998-9144-4.ch003
96. Arbab S, Ullah H, Khan MI, Khattak MN, Zhang J, Li K, Hassan IU (2022) Diversity and distribution of thermophilic microorganisms and their applications in biotechnology. J Basic Microbiol 62:95–108. https://doi.org/10.1002/jobm.202100529
97. Islam MT (2021) Enzymatic wet processing. Sustain Pract Text Indust 87. https://doi.org/10.1002/9781119818915.ch4
98. Rashid MR, Rahman MF (2020) Study on fabric and seam strength loss of denim trousers for different washing treatments. J Text Sci Technol 6(03):114. https://doi.org/10.4236/jtst.2020.63009
99. De Souza LJ, Immich APS, de Araújo PHH, de Oliveira D (2022) Cellulase immobilized on kaolin as a potential approach to improve the quality of knitted fabric. Bioprocess Biosyst Eng 45:1–10. https://doi.org/10.1007/s00449-021-02686-5
100. Gupta S, Kelkar-Mane V (2022) Use of uni-enzyme on cotton knitted fabric and its comparison with the commercial formulation. India J Fibre Text Res 46(4):349–353. http://op.niscair.res.in/index.php/IJFTR/article/view/ 46218/465480311
101. Khan M, Singh T, Pal DB, Khan S, Ahmad S, Jandrajupalli SB, Srivastava N (2022) Enhanced production of bacterial hydrolytic endoglucanase enzyme using waste leaves of water hyacinth and its thermal stability under the influence of TiO2 nanoparticles. Biomass Conversion Bioref 1–7. https://doi.org/10.1007/s13399-022-02421-4
102. Rather AH, Khan RS, Wani TU, Beigh MA, Sheikh FA (2022) Overview on immobilization of enzymes on synthetic polymeric nanofibers fabricated by electrospinning. Biotechnol Bioeng 119(1):9–33. 1002/bit.27963
103. Sivasankar P, Poongodi S, Sivakumar K, Al-Qahtani WH, Arokiyaraj S, Jothiramalingam R (2022) Exogenous production of cold-active cellulase from polar Nocardiopsis sp. with increased cellulose hydrolysis effciency. Archiv Microbiol 204(4):1–9. https://doi.org/10.1007/s00203-022-02830-z
104. Amengual NG, Csarman F, Wohlschlager L, Ludwig R (2022) Expression and characterization of a family 45 glycosyl hydrolase from Fomitopsis pinicola and comparison to Phanerochaete chrysosporium Cel45A. Enzyme Microb Technol 156:110000. https://doi.org/10.1016/j.enzmictec.2022.110000
105. Saif A, Cuccurullo A, Gallipoli D, Perlot C, Bruno AW (2022) Advances in enzyme induced carbonate precipitation and application to soil improvement: a review. Mater 15(3):950. https://doi.org/10.3390/ma15030950
106. Bajpai P (2022) Cellulases in the Biofuel Industry. Elsevier
107. Larsson PT, Stevanic-Srndovic J, Roth SV, S¨oderberg, D. (2022) Interpreting SAXS data recorded on cellulose rich pulps. Cellulose 29(1):117–131
108. Rajulapati V, Dhillon A, Kumar Gali K, Katiyar V, Goyal A (2020) Green bioprocess of degumming of jute fbers and bioscouring of cotton fabric by recombinant pectin methylesterase and pectate lyases from Clostridium thermocellum. Proc Biochem 92:93–104. https://doi.org/10.1016/j.procbio.2020.02.024

109. Rani A, Saini KC, Bast F, Varjani S, Mehariya S, Bhatia SK, Funk C (2021) A review on microbial products and their perspective application as anti microbialagents. Biomol 11(12):1860. https://doi.org/10.1016/j.bprint.2020.e00116
110. Saini N (2023) Futures of Queer Well-Being in India

Economic-Related Aspects of Pectinase in the Textile Industry

Gulrukh Khan, Syeda Tatheer Fatima, Rehana Masood, and Aiman Umar

Abstract Textiles have experienced wet chemical procedures and protocols since old age. Human creativity imagination and technique are apparent in textile goods all over the ages. With the passage of time and technological advancement, enzymes are used in place of harsh hazardous chemical processes. They are non-toxic and help in reducing pollution. The specific textile enzymes used are pectinase, laccase, amylase, and catalase. A class of enzymes known as pectinases breaks down pectin compounds. The three main forms of pectinases that are important to the food industry are pectin esterases, depolymerizing pectinases and protopectinases. Pectin esterases remove the methoxyl residue from pectin while the other two enzyme convert the indissoluble protopectin into dissolvable pectin. Pectin substances are categorized into four main classes: protopectins, pectinic acids, pectins, and pectic acid. Protopectins are not soluble in water, whereas three of pectins either fully or moderately solvable. Pectinases are extensively found in the universe and are formed in different microbes, including prokaryotes and eukaryotes. Various microbes use different fermentation techniques like solid-state and submerged for pectinase production. The generation of pectinase depends critically on the optimalization of critical bioreactor factors and factors, including environment, pH, and duration of fermentation. Because microbial pectinase has applications in many industries, there is a growing global demand for its commercial production. There is widespread use of pectinases in the culinary, agricultural, environmental, and pharmaceutical industries. Textiles are necessary for daily life, and the fashion industry shows a substantial part in world economy. The making and consumption of textiles worldwide are growing in tandem with the global population. The textile sector is thought to be worth 1.1 trillion euros and hires more than three hundred million employees [33]. Global eco-friendly textile industry

Gulrukh Khan, Syeda Tatheer Fatima and Rehana Masood—Equal contribution.

G. Khan · S. T. Fatima · R. Masood (✉)
Department of Biochemistry, Shaheed Benazir Bhutto Women University, Peshawar 25000, Pakistan
e-mail: rehana.masood@sbbwu.edu.pk

R. Masood · A. Umar
Department of Food and Nutrition, Shaheed Benazir Bhutto Women University, Peshawar 25000, Pakistan

M. Arshad (ed.), *Enzymes in Textile Processing: A Climate Changes Mitigation Approach*, SDGs and Textiles, https://doi.org/10.1007/978-981-97-8058-7_7

is anticipated to grow with annual average growth rate of 7.6% between 2023 and 2030, translating into a projected value of 77.17 billion USD. Reducing pollution and using more environmentally friendly alternatives instead of harsh chemicals is possible when using textile-found enzymes like pectinase. The textile industry is keen to use this sanitary method. Both water and energy consumption are decreased when pectinase is used. Since pectinase is non-toxic, the water can be recycled after cleansing. Energy efficiency boosts the economy and increases profitability. Reducing workplace fatalities and cancerous substances eventually leads to less medical expenses, higher productivity, and decreased absenteeism. Along with the consumption of pectinase from microbial sources, investigations should direct on engineered proteins to generate new powerful and flexible pectinases, as well as the improvement of manufacturing procedures utilizing novel microbes. Subsequent investigations into pectinolytic enzymes ought to focus on underlying mode of action governing enzyme release in addition to in what way distinct pectinolytic enzymes interact with various pectic substrates. This chapter tries to cover the introduction, classification, substrate, and textile industry applications of pectinase. It also summarized the bioprocessing processes of the textile industry using the pectinase enzyme. Further, the eco-friendly and economic-related aspects of pectinase are also briefly discussed.

Graphical abstract

Keywords Sustainability · Eco-friendly · Pectinolytic enzymes · Pectin esterases

1 Introduction

1.1 Enzymes

Enzymes are biomolecules that can speed up chemical reactions without altering their original composition. They are present in small amounts in the cells of living organisms. After their finding in the end of 1800s, enzymes are utilized extensively. The enzyme utilization is greatly encouraged in the edible, chemical, pharmaceutical, and textile industries over chemical catalysts due to customized enzyme development having increased accuracy, elevated catalytic efficiency, and selectivity [32]. In light of these significant characteristics, the need for commercial enzymes has increased rapidly with constant research and development to enhance output with minimum resource consumption [46]. Initially, the fungal enzymes were employed in industrial processes, later on, the bacterial enzymes were introduced too by Boidin and Effront [3, 4, 41]. Microbes have enormous potential and they produce a variety of enzymes that have been used for commercial purposes for many years [17]. They are the major source of most of the industrial enzymes demand because of their incredible properties like fast growth, short life spans, and easy genetic manipulation. That is why these enzymes are provided, organized, and sold by a small number of commercial companies.

1.2 Pectinase

Including industrial significant enzymes, pectinases are predominantly significant because of their numerous applications in various crucial sectors like edible, textile, beverage, paper, and biofuel. The demand for microbes based pectinases has occasionally risen above 25% among the global food and industrial enzymes [26].

The industrial applications of pectinase were witnessed in the 1930s. They were also a few of the first enzymes utilized in households, mostly for the preparation of wine and fruit juice [18]. Pectinases associated with the polysaccharides family are also known as pectic enzymes. They support plants in their cell wall growth, modifying of tissue in the time of maturity and storage, and recycling and degradation of plant waste products in an environmentally friendly manner. Pectinases are widely utilized in the food sector, especially for the extraction and clarification of fruit juice [7] (Fig. 1).

2 Pectic Substance

Pectic substances are chemically complex polysaccharides of colloidal acid linked by α (1–4) bonds, containing a backbone of galacturonic acid residues. The pendant chains of the molecular pectin are composed of L-rhamnose, arabinose, galactose, and xylose. The carboxyl residues of galacturonic acid are entirely or incompletely neutralized by sodium, potassium, and ammonium ions, which are esterified by methyl groups [18]. The American Chemical Society classified pectic substances into four main categories: protopectins, pectinic acids, pectins, and pectic acids. Protopectins are not soluble in water, whereas three of the pectins are either fully or moderately soluble [36].

3 Classification of Pectinase

Pectinases are characterized as diverse enzymes that break pectic materials from microbial and plant sources. These enzymes are responsible for catalyzing the synthesis of pectin-associated materials by microorganisms and plants. Fungal cultures are used to obtain most commercial enzymes. Pectinase has a role in the abscission process, senescence, fruit ripening, cell proliferation, and disease in addition to the metabolism of the cell wall. Since the 1970s, pectinases have been produced from microbes on industrial scale, primarily fungi. Peptinase enzymes are categorized as hydrolase enzymes by the International Union of Biochemistry and the Enzyme Commission [13].

3.1 Protopectinase

Pectic compounds have an insoluble component known as protopectin, which must be broken down using additional solvents. Subtype of pectinase that degrade the unsolvable protopectin associated with unripe fruits and produces highly polymerized soluble pectin is called protopectinase. It is also known as pectinosinase. Protopectinase converts insoluble protopectin into vastly polymerized solvable pectin. The quantity of pectin-associated material free from protopectin can be estimated by checking the activity of protopectinase, which is evaluated with the carbazole sulfuric acid method [15].

3.2 Pectin Methylesterases

Methoxyl ester is among the principal moiety in the configuration of pectin. Pectin methylesterases are accountable for eliminating this methoxyl group from pectic substances, which eventually marks in the making of methanol and pectic acid. Pectin methylesterase is a carboxylic acid esterase belongs to hydrolase group of enzymes. It is also known as pectin pectylhydrolase, pectinesterase, pectin demethoxylase, pectase, and pectolipase [28].

Pectin methylesterase made by plants acts either at a non-reducing sides or adjacent free carboxyl group and continues along the molecule by a single mechanism, whereas pectin methylesterase extracted by fungal species acts as a multistep system in the elimination of a methyl group in a random action [6]. When water is added, pectin methylesterases the methyl ester link at the α1-4 D galacturonosyl subunit, producing methanol and the negatively charged pectate polymer. Pectate was in the end formed by pectin methylesterase's action in pectic compounds. Pectinase from microorganisms is known for its industrial applications.

3.3 Polygalacturonases

As a member of the pectinase class, polygalacturonase is also known as depolymerase due to its role in the depolymerization process. The pectinolytic enzymes known as polygalacturonases are responsible for the hydrolytic cleavage of the polygalacturonic acid chain in the presence of water. Thus, its role is to breakdown the alpha1-4 glycoside link between galacturonic subunits. There are two distinct categories for polygalacturonases. They are known as exo- and endo-polygalacturonases. Oligogalacturonic acids are released via the hydrolysis of polygalacturonic acids by endopolygalacturonase. Mono-galacturonate is released when pectic acids are hydrolyzed by exo-polygalacturonase [38]. The pectinolytic-depolymerase enzyme known as polygalacturonases hydrolyzes the polygalacturonic acid chain at the $1 \longrightarrow 4\ \alpha$ glycosidic linkage position while inducing a water molecule. The pectinase enzyme family, which includes polygalacturonases, is the one that has been investigated the most because of its biological, technical, and functional roles in food processing industries and the relationship between fungi and plants [39].

4 Substrate for Pectinase Production

Aspergillus fumigatus demonstrated the greatest pectinase activity of 589.0–0.36 U/ml in wheat bran when compared to other organic substrates such as malt sprout, wheat and rice bran, pomegranate, lemon, banana, and orange. Peels from *Solanum tuberosum* showed the maximum polygalacturonase production (437.5 U/ml) when

Bacillus firmus was allowed to ferment the waste material for 92 h at 37 °C [10]. Out of the wastes from cashew, banana, pineapple, and grape, grape waste showed the greatest amount of pectinase activity, with a maximum level of 0.35 U/ml; at 40 °C, pineapple waste came in second (30.0 U/ml). Pectin is reported to be present in significant amounts in oranges, kinnow, khatta, lime, lemon (Galgal), Malta, sweet orange, and other citrus fruits. The following fruits have notable quantities of pectin with a great crystallizing grade: tomato, garden egg, cashew apple, papaya, guava, mango, avocado pear, and papaya [2].

4.1 Choice of Substrate

The selection of substrate for enzyme biosynthesis is processed on the subsequent criteria. First of all, they should be low-cost agricultural and industrial effluent that is readily available. Secondly, in contrast to other substrates, they should be easy to store. Thirdly, they should be able to cope with any adverse effects of various environmental elements, such as temperature fluctuations, variations in the weather from day to night, and changes in the seasons [31].

5 Pectinase Production and Microorganisms

Enzymes from microbes are better than those from plants or animals. These microbes produce pectinase with the utilization of pectin as a carbon source. Numerous microorganisms break down pectins to form a widespread chemicals and enzymes further utilize in various industrial applications. A wide variety of significant microbes are capable of degenerating pectin compounds to produce pectinases.

5.1 Pectinolytic Fungi

Pectinolytic enzymes associated with fungal species help in the breakdown of pectic substances. *Aspergillus, Penicillium, Trichoderma, Mucor*, and *Yarrowia* fungal species are the most frequently used in the synthesis of pectinase. They are important both in submerged and solid-state fermentation that leads to several compounds of commercial importance [20].

5.2 Pectinolytic Yeast

Pectin-containing materials can be degraded by yeast species such as *Saccharomyces fragilis*, *Saccharomyces thermantitonum*, *Torulopsis kefyr*, and *Candida pseudotropicalis* for the production of pectinase [25].

5.3 Pectinolytic Bacteria

Pectin is produced by a wide range of bacterial species, including Micrococcus, *Pseudomonas fluorescens*, Bacillus, and Erwinia species. Following research investigating pectinolytic properties, Bacillus strains were determined to be the most efficient producers of enzymes among various bacterial isolates [19].

6 Industrial Application of Pectinase

Several industrial applications of pectinase contribute to its enormous global demand. It has been employed in diverse conventional industrial processes such as the extraction of oil, tea and coffee, textiles, plant fiber processing, wastewater treatment, etc. Figure 2 provides a summary of various applications.

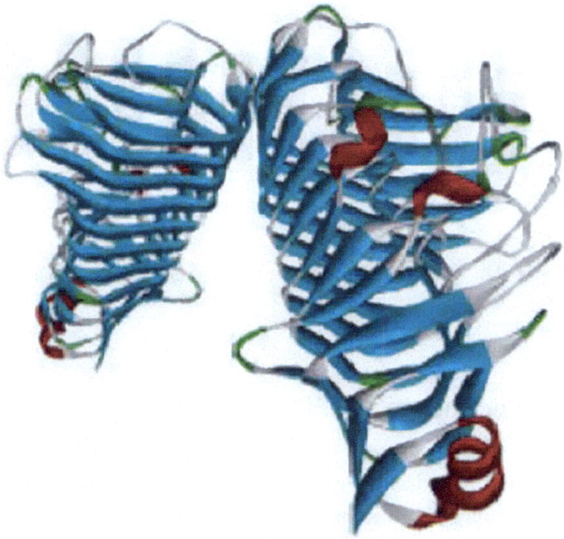

Fig. 1 Three-dimensional structure of pectinase enzyme

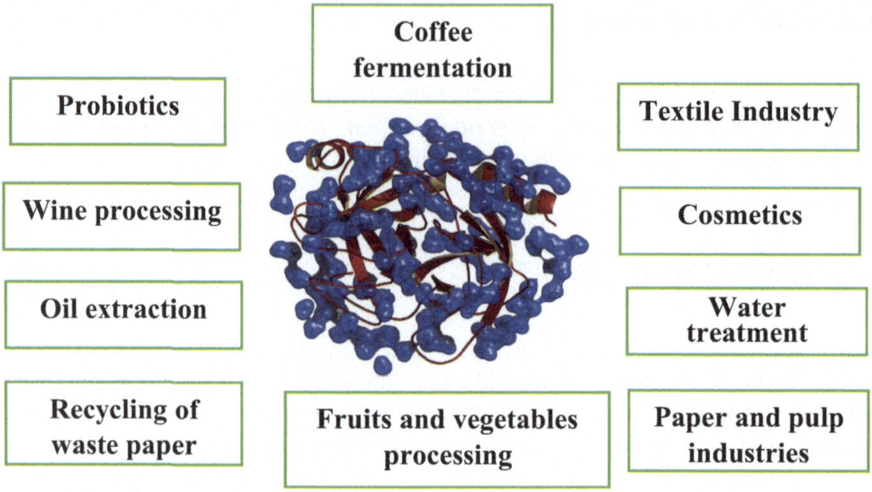

Fig. 2 Summary of Pectinase in different industries [35–37]

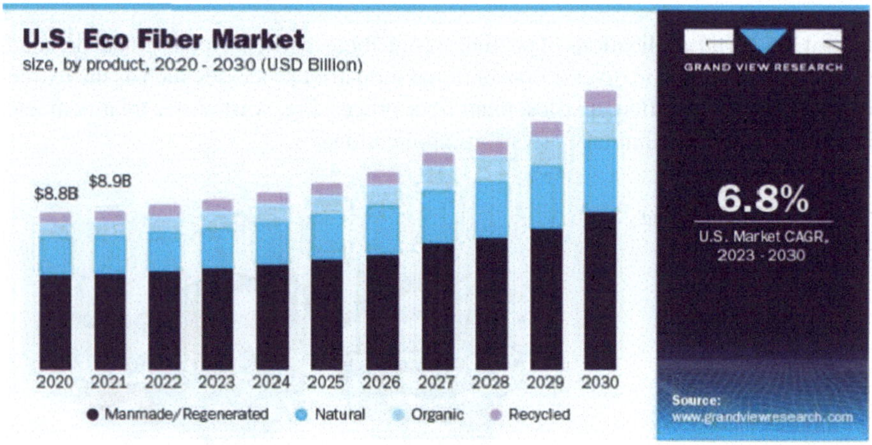

Fig. 3 US Eco-fiber market by product Source: Grand View Research, Market Analysis Report 2021. Available online: https://www.grandviewresearch.com/industry-analysis/eco-fiber-market (accessed on 10 April 2023) [34]

7 Pectinases and Textile Industry

The textile industry uses a variety of chemical agents in its wet operations. These chemicals contaminate effluents after use. A few of them have corrosive properties that could damage the substrate and machinery. An environmentally friendly production process can be ensured by implementing enzymatic processes. The significant waste pollution that traditional textile finishing produces has sparked research on

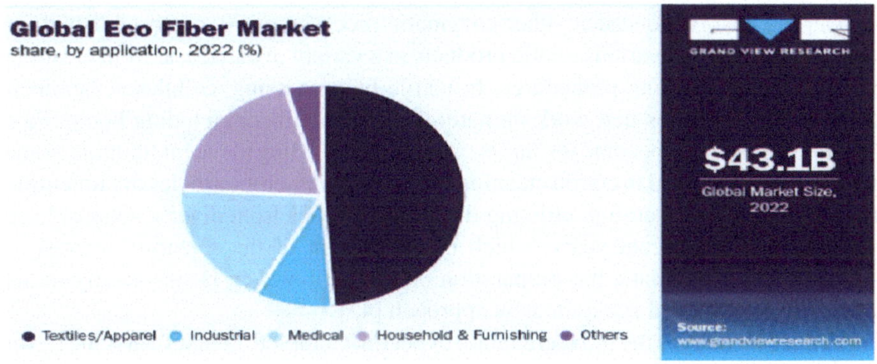

Fig. 4 Global Eco fiber market by application Source: Grand View Research, Market Analysis Report 2021. Available online: https://www.grandviewresearch.com/industry-analysis/eco-fiber-market (accessed on 10 April 2023) [34]

Fig. 5 Economic aspects of pectinase in the textile sector

the application of enzymes in wet textile processes. Amylase's ability to hydrolyze starch provides the basis for one of the oldest technologies still in use today. The benefit of these enzymes is that they are particular to starch and can eradicate it, omitting damage of support fabric [9].

Over the past two decades, other enzymatic processes were also established for the wet processing of various textile products in a variety of activities, from cleaning preparations to finishing procedures. In textile bioprocessing, cellulases, hemicelluloses, and pectinases that work on natural cellulosic fibers including hemp, flax, jute, and cotton, have become the target enzymes. According to recent findings, some enzymes may be useful in cotton cleaning processes. Numerous articles documenting the results of bio-preparation utilizing different enzymes from diverse sources have been published in recent years, which is a reflection of the scientific interest in this technique. Enzymatic bio-preparation of cotton, however, is still mostly in the development stage and is a quite new approach [43].

Pectinases catalyze the biodegradation of pectin. Complete breakdown is the result of a concerted effort by many different enzymes with varying levels of activity. The interactions between these enzymes are synergistic. Pectin esterase is a non-depolymerase in a system of pectin-degrading enzymes. This enzyme decreases the esterification level by catalyzing the breakage of the ester links in polygalacturonans, a reaction that produces free methyl and carboxyl groups. Polygalacturonase catalyzes the breakdown of alpha 1,4 links in pectic polymer chains. Polygalacturonase catalyzes the breakage of alpha 1,4 links in pectic polymer chains, releasing water at the end of the chain, while endopolygalacturonase functions within the chain [24].

8 Bioprocessing of Textile Industry

The use of enzymes by humans first appeared spontaneously among prehistoric groups, not only for the manufacture of food and drinks but also for tanning hides. There are about 4,000 identified enzymes, and 200 are used by different industries. In the processing of textiles, the use of enzymes is considered to be a useful way to replace traditional, harsh chemical products, conserve energy and water, and significantly reduce the impact on the environment and harm to fibers. Nearly all steps of textile manufacture use enzymatic reactions. Desizing, bio-preparation, bio-finishing, and process-time savings are all facilitated by enzymes (Table 1).

Table 1 Summary of processes and their respective enzymes

Enzymes used in textile industry	
Name of process	Type of enzyme used
Desizing	Amylase
Bioscouring	Pectinase
Peroxide killing	Catalase
Denim bleaching	Laccase
Biopolishing	Cellulase
Anti-shrink and Anti-felting on wool	Protease

8.1 Desizing

Desizing is the process of taking the protective layer based on starch off a warp yarn. Chemical agents such as weak oxidizing agents and inorganic acids were used in the past, but because of their non-specificity and poor reaction control, the cellulosic material becomes weaker and gets destroyed [22, 44]. The Bacillus species produces hydrolytic enzyme amylase, which can breakdown the starch into readily removable dextrin and maltose monomers that can be dissolved in hot water or washed off. Since amylase breaks down starch, it does not damage the cellulosic fabric, making it a better option for strength retention than chemical desizing. A pH of 5.5–6.5 and a temperature range of 30–60 °C are excellent for amylase desizing. Water is also saved because after desizing, only one washing is required to remove any leftover chemicals. Together, amylase and neutral cellulase improve stiffness, dye absorption, and wettability. Additional advantages include softer fabric, no adverse effects on equipment, and reduced risk when handling enzymes [5].

8.2 Bioscouring

Cell walls of cellulosic fibers, which are used to make fabric, contain non-cellulosic contaminants such as proteins, pectins, waxes, lipids, minerals, and natural colorants. The aforementioned contaminants cause a filthy, uneven look and poor wettability by reducing the productivity of dyeing and finishing. These non-cellulosic contaminants are removed from cotton's surface through scouring, and sodium hydroxide is a common strong alkali utilized in this procedure [34]. Bioscouring is the process of removing pectin and wax from cotton fabric using microbial enzymes such as cellulases, pectinases, proteases, and cutinases. The wall beneath the cuticle is weakened by the cellulase that *Clostridium thermocellum* and *Trichoderma reesei* produce [40]. *Aspergillus niger* and *Tetracladium sp.* provide pectinase, which facilitates the random hydrolysis of 1,4-D-galactosiduronic links in pectin compounds. This process breaks down and removes waxes, non-cellulosic contaminants, and pectin from cotton. The cotton fibers' cuticle is weakened as a result. The fact that the enzymes used in bioscouring, do not break the cellulose backbone, which greatly minimizes fiber degradation, is another advantage of this method [14]. Compared to conventional scouring, bioscouring offers many benefits, such as being more affordable and safe for the environment, as well as requiring fewer chemicals overall, and posing less of a danger of accidents. Bioscouring has much lower total dissolved solids (TDS), chemical oxygen demand (COD), and biological oxygen demand (BOD) values than alkaline scouring (100%) does. Furthermore, following bioscouring, the finishing is substantially softer than following alkaline scouring. Furthermore, there are no negative consequences of bioscouring on the fabric, the environment, or the health of the workers [1].

The pectinases to be used for bioscouring are selected according to their compatibility with pH and temperature, in addition to the required treatment duration, end-product quality, water absorbency, whiteness, and residual pectin [16].

8.3 *Biobleaching*

To make cotton fibers white and let them be subsequently dyed to suit consumer preferences, the natural color of the flavonoids is removed through bleaching. In traditional bleaching, hydrogen peroxide (H_2O_2), chlorine, and sodium hypochlorite are all utilized. Overuse of these two compounds has the potential to react with fibers and slow down polymerization, which might be quite harmful [33].

In biobleaching, enzymes are used in place of these chemicals, reducing the amount of water needed to remove the hydrogen peroxide and minimizing harm to the fibers. A variety of enzymes, such as laccases, amoglucosidases, glucose oxidases, and pectinases, are used in this procedure to oxidize flavonoids, which enhances the hue [30].

8.4 *Biopolishing*

Polishing is the last step in improving the fabric's quality. The cellulose fibers' propensity to pill and their fuzzy appearance are removed by eliminating the cotton microfibrils. When acidic cellulases from Aspergillus, Penicillium, and Trichoderma sp. are used, the process is referred to as "biopolishing". Biopolishing creates a smoother yarn surface by hydrolyzing pills and breaking off fiber to create a clean, glossy, softer, and cooler [23].

8.5 *Biowashing of Denim*

Denim is a type of cotton that is thick and absorbs dye on its surface. In the past, denim was treated using pumice stones composed of potassium permanganate and sodium hypochlorite. These methods have several shortcomings, including back-staining and fabric deterioration. These issues accompanied with massive extents of pumice pebbles required, led to the usage of cellulases in denim bathing. During bio-stone washing, cellulases release the indigo dye. One benefit is the tiny quantity of enzyme needed. Laccase employed to wash jeans [29].

8.6 Plants Bast Fibers Degumming

Recent usage of pectinolytic enzymes is in the removing of gum from plant fibers, including hemp, flax, jute, sunn hemp, and ramie. The fibers are not harmed by the enzymatic processing, and more significantly, it is not only energy-efficient but also environmentally friendly. Plant fibers are long, narrow, lignified sclerenchymatous cells with strong walls. Because they are dead, their only purpose is to provide the plant body strength and rigidity. Based on their source and structure, plant fibers are characterized into three main groups e.g. cotton (Gossypium sp.); soft or strong woody fibers, which form in groups outside of the xylem in the cortex, phloem, or pericycle; and hard or structural fibers, which are supporting and conductive fibro-vascular bundles, mostly exist in monocotyledonous plants, such as Manila hemps (*Musa textilis*), such as sunn hemp and ramie [8].

The flax plant's stalk yields flax fiber, also known as bast fiber. The fiber's lumen is tiny, and its primary constituents are pectins (25–30%) and cellulose (70–75%). The fibers of flax are often straight, silky, and glossy. However, the resulting yarn has a higher tensile resilience as compared to cotton yarn. Flax is picked out prior decolorizing and tinting. After calendaring, beetling (pumping the fabric with woody hits), stentering, and damping are done to modify the moisture content [27].

Ramie fibers are a great natural textile resource, however, to meet textile standards, degumming is required. Decorticated ramie fibers include 20–30% ramie gum, which is typically composed of pectin and hemicelluloses. The fibers are densely covered in a sticky material that makes the procedure of extracting the fibers from the secondary phloem challenging. The degummed and processed ramie fiber is used to make "grass cloth," or "Chinese linen," a fabric that is used for clothes [47].

Sunn hemp fibers are more resilient than jute and have a higher tensile strength. Fiber strands are typically 1.2–1.5 m in length, glossy with a moderate level of resistance to 413 dampness and microbes. The harvested crop is left for sun drying, causing stem to partially bleach and the leaves to fall off. After the shoots enfeebled from maintained in transparent, stationary, or flowing water for retting, the fibers are pared by fragmenting the bottom margins. After stripping, the fibers are washed and allowed to air dry [35].

Degumming is the process of removing thickly coated, non-cellulosic gummy material from the cellulosic portion of plant fibers. This process is required before the fibers may be used in industrial settings. There are various forms of degumming. In dew retting, plant pectic compounds are attacked by bacterial or fungal enzymes as the shoots are laid out on the land. The procedure's affordability and ease of use are its main advantages; yet, environmental variables like moisture content and temperature affect how much degumming occurs. The primary drawback of aforementioned procedure is likelihood of fiber deterioration due to cellulase enzymes released by the microbiota interacting with the fibers [21].

In container retting, anaerobic microflora that can break down pectic materials is used to degumming fibers in sizable tanks filled with warm water. The thick layer of mucilage, waxes, and pectin that is left on the treated strong woody fibers is chemically detached during the traditional degumming procedure. The degumming of bast fibers is accomplished on a commercial scale by applying diluted lye solutions to the crude fibers. The fibers are soaked for 24 h, then they are heated for 1–4 h, washed, neutralized, cleaned, and centrifuged again and again. To keep fibers from becoming brittle, they are then treated with softeners like glycerin, wax, soap, etc. after being dried over a charcoal fire. The cleaned fibers go through additional processing and grading. In addition to consuming a lot of energy, the chemical processing of fibers is thought to result in hazardous, non-biodegradable, and polluted wastewater as well as major environmental risks and biological disruptions. Here is an immediate requirement to create alternative, eco-friendly techniques for degumming and handing of strong woody fibers due to the shortage of energy supplies, rising energy costs, and environmental degradation beyond bearable levels [16].

The inadequacies in current conventional techniques make it increasingly necessary to effectively harness the biochemical activities of microorganisms for degumming operations. The sticky stuff in plant fibers can be eliminated by using enzymes like pectinases or a mixture of pectinases and xylanases, which break down polysaccharides. Fiber is processed quickly, and degumming is considerably more focused. Pectin makes around forty percent of the dehydrated mass of cambium of plant, hence pectinases are thought to be key players in the digestion of these fibers. Pectinases break down the pectin found in the intermediate lamella and main cell wall, which helps with the degumming, maceration, and retting of jute, flax, hemp, and ramie bast fibers [11, 16].

8.7 Plant Fibers Retting

Jute and flax have been treated (retting) with pectinases to remove pectins and separate the fibers. It has recently been discovered that applying cellulases to corchorus, linen, and cellulosic fiber enhances the instinctive properties of the yarn, improving malleability while maintaining a decent tensile strength. A long-standing technique that dates back to the dawn of civilization is microbial retting. In conventional soaking, hybrid microbial populace, primarily slimy decaying microbes were used. Retted fibers include coconut hulls, corchorus fibers, and linen. Microbial pectinases, or pectin-depolymerizing enzymes, liberate cotton fibers from fasciculi during this process. Although enzymatic retting is quicker, easier to control, and emits less odors than conventional soaking, further research, and development are needed to surpass traditional techniques. Industrial enzymes like pectinases, cellulases, hemicellulases along with polysaccharides consumed in flax at different concentrations in comparison with conventional soaking techniques. Plant phloem-derived woody bast fibers macerate due to pectinase enzyme released by slimy decaying microbes. These long, robust, and typically stiff fibers are utilized in manufacturing rope, carpet, and a variety of fabrics. The surfaces of blended golden fiber and cloth fiber have been softened and smoothed using pectinase along with cellulases and xylanases. Fiber quality is enhanced by the compound treatment of base along with enzymes. Efficiency can be increased and the enzyme dose rate lowered by achieving the ideal balance of enzymes.

Some basic research on the enzymatic retting process has been started in recent years. These make use of refined enzymes on specified substrates and characterize the end products. For flax retting, a pectinase derived from *Rhizomucor pumilis* was utilized. To achieve the highest possible strength in the yarn made from retted linseed, just a minor portion of fiber bundle's pectins must be hydrolyzed [12].

9 Effluent Treatment

The unregulated liberation of pigment-defiled effluents into water reservoirs has resulted in a significant increase in water pollution in recent decades. An additional factor contributing to the rise in water pollution is the expansion of the textile manufacturing and processing sectors. The dyes have a major detrimental influence on aquatic species because of their carcinogenic and mutagenic properties. Because of the extraordinarily increased accumulation of dyes (10–200 mg/L) and secondary toxic wastes such as wetting agents, electrolytes, cleansing agents, oxidizing agents, and lubricants, the textile discharge cannot decompose biologically. For this reason, it is essential to properly handle textiles before releasing them into the environment [45].

The current physiochemical dye decolorization procedures have several limitations, such as labor costs, expensive chemicals, and the creation of sludge as a byproduct that needs to be disposed of safely. As a result, they are not suitable for widespread use in impoverished countries. Because of their many advantages, such as being affordable, environmentally benign, and producing no hazardous byproducts enzymatic technologies are becoming more and more popular. The process of bioremediation uses biomass, including microbes, saprophytes, seaweeds, and infact plants, to remove reagent contaminants from an environment [37].

A variety of artificial pigments found in textile wastes, including reactive dyes like orange-16, 107, 19, rancid fast blue, and procion brilliant blue-hgr, can be neutralized by peroxidases and ligninases derived from bacteria and fungi that cause white rot, such as *Serratia marcescens* and *Phanerochate chrysosporium* [42].

10 Economic Aspects of Eco-Textile

Daily existence relies on textiles, and the clothing sector is an key element of the world financial management. Global textile production and consumption are rising together with the human population. The textile industry employs over 300 million people and is estimated to be worth 1.1 trillion euros [33]. During the period of 2023–2030, with an average annual rate of 7.6% the global marcatus for eco-friendly textiles is anticipated to grasp 77.17 billion USD by 2030. The negative impacts of consuming synthetic fibers are the main cause of the market's growth for eco-friendly textiles. The eco-fiber sector's goods and applications are depicted in the figure. Additionally, the eco-textile industry led the Asia Pacific region in year 2022 having a profit fourteen billion and is estimated to increase by 9.2% of average annual rate. It is anticipated that in the future, major textile economies like China, India, Japan, and Australia will experience a similar problem [34] (Figs. 3, 4).

11 Economic Aspect of Pectinase in Eco-Textile

Using enzymes found in textiles, such as pectinase, instead of harsh chemicals is more environmentally friendly and helps to reduce pollution. The textile sector is eager to adopt this hygienic technique. Using pectinase reduces energy use and water usage as well. Pectinase is non-toxic, therefore after cleaning, the water can be recycled. Energy conservation has a positive economic impact, increasing profits. Protecting workers from deadly and cancer-causing substances eventually results in lower absenteeism, lower medical costs, and increased productivity (Fig. 5).

12 Conclusion

Environmental pollution is an important consideration in the industrial sector including the textile sector throughout the world. As a result of increased awareness of the toxic nature of textile effluents, various aspects are under thought. Enzyme utilization allows the textile industry to develop a new bio-textile sector with eco-friendly fiber with less toxic effluents. Pectinases seem to be a crucial enzyme in many future processes of various industries like textile, fruit, oil extraction, coffee, and tea. Fruit processing and textile industries are two industries that pay close attention to pectinases. The textile sector is thought to be worth 1.1 trillion euros and employs approximately 300 million people. By 2030, it is anticipated that the global market for eco-friendly textiles will grow to 77.17 billion USD. Application of pectinases with other enzymes like amylase, cellulases, and proteases can perform various steps of eco-textile processes with high efficiency and specificity. Furthermore, it works under moderate environment that ultimately eliminate the usage of ruthless hazardous chemical substances, thus contributing to energy saving, reduced water waste loss, improved work and employee safety, and enhanced eco-fiber quality that contribute to progress in the economic sector. The substantial creation or upgrading of enzymes for industrial uses has been deemed necessary for pectinolytic enzymes, as the vast body of research indicates. Experimental investigations should concentrate on modified targeted protein to produce new resilient adaptable pectinolytic enzymes along improved yield systems using novel microbes, after fulfilment of following approach for the utilization of bacterial pectinase. Future studies on pectinolytic enzymes should concentrate on comprehending the molecular processes.that control enzyme release and how different pectinolytic enzymes interact with different pectic substrates.

References

1. Al-Dhabi NA, Esmail GA, Ghilan A-KM, Arasu MV, Duraipandiyan V, Ponmurugan K (2020). Characterization and fermentation optimization of novel thermo stable alkaline protease from Streptomyces sp. Al-Dhabi-82 from the Saudi Arabian environment for eco-friendly and industrial applications. J King Saud Univ-Sci, 32(1), 1258–1264
2. Anuradha K, Padma PN, Venkateshwar S, Reddy G (2014) Selection of nutrients for polygalacturonase production by Aspergillus awamori MTCC 9166 using Plackett-Burman design
3. Arshad M, Adil M, Sikandar A (2014) Hussain T (2014) Exploitation of meat industry by-products for biodiesel production: Pakistan's perspective. Pakistan J Life Soc Sci. 12:120–125
4. Arshad M, Ansari AR, Qadir R, Tahir MH, Nadeem A, Mehmood T, Alhumade H, Khan N (2022) Green electricity generation from biogas of cattle manure: an assessment of potential and feasibility in Pakistan. Frontiers in Energy Research 25(10):911485
5. Aydemir D, Gecili F, Özdemir N, Ulusu NN (2020) Synthesis and characterization of a triple enzyme-inorganic hybrid nanoflower (TrpE@ ihNF) as a combination of three pancreatic digestive enzymes amylase, protease and lipase. J Biosci Bioeng 129(6):679–686
6. Bano I, Ghauri MA, Arshad M, Yasin T, Younus M (2017) Bioactivity of variant molecular weight chitosan against drug-resistant bacteria isolated from human wounds. Microbial Drug Resistance 1;23(8):958–65.
7. Bano I, Arshad M (2018) Climatic changes impact on water availability. Perspect Water Usage Biofuels Prod: Aquat Contam Clim Chang 2018:39–54
8. Brühlmann F, Leupin M, Erismann KH, Fiechter A (2000) Enzymatic degumming of ramie bast fibers. J Biotechnol 76(1):43–50
9. Cavaco-Paulo A, Gubitz G (2003) Textile processing with enzymes (Vol. 29). Woodhead publishing
10. Deepak V, Kalishwaralal K, Ramkumarpandian S, Babu SV, Senthilkumar S, Sangiliyandi G (2008) Optimization of media composition for Nattokinase production by Bacillus subtilis using response surface methodology. Biores Technol 99(17):8170–8174
11. Dijkstra AJ (2010) Enzymatic degumming. Eur J Lipid Sci Technol 112(11):1178–1189
12. Foulk J, Akin D, Dodd R (2008) Influence of pectinolytic enzymes on retting effectiveness and resultant fiber properties. BioResources 3(1):155–169
13. Geetha M, Saranraj P, Mahalakshmi S, Reetha D (2012) Screening of pectinase producing bacteria and fungi for its pectinolytic activity using fruit wastes. Int J Biochem Biotech Sci 1:30–42
14. Gururaj P, Khushbu S, Monisha B, Selvakumar N, Chakravarthy M, Gautam P, Nandhini Devi G (2021) Production, purification and application of Cutinase in enzymatic scouring of cotton fabric isolated from Acinetobacter baumannii AU10. Prep Biochem Biotechnol 51(6):550–561
15. Hassan B, Ali S (2016) A review on biotechnological impact of pectinases in industries. J Pharm, Chem Biol Sci 1:16
16. Hoondal G, Tiwari R, Tewari R, Dahiya N, Beg Q (2002) Microbial alkaline pectinases and their industrial applications: a review. Appl Microbiol Biotechnol 59:409–418
17. Jayani RS, Saxena S, Gupta R (2005) Microbial pectinolytic enzymes: a review. Process Biochem 40(9):2931–2944
18. Kashyap D, Vohra P, Chopra S, Tewari R (2001) Applications of pectinases in the commercial sector: a review. Biores Technol 77(3):215–227

19. Kavuthodi B, Sebastian D (2018) Review on bacterial production of alkaline pectinase with special emphasis on Bacillus species. Biosci. Biotechnol. Res. Commun 11:18–30
20. Kumari B, Lalitha R, Sudhakar P (2013) Studies on isolation, purification and molecular identification of pectinase producing bacteria. Int J Adv Res 1(5):204–212
21. Lyu P, Zhang Y, Wang X, Hurren C (2021) Degumming methods for bast fibers—A mini review. Ind Crops Prod 174:114158
22. Madhu A, Chakraborty JN (2017) Developments in application of enzymes for textile processing. J Clean Prod 145:114–133
23. Mojsov K (2011) Application of enzymes in the textile industry: a review
24. Mojsov K (2012) Biotechnological applications of pectinases in textile processing and bioscouring of cotton fibers
25. Naumov G, Shalamitskiy MY, Naumova E (2016) New family of pectinase genes PGU1 b–PGU3b of the pectinolytic yeast Saccharomyces bayanus var. uvarum. Dokl Biochem Biophys
26. Oumer OJ, Abate D (2018) Screening and molecular identification of pectinase producing microbes from coffee pulp. BioMed research international
27. Pandey R, Jose S, Basu G, Sinha MK (2021) Novel methods of degumming and bleaching of indian flax variety tiara. Journal of Natural Fibers 18(8):1140–1150
28. Patidar MK, Nighojkar S, Kumar A, Nighojkar A (2018) Pectinolytic enzymes-solid state fermentation, assay methods and applications in fruit juice industries: a review. 3 Biotech, 8(4), 199
29. Pazarlıoğlu NK, Sariişik M, Telefoncu A (2005) Laccase: production by Trametes versicolor and application to denim washing. Process Biochem 40(5):1673–1678
30. Pereira L, Bastos C, Tzanov T, Cavaco-Paulo A, Gübitz GM (2005) Environmentally friendly bleaching of cotton using laccases. Environ Chem Lett 3:66–69
31. Phutela U, Dhuna V, Sandhu S, Chadha B (2005) Pectinase and polygalacturonase production by a thermophilic Aspergillus fumigatus isolated from decomposting orange peels. Braz J Microbiol 36:63–69
32. Pires-Cabral P, Da Fonseca M, Ferreira-Dias S (2010) Esterification activity and operational stability of Candida rugosa lipase immobilized in polyurethane foams in the production of ethyl butyrate. Biochem Eng J 48(2):246–252
33. Rahman M, Hack-Polay D, Billah MM, Un Nabi MN (2020) Bio-based textile processing through the application of enzymes for environmental sustainability. Int J Technol Manag Sustain Dev 19(1):87–106
34. Rajulapati V, Dhillon A, kumar Gali, K., Katiyar, V., & Goyal, A. (2020) Green bioprocess of degumming of jute fibers and bioscouring of cotton fabric by recombinant pectin methylesterase and pectate lyases from Clostridium thermocellum. Process Biochem 92:93–104
35. Rani, K., Ahirwar, M., & Behera, B. (2020). Comparative analysis of alkaline and enzymatic degumming process of hemp fibers. J Inst Eng (India): Ser E, 101, 1–10
36. Ribeiro DS, Henrique SM, Oliveira LS, Macedo GA, Fleuri LF (2010) Enzymes in juice processing: a review. Int J Food Sci Technol 45(4):635–641
37. Sen A, Kapila R, Chaudhary S, Nigam A (2021) Biotechnological applications of microbial enzymes to replace chemicals in the textile industry-a review. Textile Association 82(2):68–73
38. Sharif N, Munir N, Hasnain M, Naz S, Arshad M (2021) Environmental impacts of ethanol production system. Sustain Ethanol Clim Chang: Sustain Assess Ethanol Distill. 2021:205–223
39. Sharma N, Rathore M, Sharma M (2013) Microbial pectinase: sources, characterization and applications. Rev Environ Sci Bio/Technol 12(1):45–60
40. Singh A, Varghese LM, Battan B, Patra AK, Mandhan RP, Mahajan R (2020) Eco-friendly scouring of ramie fibers using crude xylano-pectinolytic enzymes for textile purpose. Environ Sci Pollut Res 27:6701–6710
41. Tabssum F, Ali SS (2018) Screening of pectinase producing gram positive bacteria: isolation and characterization. Punjab Univ J Zool 33(1):11–15
42. Verma P, Madamwar D (2003) Decolourization of synthetic dyes by a newly isolated strain of Serratia marcescens. World J Microbiol Biotechnol 19:615–618

43. Vigneswaran C, Ananthasubramanian M, Kandhavadivu P (2014) Bioprocessing of textiles. Elsevier
44. Wu XT, Ansari AR, Pang XX, Li HZ, Zhang ZW, Luo Y, Arshad M, Song H (2019) Visfatin plays a significant role in alleviating lipopolysaccharide-induced apoptosis and autophagy through PI3K/AKT signaling pathway during acute lung injury in mice. Arch Immunol Ther Exp 1(67):249–261
45. Yuan H, Chen L, Cao Z, Hong FF (2020) Enhanced decolourization efficiency of textile dye Reactive Blue 19 in a horizontal rotating reactor using strips of BNC-immobilized laccase: Optimization of conditions and comparison of decolourization efficiency. Biochem Eng J 156:107501
46. Yücel Y, Demir C, Dizge N, Keskinler B (2011) Lipase immobilization and production of fatty acid methyl esters from canola oil using immobilized lipase. Biomass Bioenerg 35(4):1496–1501
47. Zheng L, Du Y, Zhang J (2001) Degumming of ramie fibers by alkalophilic bacteria and their polysaccharide-degrading enzymes. Biores Technol 78(1):89–94

Role of Laccases to Achieve Net Zero Carbon Emissions

Sadia Batool, Waqas Haider, Mazhar Abbas, Iram Liaqat, Aqsa Mumtaz, Iram Saba, Muhammad Farooq, Sher Zaman Safi, Syed Shatir A. Syed-Hassan, and Muhammad Arshad

Abstract Biotechnology has provided a number of promising new avenues for the urgently needed fight against climate change and carbon emissions. Many are curious in laccases because of their potential role in achieving net zero carbon emissions; these enzymes are versatile and have unique catalytic properties. As a kind of multicopper oxidase, laccases oxidise a wide range of organic substrates with remarkable efficiency. Their versatility in catalysing the oxidation of aromatic pollutants and phenolic compounds makes them valuable in several industrial processes, such as bioremediation, biofuel production, and wastewater treatment. By facilitating the breakdown of recalcitrant pollutants and lignin-derived materials, laccases offer a sustainable alternative to conventional, energy-intensive methods, thus reducing overall carbon footprints. Moreover, the integration of laccase-based systems in bioconversion processes holds immense promise for achieving net zero carbon emissions. Through synergistic partnerships with microbial consortia or genetically engineered organisms, laccases can enhance the efficiency of biomass degradation and

S. Batool · W. Haider · M. Abbas · A. Mumtaz · M. Arshad (✉)
Biochemistry Section, Jhang-campus, University of Veterinary and Animal Sciences Lahore, Lahore, Pakistan
e-mail: muhammad.arshad@uvas.edu.pk

I. Liaqat
Microbiology Lab, Department of Zoology, Government College University Lahore, Lahore, Pakistan

I. Saba
Department of Chemistry, Government College, Woman University, Sialkot, Pakistan

M. Farooq
Department of Clinical Sciences, Jhang-campus, University of Veterinary and Animal Sciences Lahore, Lahore, Pakistan

S. Z. Safi
Faculty of Medicine, Bioscience and Nursing, MAHSA University, Petaling Jaya, Selangor, Malaysia

S. S. A. Syed-Hassan
School of Chemical Engineering, College of Engineering, Universiti Teknologi MARA, 40450 Shah Alam, Malaysia

M. Arshad (ed.), *Enzymes in Textile Processing: A Climate Changes Mitigation Approach*, SDGs and Textiles, https://doi.org/10.1007/978-981-97-8058-7_8

biofuel synthesis, thereby displacing fossil fuel dependency and mitigating greenhouse gas emissions. Furthermore, laccase-mediated processes enable the valorization of lignocellulosic feedstocks, transforming waste streams into valuable bioproducts and bioenergy sources. In the context of sustainable agriculture and forestry practices, laccases play a pivotal role in promoting carbon sequestration and soil health. By facilitating the decomposition of lignin-rich residues and promoting humus formation, laccases contribute to the long-term storage of carbon in terrestrial ecosystems, mitigating the impacts of deforestation and land degradation. In addition, the use of laccase-mediated bioremediation techniques provides cost-efficient methods to restore polluted areas and reduce environmental pollution. This helps to improve the ability of ecosystems to recover and supports efforts to conserve biodiversity. Therefore, adopting laccase-based technology is a fundamental aspect of the worldwide shift towards a sustainable and low-carbon civilization.

Graphical Abstract

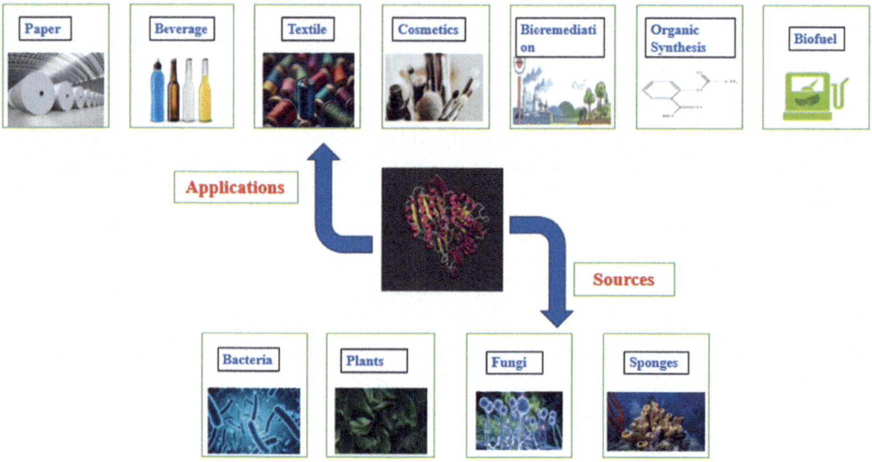

Keywords Laccases · biochemistry · Classification · Zero carbon emissions · Circular economy · Applications

1 Introduction

The intricate protein molecules created by living cells that are frequently referred to as biocatalysts are called enzymes. Both the reactions they catalyse and the reactants are extremely selective. Usually, an enzyme catalyses one chemical reaction, or a group of related reactions. Compared to uncatalysed reactions, side reactions that lead to the inefficient production of byproducts are uncommon in enzyme-catalysed reactions. Enzymes can alternatively be described as organic, liquid, and colloidal catalysts that are made by live cells yet have the ability to function separately from the cells [4,

17]. Cyanobacteria, which began releasing photosynthetic oxygen into the Earth's atmosphere 2.45 billion years ago, are responsible for the elevated oxygen levels in the biosphere. The FeII (ferrous ion) undergoes a gradual transformation into FeIII (ferric ion) due to its solubility in water. FeIII has limited solubility in water and is minimally bioavailable. Aerobic bacteria have evolved complex mechanisms for acquiring iron (siderophores) and regulating its use in response to the presence of oxygen in the environment. This is because the availability of oxygen in the air has exerted selective pressure on the evolution of bacteria to find metals with high redox potentials, similar to iron, that are readily accessible for their biological processes. As a result, copper (CuII/CuI) and manganese (MnIII/MnII) became important, particularly in aerobic organisms, where they performed a function that was similar to, but distinct from, that of iron (FeIII/FeII). Copper serves various biological functions, including its interaction with proteins to facilitate the formation of specific structures and enable metabolic processes [3].

Proteins that contain copper, nearly all of which are extracellular, are found in nature in large quantities and are involved in redox activities such as electron(s) transfer, oxygen transport, and activation. This class of multicopper oxidases uses oxygen as an electron acceptor to oxidize a variety of substrates. Reactive oxygen species (ROS), which are dangerously partly reduced O_2 molecules, are not released when O_2 is reduced to H_2O by multicopper oxidases. MCOs comprise laccases a broad family of copper oxidases include Fet3p, bilirubin oxidase, ascorbate oxidases, ceruloplasmin, and phenoxazinone synthase. All multicopper oxidase amino acid sequences are contained in a tiny, 10–20 kDa, cupredoxin-like domain and have very simple three-dimensional structures that are mainly made up of beta sheets and twists. They are mostly involved in electron transport [23, 90]. Recently, the structure of laccases has been examined many times by using crystallography. But the majority of the proteins that have been characterized are bacterial or fungal in origin. It is possible that the purification procedures for plant and animal laccases are so intricate and their small quantities contribute to their difficulty in crystallization. However, it has been demonstrated that despite their diverse range of substrates and extensive taxonomic distribution, laccases' molecular architecture is common for multicopper oxidase [38].

Unlike lignolytic peroxidases, which emerged in the evolutionary timeline shortly before fungi acquired the capacity to break down lignin [36], laccases are most likely among the first metalloproteomes because they were initially seen in organisms during the early phases of evolution. Consequently, the lignolytic function of these most commonly researched enzymes in the MCOs group is more likely the consequence of substrate specialization and their secondary role [74, 90]. MCOs can be classified into three categories based on the presence of the cupredoxin domain: two-domain, three-domain, and six-domain enzymes. Most laccases consist of three domains, each comprising 500 amino acids. Recently discovered bacterial laccases consist of only two domains, comprising 200 amino acids. The human LACC1 gene shows minimal divergence from mammalian proteins, as its C-terminal region shares homology with both laccase and bacterial multicopper polyphenol oxidase [11]. Although fungal and bacterial laccases have different amino acid sequences, their molecular structures are similar, and their active site shape is well conserved overall [73]. Additionally, it is noteworthy that both bacteria and animals possess two and three domains of these enzymes. Laccases from the Ascomycota phylum exhibit higher similarity to laccases found in plants and bacteria, which share a closer evolutionary relationship with insects than fungal enzymes. The laccases discovered in Monoraphidium, a type of green algae, and Cryptococcus, pathogens that impact both animals and plants, are categorized distinctively from the enzymes present in fungi and plants. It is hypothesized that this enzyme originated from a single ancestral protein and subsequently underwent differentiation, leading to the emergence of distinct laccases with diverse structures and functions in various organisms.

2 Biochemistry and Molecular Nature of Laccase

Laccases are basically oxidoreductases in nature which catalyse the monoelectronic oxidases from their substrates. Although various laccases from different species have different structure but commonly laccases have four centers in its active sites and classified as type I, type II and type III. Central core of enzyme binds to oxygen and reduces into water molecule. Other than type I copper center all of them binds oxygen and catalyse the reaction. In type I, single copper bonded to two histidine and one single cysteine amino acids. As earlier discussed, the variability also presents relevant to origin of enzyme. Type III laccases contain two central copper that bounded to three histidine residues via hydroxide. Various categories of anions inhibit the activity of laccase enzyme like cyanide remove all of copper from enzyme active site [2].

Laccase is a glycosylated monomer or homodimer protein that is often found in bacteria and fungi with fewer carbohydrates (10–25%) than in plant enzymes. The substance made of carbohydrates includes monosaccharides like glucose, hexoamines, fucose, arabinose, mannose and galactose. Most laccases on SDS-PAGE have mobilities that match molecular weights between 60 and 100 kDa, of which glycosylation may be responsible for 10–50% of the mobilities. One of the main parts of the carbohydrate that is linked to laccase is mannose. In laccase, glycosylation governs secretion, activity, sensitivity to proteolysis, retention of copper, and thermal stability [72]. Although intracellular laccases have been found in a number of fungi and insects, the majority of laccases that have been investigated are extracellular proteins. Plant laccases typically have a wide range of isoelectric points (pI) starting from 9, whereas fungal laccases generally have pI values ranging from 3 to 7. The enzymes differ primarily in their optimal pH ranges. The optimal pH range for fungal enzymes is 3.6–5.2, whereas *Rhus vernicifera* laccase has an optimal pH range of 6.8–7.4. The low pH optima of fungal enzymes may be attributed to their adaptation to acidic environments, whereas the intracellular pH optima of plant laccase are closer to the physiological range. Hence, the variations in physiological activity may account for the disparities in pH optima. Furthermore, these enzymes exhibit unique functionalities in addition to their pH variations. Plant enzymes are involved in the synthesis of lignin, while fungal enzymes help in the removal of harmful phenols from the fungi's natural environment [16, 76]. Purified laccases exhibit an electronic absorption peak at approximately 600 nm, resulting in a characteristic blue colour. When laccase is inactive, its UV–visible spectra typically exhibit a single peak at approximately 330 nm and two peaks at around 280 and 600 nm. The absorbance at 280 and 600 nm typically ranges from 14 to 30. Moreover, the absorbance ratio at 330 to 600 nm typically falls within the range of 0.5–2. Most laccases have four copper atoms per monomer in the holoenzyme, but the Phlebia laccase has two copper atoms and one pyrroloquinoline quinone atom as its prosthetic group. The copper atoms can be classified into three distinct groups using electron paramagnetic resonance (EPR) and UV/visible spectroscopy [46].

Laccase selectively acts on the phenolic components of lignin, resulting in the cleavage of aryl–alkyl bonds, CαCβ bonds, and oxidation of the Cα position. Laccase catalysis encompasses the following processes:

- The process involves the reduction of type 1 copper by a reducing substrate.
- Internal electron transfers from types 1–3 copper.
- Oxygen is reduced to water at the types 2 and 3 copper site.

The oxidation of a reducing substrate by laccase usually leads to the loss of one electron, resulting in the formation of a free cation radical. The radical is frequently unstable and can undergo no enzymatic reactions, including hydration, disproportionation, or polymerization. Alternatively, further oxidation of phenol may occur through laccase catalysis, leading to the formation of quinone. The regulation of type 1 copper electron transport is primarily influenced by changes in redox potential. A faster rate of substrate oxidation can be achieved by either reducing the substrate oxidation potential or increasing the laccase redox potential [29, 69]. According to laccases, similar to pharmacologists, have a predilection for polymerizing lignin by connecting the phenoxy radicals generated from the oxidation of lignin's phenolic groups. This characteristic renders them similar to other phenol-oxidizing enzymes. By introducing a mediator like ABTS, laccase may expand its spectrum of substrates to include non-phenolic constituents of lignin (Fig. 1).

(1)

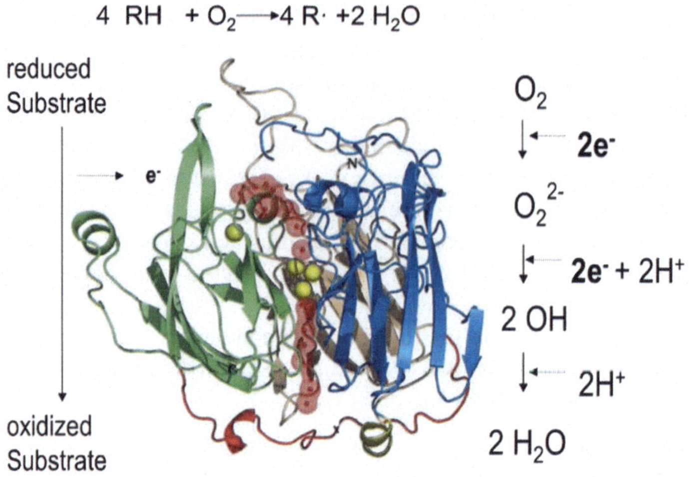

Fig. 1 Laccases as multicopper oxidases

3 Classification of Laccases

On behalf of catalytic activity, laccases have been divided into three different classes.

- Catechol oxidase or o-diphenol: oxygen oxidoreductase
- Laccase or p-diphenol: oxygen oxidoreductase
- Cresolase or monophenol monooxygenase

Hence, the capacity of a substrate to selectively engage with specific enzymes can be employed for their differentiation. Defining laccase based solely on substrate specificity is challenging because laccase and tyrosinase have overlapping substrate ranges. Catechol oxidases, also known as tyrosinases, exhibit O-diphenol and cresolase activity by oxidizing L-tyrosine. Laccases exhibit enzymatic activity towards ortho and para-diphenols, displaying a preference for the para-diphenol group. Laccases selectively oxidise syringaldazine, whereas tyrosinases exhibit specific cresolase activity. This enzyme demonstrates dual functionality, acting as both a tyrosinase and a laccase. Moreover, laccases display a wide range of substrate specificity and demonstrate a significant lack of specificity towards the initial substrate [88]. The variation in substrate ranges among different species presents an additional challenge in defining laccase based on substrate. Neurospora crassa laccase selectively oxidises phloroglucinol, the sole para- and ortho-diphenol substrate. Pyricularia oryzae laccase exhibits a preference for phloroglucinol as a substrate over other substituted monophenols. Partially oxidized meta-substituted phenols are observed when laccases from *Trametes versicolor* and *Cerrena unicolor* are used. The oxidation of phenols with substituents in the para position is facilitated by laccase from *Cerrena unicolor*, while the oxidation of phenols with substituents in the ortho position is facilitated by laccase from *Trametes versicolor* [14, 37].

4 The Multifunctional Catalysts: Laccase

Laccases possess two physiologically relevant characteristics: a broad spectrum of substrates and the utilization of molecular oxygen as the final electron acceptor. In the process of laccase-catalysed oxidation, the first electron acceptor is copper T1, which is located in the cavity close to the surface of the enzyme. The step that limits the pace of the processes catalysed by laccase is the reduction of copper T1. The range of substrates that laccase can act with is limited to molecules that contain phenolic moieties because of the relatively low values of the T1 redox potential. Laccase catalyses the oxidation of phenolic compounds, resulting in the formation of phenoxy radicals. These radicals may then undergo either radical rearrangement or polymerization via coupling reactions. Redox reversibility occurs when a particular substrate undergoes oxidation, however this is dependent on the stability of phenoxyl radicals [32]. Radical-based coupling or redox recycling of phenolic substrates expands the variety of laccase substrates, since these substrates also serve as mediators. Laccases are categorized as either low- or high-redox-potential enzymes, depending on their redox potential. High-redox-potential laccases are mostly present in fungus, whereas low-redox-potential enzymes may be found in bacteria, plants, and insects [59]. Laccases catalyse both anabolic and catabolic biochemical reactions. Fungal laccases exhibit catabolic processes, such as the breakdown of lignin and humus. The transformation of soil organic matter occurs during morphogenesis through anabolic reactions facilitated by laccases. The reactions encompass various processes, including the production of polymeric colours, the clerotization of cuticles, the synthesis of polyavonoid compounds, lignification, and humidification. Laccases, found in plants, bacteria, and insects, facilitate thermodynamically favourable radical coupling reactions in anabolic processes, eliminating the requirement for supplementary chemicals [55].

These reactions often include the utilization of low molecular weight phenolic compounds as substrates, leading to the formation of various polymeric products such as pigments, lignins, polyavonoids, and humus. Purified laccases from plant tissues are used as catalysts in the radical-based polymerization of lignin and polyavonoid synthesis. This process is seen in Eucalyptus lignin or during the production of seed coats in *Arabidopsis thaliana.* The substrates employed by these laccases include monoolignols and flavonoids. *Populus trichocarpa*, a kind of plant, has oligomeric products that are covalently bound in a random manner inside its cell walls. It has been revealed that laccases connected to the insect cuticle matrix exhibit a comparable radical-based coupling. Protein cross-linking processes mediated by laccase were shown to cause cuticle sclerotization in *Manduca sexta* and *Tribolium castaneum.* Catechol, N-acetyldopamine, and N-β-alanyldopamine are often found substances in cuticles that undergo the process of clerotization. Laccase catalyses the conversion of these substrates into quinones, which are subsequently involved in the radical coupling of the histidine residues present in the proteins found in the cuticles [41]. In order to achieve cuticle coloration, hardness, water resistance, and defence against environmental stressors, it is necessary for these low molecular weight substrates to undergo cross-linking with the protein-based matrix. Melanins, the main pigments synthesized by laccase, are derived from both external substrates such as homogentisic acid and dihydroxyphenylalanine, and endogenous substrates like dihydroxynaphthalene. *Aspergillus fumigatus* and *C. neoformans* are two bacteria that catalyse the polymerization of precursors to produce melanin pigment [15, 83]. Free radicals are generated by laccases during catabolic processes, and they oxidise target substances such as lignin and humus. Fungal laccases commonly produce the initial class of radicals from naturally existing methoxyhydroquinones. The radicals initiate the Fenton reaction, resulting in the generation of multiple reactive oxygen species (ROS). The second class of radicals is generated through the oxidation of naturally occurring low molecular weight redox mediators, often derived from oxidized target substrates. Phenolic lignin units release small phenolic residues with oxidized side chains during preferential oxidation. In addition to the phenolic hydroxyl group, these substances may also contain other functional residues such as methoxyl, amine, ketone, aldehyde, or carboxyl groups. Mediator-based oxidations primarily occur in basidiomycetes, which are efficient decomposers of lignocellulose [13]. Laccases modify the surface of natural lignin, without depolymerizing the lignin itself. Regrettably, the precise mechanisms of attack and the enzymatic characteristics of laccases in relation to lignin alteration remain uncertain. This shortage of analytical methodologies is the reason. Laccase has the ability to break the Cα–Cβ bonds and the Cα–aryl bonds in model phenolic compounds of lignin. This process occurs without a mediator. The hypothesis suggests that laccase can only break the bonds between non-phenolic subunits of lignin when a mediator is present [71]. Small molecular weight compounds such as vanillin, acetosyringone, p-coumaric acid, p-cinnamic acid, sinapic acid, syringaldehyde, 4-hydroxybezylic alcohol, and 3-hydroxyantharnillic acid serve as laccase redox mediators. These compounds are derived from natural lignin degradation, plant phenolic secondary metabolites, or extracellular fungal metabolites. Fungal laccase catalyses the oxidation of phenolic compounds to phenoxyl radicals. These radicals can then oxidise non-phenolic lignin residues through the hydrogen-abstraction mechanism [26]. Phenolic chemicals with similar properties have been detected in bacterial cultures capable of lignin degradation. Cultures of *Aneurinibacillus aneurinilyticus, Pseudonomas putida*, Bacillus sp., Streptomyces sp., and Paucimobilis sp. all exhibit benzaldehyde with hydroxyl, methoxy, or trimethoxy substitutions. Bacillus sp. and *P. putida* cultures have been found to contain cinnamic acid with hydroxy and methoxy substitutions. The presence of natural laccase mediators during in vivo wood decay has not been proven. However, certain phenolic chemicals that are associated with lignin breakdown have the potential to act as mediators when used in vitro. When studying the solitary activity of laccase, it is difficult to eliminate the potential interference from natural mediators. These mediators are phenolic chemicals that occur naturally and are derived from a lignin polymer [59].

5 The Features and Biological Roles of Polyphenol Oxidase

Enzymes that have comparable catalytic activities and molecular structures resembling laccases are classified under the MCO family. Polyphenol oxidases (PPOs) are a kind of enzymes that use molecular oxygen as a final electron acceptor to facilitate the oxidation of phenols into quinones. Plant phenolic compounds, often known as PPOs, are essential oxidases in biological systems. They have several functions including defence mechanisms, metabolic activities, polymerization, and detoxification. During the process of lignification in higher plants, a critical step involves the formation of chemical bonds between phenolic precursors. Tyrosinases, catechol oxidases, and laccases are categorized as three separate groups of PPOs, distinguished by their substrate specificities and reaction mechanisms [9, 81]. Tyrosinases, which are a kind of monophenol monooxygenase, facilitate the process of converting diphenols into their respective o-quinones. Additionally, they are responsible for the hydroxylation and oxidation of monophenols, such as tyrosine, p-cresol, and p-coumaric acid, into o-diphenols. Catechol oxidases, also known as o-diphenol oxidases, specifically catalyse the oxidation of o-diphenols to o-quinones, which is their exclusive reaction. They are classified as 1,2-benzenediol: oxygen oxidoreductases. Each PPO enzyme does not possess the capacity to hydroxylate monophenol and has a substrate range that overlaps with other enzymes. The tyrosinases and cresolase enzymes, which are identical, may be differentiated based on their respective places of origin. Tyrosinases, also known as PPOs, are enzymes present in bacteria, fungi, and humans. However, in plants, they are referred to as cresolases. Mammalian enzymes are often located inside specialized melanocytes, whereas plant PPOs are mostly found in the thylakoid membranes of chloroplasts. At times, polyphenol oxidases are discharged outside of cells [22]. PPOs oxidize flavonoids, which control every element of plant life, and other phenolic or polyphenolic chemicals. These proteins have a major role in plant defense against biotic and abiotic stressors since they are mostly in charge of enzymatic browning responses [57]. Melanin is produced as a protective reaction to tissue injury in order to defend against microbial infections. Melanin polymers may be used to reduce the capacity of insects to absorb plant proteins or to create a hostile environment for invaders. They can be involved in the formation of protective barriers or in the modification of proteins caused by PPOs [85]. Plants not only possess wound-healing abilities, but also employ defence mechanisms against pathogens and herbivorous insects through the oxidation of phenols to quinones. Quinones hinder herbivores' protein digestion by forming covalent bonds with leaf proteins. The synthesis of melanin improves the palatability and digestibility of plant tissues, while also enhancing the cell wall's resistance to disease and insect attack. Other enzymes, including quinone isomerase, quinone oxidase, and quinone methide isomerase, play a role in facilitating cross-linking interactions between quinonoid-reactive intermediates and cuticular components during the sclerotization of insect cuticle in the wound healing process. The process by which capsules enclose parasites and parasitoids is similar [80]. Quinones and other highly reactive substances, such 5,6-dihydroxyindole, may immediately eliminate microbial pathogens and parasitoids. They are also more dangerous to herbivores compared to the original phenolic substrates. Multiple peroxidase-oxidases (PPOs) work together and carry out a range of physiological functions in plants that are both overlapping and varied. These functions include redox metabolism, wound healing responses, defence against pathogens or insects, production of lignin and suberin, and cross-linking of cell wall components [30]. PPOs are diverse polyphenolic polymers that are widely found in all living organisms and serve as the main enzymes in melanogenesis. Melanin is the chromatic substance responsible for determining the skin, eye, and hair colour of animals, as well as providing crucial protection against ultraviolet radiation. Melanin is present in fish, amphibians, and reptiles. Melanin has a role in the process of wound healing, the activation of defence systems, and the hardening of the outer layer of the cuticle in insects. The innate immunity and cellular hemostasis of insects are linked with melanogenesis. Tyrosinase and laccase enzymes are involved in the production of melanin pigments in bacterial and fungal cells. However, melanogenesis is only seen at certain phases of fungal cell growth, the formation of fruiting bodies, and wound healing [78]. Fungal melanin is ubiquitously present and is located in the cell wall rather than in specialized subcellular organelles such as animal melanosomes. Typically, melanin granules are formed on the cell wall's surface, where they are likely cross-linked with polysaccharides. This occurs when melanin precursors are released and oxidized outside the cell. Melanin contributes to the coloration of fungal cell walls and their capacity to resist hydrolytic enzymes. The same mechanism that operates in bacterial cells and spores likely also serves as a defence against harmful heavy metals and provides protection against environmental stresses such as harmful UV radiation and reactive oxygen species (ROS). It has been suggested that in soil conditions, plant phenolic compounds undergo polymerization and detoxification via the participation of extracellular polyphenol oxidases in bacteria [19]. Fungal laccases are believed to have a catabolic function in the degradation of natural polymers such as lignin, potentially working alongside other lignolytic enzymes. The enzymes can be categorized into two groups: lignin-degrading auxiliary enzymes and heme peroxidases (PODs). Heme peroxidases encompass dye-decolorizing peroxidase, manganese peroxidase, lignin peroxidase, and versatile peroxidase. Aryl-alcohol oxidase, glyoxal oxidase, pyranose 2-oxidase, glucose dehydrogenase, cellobiose dehydrogenase, and heme-thiolate halo peroxidases are accessory enzymes associated with lignin degradation. The successful linkage of lignin and polysaccharide anabolism is facilitated by the conversion of oxygen to H_2O_2, a requirement for peroxidases [36]. Cellobiose dehydrogenases enhance this association by utilizing electrons derived from the oxidation of cellobiose to cellobionolactone. This process concurrently augments the supply of redox mediators for laccases, enabling the reduction of phenoxy radicals, cations of transition metals (such as FeIII, CuII, and MnIII), or quinones [77].

6 Series of Laccase Genes

Studies on the characterization of fungal laccases concur with the multiplexity of genes and the production of many isoenzymes in various species. Laccase gene families include those found in *Trametes villosa* (codified by the lcc1 and lcc2 genes), *Trametes sanguinea* (five laccases; lcc1–5), *Rhizoctonia solani* (four laccases; lcc1-4), Trametes sp. I62 (three laccases; lccA–C) in Trametes sp. AH28-2 and *Gaeumannomyces graminis* two laccases; Moreover, Pleurotus has been found to have laccase gene families. Isolated four genes (lac1-4) from other identified two genes (pel3-4) from *Pleurotus eryngii*, and Poxa1a, Poxa1b, PoxC, Poxa2, Poxa3a, and Poxa3b) from *Pleurotus ostreatus* [63]. Understanding the physiological relevance of various laccase isoforms produced in varied organisms has led to an increase in interest in the research of laccase gene expression regulation mechanisms in recent years. Moreover, this understanding can be applied to enhance industrial procedures. The amount of nutrients (N/C ratio), the culture environment, the fungal developmental stage, and the addition of inducing chemicals to the culture media all have a significant impact on the synthesis and secretion of laccases. Factors such as lignin or its derivatives, carbon and nitrogen supplies, and nitrogen and ion control can work in concert or in opposition to each other to affect the expression of laccase. Few publications indicate a complex system of cis regulatory mechanisms for laccase, despite the lack of comprehensive research on these systems [65]. Some authors argue that the presence of several genes that produce various versions of the same enzyme leads to a variety of physiological activities that laccases acquire over their life cycle. The presence of functional clusters in the evolutionary trees generated by neighbour-joining in the laccases of Batrachomycete fungus provides support for this assertion. After the recent release of the P. ostreatus genome, it was discovered that there are new laccase genes (lacc3, lacc5, lacc7, and lacc8) that code for previously unknown laccases. This has made the understanding of the laccase gene family more complicated, as there are now a total of 11 members in this organism. This raises questions about the specific functions of different laccase isoforms found in Ascomycete and Basidiomycete fungi [32].

7 Classification of Laccases on Base of Origin

See Table 1.

7.1 *Bacterial Laccases*

It has been reported that prokaryotes possess MCOs that exhibit laccase activity. The enzymes are commonly known as polyphenol oxidases, multicopper oxidases,

Table 1 Classification of laccases on base of origin

Source of laccases	Description
Fungi	Laccases are widely distributed in fungi, with species like *Trametes versicolor, Pleurotus ostreatus*, and *Cerrena* unicolor being well-known producers of laccase enzymes. These fungi are often used in industrial processes for the production of laccases due to their high enzyme yields and stability
Plants	While laccases are less common in plants compared to fungi, they do occur. Some plants, especially those with lignified tissues like wood, contain laccases. Examples include certain tree species, such as poplar and pine
Bacteria	Laccases have been identified in various bacterial species, including *Bacillus, Streptomyces*, and *Pseudomonas*. Although bacterial laccases are less common than fungal laccases, they are of interest for their potential in biotechnological applications due to their unique properties
Insects	Laccases are found in the saliva and cuticles of certain insects. For example, laccases play a role in cuticle sclerotization and melanization in insects like beetles. They are also involved in the defense mechanisms of some insects against pathogens
Marine organisms	Laccase-like enzymes have been identified in marine organisms such as algae, mollusks, and sponges. These enzymes have attracted attention for their potential biotechnological applications and their role in marine ecosystem processes
Genetic engineering	With advances in genetic engineering and biotechnology, laccase genes can be inserted into various expression hosts such as bacteria, yeast, and plants to produce recombinant laccases. These engineered systems offer opportunities for customized enzyme production and enhanced enzyme properties

or laccase-like enzymes due to differences in their catalytic mechanism. All of them catalyse the oxidation of common laccase substrates. *Azospirillum lipoferum*, a non-motile bacterium found in plant roots, was found to possess the initial prokaryotic protein exhibiting polyphenol oxidase activity (LMCO). The action of isolated laccase, a multimeric enzyme, leads to the production of a dark-brown pigment [27]. Prokaryotic laccase is produced by various soil and aquatic bacteria, including gram-positive and gram-negative bacteria from the phyla α- and γ-proteobacteria, Firmicutes, Cyanobacteria, Aquificae, Deinococcus-Thermus, and Archaea. *Bacillus pumilus, B. subtilis, B. licheniformis, S. lavendulae, S. griseus, Escherichia coli, P. syringae, Thermus thermophilus, Sinorhizobium meliloti, Oscillatoriaboryana, Haloferaxvolcanii*, and *Marinomonas mediterranea*, as well as species inhabiting extreme environments, have been documented to possess laccase enzymes. The cellular localization of laccase in prokaryotes exhibits significant variation across different species [12]. The enzyme's physiological function appears to depend on the presence of inducing substrates and the growth phase. The majority of intracellularly expressed laccases are found in bacteria such as *B. subtilis, S. meliloti, T. thermophilus*, and *M. mediterranea*. Bacterial cells must develop strategies to manage the intracellular presence of laccase due to its potential hazardous effects. The hypothesis suggests that laccase-positive cells adapt to endogenous reactive quinones generated by laccases through reorganization of the electron transport system. Several bacilli and filamentous actinomycetes have been found to localise laccase extracellularly. Laccase-related genes have been identified in important human pathogens such as E. Coli, *Bordetella pertusis, P. aeruginosa, Campylobacter jejuni, Yersinia pestis,* and *Mycobacterium leprae.* The laccase activity and melanin production are likely contributing factors to the virulence of these species [12]. The CotA protein is a well-known bacterial laccase located in the outer coat of the endospore of *B. subtilis* and other Bacillus species. The EPR spectra of the 65 kDa CotA, obtained from an overproducing *E. coli* strain, exhibit characteristic features of the blue multicopper oxidase family. The CotA structure has been clarified using comparative modelling. The enzyme possesses two buried copper centres (T2 and T3) and one surface-exposed copper centre (T1), exhibiting all the typical features of a fungal laccase [50]. The crystal structure analysis reveals that the enzyme has a bigger hypothetical substrate-binding chamber in comparison to laccases found in plants or fungi [5]. The biochemical properties of CotA are substantially modified by ligand alterations in the T1 region, resulting in a detrimental impact on copper coordination. The protein has remarkable thermostability, maintaining its structural integrity for almost two hours when exposed to a temperature of 80 °C. The examination of the crystal structure indicates that the enzyme has a comparatively bigger hypothetical cavity for binding substrates, in contrast to laccases found in plants or fungi. Alterations in the biochemical characteristics of CotA occur as a result of mutations in the ligand of the T1 site, which have a deleterious impact on copper coordination. The protein has exceptional thermostability, maintaining its structural integrity for around two hours at a temperature of 80 °C. The CotA recombinant laccase shows the highest level of activity when oxidizing syringaldazine (SGZ) at a pH of 3.0 or below, and when oxidizing 2,20-azino-bis (3-ethylbenzothiazoline-6-sulphonic acid) (ABTS) at a pH of 7.0. Nevertheless, the genome of filamentous *S. coelicolor* harbours the SLAC protein. This protein is an example of laccases with two domains, which are known for their unique protein structure and exceptional stability. The SLAC enzyme has a pH optimum of 9.4 for the oxidation of 2,6-dimethoxyphenol (DMP). Nevertheless, the enzyme produced via recombinant techniques still exhibits the characteristic paramagnetic characteristics that are often reported in laccases [48]. Prokaryotic laccases exist in various forms, including monomers, trimers, and tetramers, with sizes ranging from 32 to 180 kDa. The glycosylation of bacterial MCOs has been insufficiently investigated, despite the abundant evidence of prokaryotic protein glycosylation. Additionally, there is a scarcity of reports on the carbohydrate composition of bacterial laccases. Insufficient data is available on the electrochemistry of prokaryotic MCOs. Low-redox potential enzymes, such as bacterial and plant laccases, have a redox potential at the T1 site (E00T1) that is below 460 mV compared to NHE. Bacterial polyphenol oxidases exhibit greater thermal stability and enzymatic activity at elevated pH levels compared to fungal laccases. Several Streptomyces laccases displayed a high alkaline pH optimum, with values of 8.5 and 9 for ABTS and DMP, respectively. The highest documented temperature optimum was 85 °C [56]. Laccase-producing bacterial MCOs are known as "moonlighting proteins" or multifunctional enzymes that can perform a variety of tasks depending on where in the cell they are found. A plausible mechanism for the switching of multifunctional enzymes has been proposed. Prokaryotic laccases' potential uses have previously been discussed, but it is still unclear what their actual biological function is in vivo. The majority of the bacterial proteins that have been found thus far are rarely directly connected to the breakdown of lignin [20]. The function and purpose of this phenomenon have not been definitively established. Currently, extensive research focuses on the efficacy of bacterial laccases in lignin degradation. A recent proposal suggests modifying the lignin polymer to allow other enzymes to access cellulose and hemicellulose. Laccase-related gene products are commonly linked to stress resistance and play key roles in sporulation, morphogenesis, metal oxidation, and pigmentation of cells and spores. The Azospirillum laccase is known to participate in electron transport, lignin metabolism, utilization of plant phenolic compounds, and cell pigmentation. The abilities of Azospirillum sp. may be linked to its competitiveness in the rhizosphere and are important for the colonization of plant roots, especially in environments with fluctuating oxygen levels [27]. Streptomyces laccase has been proposed to be involved in morphogenesis, pigmentation, lignocellulose degradation, bacterial interactions, and antibiotic production. The majority of laccases in Bacillus are present in the outer coat of the spore, which provides protection to endospores against various stressors. The cotA gene is responsible for the production of a melanin-like polymer known as brown spore pigment, which serves as a shield against ultraviolet radiation. The observation of CotA's protective effect in a laccase-positive strain of B. subtilis, which exhibits greater resistance compared to cotA-deficient spores, supports this theory. Brown-pigmented Bacillus sp. HR03 spores display high resistance to UVA, UVC, and hydrogen peroxide. The resistance of Bacillus to harsh environments is attributed to its spore pigments, with laccase playing a key role in their synthesis. The involvement of plant laccases in cell wall formation can be compared to their role in the cross-linking of spore coat proteins, as indicated by the presence of tyrosine to di-tyrosine crosslinks in B. subtilis spores [64]. While B. subtilis is capable of oxidizing MnII, its CotA protein does not participate in metal oxidation, unlike CumA, the sporecoat protein found in Pseudomonas sp. Laccase and CumAisan MCO exhibit similarities. The presence of MnII ions can enhance cell viability by facilitating the oxidation process. *Bacillus halodurans*, which produces laccases, is also resistant to CuII toxicity. In addition, laccase-like coding genes have been found in *Pseudomonas syringae* (CopA) and *Escherichia coli* (PcoA and CueO, previously known as YacK). Bacterial copper resistance relies on pseudo-laccases, which exhibit structural similarities to multicopper oxidases. The yacK gene encodes a putative multicopper oxidase with ferroxidase and phenol oxidase activity and expected copper binding centres. The enzyme exhibits significant activity against siderophores utilized by *E. coli* for FeII absorption, thereby diminishing *E. coli*'s susceptibility to copper [40]. The T1 copper active site entrance in *E. coli* CueO is partially obscured by a Met and His-rich region. The binding of CuII may result in the availability of additional Cu-binding sites in this region, potentially leading to a structural alteration of the active site. The presence of Pseudomonas CopA is necessary for the full expression of copper resistance in these bacteria. Additionally, it shows a moderate, yet noteworthy, sequence similarity with MCO proteins. The potential of laccase in cyanobacteria for bioremediation has been investigated, suggesting its role in protecting the cell from unfavourable environmental conditions. However, there is a limited amount of published research on cyanobacterial laccases [1].

7.2 Plant Laccases

The sources of laccase include gymnosperms, angiosperms, and extracellular phenol oxidase from *Tetracystis aeria* or green algae. Recent studies have identified the latter as laccase based on the characteristics and substrate specificity of the purified enzyme. Laccases have been identified in various plant species, including Anacardiaceae members, Zea mays, *Pinus taeda, Acer pseudoplatanus, Nicotiana tabacum, P. trichocarpa, Liridendron tulipifera, Lolium perenne, A. thaliana, Oryza sativa, Saccharum ocinarum, Brassica napus,* and *Brachypodium distachyon.* The Chinese lacquer tree, scientifically known as *Rhus vernicifera,* was the first species in which laccase was detected by Yoshida. Ten years later, Gabriel Bertrand isolated this enzyme from members of the Anacardiaceae family, including *Mangifera indica, Schinus molle, Pistacia palaestina,* and *Pleiogynium timoriense* [82]. The majority of laccase was found in these representatives' resin ducts. According to Bligny and Douge, laccase was produced and secreted from *A. pseudoplatanus* cultures. Later, the enzyme was found in the xylem tissues of *P. taeda, P. euramericana,* and *N. tabaccum.* Additionally, the enzyme was extracted from *Aesculus parviflora* leaves [89]. Plant species vary in the quantity of laccase isoforms they possess. *P. trichocarpa*'s xylem expresses a total of five laccases, whereas *P. taeda*'s xylem tissues express a total of eight laccases. In contrast, *A. thaliana*'s xylem tissues express a total of seventeen laccase genes. Plant laccases generally have a molecular weight ranging from 60 to 130 kilodaltons with an average composition of 500–600 amino acids. The enzymes exhibit a high degree of glycosylation, ranging from 25 to 45%. They function well within a pH range of 5–7, and their isoelectric point (pI) falls between 5 and 9.6. Plant laccases have a very low redox potential of around 430 mV compared to the standard hydrogen electrode (NHE) due to their copper type I (T1) configuration. Plant laccases has the ability to oxidise substrates independently, without the need for mediators [67]. Plant laccases are reported to be involved in the following processes.

- Lignification, which maintains the mechanical rigidity and structure of the cell wall.
- Responses of plants to defense mechanisms and environmental stresses.
- Wound healing.
- Iron metabolism.
- Polymerization of phenolic compounds.
- Plant laccases have the ability to oxidise lignin precursors, indicating their involvement in the process of lignification in the plant cell wall. Laccase enzymes are released in the apoplast, where they facilitate the regeneration of injured plant

tissues and the production of lignin. The process of stem lignification is facilitated by A. thaliana laccase, an enzyme that is present in several organs and tissues of Populus sp., including the seed coat. The enzyme's existence inside the resin ducts of Anacardiaceae indicates its role as a protective mechanism against bacterial and fungal infiltration, herbivores, and predators [51]. The TT10 gene, responsible for synthesizing antimicrobial quinones, was shown to have a role in creating an antipathogenic barrier in Arabidopsis during seed storage and germination. Laccases expressed in the seed-specific group in A. thaliana participate in the synthesis of insoluble brown polymers that likely have a protective role and contribute to the polymerization of flavonoids in the seedcoat. Green algal laccases have a significant function in the production of compounds that absorb UV radiation and in the formation of polymers associated with the cell wall. They also play a role in removing phenolic compounds found in both land and water environments, as well as in the breakdown of lignocellulosic substrates to obtain nutrients. Algae laccases have the capacity to have a substantial impact on the conversion of naturally occurring and man-made aromatic compounds, including recognized environmental contaminants [62].

7.3 Fungal Laccases

Most of our understanding of fungal laccases is derived from species mostly discovered in the Ascomycota and Basidiomycota phyla. Fungal laccases assist in the degradation of lignocellulose polymers, as well as in defence, protection, virulence, pathogenicity, pigmentation, and sporulation activities. The main function of fungal laccase is to facilitate the biodegradation of lignocellulose, hence contributing to the carbon cycle in the biosphere. Due to the lignin polymer's exceptional resistance to chemical and biological degradation, the natural decomposition of wood is a challenging and protracted process. Fungal laccases, namely those present in white-rot species, have a significantly elevated redox potential of around 800 mV compared to NHE. This facilitates the more convenient extraction of electrons from substrates, which may also function as redox mediators when the laccase targets lignin [62]. Fungal laccase has the capacity to break down chemical bonds in phenolic lignin model compounds without requiring mediators. The breaking of covalent bonds by laccase in non-phenolic lignin subunits is only possible when mediators are used. In addition, when a fungal infection occurs, natural compounds derived from plants such as gallic or ferulic acids may be found in fungal cells and stimulate the activation of the laccase gene [31]. Litter breakdown is the crucial step in the intricate process of nutrient recycling, which involves many fungal species and a quick sequence of protrophic organisms. Ascomycota and Basidiomycota are the predominant species that possess extracellular enzymes. However, other fungi, such as Zygomycota and Glomeromycota, are also recognized to exist. Fungi that decompose litter have been discovered to possess laccase genes, whereas species inhabiting forest floors with high lignin content exhibit 5–10 times more basidiomycete laccase genes compared to those in low lignin settings. It is noteworthy that watery fungal species were previously unable to degrade lignin. Nevertheless, the taxa of ascomycota, chytridiomycota, and oomycota play a crucial role in the decomposition of litter and possess the ability to produce laccase. Two segments of genes encoding laccase were found in pure cultures of *Clavariopsis aquatica*, suggesting that laccase is an enzyme linked with cells [49].

Humic substances are a notable kind of carbon-based material that occurs naturally. They are dark-coloured organic compounds that are formed by chemical and biological processes, mostly from plant waste but sometimes from animal and human waste. Phenols, quinones, carbohydrates, and larger molecules such as lignins, polysaccharides, melanins, and cutins are the main components of humic substances. Microorganisms play a crucial role in the creation, decomposition, alteration, and eventual conversion of humic compounds into minerals [33]. The laccase produced by *Chaetomium thermophilum* creates water-soluble polymers that consist of hydrophobic acids. This is an important stage in the process of humification. Conversely, enzymes derived from the Basidiomycota are mainly responsible for catalysing the decomposition and conversion of HS. The transformation route of humification vs HS degradation is determined by the availability of substrates and reaction factors such as pH, humidity, and the presence of co-substrates. The participation of several laccase isoenzymes may be essential for augmenting the efficiency of HS breakdown or conversion, since pH plays a pivotal impact. The correlation between the amount and diversity of bacterial and fungal laccase activity in arable subtropical soil indicates that bacterial laccase activity is the dominant factor [28]. Laccase secretion is considered to be one of the primary ways in which fungi respond to hostile situations, such as the presence of other microorganisms, xenobiotics, metals, poisons, and biologically active chemicals. In addition to phenolic chemicals, fungal laccases are capable of oxidizing non-phenolic substrates such as aromatic amines, polycyclic aromatic hydrocarbons, synthetic colours, antibiotics, and other substrates that are not easily recognized by laccases. Laccase plays a crucial role in the active defence mechanism of fungi and serves as a valuable enzymatic resource for eliminating naturally occurring or man-made toxins present in the environment [68]. One example of this enzyme's function in defense against antibiotics like 2,4-diacetylphloroglucinol (2,4-DAPG) is BotrytiscinereaBcLCC2laccase. Tannic acid, the redox mediator, is the only substance that causes 2,4-DAPG to degrade. It is interesting to note that in *Pycnoporus cinnabarinus*, laccase functions as a catalyst during the synthesis of cinnabarinic acid. This fungus may be shielded from microbial predators by the antibacterial substance cinnabarinic acid. Key enzymes in the metabolism of melanin are laccases. Numerous melanin pigments have antibacterial properties and function as virulence factors, which aid in the survival of fungal species. They can secrete into the environment and are found in bacterial endospores, fungal spores, conidia or within cell walls [60]. Research has investigated the correlation between *C. neoformans* melanization and its ability to cause infection in humans. Cryptococcus, a fungus with yeast-like properties, is responsible for causing meningoencephalitis (cryptococcosis), a condition that closely resembles cryptococcal pneumonia. Additionally, it may also cause infections in infants. The expression of a single CNLAC1 gene is induced in a C. neoformans culture by glucose deprivation, acidification, and reduced temperature. Laccase has been identified as the main diphenol oxidase responsible for the creation of the colour generated by Candida neoformans when an external substrate is present. This substance is described as a chemical that resembles melanin. Catecholamine, dopamine, adrenaline, and noradrenaline are melanin-like precursors that are present in the central nervous system of mammals. These substances may contribute to the neurotropism of *C. neoformans* [45]. Additionally, C. neoformans' laccase increases virulence by converting FeIII to FeII, which prevents macrophages from undergoing an oxidative burst and, as a result, compromises innate immunity. Because *C. neoformans* laccase is closely linked to the cell wall, it can readily participate in the body's defense against the harmful hydroxyl radicals that macrophages produce [47]. *Talaromyces marneffei* infection of AIDS patients raises the possibility that laccase contributes to the pathogen's virulence. The generation of a red soluble pigment by *T. marneei* mold and yeast forms is linked to a potential mechanism. Single, double, triple, and quadruple deletions of the genes encoding the lacA, lacB, lacC, and pbrB laccases were obtained in order to comprehend the role of laccase in *T. marneei* virulence. The sole conidia from the quadruple mutant exhibited heightened susceptibility to antifungal agents and were susceptible to being phagocytosed and eliminated by a monocyte cell line of THP-1 [87]. These findings imply that *T. marneffei's* laccase increases the pathogen's resistance to the host's immune system. The reduction of immune recognition by the pigment or other laccase products may disrupt monocyte signalling pathways. The genome of Fonsecaea sp., a member of the class of clinically significant black yeast-like fungi, also contains laccase genes [58]. *Colletotrichum Gloeosporioides* is the primary causative agent of *Mangifera indica* fruit fever. Affected lac1 mutants (LAC1) have decreased aerial mycelial mass and radial growth rates, are less pigmented, and hardly ever produce conidia. In virulence tests on mango leaves and fruits that are both wounded and unwounded, the lac1 mutants also exhibit reduced virulence. Laccases in *Heterobasidion annosum* may function as a virulence factor during interactions with *P. sylvestris* seedlings, as reported by [44]. The fungi *T. atroviride* and *T. harzianum*, which are closely related phylogenetically, produce green and yellowish spores, respectively, but laccase activity is only seen during the formation of the darker green spores. The laccase activity is associated with the spore surface and is linked to the melanin-like polymer present in the cell wall or in the periplasm of the spore capsules. According to this study, spore-associated laccases play a role in the synthesis of melanin, a pigment that shields spores from heat, UV light, and hydrolytic enzymes [35].

8 Zero Carbon Emission

Zero emission refers to the absence of carbon dioxide and other greenhouse gases being discharged into the environment. This objective is often linked to endeavours aimed at reducing the impact of climate change and shifting towards a more sustainable and ecologically sound energy system. Human activities, such as the combustion of fossil fuels for energy, deforestation, and industrial operations, are the main sources of carbon emissions [39]. Net zero carbon emission can be achieved by the following.

8.1 Renewable Energy Sources

Adopting renewable energy sources like solar, wind, hydro, and geothermal may facilitate the production of power without releasing carbon dioxide. These sources use energy from the natural environment and have less environmental effects [61].

8.2 Energy Efficiency

Enhancing energy efficiency in buildings, transportation, and businesses has the potential to decrease the total energy need and hence decrease carbon emissions. This entails the use of energy-efficient technology and processes [34].

8.3 Carbon-Capture and Storage (CCS)

CCS systems are designed to collect carbon dioxide emissions from industrial processes and power plants prior to their discharge into the environment. Subsequently, the CO_2 that has been caught may be conveyed and stored under the Earth's surface [70].

8.4 Afforestation and Reforestation

Planting trees and restoring forests facilitate the absorption of carbon dioxide from the atmosphere by means of photosynthesis. Forests function as carbon sinks, capturing and storing carbon, so decreasing the total concentration of carbon in the atmosphere [53].

8.5 Electrification of Transportation

Transitioning from conventional internal combustion engine cars to electric vehicles (EVs) may substantially decrease carbon emissions in the transportation industry, particularly if the power used to recharge the EVs is derived from renewable sources [18, 54].

8.6 Circular Economy

Adopting a circular economy approach promotes recycling, reuse, and sustainable consumption, reducing the need for the extraction of raw materials and the associated carbon emissions from manufacturing processes [54].

9 Applications of Laccases

Laccases belongs to polyphenol oxidase, and due to potential to catalyse oxidation of different substrates, it has wide range of applications in different fields, due to which it helps to gain zero carbon emission. Few applications of laccases are given here (Table 2).

- Bioremediation
- Pulp and paper industry
- Textile industry
- Biofuel production
- Food and beverages industry
- Biosensors
- Biomedical application
- Nanotechnology

9.1 Bioremediation

Laccases have the ability to break down a range of environmental contaminants, including as phenolic compounds, dyes, and aromatic pollutants found in industrial effluent. They accelerate the process of transforming these contaminants into forms that are less harmful or completely harmless. Laccases have exceptional efficacy in the degradation of phenolic chemicals, which are prevalent contaminants found in industrial effluent. Phenolic compounds often arise from the release of industrial waste, and laccases facilitate the transformation of these chemicals into less harmful forms [7]. They have the ability to remove colour from artificial dyes, which are

Table 2 Applications of laccases

Application of laccases	Description
Bioremediation	Laccases can degrade various environmental pollutants, including phenolic compounds, aromatic hydrocarbons, and dyes, by oxidizing them into less toxic forms
Pulp and paper industry	Laccases are employed in the bleaching process of pulp and paper production as an eco-friendly alternative to chemical bleaching agents
Textile industry	Laccases are utilized for enzymatic bleaching of textiles, replacing harsh chemical bleaching agents
Biofuel production	Laccases play a role in lignin degradation, which is a key step in the conversion of lignocellulosic biomass into biofuels such as bioethanol and biodiesel
Food and beverages industry	Laccases are employed in various food processing applications, including wine and juice clarification, where they aid in the removal of phenolic compounds and haze-forming substances
Biosensors	Laccases are utilized in biosensor technology for the detection of various analytes, including phenolic compounds, pesticides, and environmental pollutants
Biomedical application	Laccases have potential applications in biomedical research and therapy, including wound healing, tissue engineering, and biosensing in medical diagnostics
Nanotechnology	Laccases have been employed in nanotechnology for the synthesis and modification of nanoparticles

widely used in sectors such as textiles, leather, and paper. Laccases catalyse the oxidation of dyes, causing the disintegration of intricate dye compositions. This process effectively eliminates colour and detoxifies the wastewater. They have the ability to undergo oxidation reactions with aromatic pollutants, such as polycyclic aromatic hydrocarbons (PAHs) and other aromatic chemicals found in polluted settings. This skill is essential in the process of restoring locations that have been polluted with substances derived from petroleum or industrial waste. Wastewater treatment facilities might use them as a supplementary measure to conventional treatment procedures. They aid in the elimination of stubborn organic contaminants that cannot be readily broken down by traditional biological treatment methods [24] (Fig. 2).

9.2 Pulp and Paper Industry

Laccases are employed in the bleaching process of pulp and paper production. They can replace or complement traditional chemical bleaching agents, reducing the environmental impact of the industry. Laccases are used in the biological bleaching of pulp as an alternative to chemical bleaching agents. Enzymatic bleaching helps

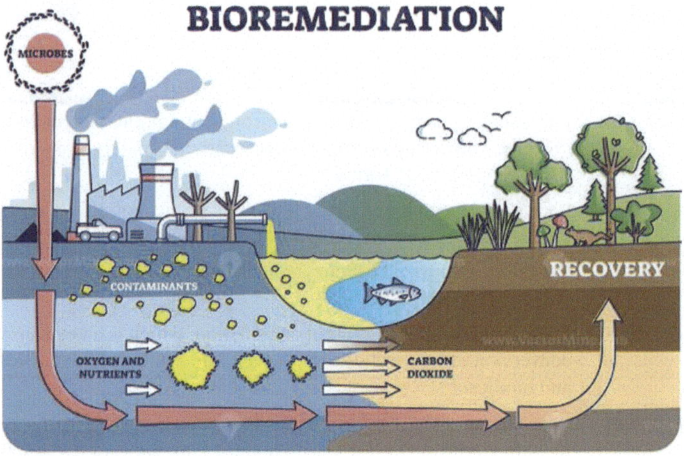

Fig. 2 Role of laccases in bioremediation

reduce or eliminate the need for chlorine-based chemicals, minimizing the environmental impact of the bleaching process. It participates in the delignification of pulp. They oxidize and break down lignin, which is a complex and rigid polymer present in wood. By selectively targeting lignin, laccases help open up the structure of the wood fibres, making them more accessible for subsequent processing. It reduces the need for harsh chemicals, resulting in a decreased environmental impact. This aligns with the industry's efforts to adopt more sustainable and environmentally friendly practices. Enzymatic bleaching with laccases can lead to improved pulp quality. The process is selective, preserving cellulose while removing lignin and other impurities. This results in pulp with better strength properties and brightness. Laccase-mediated bleaching is often carried out under milder conditions compared to traditional bleaching processes. This can lead to energy savings as it requires lower temperatures and less chemical input. Laccases efficiently remove colour from the pulp by oxidizing chromophores, contributing to the production of bright and white paper [84].

9.3 Textile Industry

Laccases are used in the textile sector to achieve eco-friendly bleaching and colour removal of textiles. They have the ability to efficiently remove colour from dyes and other pigments. It is used for the enzymatic whitening of fabrics. They have the ability to substitute or enhance conventional chemical bleaching agents, such as hydrogen peroxide or chlorine-based chemicals. Enzymatic bleaching with laccases is a more ecologically sound approach, since it minimises the use for aggressive chemicals [8].

They exhibit high efficacy in the process of decolorizing textile colours. They accelerate the chemical reaction that causes the oxidation of chromophores, resulting in the decomposition of dye molecules and the elimination of colour from textiles. This is especially beneficial for the treatment of wastewater generated by textile dyeing operations. It aids in the process of biodegrading textile dyes by decomposing intricate dye complexes into less complicated and less harmful chemicals. The enzymatic technique is more environmentally friendly in comparison to traditional chemical procedures. It enhances the colour durability of coloured materials. The enzymatic treatment enhances the fixation of dye molecules to the cloth, hence improving the colour permanence of textiles. It has the capability to be used for the treatment of effluents produced during the process of textile manufacturing. Laccases facilitate the degradation of dyes and other contaminants in wastewater, hence aiding in the mitigation of environmental pollution linked to the textile industry. Laccases, unlike aggressive chemical bleaching treatments, exhibit selectivity in their action and do not cause harm to the textile fibres. This process aids in maintaining the physical and mechanical characteristics of the textiles, so guaranteeing the excellence of the end result. Laccases are used in textile processing to decrease the total chemical burden. This aligns with the textile industry's growing emphasis on sustainability, eco-friendly chemistry, and minimizing environmental harm [6, 25].

9.4 Biofuel Production

Laccases are involved in the breakdown of lignocellulosic biomass, facilitating the liberation of fermentable sugars that may be used for the manufacture of biofuels. This application is pertinent in the context of sustainable energy generation. Laccases play a vital role in the degradation of lignin, which is an intricate and inflexible polymer found in the cellular walls of plants. Lignin, a significant constituent of lignocellulosic biomass, poses a significant obstacle to the effective retrieval of fermentable sugars for the manufacture of biofuels. Laccases alter the structure of lignin by oxidizing it, hence increasing its susceptibility to subsequent enzymatic hydrolysis or chemical reactions. This process enhances the liberation of sugars from cellulose and hemicellulose, which may then undergo fermentation to produce biofuels. Laccases are often used in conjunction with other enzymes, such as cellulases and hemicelluloses, to enhance the efficacy of enzymatic hydrolysis. Laccases facilitate the decomposition of intricate formations, hence enhancing the accessibility of other enzymes to cellulose and hemicellulose, which may then be hydrolysed into fermentable sugars. Laccases are used in bio-electrochemical systems to enhance electron transport during substrate oxidation. These findings may be used to advance the technology of microbial fuel cells (MFCs) or enzymatic biofuel cells, enabling the direct conversion of biomass-derived substances into electrical energy [42].

9.5 Food and Beverages Industry

Laccases are used for the purpose of stabilizing and clarifying liquids, such as wine and beer. They may assist in the elimination of unwanted substances and enhance the overall quality of the end product. Laccases are used in organic synthesis to manufacture high-quality chemicals and intermediate compounds used in the pharmaceutical industry. They have the ability to accelerate reactions at low temperatures, eliminating the need for strong chemicals. It may be used in food processing to alter the characteristics of certain components. For instance, they may be used to modify the composition of proteins, resulting in modifications to the texture and uniformity of food items. Phenols, which are unwanted chemicals, are eliminated in the wine business using these substances. Laccases aid in the oxidation process to eliminate these chemicals, which may have an adverse effect on the flavour and overall quality of wine. They are used to improve the colour stability of fruits and vegetables. They have the ability to accelerate the oxidation of phenolic chemicals that are responsible for the browning processes, therefore avoiding the discoloration of freshly cut fruits and vegetables. Within the dairy business, they have the capacity to enhance the processing of milk and dairy products. They participate in the oxidation process of polyphenols, which may have an impact on the taste and longevity of dairy products. They are used in the processing of coffee and tea to decrease the concentrations of polyphenols. The use of this enzymatic treatment might lead to enhanced flavour and superior quality of the end goods [52].

9.6 Biosensors

Laccases are used in biosensors to detect a wide range of chemicals. These biosensor devices may be created by immobilizing them on various surfaces to detect phenolic chemicals, contaminants, or particular biomolecules. It accelerates the process of oxidizing various substances, such as phenols, aromatic amines, and other organic molecules, by converting molecular oxygen into water. Laccases catalyse redox processes that produce an electric current, which may be quantified and harnessed for biosensor purposes. They participate in electron transfer mechanisms, in which electrons are moved from the substrate to the enzyme and then to an electron acceptor. Laccases in biosensors enable the transmission of electrons from the substrate to an electrode surface, leading to a detectable electrical signal. Laccases are very efficient in identifying phenolic chemicals, and biosensors that use laccase activity may be specially engineered to quantify phenols. Laccases are recognized for their robustness in several environments, including a broad spectrum of pH levels and temperatures. Due to their stability, they are well-suited for integration into biosensors, since they are able to retain their activity for long periods of time. They may be readily immobilized on various electrode surfaces, enabling the development of durable and reusable biosensors. Immobilizing the enzyme increases its lifespan and makes it easier to utilise in continuous monitoring applications [86].

9.7 Biomedical Applications

Laccases have potential applications in the biomedical field, including the development of anti-tumour agents and biosensors for medical diagnostics. It has been explored for their potential in wound healing applications. They can assist in the cross-linking of proteins, promoting the formation of a stable extracellular matrix, and thereby aiding tissue repair. It exhibits antimicrobial properties, which make them useful in combating infections. They can be employed in the development of antimicrobial coatings for medical devices, such as catheters or implants, to reduce the risk of infections. It can be utilized in drug metabolism studies [10]. They can catalyse the oxidation of certain drugs, providing insights into their metabolic pathways. Additionally, laccases can be employed in bioconjugation reactions, facilitating the attachment of drugs to carrier molecules for targeted drug delivery. It plays a role in the degradation of various pharmaceutical compounds and other organic pollutants. This ability can be exploited in the treatment of pharmaceutical waste and in the development of environmentally friendly processes for drug manufacturing. It has been investigated for their potential role in cancer therapy. They can be used in combination with certain compounds to generate cytotoxic products that selectively target cancer cells. This approach is being explored as a potential strategy for cancer treatment. It can be incorporated into diagnostic assays and imaging techniques. Their catalytic activity can be used to amplify signals in detection methods, enhancing the sensitivity of diagnostic tests [75].

9.8 Nanotechnology

Laccases have the potential to be used in the process of creating and altering nanoparticles. This application is crucial for the advancement of nanomaterials with a wide range of characteristics. It has the capability to be used for the eco-friendly production of nanoparticles. They enable the conversion of metal ions into nanoparticles by means of a bio-reduction process. This environmentally conscious strategy serves as a substitute for traditional chemical processes and produces nanoparticles that possess consistent size and shape. Catalysts may be used in the synthesis of a wide range of nanomaterials, such as metal oxides and organic–inorganic hybrid materials. Their catalytic activity is used to regulate the dimensions and structure of nanoparticles, enabling meticulous manipulation of nanomaterials for targeted purposes. It may be used to modify the surface of nanoparticles. By altering the surface of nanoparticles using particular biomolecules, such as proteins or peptides, their characteristics may be customized to suit specific purposes in the fields of medicine, sensing, or environmental remediation. It contributes to the advancement of nanobiotechnological gadgets. They have the ability to be included into biosensors and biofuel cells in order to improve their efficiency. Researchers have investigated the use of nanobiosensors that rely on laccase enzymes to detect a range of substances, such as phenolic chemicals and environmental contaminants. Laccases has the capacity to break down certain nanoparticles within the realm of environmental nanotechnology. The use of this characteristic may be employed for the environmentally conscious disposal of nanoparticles, hence mitigating their possible ecological consequences [79].

10 Role of Laccases in Obtaining Zero Carbon Emission

Laccases participate in the process of breaking down lignin, which is an intricate organic polymer present in the cell walls of plants. Laccases aid in the degradation of lignin, so facilitating the bioremediation of waste materials and contaminants, including those generated by the pulp and paper sector. This procedure facilitates the mitigation of the environmental consequences of industrial operations and encourages the reuse of organic materials via recycling. It contributes to the synthesis of biofuels derived from lignocellulosic biomass. Lignocellulosic resources, including agricultural leftovers and wood refuse, may be transformed into biofuels using enzymatic procedures that utilise laccases. Biofuel production provides a renewable and sustainable substitute for conventional fossil fuels, therefore aiding in the reduction of carbon emissions [21]. In the textile industry, laccases are used to bleach and colour textiles without the need for environmentally harmful chemicals. This enzyme-based approach minimizes the use of toxic substances, reduces water consumption, and lowers the carbon footprint associated with conventional textile processing [5, 66]. It is used in wastewater treatment procedures to eliminate different contaminants, such as phenolic compounds and dyes. Laccases facilitate the degradation of these contaminants, so aiding in the purification of industrial wastewater. This process helps minimise the environmental consequences of wastewater discharge and prevents the release of toxic compounds into natural water sources. Laccases are used in the production and alteration of bio-based products. This encompasses the process of chemically bonding proteins and polymers together to produce eco-friendly materials that may be used in many sectors such as packaging, building, and other industries. The advancement of bio-based materials aids in diminishing reliance on carbon-intensive resources, hence promoting a more sustainable and low-carbon economy. Laccases function as biocatalysts in several industrial processes, facilitating the synthesis of high-quality chemicals and medicines while minimizing the need for conventional chemical synthesis approaches. Utilizing laccases in biocatalysis may result in enhanced energy efficiency and a reduction in the total carbon emissions associated with chemical manufacture [43].

11 Conclusion and Future Perspective

In conclusion, laccases are an essential component in the process of achieving carbon neutrality on a worldwide scale because they act as biocatalysts. The diverse enzymatic characteristics of the enzymes have the potential to be beneficial to a variety of sectors, including the generation of biofuels and the treatment of wastewater. In the process of oxidizing a wide variety of substrates, laccases demonstrate a high level of efficiency and selectivity. This enables the degradation of pollutants that are persistent and the synthesis of compounds that are of great scientific value. As a consequence, they offer an alternative that is better for the environment than conventional chemical processes, which use a substantial amount of energy and do damage to the environment. The industrial uses of laccases have shown encouraging results, which indicates that they have the potential to revolutionise traditional processes and assist the establishment of an economy that is carbon neutral. Laccase-based technologies provide scalable solutions for a variety of applications, including bioremediation and the textile industry, among others. Both the concepts of green chemistry and the principles of the circular economy are associated with these technologies. By using enzymes' natural potential for more effective resource utilization and waste reduction, businesses, nonprofit organizations, and people may lessen their influence on the environment.

In the future, the incorporation of laccases into standard industrial processes has the potential to significantly contribute to the achievement of carbon neutrality on a global scale. It is vital to undergo ongoing research and development in order to increase the substrate specificity, stability, and catalytic activity of laccases. This is required in order to boost the practical uses of laccases and the economic feasibility of developing them. Without a doubt, significant developments in protein engineering, enzyme immobilization techniques, and bioprocess optimization are required in order to make full use of laccases as catalysts for innovation that is driven by sustainability. The recycling of biomass and the use of laccase-mediated biotransformations have both been made possible by integrated bioprocessing platforms and biorefineries. Through the transformation of organic waste and lignocellulosic materials into biofuels, bioplastics, and biochemicals, laccase-enabled biorefineries not only contribute to the achievement of carbon neutrality but also foster economic growth and rural development.

It has been shown that laccases have the ability to reduce pollution, combat climate change, and serve a variety of specialized industrial uses. It is possible that the catalytic capabilities of these enzymes will make it possible for scientists to create fresh ways for the restoration of ecosystems, the remediation of contaminated places, and the removal of carbon dioxide from the atmosphere. The incorporation of laccase-producing microorganisms into bioremediation consortiums has the potential to improve the efficiency and longevity of remediation processes, which in turn is beneficial to the health of both humans and the environment. In the midst of the problems that the twenty-first century presents, laccases will play an increasingly vital part in the effort to achieve carbon neutrality. It is imperative to engage in interdisciplinary collaboration and make deliberate investments in research and development in order to fully harness the capabilities of these exceptional enzymes in order to effectively tackle the interrelated issues of social justice, economic prosperity, and environmental sustainability. This is the only way to effectively tackle these issues. The implementation of biocatalysis and bioprocessing principles has the potential to bring about a future that is marked by resiliency, regeneration, and inclusion for future generations.

References

1. Afreen S, Anwer R, Singh R, Fatma T (2018) Extracellular laccase production and its optimization from Arthrospira maxima catalyzed decolorization of synthetic dyes. Saudi J Biol Sci 25(7):1446–1453
2. Alcalde M (2007) Laccases: biological functions, molecular structure and industrial applications. In: Industrial enzymes: structure, function and applications. Springer, pp 461–476
3. Andrews SC, Robinson AK, Rodríguez-Quiñones F (2003) Bacterial iron homeostasis. FEMS Microbiol Rev 27(2–3):215–237
4. Arshad M, Adil M, Sikandar A, Hussain T (2014) Exploitation of meat industry by-products for biodiesel production: Pakistan's perspective. Pakistan J Life Soc Sci 12:120–125
5. Arshad M (ed) (2017) Perspectives on water usage for biofuels production: aquatic contamination and climate change, 24 November 2017. Springer
6. Arshad M, Bano I, Younus M, Khan A, Rahman A (2018) Health concerns associated with biofuel production. In: Perspectives on water usage for biofuels production: aquatic contamination and climate change, pp 97–105
7. Arshad M, Hussain T, Chaudhry N, Sadia H, Aslam B, Tahir U, Abbas M, Qureshi N, Nazir A, Rajoka MI, Iqba M (2019) Enhancing profitability of ethanol fermentation through gamma ray mutagenesis of saccharomyces cerevisiae. Pol J Environ Stud 28(1)
8. Arshad M (ed) (2020) Sustainable ethanol and climate change: sustainability assessment for ethanol distilleries, 13 November 2020. Springer Nature
9. Arshad M, Ansari AR, Qadir R, Tahir MH, Nadeem A, Mehmood T, Alhumade H, Khan N (2022) Green electricity generation from biogas of cattle manure: an assessment of potential and feasibility in Pakistan. Front Energy Res 25(10):911485
10. Arshad M (ed) (2023) Climate changes mitigation and sustainable bioenergy harvest through animal waste: sustainable environmental implications of animal waste, 28 April 2023. Springer Nature
11. Assadi G, Vesterlund L, Bonfiglio F, Mazzurana L, Cordeddu L, Schepis D, Mjösberg J, Ruhrmann S, Fabbri A, Vukojevic V (2016) Functional analyses of the Crohn's disease risk gene LACC1. PLoS One 11(12):e0168276
12. Bains J, Capalash N, Sharma P (2003) Laccase from a non-melanogenic, alkalotolerant γ-proteobacterium JB isolated from industrial wastewater drained soil. Biotech Lett 25:1155–1159
13. Bajpai P (1999) Application of enzymes in the pulp and paper industry. Biotechnol Prog 15(2):147–157
14. Bano I, Ghauri MA, Arshad M, Yasin T, Younus M (2017) Bioactivity of variant molecular weight chitosan against drug-resistant bacteria isolated from human wounds. Microbial Drug Resist 23(8):958–965
15. Bano I, Arshad M (2018) Climatic changes impact on water availability. In: Perspectives on water usage for biofuels production: aquatic contamination and climate change, pp 39–54

16. Benfield G, Bocks SM, Bromley K, Brown B (1964) Studies of fungal and plant laccases. Phytochemistry 3(1):79–88
17. Bhatia S (2018) Introduction to pharmaceutical biotechnology. Volume 2: Enzymes, proteins and bioinformatics. IOP Publishing
18. Blynn KKM (2018) Accelerating bus electrification: enabling a sustainable transition to low carbon transportation systems. Massachusetts Institute of Technology
19. Bollag J, Shuttleworth K, Anderson D (1988) Laccase-mediated detoxification of phenolic compounds. Appl Environ Microbiol 54(12):3086–3091
20. Brown ME, Chang MC (2014) Exploring bacterial lignin degradation. Curr Opin Chem Biol 19:1–7
21. Cañas AI, Camarero S (2010) Laccases and their natural mediators: biotechnological tools for sustainable eco-friendly processes. Biotechnol Adv 28(6):694–705
22. Chisari M, Barbagallo RN, Spagna G (2007) Characterization of polyphenol oxidase and peroxidase and influence on browning of cold stored strawberry fruit. J Agric Food Chem 55(9):3469–3476
23. Choi M, Davidson VL (2011) Cupredoxins—a study of how proteins may evolve to use metals for bioenergetic processes. Metallomics 3(2):140–151
24. Christopher JG, Kumar G, Tesema AF, Thi NBD, Kobayashi T, Xu K (2016) Bioremediation for tanning industry: a future perspective for zero emission. Management of hazardous wastes. INTECH, pp 91–102
25. Couto SR, Toca-Herrera JL (2006) Lacasses in the textile industry. Biotechnol Mol Biol Rev 1(4):115–120
26. d'Acunzo F, Galli C, Gentili P, Sergi F (2006) Mechanistic and steric issues in the oxidation of phenolic and non-phenolic compounds by laccase or laccase-mediator systems. The case of bifunctional substrates. New J Chem 30(4):583–591
27. Diamantidis G, Effosse A, Potier P, Bally R (2000) Purification and characterization of the first bacterial laccase in the rhizospheric bacterium Azospirillum lipoferum. Soil Biol Biochem 32(7):919–927
28. Feng S, Su Y, Dong M, He X, Kumaresan D, O'Donnell AG, Wu J, Chen X (2015) Laccase activity is proportional to the abundance of bacterial laccase-like genes in soil from subtropical arable land. World J Microbiol Biotechnol 31:2039–2045
29. Flickinger MC, Drew SW (1999) Encyclopedia of bioprocess technology: fermentation, biocatalysis, and bioseparation
30. Fuerst EP, Okubara PA, Anderson JV, Morris CF (2014) Polyphenol oxidase as a biochemical seed defense mechanism. Front Plant Sci 5:689
31. Gianfreda L, Xu F, Bollag J-M (1999) Laccases: a useful group of oxidoreductive enzymes. Bioremediat J 3(1):1–26
32. Giardina P, Faraco V, Pezzella C, Piscitelli A, Vanhulle S, Sannia G (2010) Laccases: a never-ending story. Cell Mol Life Sci 67:369–385
33. Grinhut T, Hadar Y, Chen Y (2007) Degradation and transformation of humic substances by saprotrophic fungi: processes and mechanisms. Fungal Biol Rev 21(4):179–189
34. Guseva T, Shchelchkov K, Sanzharovskiy A, Molchanova Y (2019) Best available techniques, energy efficiency enhancement and carbon emissions reduction. Int Multidisc Scie GeoConf: SGEM 19(5.1):63–70
35. Hölker U, Dohse J, Höfer M (2002) Extracellular laccases in ascomycetes Trichoderma atroviride and Trichoderma harzianum. Folia Microbiol 47:423–427
36. Janusz G, Pawlik A, Sulej J, Świderska-Burek U, Jarosz-Wilkołazka A, Paszczyński A (2017) Lignin degradation: microorganisms, enzymes involved, genomes analysis and evolution. FEMS Microbiol Rev 41(6):941–962
37. Jolivalt C, Raynal A, Caminade E, Kokel B, Le Goffic F, Mougin C (1999) Transformation of N′, N′-dimethyl-N-(hydroxyphenyl) ureas by laccase from the white rot fungus Trametes versicolor. Appl Microbiol Biotechnol 51:676–681
38. Kallio JP, Gasparetti C, Andberg M, Boer H, Koivula A, Kruus K, Rouvinen J, Hakulinen N (2011) Crystal structure of an ascomycete fungal laccase from Thielavia arenaria–common structural features of asco-laccases. FEBS J 278(13):2283–2295

39. Karlsson I, Rootzén J, Johnsson F (2020) Reaching net-zero carbon emissions in construction supply chains–analysis of a Swedish road construction project. Renew Sustain Energy Rev 120:109651
40. Kim C, Lorenz WW, Hoopes JT, Dean JF (2001) Oxidation of phenolate siderophores by the multicopper oxidase encoded by the Escherichia coli yacK gene. J Bacteriol 183(16):4866–4875
41. Kramer KJ, Kanost MR, Hopkins TL, Jiang H, Zhu YC, Xu R, Kerwin J, Turecek F (2001) Oxidative conjugation of catechols with proteins in insect skeletal systems. Tetrahedron 57(2):385–392
42. Kudanga T, Le Roes-Hill M (2014) Laccase applications in biofuels production: current status and future prospects. Appl Microbiol Biotechnol 98:6525–6542
43. Kunamneni A, Plou FJ, Ballesteros A, Alcalde M (2008) Laccases and their applications: a patent review. Recent Pat Biotechnol 2(1):10–24
44. Kuo H-C, Détry N, Choi J, Lee Y-H (2015) Potential roles of laccases on virulence of Heterobasidion annosum ss. Microb Pathog 81:16–21
45. Lee SC, Dickson DW, Casadevall A (1996) Pathology of cryptococcal meningoencephalitis: analysis of 27 patients with pathogenetic implications. Hum Pathol 27(8):839–847
46. Leontievsky A, Myasoedova N, Baskunov B, Golovleva L, Bucke C, Evans C (2001) Transformation of 2, 4, 6-trichlorophenol by free and immobilized fungal laccase. Appl Microbiol Biotechnol 57:85–91
47. Liu L, Tewari RP, Williamson PR (1999) Laccase protects Cryptococcus neoformans from antifungal activity of alveolar macrophages. Infect Immun 67(11):6034–6039
48. Machczynski MC, Vijgenboom E, Samyn B, Canters GW (2004) Characterization of SLAC: a small laccase from Streptomyces coelicolor with unprecedented activity. Protein Sci 13(9):2388–2397
49. Martin C, Corvini PF, Vinken R, Junghanns C, Krauss G, Schlosser D (2009) Quantification of the influence of extracellular laccase and intracellular reactions on the isomer-specific biotransformation of the xenoestrogen technical nonylphenol by the aquatic hyphomycete Clavariopsis aquatica. Appl Environ Microbiol 75(13):4398–4409
50. Martins LO, Soares CM, Pereira MM, Teixeira M, Costa T, Jones GH, Henriques AO (2002) Molecular and biochemical characterization of a highly stable bacterial laccase that occurs as a structural component of the Bacillus subtilis endospore coat. J Biol Chem 277(21):18849–18859
51. Mayer AM, Staples RC (2002) Laccase: new functions for an old enzyme. Phytochemistry 60(6):551–565
52. Mayolo-Deloisa K, González-González M, Rito-Palomares M (2020) Laccases in food industry: Bioprocessing, potential industrial and biotechnological applications. Front Bioeng Biotechnol 8:222
53. Ménard I, Thiffault E, Kurz WA, Boucher J-F (2023) Carbon sequestration and emission mitigation potential of afforestation and reforestation of unproductive territories. New Forest 54(6):1013–1035
54. Meys R, Kätelhön A, Bachmann M, Winter B, Zibunas C, Suh S, Bardow A (2021) Achieving net-zero greenhouse gas emission plastics by a circular carbon economy. Science 374(6563):71–76
55. Mikolasch A, Schauer F (2009) Fungal laccases as tools for the synthesis of new hybrid molecules and biomaterials. Appl Microbiol Biotechnol 82:605–624
56. Miyazaki K (2005) A hyperthermophilic laccase from Thermus thermophilus HB27. Extremophiles 9:415–425
57. Molitor C, Mauracher SG, Rompel A (2016) Aurone synthase is a catechol oxidase with hydroxylase activity and provides insights into the mechanism of plant polyphenol oxidases. Proc Natl Acad Sci 113(13):E1806–E1815
58. Moreno LF, Feng P, Weiss VA, Vicente VA, Stielow JB, de Hoog S (2017) Phylogenomic analyses reveal the diversity of laccase-coding genes in Fonsecaea genomes. PLoS One 12(2):e0171291

59. Munk L, Sitarz AK, Kalyani DC, Mikkelsen JD, Meyer AS (2015) Can laccases catalyze bond cleavage in lignin? Biotechnol Adv 33(1):13–24
60. Nosanchuk JD, Casadevall A (2003) The contribution of melanin to microbial pathogenesis. Cell Microbiol 5(4):203–223
61. Obobisa ES (2022) Achieving 1.5 C and net-zero emissions target: the role of renewable energy and financial development. Renew Energy 188:967–985
62. Otto B, Schlosser D (2014) First laccase in green algae: purification and characterization of an extracellular phenol oxidase from Tetracystis aeria. Planta 240:1225–1236
63. Palmieri G, Cennamo G, Faraco V, Amoresano A, Sannia G, Giardina P (2003) A typical laccase isoenzymes from copper supplemented Pleurotus ostreatus cultures. Enzyme Microb Technol 33(2–3):220–230
64. Pandey N, Aronson A (1979) Properties of the Bacillus subtilis spore coat. J Bacteriol 137(3):1208–1218
65. Piscitelli A, Giardina P, Lettera V, Pezzella C, Sannia G, Faraco V (2011) Induction and transcriptional regulation of laccases in fungi. Curr Genomics 12(2):104–112
66. Piscitelli A, Pezzella C, Giardina P, Faraco V, Sannia G (2010) Heterologous laccase production and its role in industrial applications. Bioeng Bugs 1(4):254–264
67. Polak J, Jarosz-Wilkołazka A (2007) Reakcje katalizowane przez lakazę–mechanizm i zastosowanie w biotechnologii. Biotechnologia 4(79):82–94
68. Polak J, Jarosz-Wilkolazka A (2012) Structure/Redox potential relationship of simple organic compounds as potential precursors of dyes for laccase-mediated transformation. Biotechnol Prog 28(1):93–102
69. Qasim M, Bukhari SA, Ghani MJ, Masoud MS, Huma T, Arshad M, Haque A, Ibrahim Z, Javed S, Rajoka MI (2016) Relationship of oxidative stress with elevated level of DNA damage and homocysteine in cardiovascular disease patients. Pak J Pharm Sci 2:29
70. Rackley SA (2017) Carbon capture and storage. Butterworth-Heinemann
71. Rochefort D, Leech D, Bourbonnais R (2004) Electron transfer mediator systems for bleaching of paper pulp. Green Chem 6(1):14–24
72. Rogalski J, Wojtas-Wasilewska M, Apalovič R, Leonowicz A (1991) Affinity chromatography as a rapid and convenient method for purification of fungal laccases. Biotechnol Bioeng 37(8):770–777
73. Ruijssenaars H, Hartmans S (2004) A cloned Bacillus halodurans multicopper oxidase exhibiting alkaline laccase activity. Appl Microbiol Biotechnol 65:177–182
74. Sakurai T, Kataoka K (2007) Basic and applied features of multicopper oxidases, CueO, bilirubin oxidase, and laccase. Chem Rec 7(4):220–229
75. Sharma A, Jain KK, Jain A, Kidwai M, Kuhad R (2018) Bifunctional in vivo role of laccase exploited in multiple biotechnological applications. Appl Microbiol Biotechnol 102:10327–10343
76. Sharif N, Munir N, Hasnain M, Naz S, Arshad M (2021) Environmental impacts of ethanol production system. In: Sustainable ethanol and climate change: sustainability assessment for ethanol distilleries, pp 205–223
77. Sinsabaugh RL (2010) Phenol oxidase, peroxidase and organic matter dynamics of soil. Soil Biol Biochem 42(3):391–404
78. Solano F (2014) Melanins: skin pigments and much more—types, structural models, biological functions, and formation routes. New J Sci 2014:1–28
79. Srinivasan P, Thiyagarajan P, Yuvarajan R, Selvankumar T (2021) Laccase enzyme in nanotechnology and their environmental applications. Int J Adv Sci Eng 8(2):2210
80. Sugumaran M (2002) Comparative biochemistry of eumelanogenesis and the protective roles of phenoloxidase and melanin in insects. Pigment Cell Res 15(1):2–9
81. Taranto F, Pasqualone A, Mangini G, Tripodi P, Miazzi MM, Pavan S, Montemurro C (2017) Polyphenol oxidases in crops: biochemical, physiological and genetic aspects. Int J Mol Sci 18(2):377
82. Thurston CF (1994) The structure and function of fungal laccases. Microbiology 140(1):19–26

83. Tsai H-F, Wheeler MH, Chang YC, Kwon-Chung K (1999) A developmentally regulated gene cluster involved in conidial pigment biosynthesis in Aspergillus fumigatus. J Bacteriol 181(20):6469–6477
84. Virk AP, Sharma P, Capalash N (2012) Use of laccase in pulp and paper industry. Biotechnol Prog 28(1):21–32
85. Walker JR, Ferrar PH (1998) Diphenol oxidases, enzyme-catalysed browning and plant disease resistance. Biotechnol Genet Eng Rev 15(1):457–498
86. Wang Y, Chen Z-H (2019) Bioinformatics and enzymatics investigation of Trametes laccase for optical biosensing application. J Mater Sci 54(6):4970–4983
87. Williamson PR (2016) Role of laccase in the virulence of Talaromyces marneffei: a common link between AIDS-related fungal pathogens? Virulence 7(6):627–629
88. Wood D (1980) Production, purification and properties of extracellular laccase of Agaricus bisporus. Microbiology 117(2):327–338
89. Wosilait WD, Nason A, Terrell AJ (1954) Pyridine nucleotide-quinone reductase: II. Rôle in electron transport. J Biol Chem 206(1):271–282
90. Wu XT, Ansari AR, Pang XX, Li HZ, Zhang ZW, Luo Y, Arshad M, Song H (2019) Visfatin plays a significant role in alleviating lipopolysaccharide-induced apoptosis and autophagy through PI3K/AKT signaling pathway during acute lung injury in mice. Arch Immunol Ther Exp 1(67):249–261

Catalases: Role in Pollution Reduction Caused by Dyes

Aijaz Ali Otho, Sakib Hussain Laghari, Mujahid Mehdi, Sohail Shahzad, Sarfaraz Ahmed Mahesar, Rabia Asma Memon, Saeed Akhter Abro, Altaf Ahmed Simair, Abdul Hameed Kori, and Saeeda Rind

Abstract The textile industry annually releases a substantial quantity of dyes into liquid effluents, posing environmental risks due to their high toxicity, carcinogenicity, and xenobiotic nature. The imperative for eco-friendly dye removal methods has prompted exploration into various approaches, including physico-chemical and biological methods. Among the biological methods, enzymes such as catalase, peroxidases, manganese peroxidases, lignin peroxidases, laccases, micro peroxidase-11, polyphenol oxidases, and azoreductases are commonly employed. However, the extraction of natural enzymes is laborious and time-consuming. To overcome these challenges, researchers have developed enzyme-mimicking nanomaterials (NMs) as alternatives to natural enzymes. Nano-enzymes (NZs) exhibit advantages such as low cost, ease of bulk preparation, high stability, biocompatibility, and resilience to harsh conditions. This chapter explores carbon-based NMs, widely utilized in NZ production for their well-defined electronic and geometric structures, enabling efficient mimicry of highly evolved catalytic centers in natural enzymes. Carbon dots

A. A. Otho (✉) · R. A. Memon · S. A. Abro
Institute of Plant Sciences, University of Sindh, Jamshoro 76080, Pakistan
e-mail: aijazotho@gmail.com

S. H. Laghari · S. A. Mahesar · A. H. Kori
National Centre of Excellence in Analytical Chemistry, University of Sindh, Jamshoro 76080, Pakistan

M. Mehdi
Department of Textile Design, Faculty of Design, Aror University of Art, Architecture, Design and Heritage, Sukkur 65200, Pakistan

S. Shahzad
Department of Chemistry, University of Sahiwal, Sahiwal 57000, Pakistan

A. A. Simair
Department of Botany, Government College University Hyderabad, Hyderabad 71000, Pakistan

S. Rind
State Key Laboratory of Electroanalytical Chemistry, Changchun Institute of Applied Chemistry, Chinese Academy of Sciences, Changchun, Jilin 130022, P.R. China

School of Applied Chemistry and Engineering, University of Science and Technology of China, Hefei, Anhui 230026, P.R. China

M. Arshad (ed.), *Enzymes in Textile Processing: A Climate Changes Mitigation Approach*, SDGs and Textiles, https://doi.org/10.1007/978-981-97-8058-7_9

(CDs) or quantum dots (QDs) within these NMs also demonstrate the scavenging ability of free radicals, contributing to the production of azo compounds. Notably, CDs exhibit superior reactive oxygen species (ROS) scavenging compared to other compounds, showcasing their potential in environmental remediation.

Graphical Abstract

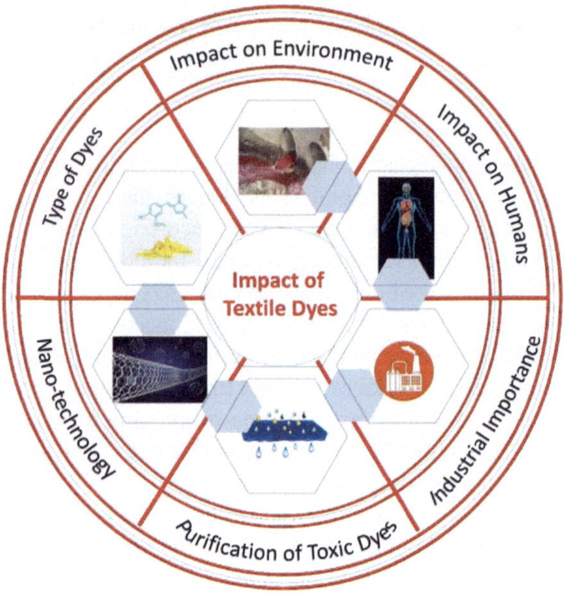

1 Introduction

The fastest growth of industrialization and population has increased water pollution globally. Water is a crucial element for life on earth as industries, agriculture and the general population are using water and releasing various types of pollutants in the wastewater [150]. However, industrial discharge, agriculture practices and human beings play a vital role in the increasing pollutant ratio in the environment [132]. Almost every year, more than 100,000 toxic chemicals are excluded from the environmental origin, due to their increasing manufacture, consumption and discarding [111]. The water pollution caused by waste sewage from the textile industry is a serious concern because it may contain toxic chemicals, thus, the human beings may be affected by those toxic chemicals [73, 77]. Pollutants are classified as pririty pollutants (PPs) and emerging pollutants (EPs). The PPs are a group of substances which may include pesticides [50], metals, phthalates [97], polycyclic aromatic hydrocarbons (PAHs) [160] and endocrine disruptors [124], whereas emerging pollutants (EPs) may include dyes [96], pharmaceuticals and personal care products (PPCPs)

[91], hormones [40], plasticizers [129] and fire retardants [53]. Among them, pollutants such as organic dyes from EPs are the vital contaminants form the factory waste effluents [86]. Dyes defined as synthetic or natural Chroma that are water-miss, solvable and have potential use in various textile industries. A higher amount of dyes lost to liquid effluents from the textile industry every year. The dyes are highly toxic, carcinogenic and xenobiotic for living organisms [83, 92, 131]. The large user of dye is the textile industry. The dyes present in wastewater may inadequately affect the intense colour of aqueous media due to which sunlight cannot penetrate the interior aqueous body because the textile pollutant is dark in colour [30]. There are more than 100,000 dyes available commercially of which azo dye is present about 70% by weight [66]. Azo and triphenylmethane dyes are broadly used in the textile industry. Every year, over 1 million dyes are produced, among which 50% are textile dyes [26]. The number of dyes belonging to the azo dyes family is well known for the non-biodegradability in aerobic conditions [161]. According to various reports, the textile industry may consume 1.8 million litres of water every single day to produce 8000 kg of fabrics [43]. Azo dyes are highly colored, quickly apparent and this may lead to significant environmental problems. Almost 18–20% of the textile industry's polluted water comes from dyeing and finishing. An azo dye may cause bladder cancer to splenetic sarcomas in human beings [142]. The dyes treatment can be done by numerous methods from the effluents such as physio-chemical and biological methods. Physio-chemical methods may include adsorption, ion exchange, chemical precipitation, coagulation-flocculation, oxidation, electrochemical treatment and membrane filtration. Biological methods may include bacteria, fungi, algae, plants and enzymes. Among them, the biological treatments were carried out by using catalase enzyme to remove or biodegrade the dye from the wastewater. Several biological enzymes are available including catalase, peroxidases, manganese peroxidases, lignin peroxidases, laccases, micro peroxidase-11, polyphenol oxidases, and azoreductases to remove or degrade the dyes from the wastewater but these natural enzymes possess few drawbacks or limitations in the application of dye removal or bioremediation. Natural enzymes have limitations because of their origin from biological sources, they are expensive in terms of their purification and extraction process. Natural enzymes are unstable, sensitive and difficult to store for a longer time; if they are stored for a long time, then they may lose their activity. To overcome these limitations, researchers have developed enzymes mimicking nanomaterials (which are alternatives to the natural enzymes. Nano-enzymes (may have low cost, ease of preparation in bulk, are highly stable, biocompatible, and can tolerate harsh conditions, promising shape, size, composition and large surface area. Several NZs with different compositions were utilized for the degradation of dyes from the waste effluent. Nano-carbons may have covalently bonded carbon atoms with various hybridization states (sp, sp^2, and sp^3), which may deliver a great morphology and structures. The Nano-carbons emerged as mimetic in the field of NZs due to their both catalytic and enzymatic properties. The large surface area of carbon nano-enzymes (CNZs) may enhance their active sites accelerate the electron transfer in the catalytic reaction and help them to work as the direct proxy to natural enzymes. In this chapter, the artificial

carbon-based and other catalase-like nano-enzymes (CAT-NZs) has been discussed for the removal of dye from wastewater.

2 Classification of Dyes

The dyes are complex organic compounds which are used by a variety of industries to impart color to their products. Natural as well as synthetic dyes are of great interest since they play a vital role in our everyday life. The dye used for printing of textiles, paper, leather or other material applications from which 70% of dyes are not fixed to the fibre [26]. Dyes are carcinogenic and allergic in nature [4]. The more important group of dye molecules is chromophore. However, the dyes are organic compounds which may contain 3 important groups in their molecules: (1) chromophore, (2) auxochrome, and (3) matrix [38]. The active sites of chromphore are crucial of the dye structure, summarized as the dimensional localization of atoms absorbing light energy. The chromophore may consist of group of atoms—the most frequent groups are azo (–N=N–), nitro ($–NO_2$), nitroso (–N=O), thiocarbonyl (–C=S), carbonyl (–C=O), ethene ($CH = CH$), sulfonate ($–SO_3H$), and hydroxyl (OH) along with the alkenes (–C=C–). The absorption peaks exhibited by the chromophore are due to the stimulation of electrons within a molecule and it becomes chromogenic [68]. Besides this, auxochrome is a molecule when a chromogenic molecule can achieve dyeing solely through the addition of a group of atoms. The auxochromic groups responsible for the fixation of dyes and have the potential to modify the color of the dye. They are fundamentally alkaline (NH_2, NHR and NR_2) or acidic (COOH, SO_3 and OH) [103]. The rest of the molecules are related to the matrix which is the third part of the dye. The functional groups present in dye like carboxylic, amine, and azo groups may pose challenges for treatment using conventional methods. The correlation of chromphoric, axuochromic and conjugated structures is like benzene ring, anthracene, and perylene [9]. The enhancement in aromatic nuclei also enhances the double bond conjugation. The reduction in the energy of π-bond may lead to an increase in the influence of π or n electrons, which can lead to a shift towards longer wavelengths [153]. Similarly, when an auxochrome group (such as amino, hydroxyl, alkoxy, etc.) is positioned on a conjugated aromatic system, it can result in the extension of the π-system conjugation. This, in turn, leads to hyperconjugation and the absorption of longer wavelengths by the molecule, resulting in a darker color [122]. The broad range of textile dyes may include soluble (acid, base, reactive, direct) and insoluble dyes (vat, sulfur, disperse, pigment). The classification of dyes became mandatory because of universal fabrication of dyes. Dyes are classified based on chemical composition such as azo, anthraquinone, nitro, nitroso, indigoid, cyanine, phthalocyanine and triphenylmethane [27].

3 Soluble Dyes

3.1 Acidic Dyes

The dyeing process in the aqueous solution may utilize the acidic dyes. The pH range of acidic dyes 1.0–6.0. This type of dye utilized for the deliquescent wool and nylon about 40–50%. The sulfonic group present in acidic dye or this may show the solubility in basic media [137].

3.2 Basic Dyes

This type of dye is obtained from the organic origins which may produce the cationic pigments. These dyes are present in the form of chloride, double and oxalate salts which may contain the $ZnCl_2$. The extraordinary properties of dyes are luster and intensity. Examples are Basic Yellow 2, Basic Green 4, and Basic Blue 12 [145].

3.3 Reactive Dyes

This category of dye is soluble and can chemically bond with both cotton and wool fibers, effectively adhering to the fiber substrate. Sodium chloride (NaCl) and sodium sulphate (Na_2SO_4) are used to transfer dye to the fibre. The groups which are capable of reacting with the fibre sites are the hydroxyl (OH^-), thiol (-SH) or amine (NH_2) groups. This may include Reactive Red 3, Reactive Blue 19 [72].

3.4 Direct Dyes

The linen, rayon and cellulosic fiber may interact with this type of dyes. Direct dyes weakly bonded to the molecule fiber. When applied to the fiber, they do not quickly get dry. This type of dyes are used at the temperatures between 80 and 94 °C. This may include Direct Blue 1, Direct Blue 86, and Direct Blue 106 [76].

4 Insoluble Dyes

4.1 Disperse Dyes

The disperse dyes are water-insoluble dyes assembled by hydrophobic nature of fibers such as polyester or cellulose acetate with one or more than one atom. The dye easily vaporizes and disperses, being absorbed by hydrophobic fibers. This may include Disperse Orange 30, Disperse Violet 1, and Disperse Yellow 26 [35].

4.2 Vat Dye

These types of dyes are miscible in hot water and few miscible with sodium carbonate (Na_2CO_3). Indigo and indigotic produced from the Indigo era are the general examples of vat dyes. This may include Vat Blue 1, Vat Black 25, and Vat Green 1 [168].

4.3 Sulfur Dyes

The sulfur dyes were produced around 110,000–120,000 tons per year worldwide. This type of dye applied on the cellulose fiber, mostly 90% are black sulfur dyes. Examples are Sulfur Blue 15, Leuco Sulfur Black, and Sulfur Green 3 [130].

5 Toxicity and the Negative Impact on (Dyes Environmental Ecosystems)

The water covers around 71% of earth's surface. In the current era, the purity of lacks, high quality drinking water is a challenging task and demand of clean drinking water in enhancing at alarming rate. Around 70, 19, and 11% is used by agriculture, industry and household [61]. The waste water containing long term discharge of untreated dyes can cause the severe eco-toxicological risk to the multiple environmental ecosystem. The fertility of soil and crop fertilization life cycle is concerned with the aquatic life. The chemical oxygen demand (COD), biological oxygen demand (BOD) and total organic carbon (TOC) along with the suspended solids may reduce the photosynthesis toxicity of numerous types of intermediates, such as amino acids and plant growth factors, and can contribute to the development of resistance, bioaggregation, and are all connected with the color of the dye [114]. The residue left after the chemical used for the dye manufacturing can be evaporated and it may pollute the air and water. Respiratory and skin allergic problems might be caused after intake of dye

[42]. When azo dyes get dissolved in water, they become more toxic and metabolized by enzymes. Carnivores such as human beings and fishes affected by ingestion of dyes may have toxic effects. Heavy metals such as chromium (Cr), copper (Cu), nickel (Ni), lead (Pb), mercury (Hg), cadmium (Cd), arsenic (As), and zinc (Zn) [164] are present in the fixing, oxidizing, cross-linking agents and catalyst which are used in dying process. Accumulation of heavy metals in the food chain can lead to severe effects on both human health and plants. Gita et al. [63] carried out an toxicological study upon the reduced mortality and cytotoxicity effects on red blood cells (RBCs) and also found the reduced in the count and alteration in shape and size. Athira and Jaya [20] suggested that anthraquinone dyes may have some effects on the biochemical features of muscles of two fishes *Channa punctatus* and *Cyprinus carpio*. Tkaczyk et al. [162] showed that the dye was associated with malachite green in fish samples, with concentrations ranging between 265 and 1663 μg/kg. Sharma et al. [146] conducted a study on the toxicological effects induced by 2 ABS, along with its impact on the liver and gills of fish.

6 Toxicity and the Negative Impact of Dyes on the Humans

The poor quality of water can directly impact the health of various organs, ranging from simpler to more complex organisms. The highly toxic effects of dyes can pose a threat to human health. Industries located near processing plants contribute to a large accumulation of dyes in the soil [94]. While these dyes may be initially stabilized in the soil for a short period, there is potential for prolonged retention. Human exposure to these dyes occurs in two main ways [71]: through the consumption of contaminated food or crops grown in soil containing toxic dyes, leading to their accumulation in the human body in degraded forms. Consuming contaminated food introduces the dye into human organs, potentially causing a series of health issues, compounded by the presence of synthetic dye compounds such as benzidine and 2-naphthylamine, associated with the development of bladder cancer [57]. Additionally, dyes have been found to be more impactful on asthma due to their chromogen, which facilitates binding to hydroxyl groups on cellulose fibers. This interaction may trigger an immune response, with the dyes acting as haptens and potentially exacerbating asthma symptoms. Al-Tohamy et al. [6] showed that the textile dyes, which are highly toxic and carcinogenic, have been linked to a various type of human diseases. Textile dyes may cause diseases from dermatitis to central nervous system. Lellis et al. [93] concluded that basic Red 9 dye which is used in the textile, leather, paper, and ink industries may lead to carcinogenic effects in humans [109]. Benzidine causes cancer of the bladder in humans, due to the azo dyes intestinal microflora metabolized to their parent amines. These amine derivatives, dissimilar to their parent compounds, are quickly absorbed by the gut [151]. Sudan I dye (Solvent Yellow 14) belongs to the category of azo-lipophilic complexes frequently employed in industrial applications. Upon entering the bodies of humans and animals, it undergoes enzymatic transformation into carcinogenic aromatic amines. Bhardwaj et al. [29]

found that the oral ingestion and inhalation have occurrence of acute toxicity as well irritation to eyes and skin.

7 Industrial Importance of dyes

The dyes are highly colored, stable organic compounds that may produce the brighter colour for various Substrates like textiles, paper printing, cosmetics, plastics, waxes, leather, and pharmaceuticals are common applications for colorants [2]. There are two primary types of colorants: dyes and pigments. Dyes are characterized by properties such as a straightforward chemical structure and high stability in water and covalently bonded with substrates due to this reason dyes are more beneficial than pigments. Dyes are colored because they sponge up the light between the wavelength range of 400–700 nm in the UV–visible spectrum. Triphenylmethane dyes may have their importance because they are cheap and brilliant in color along with the shades of red, violet, blue and green [85]. The major role of this dye is to dyeing the nylon, wool, silk and cotton. Indigo is one of the most important dyes in the textile industry which has been used since 5000 years. Saffron is used as natural dye in the cosmetic and food industry. Sunset yellow (SY) is present in thousands of food and drugs [65].

8 Oxidoreductive Enzymes

Oxidoreductive enzymes derived from animals and plants, encompassing both reductive and oxidative enzymes, have the potential to produce free radicals, initiating a complex cascade of reactions. These enzymes exhibit their highest level of activity in this context [44].

However, the presence of these oxidoreductive enzyme may target the maximum in terms of concentration. These enzymes may produce a higher amount of reactive free radicals. Various types of enzymes have been used by the bacterium to decolorize the dye. Azoreductase is a prevalent reductive enzyme tasked with the degradation or decolorization of dyes. The oxidoreductive enzymes are peroxidases, laccases and catalases [108].

9 Catalases

The hydrogen peroxide (H_2O_2) is broken down by using catalase enzyme, which is well known antioxidant because the H_2O_2 metabolism commonly controlled by this type of enzyme or any distress in imbalance between antioxidants and reactive species leads to a physiological condition known as oxidative stress [79]. Catalase, among the crucial antioxidant enzymes, significantly alleviates oxidative stress by converting cellular H_2O_2 into water and oxygen [107]. Catalase generally used enzyme that present in all type of living organisms. The biggest turnover of this enzyme is that it can breakdown each molecule of the enzyme can decompose more than one million molecules of H_2O_2 [52]. In bioremediation or biodegradation, catalase stands out as a crucial enzyme. The catalase enzyme functions as an oxidoreductase, playing a crucial role in neutralizing ROS like hydrogen peroxide, commonly generated as a byproduct in aerobic conditions [104]. This enzyme protects the cell against the stress as it is antioxidant.

$$\mathbf{2H_2O_2}\ \text{Catalase} \rightarrow \mathbf{2H_2O_{+}O_2}$$

Catalase may have tetramer of at least four polypeptide chains and approximately 500 amino acids; it may also feature porphyrin heme iron groups, enabling the enzyme to interact with hydrogen peroxide [47]. The interesting thing is that it may work in the range of 4–11 pH. In the living organisms, catalase is situated within a cellular, bipolar organelle known as the peroxisome [64], this enzyme is present in various organisms, both aerobic and anaerobic. Its applications extend to diverse fields including food processing, pharmaceuticals, and the textile industry [149]. The primary emphasis of this chapter is on the utilization of catalase enzyme in the process of bioremediation for the removal of dyes, contributing to environmental cleanup. Catalase (CAT) [21], catalase-peroxidases (Kat) [121] and manganese catalases [186]. The unifunctional catalases are most efficient groups among them. The huge sub kits such as 80–90 kDa may have been in active mass center. The dual functional roles of peroxidase involve both catalytic activity and catalase-peroxidase. This combination exhibits the availability of both heme and methionine tyrosine-tryptophan, facilitating the transfer of electrons from various substrates such as heme and manganese CN– or N– [11, 183]. Subsequently, the dual functionality of catalase is demonstrated in *Bacillus pumilus* and *Scytalidium thermophilum.* In *Amaranthus Cruentus*, CATPO enzymes (catalase/phenol oxidase bifunctional) showcase both catalytic and phenolic oxidase activities, allowing the oxidation of phenolic ingredients [159].

9.1 *Catalase Mechanism of Neutralizing H_2O_2*

Various types of contaminants may exist in oxidized states and undergo biotransformation to generate free radicals, including radical thiyl (RS·), hydroxyl radical (OH·), superoxide ($O\cdot^{-}$), nitric oxide (NO·), and hydroperoxyl radical (HO*) [1]. The inherent instability and reactivity found in free radicals make them prone to radial interactions with various molecules, leading to cellular damage. The reactive oxygen species (ROS) are catalytically converted into H_2O_2 by using particular enzymes such as superoxide dismutase (SOD) [16, 67]. Heavy metals such as Fe may have a strong electron affinity for H_2O_2 and undergo a Fenton reaction with H_2O_2 to extremely reactive with OH radicals, which can attack the various biomolecules. Through a stepwise biodegradation mechanism, catalase can effectively break down the H_2O_2 molecule into H_2O and O_2. The heme, undergoing oxidation, generates H_2O_2 through a π-cationic porphyrin radical oxyferryl species have been identified in this context [101]. This sequence entails the creation of compound I (Comp I), where a proton is transferred from one O_2 of the peroxide and binds to the iron center. In the subsequent stage of Comp I, it reverts to its initial state position, accompanied by the net consumption of two H_2O_2 molecules per cycle [154].

$$H_2O_2 + \text{Enzyme}\left(\text{Por} - \text{Fe}^{\text{III}}\right) \cdot\!\rightarrow H_2O + \text{comp I}\left(\text{Por}^{+\circ} - \text{Fe}^{\text{IV}} = 0\right)$$
$$H_2O_2 + \text{Comp I}\left(\text{Por}^{+\circ} - \text{Fe}^{\text{IV}} = 0\right) \rightarrow H_2O + \text{Enzyme}\left(\text{Por} - \text{Fe}^{\text{III}}\right) + 0$$

Catalase can catalyze the peroxidase reaction in the presence of ethanol, phenol, or heavy metals. However, the compound I may go into reduction condition to form compound II. The H_2O_2 is oxidized by inactive compound III [95].

$$\text{Comp I}\left(\text{Por}^{+\circ} - \text{Fe}^{\text{IV}} = 0\right) + AH_2 \rightarrow \text{Comp II}\left(\text{Por} - \text{Fe}^{\text{IV}} - \text{OH}\right) + AH^{\circ}$$
$$\text{Comp II}\left(\text{Por} - \text{Fe}^{\text{IV}} - \text{OH}\right) + H_2O_2 \rightarrow \text{Comp III}\left(\text{Por} - \text{Fe}^{\text{III}} - O_2^{-\circ}\right) + H_2O$$

The compound oxyferryl radicals release coupled with the compound II production. Active sites of the surface present with the co factors NADPH may prevent the generated radicals by compound II by the inactivation of catalase enzyme [14, 148].

9.2 *Structure of Catalases*

Catalase is a crystallized from the bovine liver. However, in the 1980s the 3D structure was elaborated. The crystal growth experiments may utilize the catalase from the eukaryotes and prokaryotes [88]: *Escherichia coli* hydro peroxidase II, penicillium vital catalase *Saccharomyces cerevisiae*, and human erythrocytes catalase (HEC). The proteolysis and denaturation of pH thermal due to harshness and strength are resistant to the stable structure of catalase [120]. The 3D positioning folds of catalase may result in tetrameric subunits of unique structural domains—the carbon (C) and nitrogen (N) terminal arm and b-barrel domains or bundle up loop or helix region [34]. The thermal stability of catalase may enhance the non-covalent interactions by each C and N terminals. The selective length of 20Å is confirmed the X-crystallography by directed sites to the mutagenesis potentials to deliver the H_2O_2 via H_2O_2 H bonding and dipole to the active sites such as haem and it may help with the adjustment of highly level entry of substrate which is selective as potential designed for drugs [24, 59]. Interconnection of antioxidants may bring to being in the turmeric plants along with the enzyme developed as remarkable. The amino residue alteration may have got efficiency. The catalase from BLC in the presence of curcumin demonstrated the amino acids distance and size enhancement [133]. The catalase sites such as haem may readily allow the substrate permission. The b-helix protein increases in the curcumin in the different types of catalase spectra [39]. Structure of catalase stabilized by secondary coil suggests the reduction followed by catalase. The carboxylic group of aspartic and glutamic acids is responsible for creating prominent gaps in the secondary structure of proteins [143].

9.3 Catalases in Textile Dye Decolorization

A large quantity of synthetic dyes is released into the aquatic environments by the textile industry. The potential toxic effects caused by dyes in living organisms include carcinogenic and mutagenic effects. The food, cosmetics, tanneries, pharmaceuticals and cloths used the dyes during the dyeing process [113]. About 15% or 280 kT (kilotons) dyes left unfixed during the dyeing process. The wastewater contains at least 280 kT dyes. The discarded wastewater may accommodate the unfixed molecule from the local areas. The health risk issues are extraordinary to increase the environmental pollutants [23, 41]. A number of methods are available for dye removal but they may have restrictions such as greater toxicity and higher cost [127]. The dye degradation can be done by using plants, mammalian cells and algae based catalase. Kadam and co-workers demonstrated the decolorization of Methyl Orange, Malachite Green, Rubin GFL, and Red HE3B [81]. Donga and Chanda [48] studied the catalase increased activity along with highest degradation rate up to 83% and noted at a concentration of 70 mg/L for Red HE3B. Dye degradation mechanism of dye is not understood yet by using catalase. ROS and oxidative stress are produced by textile dyes. The ROS will be converted in H_2O_2 by using SOD and catalase neutralizes active dye degradation at higher stress. The dye degradation of C.I acid blue 92 rubine GFL by using plant catalase (Lemna minor L., Salvinia molesta) [128] was reported. The pH reduced from 3.4 to 1.6 demonstrated in the dye degradation into organic acids when the effect of ozone catalase on reactive black 5 dyes was compared with the COD with the degradation efficiency of 34–43 with 95% dye removal [31].

10 Removal of Organic Pollutants by Using Physio-Chemical and Biological Methods

The physio-chemical methods may include the adsorption [138], filtration [165], coagulation [155], chemical flocculation [117], electrochemical destruction [118], advanced oxidation processes (AOPs) [62]. The physio-chemical methods are cost-effective and have few drawbacks. However, the coagulation, sedimentation and flocculation techniques are explored as the most attractive approaches to remove the pollutants [74, 123], while osmosis and filtration are effective but are most expensive and may produce secondary waste as well. Besides this, the AOPs are one of the most promising sets of chemical remediation methods which are costly and lots of energy demanding methods. Biological approaches may contain less expensive and eco-friendly alternatives for removing the organic dye pollutant [22]. Several microorganisms like bacteria, fungi, yeast, and algae are employed in the bioremediation or biodegradation of dyes. Most of the dyes are removed from the wastewater through the activity of oxidoreductase enzymes found within microorganisms [28]. On the basis of microorganisms, the oxidoreductase enzyme can be classified as catalase, peroxidases, manganese peroxidases, lignin peroxidases, laccases etc. Dincer [45] used the chemical (BOD_5/COD = 0.48) 270-min time, which converted the Reactive Black5 readily. Alizadeh et al. [7] used biological method and demonstrated the dye degradation by using 200 mg L^{-1}, ZA14, which showed the highest decolorization (90%). [171] used the biological method with *Lemna minor* showing the attractive results in a reduction of both Methylene Blue 99% and Congo red 25% after a 24-day period. Sarkar et al. [135] used biological method to synthesize silver nanoparticle (Ag-NPs) which possess remarkable degradation of methylene blue and organic dyes by catalytic activity. Hassan et al. [70] used micro dilution method and revealed that Ag–Fe_3O_4 NZ (silver-iron oxide nano enzyme) revealed peroxidase like activity to eradicate H_2O_2 and degrade triarylmethane dye.

11 Sustainable Approach via Catalase Like Nano-Enzyme Bioremediation of Dye

The nanomaterials referred to the material that have a particle size ranging from 1 to 100 nm. This scale may define according to the unique physical, chemical and biological properties. The morphology of NMs may include the solid, hollow, porous, nano-sphere, nano-rods, nano-sheets, nano-fibers, nano-flowers, nano-enzymes and so on [15, 17, 98]. Among them, the scientists found that the nano-enzymes are more effective in order to have enzyme-like activities. Scrimin et al. [140] discovered the NZs in 2004. Fe_3O_4 NPs peroxidase-like activity was discovered by Yan et al. [177] in 2007. From then, NZs are rising stars and have received considerable attraction in the field of NMs. The non-metal compounds (g-C_3N_4, GO), metal compounds (Fe_3O_4, CeO_2, MnO_2, CuS, MnSe), metals (Au, Ag, Pt, Pd, Ir, Au@Pt), catalase-like enzyme-mimicking characteristics were shown by [173]. The natural enzymes are now replaced by NZs, which shows the same properties like as catalytic oxidation of substrates, alike kinetics, pH and concentration reliant. The term artificial enzymes were coined by Ronald Breslow for enzymes mimics, which is considered as very important and remarkable branch of biomimetic chemistry [58]. The NZs has turned over the role of natural enzymes because the artificial enzymes may have high stability, they are less expensive, easy to synthesis from green approaches with causing any harm to human beings health and are easy to store. NZs are more tolerant in the harsh environment, pH and temperature, and have large surface area and excellent tenability [8, 141]. From the past decades to recent time, the quick development in the nanotechnology and NZs depth research gained the attraction. NZs are actually NMs which shows the surface catalytic and enzymatic properties. The NZs, a cross breed, may have the natural enzymatic like properties. Enzyme-mimicking functionalities are possessed by NZs. The active sites in peroxidase enzymes have the protein scaffold reciprocal with the specific active sites like as heme molecule of natural enzymes [110]. The metals like Fe, Cu, Mn or Zn are covered by heme center, which are crucial for the substrate binding catalysis process produced by amino acids with fix structure. The substrate binding is combined with the stereoscopic particles. The catalytic and selective activities of natural enzymes are complex [56]. Too many suspended bonds under the co-ordination of the atoms in the NMs surface may have extra energy and instability. The energy of NZs is greater due to smaller size. The catalytic performance extraordinary with enzyme-mimicking NZs with single atom. Rare coordination structures, multivalent elements and active sites in natural enzymes due to using bio-orthogonal reactions of NZs. The 90–450 $m^2\ g^{-1}$ of fixed surface area with the particle size 2–10 nm. Catalytic activities of NPs may highly influence the higher surface area, surface atomic number and energy [33, 174]. The fraction of surface atoms is enhanced by surface atomic number and decreased particle size. Around 20–99% atoms have particle size 2–10 nm. Enhancement in the surface energy may deal with the functionalities of NPs which may enhance the atoms surface. The inner core may vary form the surface atoms to bind the crystal field environment. The surface atoms may have unsaturated linkages, which may lead to the decrease in size and fixed area. The significant chemical and catalytic activities as shown in substrates by atoms can easily react the with unstable surface energy to the surface atoms of NZs [182].

11.1 Catalase Enzyme Like Activity of NZs

The OXD, POD, CAT, SOD and nitrate reductase like activities were found in the various type of NMs. The oxidoreductase such as the H_2O_2 plays a crucial role in ROS in nature [184]. NZs with catalase like activity can be catalyzed by H_2O_2 in order to generate the water and molecular oxygen. The activity depended upon concentration of both substrate and NZs. If the concentration increases, the decomposition of rate will be higher. The process may prevent the accumulation the ROS effectively which may protect the oxidative stress of the cell. The level of oxygen or dissolved O_2 during the catalase-mimicking enzyme like activity [12, 13, 69]. The activity of NZs is dependent upon the pH and temperature. The most preferred NZs environments are neutral and alkaline. The acidic media is most preferred for catalase like activity. NZs may have larger affinity values for H_2O_2. Michaeis-Menten catalytic process is followed by catalase. The NZs, depending upon the morphology, structure and surface volume, are firmly related with catalase like activity in the pH environment. The extraordinary behavior of catalytic activity of NZs is extremely focused to some extent [19]. Generally, the activation and redox reactions may base upon the catalase mimicking NZs. The NZ surface may have a key role over the pre absorbed OH* (the sign * means species adsorbed on metal surfaces). For the synthesis of HO_2* and H_2O*, the H_2O_2 used to break down the H with pre absorbed OH*. The produced HO_2* passed H to another H_2O_2* leaving an O_2* and this will convert the H_2O_2* to H_2O* and OH* [105].

11.2 Catalase-Like Carbon NZs

Carbon-based nanomaterials are extensively employed in the synthesis of nanozymes (NZs) due to their precisely defined electronic and geometric structures, making them suitable for emulating highly advanced catalytic sites in natural enzymes. Various carbon-based NMs, such as GQDs, fullerenes, CNTs, carbon nanospheres, and CDs, have been documented for catalytic-like activities, including CAT, POD, and SOD activities [156]. Although alternative materials exist, the preference for carbon-based NMs in enzymatic mimicking, especially for catalase activities, stems from their well-defined electronic and geometric structures. CNZs hold significant potential for a broad spectrum of applications. The initial generation of abundant o_2 bubbles demonstrated the prepared graphene oxide quantum dots GOQDs were almost comparable with 4.0 ml^{-1} [54]. CDs or QDs may also have rummage ability of free radicals to produce the azo compounds. The scavenge ROS ability found in CDs is more powerful as compared to other compounds. However, the Cl and copper-doped carbon dots (CuCl-CDs) may exhibit exclusive reactive oxygen species (ROS) production ability under laser irradiation for photodynamic therapy (PDT). Additionally, their peroxidase-mimic activity can generate oxidative ·OH from H_2O_2. The transfer of electron donor and acceptor depends upon the catalytic dye degradation, whereas the greater difference in redox potential between acceptor and donor may determine the electron passage between them. Ag, Pt and Au NPs act as effective catalysts based on the intermediate redox potential value between the donor and the acceptor and may aid in the transfer of electrons, which could cause an electron relay system (Tables 1, 2 and 3).

12 Conclusion

It was conluded that the Azo and Triphenylmethane dyes are broadly used in the textile industry. Every year, over 1 million tons of dyes produced of which 50% are textile dyes. Such dangerous dyes should be removed via enviromental friendly methods such as artificial NZs. Carbon-based nanomaterials are extensively employed in crafting nanozymes due to their well-defined electronic and geometric structures, which make them more straightforward for emulating the highly evolved catalytic centers found in natural enzymes. From the cluster of carbon-based NZs, the CDs or QDs may also have rummage ability of free radicals to produce the azo

Table 1 Studies on removal or degradation of dyes by enzymes

Precursor	Source	Dye	Method	Degradation %	Reaction condition (IDC; Temp; pH; OT)	Enzymatic concentration (mM)	References
Manganese peroxidase	*P. sordida, P. chrysosporium*	Indigo	Biological	42–95	25 mg/L; 25–70 °C; 3.5–9.0; 20 days	50	[90]
Manganese peroxidases	*Trametes lacinia*	Indigo Carmine	Biological	81	200 mg/L; 28 °C; 5.0; 21 days	0.5	[55]
Manganese peroxidases	*Aspergillus flavus*	Direct red 31	Biological	92–100	0.0886 mM; 30 °C; 5.0; 5 days	0.704	[82]
Laccases	*Phosphate enzyme*	Indigo	Biological	86	5.0 mg/mL; 28 °C; 5.0–7.0; 15 days	25	[167]
Peroxidase	*Moringa* Oliefera	Reactive Blue 221, direct blue 297	Biological	75–86	303 U/mg; 40 °C; 5.0; 7–8 h	3	[158]
Manganese peroxidases	*Schizophyllum commune*	Drimaren Turquoise	Physical	95	300 mg/L; 35 °C; 4.5; 10 days	4	[125]
Laccases	*Pseudomonas sp*	Sulfonate azo	Biological	25	1,500 ppm; 28 °C; 7.0; 7 days	100	[80]
Laccases	*Quinoa husk*	Malachite green and Congo red	Biological	83	150 mg/L; 28 °C; 8.0; 72 h	40	[10]
Bacteria	Streptomyces bacillaris	Malachite green (MG)	Biological	30–40	200 mg/L; 28 °C; 7.0; 7 days	10	[3]
Azoreductase	*Armillaria sp*	Azo	Biological	90	500 mg/l; 50 °C; 7.0; 5 days	20	[136]
Lignolytic	*Halopiger aswanensis*	Azo	Biological	82–93	600 mg/L; 50 °C; 10.0 2 h	20	[36]

(continued)

Table 1 (continued)

Precursor	Source	Dye	Method	Degradation %	Reaction condition (IDC; Temp; pH; OT)	Enzymatic concentration (mM)	References
Laccases	*Mushroom*	Azo	Biological	57.2	100 g/L; 50 °C; 4.0; 2 days	0.264	[139]
Laccases	Pleurotus ostreatus	Malachite green	Biological	58.33	0.1 g/L; 50 °C; 6.0; 2 days	10	[60]
Protein	*Bovine serum*	Azo	Physical	91	2 mg/mL; 50 °C; 3.0; 2 h	100	[170]
Lipase	*Porcine Pancreas*	Azo	Physical	89	5 mg/L; 40 °C; 5.0; 240 min	10	[179]
Laccases	*Trametes versicolor*	Methylene blue	Physical	75	10 mg/L; 28 °C; 7.0; 120 min	100	[134]
Laccase	*Trichoderma atroviride*	Disazo	Physical	95	10 mg/L; 70 °C; 5.0; 21 h	50	[78]
Catalase	*B. subtilis* SK1	Azo	Biological	82	140 mg/L; 28 °C; 8.0; 3 days	40	[81]
Catalase	*Pseudomonas aeruginosa*	Azo	Biological	92	400 mgl – 1; 30 °C; 8.0; 6.5 h	5.0	[75]
Catalase	*Bacillus circulans*	Azo	Biological	98	50 mg/L; 35 °C; 7.5 3 days	10	[116]
Catalase	*Azolla filiculoides*	Azo	Biological	80	20 mg/L; 28 °C; 8.0; 20 days	10	[87]
Catalase	*Ferrioxalate*	Azo	Chemical	23	40 mg/L; 30 °C; 4.0; 5 days	40	[49]
Catalase	*Bacillus circulans*	Black B dye	Chemical	70	2340 mg/ml; 30 °C; 4.5; 24 h	20	[18]

Table 2 Waste-derived artificial catalase-like Nanozymes including their synthesis, source, enzyme-like activity and pollutant

Nano-enzyme	Source	Synthesis method	Enzyme like activity	Pollutant	References
CuO-NPs	*Cirsium arvense*	Hydrothermal	Catalase	Rhodamine B	[163]
Ag-rGO	Tea flowers	Hydrothermal	Catalase	Rhodamine B	[166]
$NiMn_2O_4$	*Euphorbia pulcherrima*	Hydrothermal	Peroxidase	Methylene blue	[144]
$MnFe_2O_4$	Calcium alginate	Co-precipitation	Laccase	Methylene blue	[152]
Au-NPs	*Pleurotus ostreatus*	Radiation-induced	Laccase	Azo	[51]
Carbon-Au nanosphere	Ginger	Hydrothermal	Peroxidase	DAPI dye	[181]
Ag–Fe_3O_4	Glucose	Ultrasonic	Catalase	Methyl orange	[175]
CeO_2@ZIF-8	Yolk	Hydrothermal	Peroxidase	Rhodamine B	[178]
Ag-Ca NCs	Ginger	Chemical precipitation	Peroxidase	Methylene blue	[106]
Fe_3O_4- QDs	Dopamine	Hydrothermal	Peroxidase	Azo	[32]
S, N-doped carbon dots	Starch	Hydrothermal	Catalase	Azo	[102]
$(BiO)_2CO_3$	*Punica Granat*	Hydrothermal	Catalase	Rhodamine B	[46]
CN-hemin	Citric acid	Pyrolysis	Peroxidase	CV	[176]
Fe_3O_4/N-PCN	Naringin	Hydrothermal	Peroxidase	Azo	[112]
$Mn_4(P_2O_7)_3$ Nano-flakes	Chitosan	Hydrothermal	Catalase	Methylene blue	[7]
Fe_3O_4/C	Glucose	Hydrothermal	Catalase	Azo	[185]
Cu co-doped IONP	*Punica Granatum*	Hydrothermal	Catalase	Methylene blue	[25]

compounds. The scavenge ROS ability found in CDs is more powerful as compared to other compounds and they may be used as alternatives to the natural enzymes.

Table 3 Benefits and drawbacks of various approaches physical, chemical and biological involved in the removal of dyes

Approaches	Benefits	Drawbacks	References
Physical approach			
Adsorption	– Large variety of dyes can be removed – Adsorbent is regenerated – Simple process	– Cannot regenerate adsorbents like activated carbon because it is expensive – Dye removal is incomplete	[84]
Ion exchange	– It is regenerated – Adsorbent never loses – Ion exchange capacity is higher	– Cost of resins high – Applicable for a few dyes – Resins may have poor hydrodynamic properties which are hard to tolerate at higher pressure	[89]
Membrane technology	– Only effective at low concentrations of dyes – Any type of dye can be removed – Efficiency of separation is high	– It is expensive – Membrane pores too expensive	[157]
Chemical approach			
Ozonation	– Effective and fast removal of dyes – It is used as a gaseous state – Water volume is not increased	– Toxic chlorinated organic produces – Expensive – Produces Carcinogenic	[126]
Advanced oxidation process	– No sludge – Ozone may degrade to oxygen – Easy and quick removal of pollutants	– Produces unwanted products – Dependent upon pH	[99]
Electrochemical	– Degradation of dyes in unusual terms – Both soluble and insoluble dyes can be removed	– Production of additional hazardous materials – Use of expensive electricity	[172]
Photochemical	– No sludge is produced – The dye removal process is efficient	– Cost-effective – Production of by-products	[5]
Fenton reaction	– No sludge production -Suitable for both soluble and in soluble dyes – Degrades all toxic pollutants from water	– VAT dyes cannot be removed – The greater amount of sludge is produced -More reaction time	[37]

(continued)

Table 3 (continued)

Approaches	Benefits	Drawbacks	References
Ultraviolet irradiation	– Nil requirement for hazardous chemicals – No sludge production	– Energy depleted – Expensive – Short recovery	[100]
Biological approach			
Microbial biomass	– Limited affinity towards microbial biomass – Great removal adsorbents	– Cannot remove all dyes – Poor capacity of dye binding and recyclability	[115]
Algae degradation	– Utilizes more dyes – Algae is inexpensive and easily available -No harmful products	– Unstable system	[180]
Fungal culture	– Degrades a variety of dye – Greater flexibility	– Long term and unstable growth – Requires nitrogen atmosphere	[169]
Microbial culture	– Fast degradation of dyes (<40 h)	– Only limited dyes are removed – Large-scale application is applicable	[147]
Enzyme degradation	– Non-toxic or carcinogenic – Fast and effective – Greater specificity of enzymes – Cheap	– Recovery is limited – Limited reusability	[119]

References

1. Acworth IN, Mccabe DR, Maher TJ (2017) The analysis of free radicals, their reaction products, and antioxidants. Oxidants, antioxidants and free radicals. CRC Press
2. Adeel S, Kiran S, Rafi S, Ayesha T, Rehman F, Gulzar T, Zuber M (2017) Radiation pretreatment: a potential novel technology to improve fastness properties of plant-derived natural dyes. In: Plant-based natural products: derivatives applications, pp 65–87
3. Adenan NH, Lim YY, Ting ASY (2022) Removal of triphenylmethane dyes by Streptomyces bacillaris: a study on decolorization, enzymatic reactions and toxicity of treated dye solutions. J Environ Manage 318:115520
4. Affat SS (2021) Classifications, advantages, disadvantages, toxicity effects of natural and synthetic dyes: a review. Univ Thi-Qar J Sci 8:130–135
5. Ahmadian M, Jaymand MJCCR (2023) Interpenetrating polymer network hydrogels for removal of synthetic dyes: a comprehensive review. Coord Chem Rev 486:215152
6. Al-Tohamy R, Ali SS, Li F, Okasha KM, Mahmoud YA, Elsamahy T, Jiao H, Fu Y, Sun J (2022) A critical review on the treatment of dye-containing wastewater: ecotoxicological and health concerns of textile dyes and possible remediation approaches for environmental safety. Ecotoxicol Environ Saf 231:113160
7. Alizadeh N, Salimi A (2023a) Simultaneous adsorption and catalytic degradation of methylene blue dye over recyclable $Mn_4(P_2O_7)_3$ nanoflakes: mechanism and efficiency. Environ Nanotechnol Monitor Manag 20:100806

8. Alsukaibi, AKD (2022) Various approaches for the detoxification of toxic dyes in wastewater. Processes 10:1968
9. Antol I, Glasovac Z, Margetic D, Crespo-Otero R, Barbatti M (2016) Insights on the auxochromic properties of the guanidinium group. J Phys Chem A 120:7088–7100
10. Ariaeenejad S, Motamedi E, Salekdeh GHJBT (2022) Highly efficient removal of dyes from wastewater using nanocellulose from quinoa husk as a carrier for immobilization of laccase. Biores Technol 349:126833
11. Arshad M, Adil M, Sikandar A, Hussain T (2014) Exploitation of meat industry by-products for biodiesel production: Pakistan's perspective. Pakistan J Life Soc Sci 12:120–125
12. Arshad M (ed) (2017) Perspectives on water usage for biofuels production: aquatic contamination and climate change, 24 November 2017. Springer
13. Arshad M, Bano I, Younus M, Khan A, Rahman A (2018) Health concerns associated with biofuel production. In: Perspectives on water usage for biofuels production: aquatic contamination and climate change, pp 97–105
14. Arshad M, Hussain T, Chaudhry N, Sadia H, Aslam B, Tahir U, Abbas M, Qureshi N, Nazir A, Rajoka MI, Iqba M (2019) Enhancing profitability of ethanol fermentation through gamma ray mutagenesis of saccharomyces cerevisiae. Pol J Environ Stud 28(1)
15. Arshad M (ed) (2020) Sustainable ethanol and climate change: sustainability assessment for ethanol distilleries, 13 November 2020. Springer Nature
16. Arshad M, Ansari AR, Qadir R, Tahir MH, Nadeem A, Mehmood T, Alhumade H, Khan N (2022) Green electricity generation from biogas of cattle manure: an assessment of potential and feasibility in Pakistan. Front Energy Res 25(10):911485
17. Arshad M (ed) (2023) Climate changes mitigation and sustainable bioenergy harvest through animal waste: sustainable environmental implications of animal waste, 28 April 2023. Springer Nature
18. Arslan I, Balcioglu IAJCT (2001) Degradation of Remazol Black B dye and its simulated dyebath wastewater by advanced oxidation processes in heterogeneous and homogeneous media. Color Technol 117:38–42
19. Ashrafi AM, Bytesnikova Z, Barek J, Richtera L, Adam V (2021) A critical comparison of natural enzymes and enzymes in biosensing and bioassays. Biosens Bioelectron 192:113494
20. Athira N, Jaya DS (2018) The use of fish biomarkers for assessing textile effluent contamination of aquatic ecosystems: a review. Nat Environ Pollut Technol 17(1):25–34
21. Baker A, Lin CC, Lett C, Karpinska B, Wright MH, Foyer CH (2023) Catalase: a critical node in the regulation of cell fate. Free Radic Biol Med 199:56–66
22. Bal G, Thakur AJMTP (2022) Distinct approaches of removal of dyes from wastewater: a review. Mater Today: Proc 50:1575–1579
23. Bano I, Ghauri MA, Arshad M, Yasin T, Younus M (2017) Bioactivity of variant molecular weight chitosan against drug-resistant bacteria isolated from human wounds. Microbial Drug Resist 23(8):958–965
24. Bano I, Arshad M (2018) Climatic changes impact on water availability. In: Perspectives on water usage for biofuels production: aquatic contamination and climate change, pp 39–54
25. Basak S, Sikdar S, Ali S, Mondal M, Roy MNJJOMS (2023) Green synthesized copper-assisted iron oxide enzyme for the efficient elimination of industrial pollutants via peroxodisulfate activation. J Mol Struct 1283:135267
26. Benkhaya S, M'Rabet S, el Harfi AJICC (2020) A review of classifications, recent synthesis and applications of textile dyes. Inorg Chem Commun 115:107891
27. Benkhaya S, M'Rabet S, Lgaz H, El Bachiri A, El Harfi A (2022) Dyes: classification, pollution, and environmental effects. In: Dye biodegradation, mechanisms techniques: recent advances, pp 1–50
28. Bharathi D, Nandagopal JGT, Ranjithkumar R, Gupta PK, Djearamane S (2022) Microbial approaches for sustainable remediation of dye-contaminated wastewater: a review. Arch Microbiol 204:169
29. Bhardwaj D, Bharadvaja NJJOWPE (2021) Phytoremediation of effluents containing dyes and its prospects for value-added products: a review of opportunities. J Water Process Eng 41:102080

30. Bhatia D, Sharma NR, Singh J, Kanwar RS (2017) Biological methods for textile dye removal from wastewater: a review. Crit Rev Environ Sci Technol 47:1836–1876
31. Bilińska L, Gmurek M, Ledakowicz SJCEJ (2016) Comparison between industrial and simulated textile wastewater treatment by AOPs–biodegradability, toxicity and cost assessment. Chem Eng J 306:550–559
32. Boruah PK, Darabdhara G, Borthakur P, Le Ouay B, Das MR (2023) Fe_3O_4 quantum dots anchored on functionalized graphene: a multimodal platform for sensing and remediation of Cr (VI). Chem Eng J 145797
33. Boukha Z, González-Prior J, De Rivas B, González-Velasco JR, López-Fonseca R, Gutiérrez-Ortiz JI (2018) Pd supported catalyst for gas-phase 1,2-dichloroethane abatement: efficiency and high selectivity towards oxygenated products. J Indus Engi Chem 57:77–88
34. Boyko KM, Baymukhametov TN, Chesnokov YM, Hons M, Lushchekina SV, Konarev PV, Lipkin AV, Vasiliev AL, Masson P, Popov VOJB (2019) 3D structure of the natural tetrameric form of human butyrylcholinesterase as revealed by cryoEM, SAXS and MD. Biochimie 156:196–205
35. Broadbent PJ, Carr CM, Lewis DM, Rigout ML, Siewers EJ, Kaveh NS (2023) Supercritical carbon dioxide (SC-CO_2) dyeing of cellulose acetate: an opportunity for a "greener" circular textile economy. Coloration Technol 139:475–488
36. Chauhan AK, Choudhury B (2021) Synthetic dye degradation using lignolytic enzymes produced from Halopiger aswanensis strain ABC_IITR by solid state fermentation. Chemosphere 273:129671
37. Chen S, Zhu M, Guo X, Yang B, Zhuo R (2023) Coupling of Fenton reaction and white rot fungi for the degradation of organic pollutants. Ecotoxicol Environ Saf 254:114697
38. Chequer FD, De Oliveira GR, Ferraz EA, Cardoso JC, Zanoni MB, De Oliveira DP (2013) Textile dyes: dyeing process and environmental impact. In: Eco-friendly textile dyeing finishing, vol 6, pp 151–176
39. Conter C, Favretto F, Dominici P, Martinez-Cruz LA, Astegno A (2023) Key substrate recognition residues in the active site of cystathionine beta-synthase from Toxoplasma Gondii. Proteins 91:1383–1393
40. Coronado-Apodaca KG, Rodriguez-De Luna SE, Araujo RG, Oyervides-Munoz MA, Gonzalez-Meza GM, Parra-Arroyo L, Sosa-Hernandez JE, Iqbal HMN, Parra-Saldivar R (2023) Occurrence, transport, and detection techniques of emerging pollutants in groundwater. MethodsX 10:102160
41. Dahlan D, Umar MIA, Sadikin SN, Ridwan J, Umar AAJOM (2022) Photoelectrical process uplift in Mg-doped-TiO_2 photoanode of dye-sensitized solar cells. Opt Mater 133:112976
42. Daniel LS, Rahman A, Hamushembe MN, Kapolo P, Uahengo V, Jonnalagadda SBJBTR (2023) The production of activated carbon from Acacia erioloba seedpods via phosphoric acid activation method for the removal of methylene blue from water. Bioresour Technol Rep 23:101568
43. Dhruv Patel D, Bhatt S (2022) Environmental pollution, toxicity profile, and physico-chemical and biotechnological approaches for treatment of textile wastewater. Biotechnol Genet Eng Rev 38:33–86
44. di Giacomo M, Zara V, Bergamo P, Ferramosca AJNRR (2020) Crosstalk between mitochondrial metabolism and oxidoreductive homeostasis: a new perspective for understanding the effects of bioactive dietary compounds. Nutr Res Rev 33:90–101
45. Dinçer AR (2020) Increasing BOD5/COD ratio of non-biodegradable compound (reactive black 5) with ozone and catalase enzyme combination. SN Appl Sci 2:736
46. Ding P, Di J, Chen X, Ji M, Gu K, Yin S, Liu G, Zhang F, Xia J, Li H (2018) S, N codoped graphene quantum dots embedded in $(BiO)_2CO_3$: incorporating enzymatic-like catalysis in photocatalysis. ACS Sustain Chem Eng 6:10229–10240
47. Długosz O, Matyjasik W, Matysik J, Szostak K, Śliwa P, Banach M (2023) Biocatalyst based on magnetic nanoparticles with Cu(II), Mn(II), Zn(II) and immobilised catalase. J Clust Sci 1–16

48. Donga S, Chanda S (2022) Recent trends in green synthesis of gold nanoparticles (AuNPs) and their biological efficacies: a mini review. Nanotechnol Sci Technol 1
49. Doumic LI, Soares PA, Ayude MA, Cassanello M, Boaventura RA, Vilar VJJCEJ (2015) Enhancement of a solar photo-Fenton reaction by using ferrioxalate complexes for the treatment of synthetic cotton-textile dyeing wastewater. Chem Eng J 277:86–96
50. Dugan ST, Muhammetoglu A, Uslu A (2023) A combined approach for the estimation of groundwater leaching potential and environmental impacts of pesticides for agricultural lands. Sci Total Environ 901:165892
51. El-Batal AI, Elkenawy NM, Yassin AS, Amin MA (2015) Laccase production by Pleurotus ostreatus and its application in synthesis of gold nanoparticles. Biotechnol Rep (Amst) 5:31–39
52. Fall I, Czerwiec Q, Abdellaoui S, Doumeche B, Ochs M, Remond C, Rakotoarivonina H (2023) A thermostable bacterial catalase-peroxidase oxidizes phenolic compounds derived from lignins. Appl Microbiol Biotechnol 107:201–217
53. Feng J, Liu L, Zhang Y, Wang Q, Liang H, Wang H, Song P (2023) Rethinking the pathway to sustainable fire retardants. Exploration. Wiley Online Library, 20220088
54. Fernández I, Carinelli S, González-Mora JL, Villalonga R, Salazar-Carballo PAJE (2023) Nickel oxide nanoparticles/carbon nanotubes nanocomposite for non-enzymatic determination of hydrogen peroxide. Electroanalysis 35:e202200192
55. Ferreira-Silva V, de Gusmão NB, Gibertoni TB, de Oliveira da Silva (2022) Trametes lactinea and T. villosa collected in Brazil can discolor indigo carmine. Acta Botanica Brasilica 36
56. Fu R, Ma Z, Zhao H, Jin H, Tang Y, He T, Ding Y, Zhang J, Ye D (2023) Research progress in iron-based nanozymes: catalytic mechanisms, classification, and biomedical applications. Anal Chem 95:10844–10858
57. Gallo M, Ferrara L, Calogero A, Montesano D, Naviglio DJFRI (2020) Relationships between food and diseases: What to know to ensure food safety. Food Res Int 137:109414
58. Garg B, Bisht T, Ling Y-CJM (2015) Graphene-based nanomaterials as efficient peroxidase mimetic catalysts for biosensing applications: an overview. Molecules 20:14155–14190
59. Geng L, Yu X, Li Y, Wang Y, Wu Y, Ren J, Xue F, Yi T (2019) Instant hydrogel formation of terpyridine-based complexes triggered by DNA via non-covalent interaction. Nanoscale 11:4044–4052
60. George J, Rajendran DS, Senthil Kumar P, Sonai Anand S, Vinoth Kumar V, Rangasamy G (2023) Efficient decolorization and detoxification of triarylmethane and azo dyes by porous-cross-linked enzyme aggregates of Pleurotus ostreatus laccase. Chemosphere 313:137612
61. Ghaderpour E, Mazzanti P, Mugnozza GS, Bozzano F (2023) Coherency and phase delay analyses between land cover and climate across Italy via the least-squares wavelet software. Int J Appl Earth Observat Geoinform 118:103241
62. Ghaly A, Ananthashankar R, Alhattab M, Ramakrishnan VVJJCEPT (2014) Production, characterization and treatment of textile effluents: a critical review. Chem Eng Process Technol 5:1–19
63. Gita S, Hussan A, Choudhury TJEE (2017) Impact of textile dyes waste on aquatic environments and its treatment. Environ Ecol 35:2349–2353
64. Guo Q, Yin C, Chen Y, Fang Z, Xiao H, He J, Huang L, Wu G, Zeng Z, Shen FJCEJ (2023) Periodic pH regulation controls the phosphate uptake-release behavior and structural evolution of layered double hydroxides. Chem Eng J 459:141584
65. Hamdy DM, Hassabo AG (2021) Various natural dyes from different sources. J Text Colorat Polym Sci 18:171–190
66. Hanafi MF, Sapawe NJMTP (2020) A review on the water problem associate with organic pollutants derived from phenol, methyl orange, and remazol brilliant blue dyes. Mater Today: Proc 31:A141–A150
67. Hashem HA, Khalil R (2023) Insight into the interaction of strigolactones, abscisic acid, and reactive oxygen species signals. In: Reactive oxygen species: prospects in plant metabolism. Springer
68. Hashemi SH, Kaykhaii M (2022) Azo dyes: sources, occurrence, toxicity, sampling, analysis, and their removal methods. In: Emerging freshwater pollutants. Elsevier

69. Hasnat A, Hasan S, Bano S, Sultana S, Ibhadon AO, Khan MZ (2023) Development of enzyme based sensors as diagnostic tools in clinic applications: a review. J Mater Chem B (29)
70. Hassan SA, Ghadam P, Ali AA, Dadvar AA, Ahmadi M (2023) Introduction to the biogenic Ag enzyme for the control and eradication of microbes and environmental pollutants. Mater Chem Phys 307:128107
71. Hataf N, Ghadir P, Ranjbar NJJOCP (2018) Investigation of soil stabilization using chitosan biopolymer. J Clean Prod 170:1493–1500
72. Hoveizavi NB, Feiz M (2023) Synthesis of novel dyes containing a dichloro triazine group and their applications on nylon 6 and wool. Dyes Pigments 212:111086
73. Islam T, Repon MR, Islam T, Sarwar Z, Rahman MM (2023) Impact of textile dyes on health and ecosystem: a review of the structure, causes, and potential solutions. Environ Sci Pollut Res Int 30:9207–9242
74. Iwuozor KO (2019) Prospects and challenges of using the coagulation-flocculation method in the treatment of effluents. Adv J Chem-Sect A 2:105–127
75. Jadhav SB, Surwase SN, Kalyani DC, Gurav RG, Jadhav JP (2012) Biodecolorization of azo dye Remazol orange by Pseudomonas aeruginosa BCH and toxicity (oxidative stress) reduction in Allium cepa root cells. Appl Biochem Biotechnol 168:1319–1334
76. Jagadeshvaran P, Parasuram S, Bose S (2023) Polymer-based nanocomposites for smart and functional textiles. Smart Funct Text 183
77. Jahan S, Singh A (2023) Causes and impact of industrial effluents on receiving water bodies: a review. Malays J Sci Adv Technol 111–121
78. Jankowska K, Su Z, Zdarta J, Jesionowski T, Pinelo M (2022) Synergistic action of laccase treatment and membrane filtration during removal of azo dyes in an enzymatic membrane reactor upgraded with electrospun fibres. J Hazard Mater 435:129071
79. Jomova K, Raptova R, Alomar SY, Alwasel SH, Nepovimova E, Kuca K, Valko M (2023) Reactive oxygen species, toxicity, oxidative stress, and antioxidants: chronic diseases and ageing. Arch Toxicol 97:2499–2574
80. Joshi SM, Inamdar SA, Jadhav JP, Govindwar SP (2013) Random UV mutagenesis approach for enhanced biodegradation of sulfonated azo dye, green HE4B. Appl Biochem Biotechnol 169:1467–1481
81. Kadam SK, Tamboli AS, Sambhare SB, Jeon B-H, Govindwar SPJEI (2018) Enzymatic analysis, structural study and molecular docking of laccase and catalase from B. subtilis SK1 after textile dye exposure. Eco Inform 48:269–280
82. Kalsoom U, Ahsan Z, Bhatti HN, Amin F, Nadeem R, Aftab K, Bilal MJPB (2022) Iron oxide nanoparticles immobilized Aspergillus flavus manganese peroxidase with improved biocatalytic, kinetic, thermodynamic, and dye degradation potentialities. Process Biochem 117:117–133
83. Karimi-Maleh H, Beitollahi H, Senthil Kumar P, Tajik S, Mohammadzadeh Jahani P, Karimi F, Karaman C, Vasseghian Y, Baghayeri M, Rouhi J, Show PL, Rajendran S, Fu L, Zare N (2022) Recent advances in carbon nanomaterials-based electrochemical sensors for food azo dyes detection. Food Chem Toxicol 164:112961
84. Katheresan V, Kansedo J, Yon JLS (2018) Efficiency of various recent wastewater dye removal methods: a review. J Environ Chem Eng 6:4676–4697
85. Khammee P, Unpaprom Y, Whangchai K, Ramaraj R (2022) Comparative studies of the Longan leaf pigment extraction as a photosensitizer for dye-sensitized solar cells' purpose. Biomass Convers Biorefinery 1–8
86. Khan MD, Singh A, Khan MZ, Tabraiz S, Sheikh JJJOWPE (2023) Current perspectives, recent advancements, and efficiencies of various dye-containing wastewater treatment technologies. J Water Process Eng 53:103579
87. Khataee AR, Movafeghi A, Vafaei F, Lisar SY, Zarei M (2013) The potential of the aquatic fern Azolla filiculoides in biodegradation of an azo dye: modeling of experimental results by artificial neural networks. Int J Phytoremediation 15:729–742
88. Khmelevtsova L, Sazykin I, Azhogina T, Sazykina MA (2020) Prokaryotic peroxidases and their application in biotechnology. Appl Biochem Microbiol 56:373–380

89. Kordbacheh F, Heidari GJMCH (2023) Water pollutants and approaches for their removal. Mater Chem Horiz 2:139–153
90. Kumar A, Arora PKJFIES (2022) Biotechnological applications of manganese peroxidases for sustainable management. Front Environ Sci 10:365
91. Kumar M, Sridharan S, Sawarkar AD, Shakeel A, Anerao P, Mannina G, Sharma P, Pandey A (2023) Current research trends on emerging contaminants pharmaceutical and personal care products (PPCPs): a comprehensive review. Sci Total Environ 859:160031
92. Kumari N, Behera M, Singh RJF, Toxicology C (2023) Facile synthesis of biopolymer decorated magnetic coreshells for enhanced removal of xenobiotic azo dyes through experimental modelling. Food Chem Toxicol 171:113518
93. Lellis B, Fávaro-Polonio CZ, Pamphile JA, Polonio J (2019) Effects of textile dyes on health and the environment and bioremediation potential of living organisms. Biotechnol Res Innov 3:275–290
94. Li K, Su H, Xiu X, Liu C, Hao W (2023a) Tire wear particles in different water environments: occurrence, behavior, and biological effects—a review and perspectives. Environ Sci Pollut Res 1–21
95. Li X, Song H, Zhang G, Zou W, Cao Z, Pan Y, Zhang G, Zhou MJJOHM (2023) Enhanced organic pollutant removal in saline wastewater by a tripolyphosphate-FeO/H_2O_2 system: key role of tripolyphosphate and reactive oxygen species generation. J Hazard Mater 457:131821
96. Lima JPP, Tabelini CHB, Aguiar A (2023) A review of gallic acid-mediated fenton processes for degrading emerging pollutants and dyes. Molecules 28:1166
97. Liu T, Ning L, Mei C, Li S, Zheng L, Qiao P, Wang H, Hu T, Zhong W (2023) Synthetic bacterial consortia enhanced the degradation of mixed priority phthalate ester pollutants. Environ Res 235:116666
98. Liu X, Verma G, Chen Z, Hu B, Huang Q, Yang H, Ma S, Wang X (2022) Metal-organic framework nanocrystals derived hollow porous materials: synthetic strategies and emerging applications. The Innovation
99. Lu H, Yang Q, Huang B, Qi J, Wang R, Zhou Q, Chen Q, Zhu L, Jin J, Kong Y (2023) Removal performance and adsorption kinetics of dyes by a Co-based metal organic framework. Microporous Mesoporous Mater 112665
100. Mahlangu OT, Mamba G, Mamba BBJJOECE (2023) A facile synthesis approach for GO-ZnO/PES ultrafiltration mixed matrix photocatalytic membranes for dye removal in water: leveraging the synergy between photocatalysis and membrane filtration. J Environ Chem Eng 11:110065
101. Maksimchuk NV, Puiggalí-Jou J, Zalomaeva OV, Larionov KP, Evtushok VY, Soshnikov IE, Solé-Daura A, Kholdeeva OA, Poblet JM, Carbó JJJAC (2023) Resolving the Mechanism for H_2O_2 decomposition over Zr (IV)-substituted lindqvist tungstate: evidence of singlet oxygen intermediacy. ACS Catal 13:10324–10339
102. Man Y, Li Z, Kong W-L, Li W, Dong W, Wang Y, Xie F, Zhao D, Qu Q, Zou W-SJMJ (2020) Starch fermentation wastewater as a precursor to prepare S, N-doped carbon dots for selective Fe (III) detection and carbon microspheres for solution decolorization. Microchem J 159:105338
103. Manikandan A, Thanrasu K, Dinesh A, Raja KK, Durka M, Almessiere M, Slimani Y, Baykal A (2022) Photocatalytic applications of magnetic hybrid nanoalloys and their nanocomposites. In: Handbook of magnetic hybrid nanoalloys and their nanocomposites. Springer
104. Mansoor S, Ali A, Kour N, Bornhorst J, Alharbi K, Rinklebe J, Abd El Moneim D, Ahmad P, Chung YS (2023) Heavy metal induced oxidative stress mitigation and ROS scavenging in plants. Plants (Basel) 12:3003
105. Matsumoto KI, Nyui M, Ueno M, Ogawa Y, Nakanishi I (2019) A quantitative analysis of carbon-ion beam-induced reactive oxygen species and redox reactions. J Clin Biochem Nutr 65:1–7
106. Mehmood Z, Ikram M, Imran M, Shahzadi A, Haider A, Ul-Hamid A, Nabgan W, Haider J, Hayat SJPB (2022) Z. officinale-doped silver/calcium oxide nanocomposites: catalytic activity and antimicrobial potential with molecular docking analysis. Process Biochem 121:635–646

107. Miller JK, Madsen FC (2023) The power of antioxidants: a critical review of oxidative stress and its impact on health. Interdisc J Agricult Environ Sci 10:6–10
108. Mohanty S, Dash S, Pradhan N, Maji SK (2023) Bio-remediation of organic dyes from wastewater by microbial colony—a short review. In: Nano-engineered materials for textile waste remediation, pp 61–104
109. Monisha B, Sridharan R, Kumar PS, Rangasamy G, Krishnaswamy VG, Subhashree S (2023) Sensing of azo toxic dyes using nanomaterials and its health effects—a review. Chemosphere 313:137614
110. Mubeen B, Hasnain A, Wang J, Zheng H, Naqvi SAH, Prasad R, Rehman AU, Sohail MA, Hassan MZ, Farhan MJC (2023) Current progress and open challenges for combined toxic effects of manufactured nano-sized objects (MNO's) on soil biota and microbial community. Coatings 13:212
111. Naidu R, Biswas B, Willett IR, Cribb J, Kumar Singh B, Paul Nathanail C, Coulon F, Semple KT, Jones KC, Barclay A, Aitken RJ (2021) Chemical pollution: a growing peril and potential catastrophic risk to humanity. Environ Int 156:106616
112. Nsabimana A, Kitte SA, Wu F, Qi L, Liu Z, Zafar MN, Luque R, Xu GJASS (2019) Multifunctional magnetic Fe_3O_4/nitrogen-doped porous carbon nanocomposites for removal of dyes and sensing applications. Appl Surf Sci 467:89–97
113. Oladoye PO, Ajiboye TO, Omotola EO, Oyewola OJ (2022) Methylene blue dye: toxicity and potential technologies for elimination from (waste) water. Res Eng 100678
114. Palanimuthu K, Subbiah U, Sundharam S, Munusamy C (2023) Spirulina carbon dots: a promising biomaterial for photocatalytic textile industry reactive red M8B dye degradation. Environ Sci Pollut Res 30:52073–52086
115. Papapostolou H, Kachrimanidou V, Alexandri M, Plessas S, Papadaki A, Kopsahelis N (2023) Natural carotenoids: recent advances on separation from microbial biomass and methods of analysis. Antioxidants (Basel) 12:1030
116. Patil NP, Bholay AD, Kapadnis BP, Gaikwad VBJJPAM (2016) Biodegradation of model azo dye methyl red and other textile dyes by isolate Bacillus circulans NPP1. J Pure Appl Microbiol 10:2793–2800
117. Patra D (2017) Reducing energy cost for wastewater treatment in the Middle East: a physiochemical prospective. In: The sustainability triangle, pp 223–244
118. Pearce C, Lloyd JR, Guthrie JT (2003) The removal of colour from textile wastewater using whole bacterial cells: a review. Dyes Pigments 58:179–196
119. Pham VHT, Kim J, Chang S, Bang D (2023) Investigating bio-inspired degradation of toxic dyes using potential multi-enzyme producing extremophiles. Microorganisms 11:1273
120. Pinheiro FC, Bortolotto VC, Araujo SM, Couto SF, Dahleh MMM, Cancela M, Neto J, Zeni G, Zaha A, Prigol M (2022) Oxidative stress response system in Escherichia coli arising from diphenyl ditelluride (PhTe)(2) exposure. Toxicol In Vitro 83:105404
121. Poljovka A, Musil M, Bednář D, Chovanová K, Bauerová-Hlinková V, Bellová J, Kohútová L, Baráth P, Zámocký MJA (2023) Comparison of fungal thermophilic and mesophilic catalase-peroxidases for their antioxidative properties. Antioxidants 12:1382
122. Poronik YM, Vygranenko KV, Gryko D, Gryko DTJCSR (2019) Rhodols–synthesis, photophysical properties and applications as fluorescent probes. Chem Soc Rev 48:5242–5265
123. Qasim M, Bukhari SA, Ghani MJ, Masoud MS, Huma T, Arshad M, Haque A, Ibrahim Z, Javed S, Rajoka MI (2016) Relationship of oxidative stress with elevated level of DNA damage and homocysteine in cardiovascular disease patients. Pak J Pharm Sci 2:29
124. Qiao X, Li X, Qi T, Liu YJW (2023) Identification of priority pollutants in groundwater: a case study in Xiong'an New Region, China. Water 15:1565
125. Raees A, Bhatti HN, Alshehri S, Aslam F, Al-Fawzan FF, Alissa SA, Iqbal M, Nazir A (2023) Adsorption potential of schizophyllum commune white rot fungus for degradation of reactive dye and condition optimization: a thermodynamic and kinetic study. Adsorpt Sci Technol
126. Rahimi M, Salehi E, Mandooie M, Khalili N (2023) Adsorption/ozonation integration for intensified ethyl acetate plant wastewater treatment: process optimization and sensitivity analysis assessment. J Indus Eng Chem 120:271–284

127. Rai A, Sirotiya V, Mourya M, Khan MJ, Ahirwar A, Sharma AK, Kawatra R, Marchand J, Schoefs B, Varjani S, Vinayak V (2022) Sustainable treatment of dye wastewater by recycling microalgal and diatom biogenic materials: Biorefinery perspectives. Chemosphere 305:135371
128. Reddy MK, Labhasetwar V (2009) Nanoparticle-mediated delivery of superoxide dismutase to the brain: an effective strategy to reduce ischemia-reperfusion injury. FASEB J 23:1384–1395
129. Rosso B, Barbante C, Gambaro A, Corami FJS W (2023) Additives, plasticizers and small microplastics (<100 μm) in wet and dry depositions from an urban area of Venice, Italy. Springer Water
130. Saad I, Ralha N, Abukhadra MR, Al Zoubi W, Ko YG (2023) Recent advances in photocatalytic oxidation techniques for decontamination of water. J Water Process Eng 52:103572
131. Saeed M, Muneer M, Haq AU, Akram N (2022) Photocatalysis: an effective tool for photodegradation of dyes—a review. Environ Sci Pollut Res 1–19
132. Saira G, Shanthakumar SJJOEM (2023) Zero waste discharge in tannery industries—an achievable reality? A recent review. J Environ Manage 335:117508
133. Samarasinghe K, Wenk C, Silva K, Gunasekera JJA-AJOAS (2003) Turmeric (Curcuma longa) root powder and mannanoligosaccharides as alternatives to antibiotics in broiler chicken diets. Asian Australas J Anim Sci 16:1495–1500
134. Sannino F, di Matteo E, Ambrosecchio M, Pirozzi D (2023) A novel approach, based on the combined action of chitosan hydrogel and laccases, for the removal of dyes from textile industry wastewaters. Gels 9:41
135. Sarkar P, Paul SK, Sahu K, Mandal S, Islam QA, Chakrabarty RJC (2023) A greener synthesis of silver nanoparticles using tea leaf and aloe vera leaf extract for catalytic organic dye degradation application. ChemistrySelect 8:e202301906
136. Sarkar S, Banerjee A, Halder U, Biswas R, Bandopadhyay R (2017) Degradation of synthetic azo dyes of textile industry: a sustainable approach using microbial enzymes. Water Conserv Sci Eng 2:121–131
137. Sarwa P, Vijayakumar R, Verma SKJOALJ (2014) Adsorption of acid red 66 dye from aqueous solution by green microalgae Acutodesmus obliquus strain PSV2 isolated from an industrial polluted site. Open Access Libr J 1:1–8
138. Sarwar B, Khan AU, Aslam M, Bokhari A, Mubashir M, Alothman AA, Ouladsmane M, Aldossari SA, Chai WS, Khoo KS (2023) Comparative study of ZIF-8-materials for removal of hazardous compounds using physio-chemical remediation techniques. Environ Res 220:115168
139. Schallemberger JB, Libardi N, Dalari B, Chaves MB, Nagel Hassemer ME (2023) Textile azo dyes discolouration using spent mushroom substrate: enzymatic degradation and adsorption mechanisms. Environ Technol 44:1265–1286
140. Scrimin P, Manea F, Houillon F B, Pasquato L, Angew (2004) Chem Int, 43, 6165
141. Sharif N, Munir N, Hasnain M, Naz S, Arshad M (2021) Environmental impacts of ethanol production system. In: Sustainable ethanol and climate change: sustainability assessment for ethanol distilleries, pp 205–223
142. Selvaraj V, Karthika TS, Mansiya C, Alagar MJJOMS (2021) An over review on recently developed techniques, mechanisms and intermediate involved in the advanced azo dye degradation for industrial applications. J Mol Struct 1224:129195
143. Seo SJ, Das G, Shin HS, Patra JK (2023) Silk Sericin protein materials: characteristics and applications in food-sector industries. Int J Mol Sci 24:4951
144. Shahanas T, Harichandran G (2023) PEG mediated $NiMn_2O_4$ nanomaterials as a nano catalyst for peroxidase mimetic activity and photocatalytic degradation. Spectrochimica Acta Part A: Mol Biomol Spectrosc 303:123212
145. Shakir K, Elkafrawy AF, Ghoneimy HF, Elrab Beheir SG, Refaat M (2010) Removal of rhodamine B (a basic dye) and thoron (an acidic dye) from dilute aqueous solutions and wastewater simulants by ion flotation. Water Res 44:1449–1461
146. Sharma K, Sharma P, Dhiman SK, Chadha P, Saini HS (2022) Biochemical, genotoxic, histological and ultrastructural effects on liver and gills of fresh water fish Channa punctatus exposed to textile industry intermediate 2 ABS. Chemosphere 287:132103

147. Sharma M, Agarwal S, Agarwal Malik R, Kumar G, Pal DB, Mandal M, Sarkar A, Bantun F, Haque S, Singh P, Srivastava N, Gupta VK (2023) Recent advances in microbial engineering approaches for wastewater treatment: a review. Bioengineered 14:2184518
148. Shi Y, Li L, Xu Z, Qin X, Cai Y, Zhang W, Shi W, Du X, Guo F (2023) Coupled internal electric field with hydrogen release kinetics for promoted photocatalytic hydrogen production through employing carbon coated transition metal as co-catalyst. J Colloid Interface Sci 630:274–285
149. Shinde RM, Ingle PU, Trivedi HR, Wasule D, Gaharwar A, Gade AK, Rai M, Manna S, Srivastava PK, Biswas JK (2023a) Bioremediation of industrial dye waste effluents aided by GIS applications: a comprehensive review. Environ Dev Sustain 1–27
150. Shinde RM, Ingle PU, Trivedi HR, Wasule D, Gaharwar A, Gade AK, Rai M, Manna S, Srivastava PK, Biswas JK (2023b) Bioremediation of industrial dye waste effluents aided by GIS applications: a comprehensive review. J Environ Dev Sustain 1–27
151. Shindhal T, Rakholiya P, Varjani S, Pandey A, Ngo HH, Guo W, Ng HY, Taherzadeh MJ (2021) A critical review on advances in the practices and perspectives for the treatment of dye industry wastewater. Bioengineered 12:70–87
152. Shojaat R, Saadatjoo N, Karimi A, Aber SJJOECE (2016) Simultaneous adsorption–degradation of organic dyes using MnFe2O4/calcium alginate nano-composites coupled with GOx and laccase. J Environ Chem Eng 4:1722–1730
153. Shostak S, Park W, Oh J, Kim J, Lee S, Nam H, Filatov M, Kim D, Choi CH (2023) Ultrafast excited state aromatization in dihydroazulene. J Am Chem Soc 145:1638–1648
154. Singh S, Rai N, Tiwari H, Gupta P, Verma A, Kumar R, Kailashiya V, Salvi P, Gautam V (2023) Recent advancements in the formulation of nanomaterials-based nanozymes, their catalytic activity, and biomedical applications. ACS Appl Bio Mater 6(9)
155. Srivastava A, Rani RM, Patle DS, Kumar S (2022) Emerging bioremediation technologies for the treatment of textile wastewater containing synthetic dyes: a comprehensive review. J Chem Technol Biotechnol 97:26–41
156. Sun Y, Xu B, Pan X, Wang H, Wu Q, Li S, Jiang B, Liu HJCCR (2023) Carbon-based nanozymes: design, catalytic mechanism, and bioapplication. Coord Chem Rev 475:214896
157. Suresh R, Rajendran S, Gnanasekaran L, Show PL, Chen W-H, Soto-Moscoso MJC (2023) Modified poly (vinylidene fluoride) nanomembranes for dye removal from water—a review. Chemosphere 138152
158. Suvathi S, Rathi R, Ravichandran K, Kavitha P, Ayyanar M, Praseetha P, Chidhambaram NJER (2023) Improved photocatalytic dye degradation and seed germination through enzyme-coupled titanium oxide nanopowder—a cost-effective approach. Environ Res 218:114973
159. Takio N, Yadav M, Yadav HS (2021) Catalase-mediated remediation of environmental pollutants and potential application—a review. Biocatal Biotransform 39:389–407
160. Thacharodi A, Hassan S, Singh T, Mandal R, Chinnadurai J, Khan HA, Hussain MA, Brindhadevi K, Pugazhendhi A (2023) Bioremediation of polycyclic aromatic hydrocarbons: an updated microbiological review. Chemosphere 328:138498
161. Tizazu S, Tesfaye G, Andualem B, Wang A, Guadie A (2022) Evaluating the potential of thermo-alkaliphilic microbial consortia for azo dye biodegradation under anaerobic-aerobic conditions: Optimization and microbial diversity analysis. J Environ Manage 323:116235
162. Tkaczyk A, Mitrowska K, Posyniak AJSOTTE (2020) Synthetic organic dyes as contaminants of the aquatic environment and their implications for ecosystems: a review. Sci Total Environ 717:137222
163. Ur Rehman K, Khan AU, Tahir K, Nazir S, Albalawi K, Hassan HM, Alabbad EA, Refat MS, Al-Shehri HS, Aldawsari AM (2022) Facile synthesis of copper oxide nanoparticles (CuONPs) using green method to promote photocatalytic and biocidal applications. J Mol Liq 360:119453
164. Vaiano V, de Marco IJS (2023) Removal of azo dyes from wastewater through heterogeneous photocatalysis and supercritical water oxidation. Separations 10:230
165. Vasiraja N, Prabhahar RSS, Joshua AJJOCP (2023) Preparation and physio-chemical characterisation of activated carbon derived from prosopis juliflora stem for the removal of methylene blue dye and heavy metal containing textile industry effluent. J Clean Prod 397:136579

166. Vinay S, Alharthi FA, Alsalme A, Nagaraju GJJOMS (2021) Hydrothermal synthesis of Ag/rGO@ CTFE nanocomposite as a promising photocatalyst for degradation action. J Mol Struct 1228:129722
167. Wakade G, Lin S, Saha P, Kumari U, Daniell HJPBJ (2023) Abatement of microfibre pollution and detoxification of textile dye–Indigo by engineered plant enzymes. Plant Biotechnol J 21:302–316
168. Wang F, Ma W, Tang B, Zhang SJCT (2023) Synthesis and properties of poly (styrene-co-maleic acid) macromolecular reactive dyes. Color Technol 139:57–67
169. Wang H, Neal B, White B, Nelson B, Lai J, Long B, Arreola-Vargas J, Yu J, Banik MT, Dai SYJBTR (2023) Microplastics removal in the aquatic environment via fungal pelletization. Bioresour Technol Rep 23:101545
170. Wang X, Hou C, Qiu W, Ke Y, Xu Q, Liu XY, Lin Y (2017) Protein-directed synthesis of bifunctional adsorbent-catalytic hemin-graphene nanosheets for highly efficient removal of dye pollutants via synergistic adsorption and degradation. ACS Appl Mater Interfaces 9:684–692
171. Wibowo YG, Syahnur MT, Al-Azizah PS, Gintha DA, Lululangi BRG. Sudibyo (2023) Phytoremediation of high concentration of ionic dyes using aquatic plant (Lemna minor): a potential eco-friendly solution for wastewater treatment. Environ Nanotechnol Monitor Manag 20:100849
172. Włodarczyk-Makuła M, Myszograj S, Włodarczyk MJE (2023) Removal of organic micro-pollutants from wastewater in electrochemical processes. Energies 16:5591
173. Wong JKH, Tan HK, Lau SY, Yap P-S, Danquah MKJJOECE (2019) Potential and challenges of enzyme incorporated nanotechnology in dye wastewater treatment: a review. J Environ Chem Eng 7:103261
174. Wu XT, Ansari AR, Pang XX, Li HZ, Zhang ZW, Luo Y, Arshad M, Song H (2019) Visfatin plays a significant role in alleviating lipopolysaccharide-induced apoptosis and autophagy through PI3K/AKT signaling pathway during acute lung injury in mice. Arch Immunol Ther Exp 1(67):249–261
175. Xin Z, Duan K, Zhuo Q, He Q, Zhang X, Zheng C, Han X, Han T, Fu Z, Xu XJCEJ (2023) Novel nanozyme Ag/Fe_3O_4@ h-BN with peroxidase-mimicking and oxidase-mimicking activities for dye degradation, As (V) removal and detection. Chem Eng J 461:141589
176. Xu C, Zhou L, Zheng S, Yu D, Tang Y, Fan C, Lin C, Zhou D, Lin YJASS (2021) Hemin covalently functionalized carbon nanobranch with enzyme-like and photocatalytic activities for synergistic dye degradation and antibacterial therapy. Adv Sustain Syst 5:2100103
177 Yan X, Gao L, Zhuang J, Nie L, Zhang J, Zhang Y, Gu N, Wang T, Feng J, Yang D, Perrett S (2007) Nat Nanotechnol, 2, 577
178. Yang T, Liu X, Zeng Z, Wang X, Zhang P, Feng B, Tian K, Qing T (2023) Efficient and recyclable degradation of organic dye pollutants by CeO_2@ZIF-8 nanozyme-based non-photocatalytic system. Environ Pollut 316:120643
179. Yao LW, Khan FSA, Mubarak NM, Karri RR, Khalid M, Walvekar R, Abdullah EC, Mazari SA, Ahmad A, Dehghani MHJJOML (2022) Insight into immobilization efficiency of Lipase enzyme as a biocatalyst on the graphene oxide for adsorption of Azo dyes from industrial wastewater effluent. J Mol Liq 354:118849
180. Zeng G, Zhang R, Liang D, Wang F, Han Y, Luo Y, Gao P, Wang Q, Wang Q, Yu CJW (2023) Comparison of the advantages and disadvantages of algae removal technology and its development status. Water 15:1104
181. Zhang A, Pan S, Zhang Y, Chang J, Cheng J, Huang Z, Li T, Zhang C, de la Fuentea JM, Zhang Q, Cui D (2019) Carbon-gold hybrid nanoprobes for real-time imaging, photothermal/photodynamic and nanozyme oxidative therapy. Theranostics 9:3443–3458
182. Zhang H, Han A, Okumura K, Zhong L, Li S, Jaenicke S, Chuah G-KJJOC (2018) Selective hydrogenation of phenol to cyclohexanone by SiO_2-supported rhodium nanoparticles under mild conditions. J Catal 364:354–365
183. Zhang Y, Heym B, Allen B, Young D, Cole SJN (1992) The catalase—peroxidase gene and isoniazid resistance of Mycobacterium tuberculosis. Nature 358:591–593

184. Zhang Y, Zhang L, Wang M, Li P (2023) The applications of nanozymes in neurological diseases: from mechanism to design. Theranostics 13:2492–2514
185. Zhang Z, Kong JJJOHM (2011) Novel magnetic Fe_3O_4@ C nanoparticles as adsorbents for removal of organic dyes from aqueous solution. J Hazard Mater 193:325–329
186. Zhao G, Wang W, Zheng L, Chen L, Duan G, Chang R, Chen Z, Zhang S, Dai M, Yang GJIJOBM (2023) Catalase-peroxidase StKatG is a bacterial manganese oxidase from endophytic Salinicola tamaricis. Int J Biol Macromol 224:281–291

Pioneering Lipase Applications in Textile Manufacturing

Iram Liaqat, Shiza, Chand Raza, Saiqa Andleeb, Sher Zaman Safi, and Muhammad Arshad

Abstract Biocatalysts are preferred by industries over chemical catalysts. In the lipid modification business, lipases (triacylglycerol acylhydrolase) are considered the best biocatalysts. Animals, plants, and microbes all have large amounts of lipases. Bacterial lipases account for 45% of the total, followed by fungus (21%), animal (18%), plant (11%), and algae (3%). Because they are readily available and can catalyze processes such as hydrolysis, esterification, and alcoholysis, microbial lipases are the preferred option among them. Microbial lipases are now more effective when immobilized, making them useful for a variety of reactions and reducing the need to add fragrance to the immobilization procedures. Additionally, particular attention has been paid to elements like modeling, design, and process optimization that are essential for future industrial implementation and scale-up. Enzymologists have always been captivated by the distinct interfacial activation of lipases, and in the past few years, biophysicists and crystallographers have made strides toward grasping the structure–function links of these enzymes. Because of their accessibility, ability to work in a range of media, stability in organic solvents, ability to catalyze at the lipid–water interface and even in non-aqueous conditions, and ability to function in a variety of media, they are a versatile option for the food, flavor, detergent, pharmaceutical, leather, textile, cosmetic, and paper industries.

I. Liaqat (✉) · Shiza · C. Raza
Microbiology Lab, Department of Zoology, Government College University, Lahore, Pakistan
e-mail: dr.iramliaqat@gcu.edu.pk

S. Andleeb
Department of Zoology, University of Azad Jammu and Kashmir, Muzaffarabad, Pakistan

S. Z. Safi
Faculty of Medicine. Bioscience and Nursing, MAHSA University, Selangor, Malaysia

M. Arshad
Biochemistry Section, Jhang-Campus, University of Veterinary and Animal Sciences Lahore, Lahore, Pakistan

M. Arshad (ed.), *Enzymes in Textile Processing: A Climate Changes Mitigation Approach*, SDGs and Textiles, https://doi.org/10.1007/978-981-97-8058-7_10

Graphical Abstract

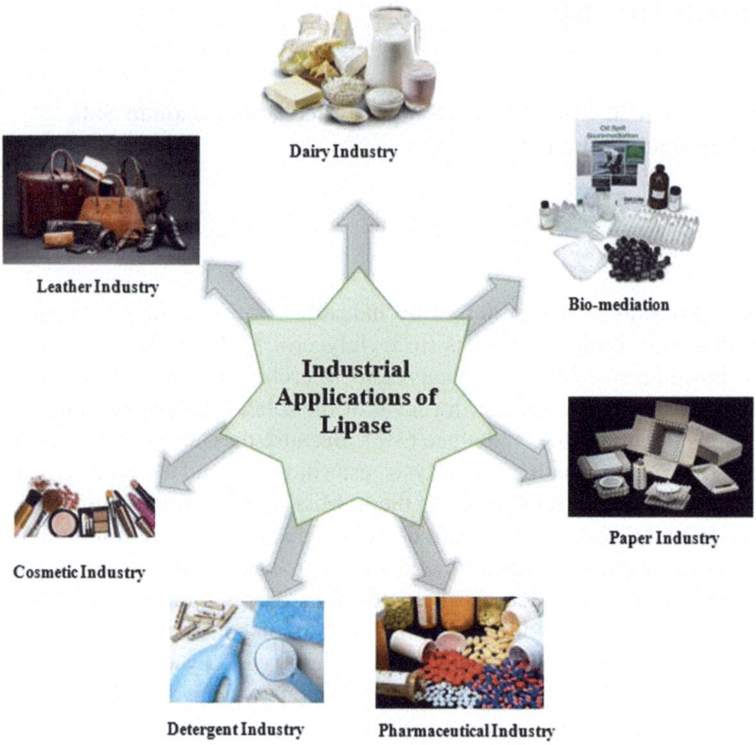

Keywords Lipases · Properties · Applications · Optimization factors · Biodegradable polymers

List of Abbreviations

FAs	Fatty acids
TAGs	Triacylglycerols
FFAs	Free fatty acids
FA	Fatty acid
TAG	Triacylglycerol
pH	Potential of hydrogen
RML	*Rhizopus miehei* Lipase
PUFA	Polyunsaturated fatty acid
HMF	Human milk fat

1 Overview of Textile Wet Processing

The textile business is among the oldest, having been around for several centuries. The design, manufacture, and distribution of yarn, fabric, and apparel are the industry's main priorities. To do this, a variety of chemical and physical processes is needed. The textile business generates significant amounts of effluents including various harmful agents and chemicals, which has drawn attention from around the world about environmental concerns. It has been claimed that the greatest substitute for these dangerous and deadly compounds is enzymes. Throughout the textile manufacturing process, enzymes such as lipase are frequently utilized. The use of enzymatic methods is particularly promising since they are low-energy, low-water, and high-quality products [37].

Lipases

Lipases are hydrolases that hydrolyze triacylglycerols (TAGs) to produce glycerol and fatty acids (FAs) (Fig. 10.1). Lipases catalyze esterification, transesterification, and inter-esterification processes in an organic medium (Fig. 10.4).The process known as interfacial activation (Fig. 10.2) occurs when enzymes are altered to function at the interfaces between biphasic systems. An aggregation, micelle, or monomolecular film made of ester molecules is the distinctive substrate [4, 62]. The substrates of lipases are diversified and can be composed by neutral lipids, phospholipids and lyso-phospholipids, or ether lipids [48, 74].

When there is an abundance of water present, lipases function at the interface between organic and aqueous solutions, facilitating the breakdown of carboxylate ester linkages into free fatty acids (FFAs) and organic alcohol (Fig. 10.3). Alcohol, acylglycerol, FA, structural specificity, TAG specificity, hetero and chiral specificities, and positional specificity are the different categories of substrate specificities of lipases. These enzymes show a high degree of stability and activity even in non-aqueous conditions.

Since most animal and microbial lipases have alkaline pH optima (pH 8–9), they do not need cofactors and can be readily immobilized on various matrices [45]. However, under certain circumstances, like the presence of salts or a particular kind

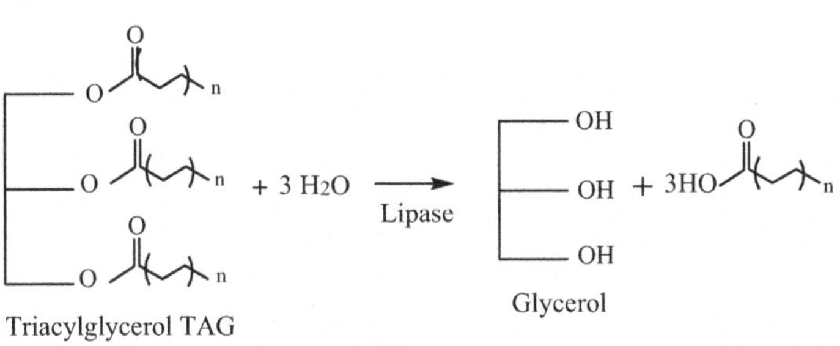

Fig. 10.1 General scheme of lipase reaction

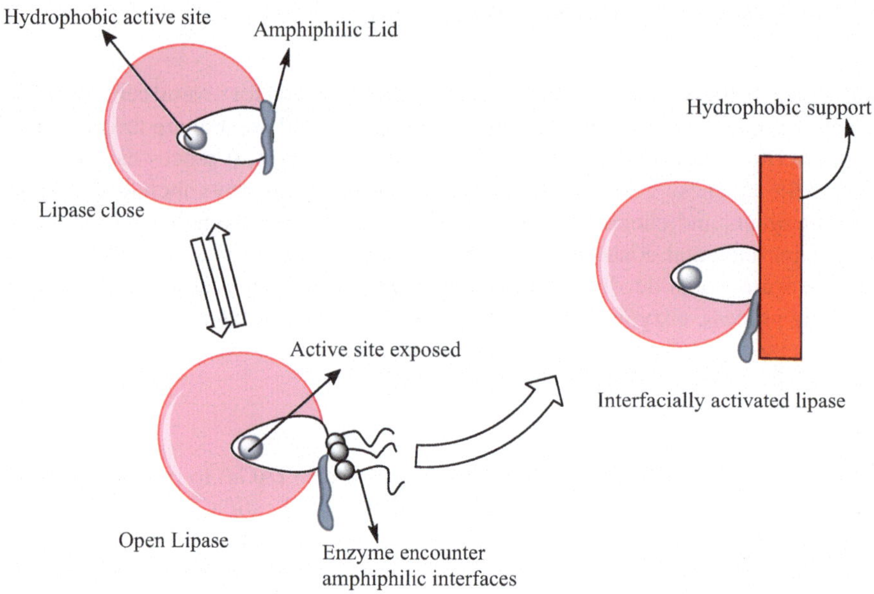

Fig. 10.2 The characteristic interfacial activation of lipases

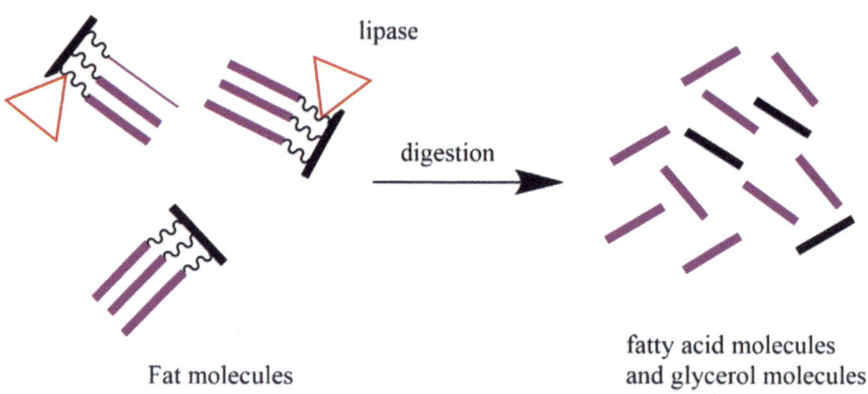

Fig. 10.3 Working of lipases [33]

of emulsifier, the optimal pH can be shifted to the acidic range [43]. In wet conditions, the hydrophobic lipase catalytic core is blocked by a polypeptide chain called a lid. When this structure comes into contact with the substrate, it moves away. Its external face is hydrophilic, while its internal face is hydrophobic [10].

Certain lipases, such *Rhizopus miehei* lipase (RML) and human pancreas lipases, display this lid. However, two approximately parallel amphipathic helices appear to surround the active region of several lipases, such as the lipase from *Geotrichum candidum*. Activation at an oil–water interface is made possible by the conformational

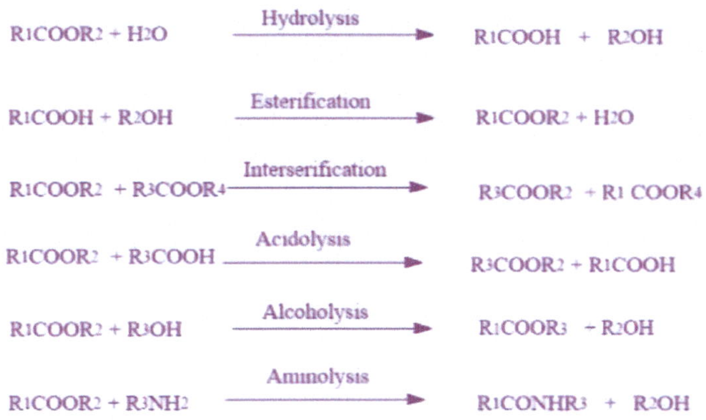

Fig. 10.4 Lipases catalyzing different reactions

shift. However, this phenomenon is not true for all lipases. Some lipases, such those from *Pseudomonas aeruginosa* and *Candida antarctica B*, do not have a lid and do not show interfacial activation [63]. Lipases are thought to be the most adaptable catalysts in lipid biotechnology because of their region-selective characteristics and enantioselective catalytic action [38, 64]. These enzymes can be used in a wide range of industrial processes, such as the synthesis and breakdown of engineering thermoplastics, the synthesis of pharmaceuticals and emulsifiers, the production of agrochemicals, cosmetics, fragrances, and short-chain alcohol-based biodiesel, as well as the hydrolysis of oils and fats to produce concentrated flavonoids [62].

2 Properties of Lipases

Lipases are monomeric proteins with a molecular weight ranging from 19 to 60 kDa.

The degree of unsaturation, the length of the fatty acid chain, and the location of the fatty acid inside the glycerol backbone all affect the physical characteristics of lipases [16, 69]. Being active in organic solvents, certain lipases catalyze a variety of helpful reactions, including esterification [19, 72]. Lipases, which are generally stable at neutral pH 7.0 or up to pH 4.0 and 8.0, demonstrated pH-dependent activity. Extracellular lipases generated by *Chromobacterium viscosum, A. niger,* and *Rhizophus sp.* are active at acidic pH levels, whereas alkaline lipases produced by *P. nitroaeducens* are active at pH 11.0 [1]. Under specific experimental settings, lipases can reverse the processes that cause esterification and interesterification when water is not present. Lipase activities can be produced without cofactors, although calcium, a divalent cation, helps to stimulate the activity [43]. Based on the specificity to a certain area, lipases were separated into two groups and identified using an acyl glycerol substrate. In the first group of lipases, solely fatty acids are released from all three glycerol sites without exhibiting regiospecificity.

The fatty acids that are released from the 1, 3, and acylglycerol sites in the second lipase group in a region-specific manner. Lipases hydrolyzed triacylglycerol to form 2-monoacylglycerol and 1,2-(2, 3)-diacylglycerols, which are free fatty acids. These characteristics make these enzymes suitable for the extraction of optically pure alcohols and esters. Therefore, an enzyme's specificity can be changed by changing the solvents' characteristics. The fragile and flexible architectures of enzymes allow any solvent to have a significant impact on their catalytic capabilities [17].

3 Classification of Lipases

See Figs. 10.3, 10.4 and 10.5.

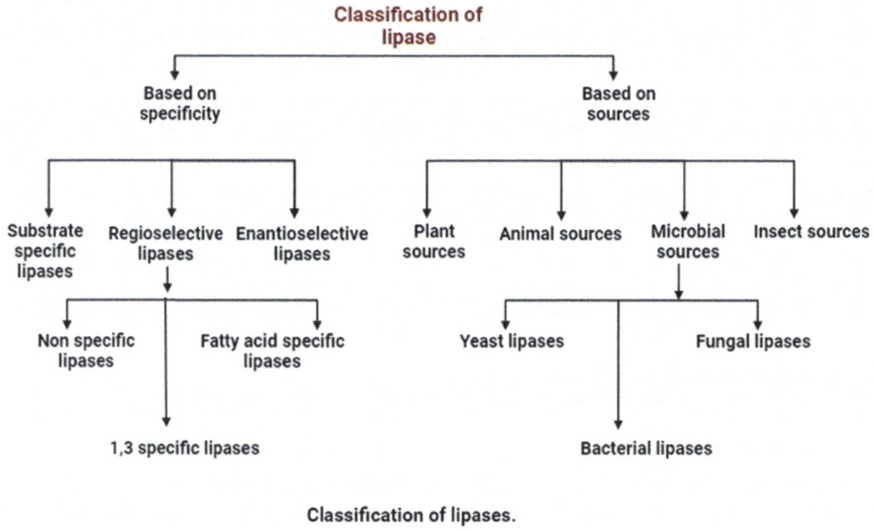

Fig. 10.5 Classification of lipases [27]

4 Types of Lipases

4.1 *Microbial Lipases*

Microbial lipases are extremely valued in a diversity of industries because of their numerous applications, including excellent substrate specificity, stability in organic solvents, activity in non-aqueous environments, and interfacial activation in micro aqueous settings. Benefits of these biocatalysts include inexpensive manufacturing costs, seasonal yield consistency, simple purification, and the ability to modify genetic material. The lipase market is anticipated to reach $590 million by 2023. Microbial lipases are used in a wide range of industries, such as dairy and baking, cosmetics, bioremediation, and biodiesel synthesis. Researchers are concentrating more on creating environmentally approachable industrial processes that consume less energy and produce less waste by retaining renewable raw materials to create biocatalysts [3, 12].

4.2 *Animal and Insect Lipases*

Lipases, which facilitate the digestion of fats and lipids, are produced by animal cells [52]. However, due to complications like culture management and product separation, animal lipases are more frequently utilized in clinical diagnosis than in commercial production. These lipases have received far less research attention than the lipases found in microorganisms and plants. Some of the lipases produced from animal tissue have been employed in various research. Since pancreatic lipase efficiently catalyzes the hydrolysis of primary alcohol esters, it has been widely employed as a research tool in the domains of biochemistry and lipid chemistry among the many animal lipases [50]. Porcine pancreatic lipases are employed in asymmetric aldol processes [13, 50], bis (indolyl) alkane [73] synthesis, and the manufacture of monoacylates of 2-Substituted (Z)-But-2-ene-1,4-diols [31]. Furthermore, it is discovered that lipases obtained from star fish can serve as an alternative to swine pancreas lipases, which are mostly used in the food processing industries. It should be mentioned that lipases can also be obtained from insect tissues, and they are mainly found in the gut or during the larval stage of insect development [8, 29, 60]. Nonetheless, research has demonstrated that lipases that are active against diacylglycerols are present throughout the body, with the exception of the alimentary canal [50], and in the flying muscles of numerous insects. Research on using gut lipase activity to produce biocontrol agents is becoming more prevalent [32, 61].

5 Industrial Applications of Lipases

See Fig. 10.6.

5.1 Biodiesel

Transesterification of different oils and fats with alcohol has led to the manufacturing of biodiesel, a viable alternative fuel that is manufactured internationally. Compared to chemical transesterification, enzymatic transesterification has numerous benefits, including little waste formation, lower energy consumption, and simple recovery of the byproduct glycerol in purified form [39]. There are several lipases known to be used in the synthesis of biodiesel, one of which being the lipase from *Burkholderia cepacia*, which catalyzes the transesterification of castor oil, an inedible oil [11]. The current state and future prospects of employing fungal lipases to produce biodiesel have been discussed by Aguieiras et al. They also demonstrate how the choice of alcohols and raw materials, the kind of biocatalyst, the usage of solvent, etc, affects the stability and activity of the lipase. Additional reviews with a specific focus are directed towards illustrating the developments in the enzymatic transesterification process for the production of biodiesel. These reviews address the important operational factors that affect the stability and activity of lipase, as well as technological solutions for the industrial application of the enzymatic process. Additionally, they highlight the significance of process parameter optimization, enzyme modification,

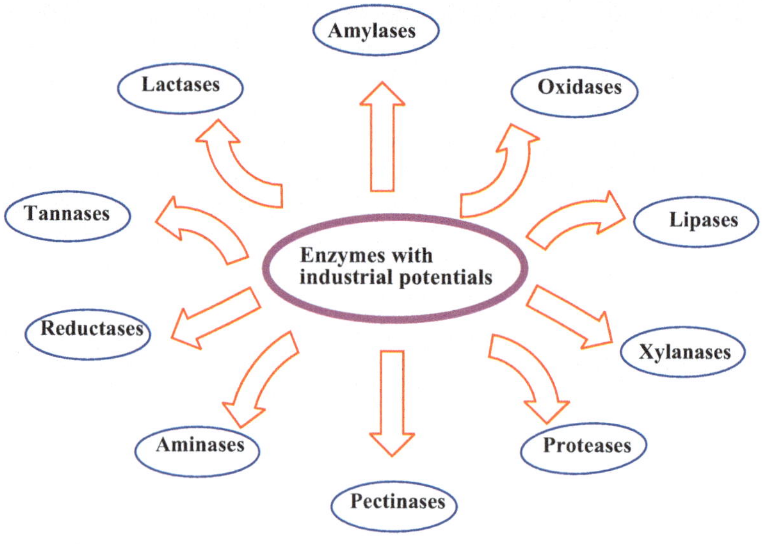

Fig. 10.6 The industrial potentialities of extensively studied enzymes in food industry

feedstock and alcohol selection, and process parameter optimization in lowering the cost of biodiesel production [25].

5.2 *Production of Biodegradable Polymers*

Another usage for lipases is in the synthesis of beneficial and biodegradable biopolymers, including polyesters, from renewable natural resources [30]. Polymerization reactions aided by lipase can be categorized into two main modes: polycondensation and ring-opening polymerization [71]. Examples of the latter are polycondensation of oxyacids or their esters and dicarboxylic acids or their derivatives with diols [30]. Several works have successfully used this polymerization process using a variety of techniques. The polymerization of bis (2,2,2-trifluoroethyl) sebacate and aliphatic diols was carried out using *Rhizomucor miehei* lipase [33, 35]. In their investigation, Mahapatro et al. discovered that Novozym 435 promoted the polymerization of 12-hydroxydodecanoic acid, resulting in a 91% conversion in high molecular weight oligomers [40]. Using immobilized lipase from *Candida rugosa*, B'odalo et al. produced polyesters of ricinoleic acid with 72% conversion. These results imply that lipases can be incorporated into polymerization operations in a variety of ways.

Understanding the molecular and atomic mechanisms underlying lipases has received a lot of research due to their wide range of applications and importance in biotechnology. There have been numerous reports of the three-dimensional structures of lipases from various different species since the BRIDGE-T lipase project (1990–1994) and up to this point. These structures provide insight into the mechanism these enzymes use during catalysis [9] (Fig. 10.7).

5.3 *Increasing the Lipophilicity of Bioactive Compounds*

Numerous bioactive substances with promising use as nutraceuticals, antioxidants, and other treatments are poorly soluble in nonpolar fluids. Adding hydrophobic groups to them is one method of altering them to make them more soluble. Then, utilizing long-chain carboxylic acids or long-chain alcohols, compounds containing hydroxyl and/or carboxyl groups can be transformed into esters. Because a lot of the chemicals are highly sensitive, lipases are good catalysts for the esterification reactions because they require mild reaction conditions [57].

The hydrophilic substrate (the bioactive component) and the hydrophobic substrate (the reagent) seen in a typical reaction combination lead to the same kinds of issues as those encountered in the acylation of carbohydrates (see preceding section). Tertiary alcohols and short-chain ketones are common solvents [39, 40]. Numerous studies have shown the application of lipases as a catalyst for the synthesis of esters that comprise two constituents with intriguing, frequently advantageous attributes

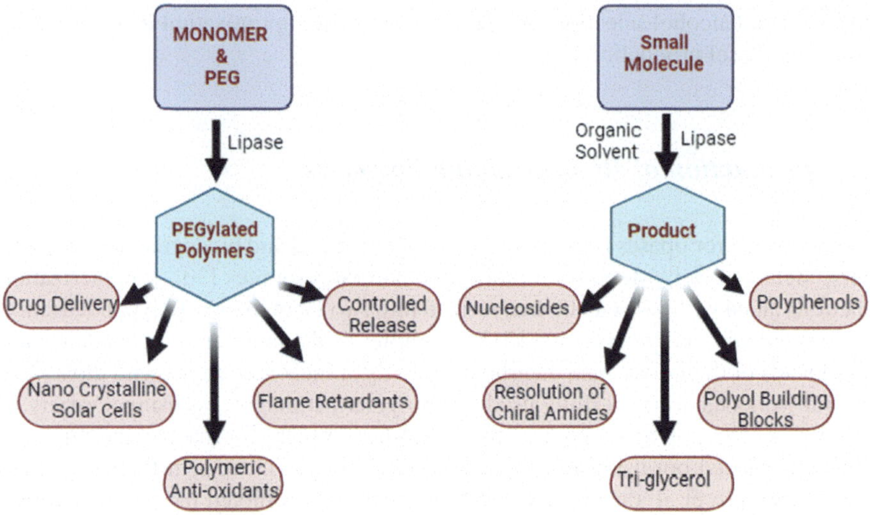

Fig. 10.7 Biodegradable polymer

for health. Rich in long-chain omega-3 fatty acids, a combination of fatty acids from fish was used to acylate the antioxidants rutin and vanillyl alcohol [41].

6 Enzyme Immobilization

The obstacles that enzymes typically encounter in harsh industrial environments—such as instability in high temperatures, organic solvents and pH ranges—have been addressed through methods of immobilization. The synthesis of undesirable byproducts, challenges with product separation, and enzyme recovery constitute a few of these obstacles, and the inability to make the entire process economically viable [3, 44, 53]. All of these issues can be resolved by immobilization, which increases the enzyme's efficiency over its free versions, produces purer products, facilitates faster product recovery, and improves the enzyme's stability, specificity, and selectivity over the course of multiple reaction cycles [47, 51, 56].

Physical techniques such as adsorption, encapsulation/entrapment, and confinement can be used to immobilize enzymes. Through these methods, reversible and relatively weak connections between the enzyme and the support are created, including hydrophobic contacts, van der Waal forces, affinity binding, and hydrogen bonding. More powerful and long-lasting chemical immobilization methods include covalent bonding and cross-linking, which ensure that an enzyme attached to a support cannot be withdrawn without endangering the structure and catalytic activity of the enzyme [15, 24].

In the food industry, for example, immobilized lipases from *Staphylococcus xylosus* and *S. warneri* are frequently used in order to create flavor esters. In order to maximize efficiency, provide easy product recovery, and allow for enzyme reusability, most bacterial lipases utilized in the commercial sector are first immobilized [66] (Fig 10.8).

Through their investigation, they found that this immobilized lipase very efficiently catalyzed the bioconversion of valeric acid into the taste ester pentyl valerate, which is connected to green apples. When *Lactobacillus plantarum* lipase was immobilized on sodium alginate, it efficiently catalyzed the extraction of beef fat and the creation of short-chain flavor esters [70]. Many immobilized enzymes have been effectively used at different phases of the biodiesel synthesis process. For example, *Pseudomonas sp.* Immobilization of ISTPL3 lipase occurred on the activation of charcoal by Khosla et al. and used in the biodiesel synthesis process.

There is still much to be discovered in this field, but lately, attention has been drawn to nanoparticle-based carriers because they have a number of benefits over traditional supports, including a large surface area, a higher capacity for enzyme

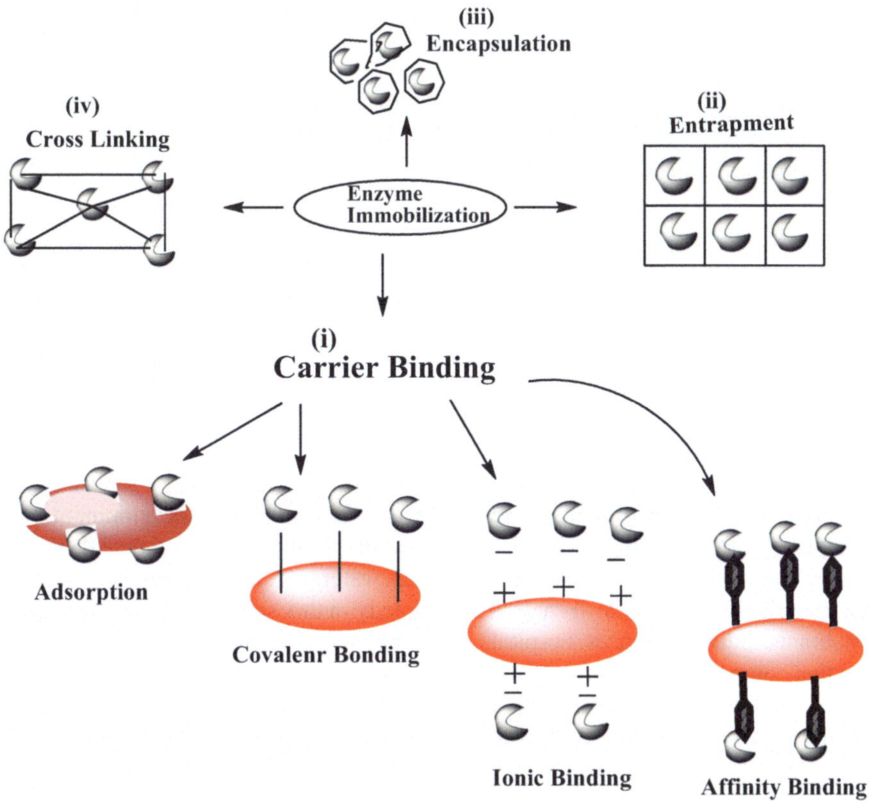

Fig. 10.8 Methods of immobilization

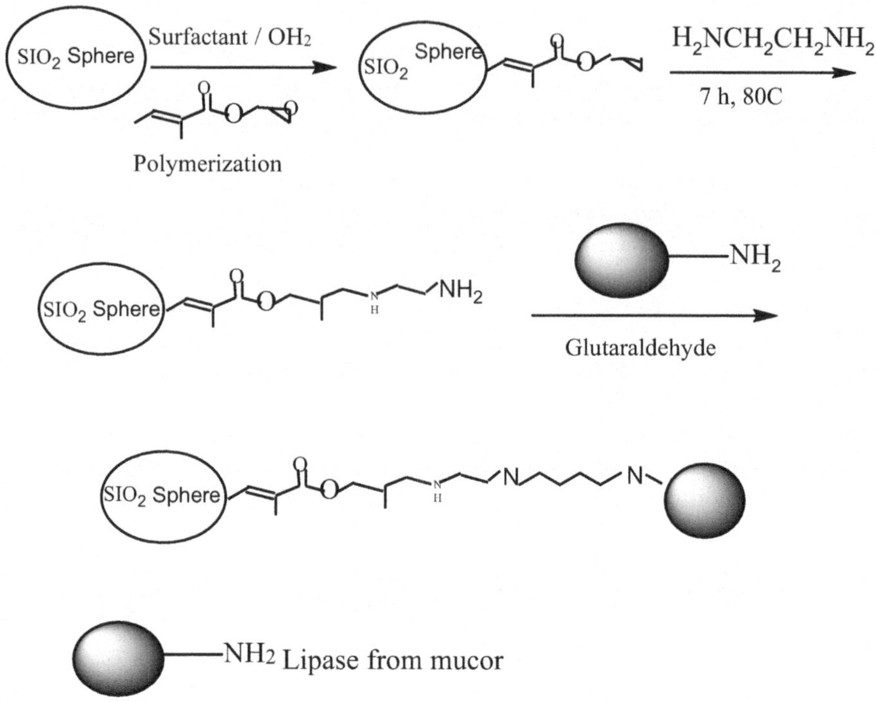

Fig. 10.9 Lipase immobilization on silica nanoparticle

adhesion, and smaller particle sizes. Additionally, studies have shown that lipases immobilized on nanostructured carriers display improved kinetic characteristics [15, 18] (Fig. 10.9).

7 Bioremediation

The hunt for bioremediation strategies to enable sustainable development was prompted by the growing concern over waste disposal and environmental preservation. The application of extracellular lipases, esterases, and proteases—enzymes that hydrolyze ester bonds—is a desirable remedy in this situation. The dairy sector, restaurants, slaughterhouses, hospitals, and health resorts can all release greasy effluents that comprise fats, oils, and proteins. These enzymes can improve the bioremediation of these wastewaters [14]. Furthermore, because of their broad range of substrate selectivity, stability in organic solvents, and absence of a cofactor need, lipases are the most often used enzymes in mediating reactions in chemical synthesis [58]. Testing was done on bacterial thermoalkalophilic lipase (BTL2) in order to produce biosensors for the detection of water pollutants [26]. In their research on the remediation of cooking oil wastes, [49] shown that treated waste oils were less harmful than untreated oils by using orange lipase for the transesterification reaction.

Lipases are widely utilized in the baking business also because they can improve bread's color and texture, stability of the dough, flour quality, lengthen the crumb structure, and make the bread softer and fresher [68]. Moreover, lipases facilitate the esterification process, which releases short-chain fatty acids and enhances the flavor and shelf life of baked goods [2, 23, 34].

8 Medical Applications

Galleria mellonella, the wax moth, discovered that bacteria were acting on *Mycobacterium tuberculosis* H3tRv. This work could be useful in identifying new drug sources. The microorganism *Candida rugosa's* lipase isolate can be used to make lovastatin, which lowers blood cholesterol [5–7, 33]. Polyunsaturated fatty acids (PUFAs) derived from plant and animal lipids, such as borage, tuna, and menhaden oil, are produced by employing microbial lipases. Mono- and di-acyl glycerides of free polyunsaturated fatty acids can be used to manufacture a range of medicinal products [65]. PUFAs are widely employed as food, medicinal, and nutraceutical additives because of their advantageous effects on metabolism. Nutraceuticals are produced using immobilized lipases [55].

8.1 Lipase in Human Milk Fat Substitutes

A variety of lipids are present in human milk fat (HMF), including stearic acids (5.7–8%), linoleic (7–14%), palmitic (20–30%), and oleic (30–35%) [59]. Palmitic acid is the main saturated fatty acid in HMF, unsaturated fatty acids, on the other hand, are located at the external positions of TAGs, whereas vegetable oils and cow's milk fat are usually esterified at the sn-2 position [64, 65]. The digestibility and intestinal absorption of food in babies are significantly influenced by the composition of HMF fatty acids. Using lipase, tripalmitin, butterfat, palm oil, palm stearin, or lard (rich in palmitic acid in the sn-2 position) can be acidolyzed to create human milk fat substitutes (HMFS) by employing free fatty acids (FFA) from diverse sources [64, 66]. Using IOI Loders Croklaan as a catalyst, Rhizomucormiehei (Lipozymc® RM) sn-1,3 selective lipase catalyzes the acidolysis of lard and soybean fatty acids to produce the commercial Betapol® product [27, 28].

8.2 Lipase in Pulp and Paper Industry

Large amounts of lignocellulosic biomass are processed annually in the pulp and paper sector. This machinery is quite varied for the production of pulp, and there are several opportunities for the introduction of microbial enzymes. Enzymes have been used in the paper industry, although primarily for processes like the modification of raw starch.

Deinking lipase has the ability to speed up the pulp's pulping rate, reduce the amount of chemicals used, reduce waste water pollution, lengthen the life of equipment, save energy and time, and lower the cost of the composite [20]. Using lipase from *Pseudomonas* sp. KWI-56, a deinking solution for ethylene oxide–propylene oxide adduct stearate improved paper whiteness and decreased persistent ink spots. Since the early 1990s, the enzymatic pitch monitoring methods—which uses lipases—has been used regularly in the large-scale paper-making process [35].

8.3 Textile Industry

Lipases are non-toxic and environmentally friendly enzymes, that are quickly becoming more popular in textile mills [15]. Lipases being crucial in textile industry remove size lubricants from cotton and denim fabrics, as they have become essential enzymes in this industry [54].

Consistent dyeing requires high fabric absorbency; lipases aid in this by minimizing cracks, streaks, and denim abrasion and improving fabric quality [33]. Lipases are utilized in the commercial domain to desize cotton and denim [42]. In textile mills, lipases provide an environmentally friendly substitute for conventional scouring techniques by aiding in the removal of wax from raw fibers to improve wettability and coloring. Lipase-catalyzed bioscouring lessens oxidative damage tso the fabric and has a lower environmental impact than chemical treatments that use surfactants and NaOH. Nevertheless, some research points to problems with particular lipase enzymes' ability to remove wax, such as *Pseudomonas mendocina* lipase. Research on bioscouring was advanced by *Bacillus licheniformis* MTCC 2618's lipase, which successfully removed wax from cotton cloth. The cotton fiber's hydrophilicity was enhanced by *B. sonorensis* bacterial lipase without causing any harm, suggesting that it could be used in the textile industry for environmentally friendly scouring [46]. By eliminating lipid barriers, *Bacillus aerius* 24 K's thermophilic lipase was used to improve the dyeability of wool fibers. Their results imply that this lipase can effectively scrub wool in a single bath, saving both water and energy [21] (Table 10.1).

Table 10.1 List of lipase uses in industry [36]

Industry	Action	Result of the application
Dairy products	Butter fat modification, cheese ripening, milk, and fat hydrolysis	Creation of flavoring agents for butter and cheese made from milk
Baked food	Enhancement of flavor	Propagation of shelf life
Drinks	Enhanced aroma	Alcoholic drinks, such as sake wine
Sauces for food	Enhancement of quality	Maysoine whippings and dressing
wholesome food	The method of transesterification	nutritious food
Fish and meat	Development of flavor	Removing fat from meat and fish products
Oil and fats	Hydrolysis and transesterification	Cocoa butter, glycerol mono and diglycerides, and margarine fatty acids
Clothing	Cutting down on biodegradable strains	Cloths for cleaning
Beauty products	The process of esterification	Bath oil, skin and sun-tan creams, etc
Surfactants in water	Replaces phospholipases in the lyso-phospholipid-producing process	Esters of polyglycerol and carbs are employed as emulsifiers in food formulation, such as sauces and ice cream, and as industrial detergents
Agricultural chemicals	The process of esterification	Among the herbicides include phenoxy-propionate
Drugs	Expolyester alcohol hydrolysis	Generate a range of intermediates utilized in the pharmaceutical industry
Fuel sectors	The process of Transesterification	Production of biodiesel
Control of pollution	Hydrolysis and transesterification of oil and grease	To hydrolyze oil and greases and remove tough stains

9 Conclusions

Lipases are amazing biocatalysts with a wide range of growing applications. Regretfully, due to high manufacturing and commercialization costs, industrialists' resistance to new and creative procedures, and the sluggish pace of research, their full potential is still undiscovered. Lipases offer a much-needed sustainable, recyclable, energy-saving substitute that works in mild temperatures and with agro-industrial wastes as substrates. This is beneficial for both the industry and the environment. The ability to create completely new, customized enzymes for our intended uses has been made possible by remarkable advancements in molecular enzymology and gene editing techniques. Every day, more and more new lipases are being cloned, described, and refined. The evaluation of metabolic pathways has quickly replaced the first use of controlled evolution of enzymes as a strategy for researching the connection between an enzyme's gene sequence and activity. There is a need for lipases designed for particular uses in important sectors because of the emergence of these new technologies. To mitigate the cost limitations, more effective production and purification techniques guaranteeing high lipase yields are required. Engineered lipases are the way of the future for molecular enzymology; hence, it is important to synthesize them and thoroughly assess and optimize their properties [3].

List of Abbreviations

FAs	Fatty acids
TAGs	Triacylglycerols
FFAs	Free fatty acids
FA	Fatty acid
TAG	Triacylglycerol
pH	Potential of hydrogen
RML	*Rhizopus miehei* Lipase
PUFA	Polyunsaturated fatty acid
HMF	Human milk fat

References

1. Abreu Silveira E, Moreno-Perez S, Basso A, Serban S, Pestana Mamede R, Tardioli PW, Sanchez Farinas C, Rocha-Martin J, Fernandez-Lorente G, Guisan JM (2017) Modulation of the regioselectivity of Thermomyces lanuginosus lipase via biocatalyst engineering for the Ethanolysis of oil in fully anhydrous medium. BMC Biotechnol 17(1):1–13
2. Adrio JL, Demain AL (2014) Microbial enzymes: tools for biotechnological processes. Biomolecules 4(1):117–139. https://doi.org/10.3390/biom4010117
3. Akram F, Mir AS, Haq IU, Roohi A (2023) An appraisal on prominent industrial and biotechnological applications of bacterial lipases. Molecul Biotechnol 65(4):521–543. https://doi.org/10.1007/s12033-022-00592-z
4. Arshad M, Adil M, Sikandar A (2014) Hussain T (2014) Exploitation of meat industry by-products for biodiesel production: Pakistan's perspective. Pakistan J Life Soc Sci. 12:120–125
5. Arshad M (ed) (2017) Perspectives on water usage for biofuels production: aquatic contamination and climate change. Springer
6. Arshad M (ed) (2020) Sustainable ethanol and climate change: sustainability assessment for ethanol distilleries. Springer Nature
7. Arshad M (ed) (2023) Climate changes mitigation and sustainable bioenergy harvest through animal waste: sustainable environmental implications of animal waste. Springer Nature
8. Arshad M, Ansari AR, Qadir R, Tahir MH, Nadeem A, Mehmood T, Alhumade H, Khan N (2022) Green electricity generation from biogas of cattle manure: an assessment of potential and feasibility in Pakistan. Front Energy Res 25(10):911485
9. Anobom CD, Pinheiro AS, De-Andrade RA, Aguieiras EC, Andrade GC, Moura MV, Almeida RV, Freire DM (2014) From structure to catalysis: recent developments in the biotechnological applications of lipases. BioMed Res Int

10. Barbosa O, Ortiz C, Berenguer-Murcia Á, Torres R, Rodrigues RC, Fernandez-Lafuente R (2015) Strategies for the one-step immobilization–purification of enzymes as industrial biocatalysts. Biotechnol Adv 33(5):435–456
11. Baron AM, Barouh N, Barea B, Villeneuve P, Mitchell DA, Krieger N (2014) Transesterification of castor oil in a solvent-free medium using the lipase from Burkholderia cepacia LTEB11 immobilized on a hydrophobic support. Fuel 117:458–462
12. Bano I, Ghauri MA, Arshad M, Yasin T, Younus M (2017) Bioactivity of variant molecular weight chitosan against drug-resistant bacteria isolated from human wounds. Microbial Drug Resistance 1;23(8):958–65.
13. Bano I, Arshad M (2018) Climatic changes impact on water availability. Perspect Water Usage Biofuels Product: Aquat Contaminat Clim Change 2018:39–54
14. Basheer SM, Chellappan S, Beena P, Sukumaran RK, Elyas K, Chandrasekaran M (2011) Lipase from marine Aspergillus awamori BTMFW032: production, partial purification and application in oil effluent treatment. New Biotechnol 28(6):627–638
15. Bilal M, Fernandes CD, Mehmood T, Nadeem F, Tabassam Q, Ferreira LFR (2021) Immobilized lipases-based nano-biocatalytic systems—a versatile platform with incredible biotechnological potential. Int J Biol Macromol 175:108–122
16. Carpen A, Bonomi F, Iametti S, Marengo M (2019) Effects of starch addition on the activity and specificity of food-grade lipases. Biotechnol Appl Biochem 66(4):607–616
17. Chandra P, Enespa S, R, Arora PK (2020) Microbial lipases and their industrial applications: a comprehensive review. Microb Cell Fact 19:1–42
18. Coelho ALS, Orlandelli RC (2021) Immobilized microbial lipases in the food industry: a systematic literature review. Critical Rev Food Sci Nutrit 61(10):1689–1703. https://doi.org/10.1080/10408398.2020.1764489
19. Damaso MCT, Salum TFC, Terzi S, Couri S (2013) Assay methods for lipase activity. Methods to determine enzymatic activity. Bentham, Rio de, pp 161–194
20. Demuner BJ, Pereira Junior N, Antunes A (2011) Technology prospecting on enzymes for the pulp and paper industry. J Technol Manag Innovat 6(3):148–158. https://doi.org/10.4067/S0718-27242011000300011
21. El-Fiky AF, Khalil EM, Mowafi S, Zaki RA, El-Sayed H (2022). A novel approach towards removal of lipid barrier from wool fibers' surface using thermophilic lipase. J Natural Fibers 19(14):9471–9485. https://doi.org/10.1080/15440478.2021.1982835
22. Enespa CP, Singh DP (2023) Sources, purification, immobilization and industrial applications of microbial lipases: an overview. Critical Rev Food Sci Nutrit 63(24):6653–6686. https://doi.org/10.1080/10408398.2022.2038076
23. Fernandes P (2010) Enzymes in food processing: a condensed overview on strategies for better biocatalysts. Enzyme Res 2010. https://doi.org/10.4061/2010/862537
24. Filho DG, Silva AG, Guidini CZ (2019) Lipases: sources, immobilization methods, and industrial applications. Appl Microbiol Biotechnol 103:7399–7423. https://doi.org/10.1007/s00253-019-10027-6
25. Ghaly A, Dave D, Brooks M, Budge S (2010) Production of biodiesel by enzymatic transesterification. Am J Biochem Biotechnol 6(2):54–76
26. Herranz S, Marciello M, Marco M-P, Garcia-Fierro JL, Guisan JM, Moreno-Bondi MC (2018) Multiplex environmental pollutant analysis using an array biosensor coated with chimeric hapten-dextran-lipase constructs. Sens Actuators, B Chem 257:256–262
27. Holm HC, Cowan D (2008) The evolution of enzymatic interesterification in the oils and fats industry. Eur J Lipid Sci Technol 110(8):679–691
28. Huang J, Zhao Q, Bu W, Zhang C, Yang Z, Zhang X, Zhang K (2020) Ultrasound-assisted hydrolysis of lard for free fatty acids catalyzed by combined two lipases in aqueous medium. Bioengineered 11(1):241–250
29. Hussain M, Khan I, Jiang B, Zheng L, Pan Y, Hu J, Ashraf A, Din AS, Waleed AA, Khan A (2023) Lipases: sources, immobilization techniques, and applications. Int J Environ, Agricult Biotechnol 8:6. https://doi.org/10.22161/ijeab.86.12

30. Kadokawa J-I, Kobayashi S (2010) Polymer synthesis by enzymatic catalysis. Curr Opin Chem Biol 14(2):145–153
31. Kawashima Y, Ezawa T, Harada T, Noguchi T, Kawasaki M, Kirihara M, Imai N (2016) Preparation of the monoacylates of 2-substituted (Z)-but-2-ene-1, 4-diols using porcine pancreas lipase. Bull Chem Soc Japan 89(2):257–267. https://doi.org/10.1246/bcsj.20150303
32. Khan S, Guo L, Maimaiti Y, Mijit M, Qiu D (2012) Entomopathogenic fungi as microbial biocontrol agent. Molecul Plant Breeding 3(7). https://doi.org/10.5376/mpb.2012.03.0007
33. Kiran S, Arshad Z, Nosheen S, Kamal S, Gulzar T, Majeed M, Jannat M, Rafique M (2016) Microbial lipases: production and applications: a review. J Biochem Biotechnol Biomater 1(2):7–20
34. Kirk O, Borchert TV, Fuglsang CC (2002) Industrial enzyme applications. Current Opinion Biotechnol 13(4):345–351. https://doi.org/10.1016/S0958-1669(02)00328-2
35. Kirk TK, Jeffries TW (1996) Roles for microbial enzymes in pulp and paper processing. ACS Publications. https://doi.org/10.1021/bk-1996-0655.ch001
36. Kishore J, Manojkumar ZC, Raghunath T (2011) Lipase biodiversity. Indian J. Sci Technol 4(8):971–982
37. Kumar D, Bhardwaj R, Jassal S, Goyal T, Khullar A, Gupta N (2021) Application of enzymes for an eco-friendly approach to textile processing. In: Environmental science and pollution research, 1–11. https://doi.org/ https://doi.org/10.1016/j.scp.2019.100178
38. Kwon CW, Park K-M, Choi SJ, Chang P-S (2015) A reliable and reproducible method for the lipase assay in an AOT/isooctane reversed micellar system: modification of the copper-soap colorimetric method. Food Chem 182:236–241. https://doi.org/10.1016/j.foodchem.2015.02.145
39. Luković N, Knežević-Jugović Z, Bezbradica D (2011) Biodiesel fuel production by enzymatic transesterification of oils: recent trends, challenges and future perspectives. Alternative Fuel 1
40. Mahapatro A, Kumar A, Gross RA (2004) Mild, Solvent-Free ω-Hydroxy Acid Polycondensations Catalyzed by Candida a ntarctica Lipase B. Biomacromol 5(1):62–68
41. Mbatia B, Kaki SS, Mattiasson B, Mulaa F, Adlercreutz P (2011) Enzymatic synthesis of lipophilic rutin and vanillyl esters from fish byproducts. J Agric Food Chem 59(13):7021–7027
42. Mehta A, Guleria S, Sharma R, Gupta R (2021) The lipases and their applications with emphasis on food industry. In Microbial biotechnology in food and health. Elsevier, pp 143–164
43. Melani NB, Tambourgi EB, Silveira E (2020) Lipases: from production to applications. Separat Purificat Rev 49(2):143–158. https://doi.org/10.1080/15422119.2018.1564328
44. Mohebrad B, Ghods G, Rezaee A (2022) Dairy wastewater treatment using immobilized bacteria on calcium alginate in a microbial electrochemical system. J Water Process Engin 46:102609. https://doi.org/10.1016/j.jwpe.2022.102609
45. Molinari F, Romano D, Villa R, Clark J (2019) Production of fine chemicals by (bio) transformation of agro-food by-products and wastes. In: Comprehensive biotechnology. Elsevier, pp 491–500. https://doi.org/10.1016/B978-0-444-64046-8.00376-1
46. Nerurkar M, Joshi M, Adivarekar R (2015) Bioscouring of cotton using lipase from marine bacteria Bacillus sonorensis. Appl Biochem Biotechnol 175:253–265. https://doi.org/10.1007/s12010-014-1259-6
47. Nimkande VD, Bafana A (2022) A review on the utility of microbial lipases in wastewater treatment. J Water Process Engin 46:102591. https://doi.org/10.1016/j.jwpe.2022.102591
48. Nomura DK, Casida JE (2016) Lipases and their inhibitors in health and disease. Chem Biol Interact 259:211–222
49. Okino-Delgado CH, Prado DZ, Facanali R, Marques MM, Nascimento AS, Fernandes CJ, Zambuzzi WF, Fleuri LF (2017). Bioremediation of cooking oil waste using lipases from wastes. PLoS One 12(10):e0186246
50. Pahoja VM, Sethar MA (2002) A review of enzymatic properties of lipase in plants, animals and microorganisms
51. Patel R, Trivedi U, Patel KC (2022) Lipases: an efficient biocatalyst for biotechnological applications. J Microbiol, Biotechnol Food Sci 11(5):e2498–e2498. https://doi.org/10.55251/jmbfs.2498

52. Patil KJ, Chopda MZ, Mahajan RT (2011) Lipase biodiversity. Indian J Sci Technol 971–982. https://doi.org/10.17485/ijst/2011/v4i8/30913
53. Rafiee F, Rezaee M (2021) Different strategies for the lipase immobilization on the chitosan based supports and their applications. Int J Biol Macromolecul 179:170–195. https://doi.org/10.1016/j.ijbiomac.2021.02.198
54. Ramesh A, Harani Devi P, Chattopadhyay S, Kavitha M (2020) Commercial applications of microbial enzymes. In: Microbial enzymes: roles and applications in industries, pp 137–184
55. Ray A (2012) Application of lipase in industry. Asian J Pharmacy Technol 2(2):33–37
56. Rios NS, Pinheiro BB, Pinheiro MP, Bezerra RM, dos Santos JCS, Gonçalves LRB (2018) Biotechnological potential of lipases from Pseudomonas: sources, properties and applications. Process Biochem 75:99–120. https://doi.org/10.1016/j.procbio.2018.09.003
57. Riva S (2002) Enzymatic modification of the sugar moieties of natural glycosides. J Mol Catal B Enzym 19:43–54
58. Saha B, Bothast R, Jordan D (2009) Enzymes, industrial. In: Encyclopedia of microbiology, pp 281–294
59. Sahin N, Akoh CC, Karaali A (2005) Lipase-catalyzed acidolysis of tripalmitin with hazelnut oil fatty acids and stearic acid to produce human milk fat substitutes. J Agric Food Chem 53(14):5779–5783
60. Sakate P, Salunkhe P (2013) Study of lipase activity during development of Chilo partellus (Swinhoe). Uttar Pradesh J Zool 33(1):61–68
61. Sandhu SS, Sharma AK, Beniwal V, Goel G, Batra P, Kumar A, Jaglan S, Sharma A, Malhotra S (2012) Myco-biocontrol of insect pests: factors involved, mechanism, and regulation. J Pathogens 2012. https://doi.org/10.1155/2012/126819
62. Santos KC, Cassimiro DM, Avelar MH, Hirata DB, de Castro HF, Fernández-Lafuente R, Mendes AA (2013) Characterization of the catalytic properties of lipases from plant seeds for the production of concentrated fatty acids from different vegetable oils. Ind Crops Prod 49:462–470
63. Schaechter M (2009) Encyclopedia of microbiology. Academic Press
64. Sharif N, Munir N, Hasnain M, Naz S, Arshad M (2021) Environmental impacts of ethanol production system. Sustain Ethanol Clim Change: Sustain Assess Ethanol Distiller 2021:205–223
65. Sharma R, Chisti Y, Banerjee UC (2001) Production, purification, characterization, and applications of lipases. Biotechnol Adv 19(8):627–662. https://doi.org/10.1016/S0734-9750(01)00086-6
66. Sharma S, Kanwar SS (2014) Organic solvent tolerant lipases and applications. Sci World J 2014. https://doi.org/10.1155/2014/625258
67. Subroto E, Nurhasanah S, Munarso SJ, Lipase immobilization by adsorption techniques on the hydrophobically modified matrix: a. https://ijettjournal.org/archive/ijett-v69i1p208
68. Tian YS, Peng RH, Xu J, Zhao W, Gao F, Fu XY, Xiong AS, Yao QH (2010) Mutations in two amino acids in phyI1s from Aspergillus niger 113 improve its phytase activity. World J Microbiol Biotechnol 26:903–907. https://doi.org/10.1007/s11274-009-0251-8
69. Tong X, Busk PK, Lange L (2016) Characterization of a new sn-1, 3-regioselective triacylglycerol lipase from Malbranchea cinnamomea. Biotechnol Appl Biochem 63(4):471–478
70. Uppada SR, Akula M, Bhattacharya A, Dutta JR (2017) Immobilized lipase from Lactobacillus plantarum in meat degradation and synthesis of flavor esters. J Genet Engin Biotechnol 15(2):331–334. https://doi.org/10.1016/j.jgeb.2017.07.008
71. Uyama H, Kobayashi S (1993) Enzymatic ring-opening polymerization of lactones catalyzed by lipase. Chem Lett 22(7):1149–1150
72. Veríssimo L, Mól P, Soares W, Minim V, Hespanhol M, Minim L (2018) Development of a bioreactor based on lipase entrapped in a monolithic cryogel for esterification and interesterification reactions. Revista Mexicana de Ingeniería Química 17(1):177–187

73. Xiang Z, Liu Z, Chen X, Wu Q, Lin X (2013) Biocatalysts for cascade reaction: porcine pancreas lipase (PPL)-catalyzed synthesis of bis (indolyl) alkanes. Amino Acids 45:937–945. https://doi.org/10.1007/s00726-013-1547-4
74. Wu XT, Ansari AR, Pang XX, Li HZ, Zhang ZW, Luo Y, Arshad M, Song H (2019) Visfatin plays a significant role in alleviating lipopolysaccharide-induced apoptosis and autophagy through PI3K/AKT signaling pathway during acute lung injury in mice. Arch Immunol Ther Exp 1(67):249–261

Perspectives Applications of Protease in Textile Industries

Iram Liaqat, Sajida Naseem, Sikander Ali, Muhammad Nauman Aftab, Muhammad Arshad, and Guojing Yang

Abstract Proteases are enzymes found in various life forms and are classified based on their source (plant, animal, bacterial, fungal, and viral) and optimal pH range. This chapter explores the different types of proteases and their applications in textile wet processing, including desizing, bioscouring, biobleaching, and biopolishing. Proteases are revolutionizing the textile industry by presenting a sustainable and multifaceted approach to fabric processing. They are preferred over the harsh chemical methods used for degumming the silk and can better unveil the hidden luster and softness of silk. They are considered as enzymatic sculptors. They further prevent the wool's rough edges from felting and getting interlocked by sculpting a smoother surface to enhance the user experience. Notably, these enzymes present a sustainable edge. Besides the surface modifications, the proteases orchestrate the harmony of functionalities that enhance the fabric quality by improving smoothness and wicking properties. The chapter mainly highlights the impact of proteases on fabric quality and environmental sustainability. Moreover, we have discussed the different types of proteases (alkaline, acidic, and neutral) and their specific applications in silk degumming, wool finishing, and stain removal. Surpassing the traditional chemical treatments proteases can be dealt with under milder conditions which provide a good environmental impact by lesser energy consumption. As scientific exploration

I. Liaqat (✉)
Microbiology Lab, Department of Zoology, Government College University, Lahore, Pakistan
e-mail: dr.iramliaqat@gcu.edu.pk

S. Naseem
Department of Zoology, University of Education, Lahore, Pakistan

S. Ali · M. N. Aftab
Ikarm-Ul-Haq Institute of Industrial Biotechnology, Government College University, Lahore, Pakistan

M. Arshad
Biochemistry Section, Jhang-Campus, University of Veterinary and Animal Sciences Lahore, Lahore, Pakistan

G. Yang
Hainan Medical University, Laboratory of Tropical Environment and Health, Haikou, Hainan, China

M. Arshad (ed.), *Enzymes in Textile Processing: A Climate Changes Mitigation Approach*, SDGs and Textiles, https://doi.org/10.1007/978-981-97-8058-7_11

unveils the deeper secrets of proteases, they are poised to inevitably redefine textile processing, weaving a future symphony of functionality, sustainability, and comfort.

Graphic Abstract

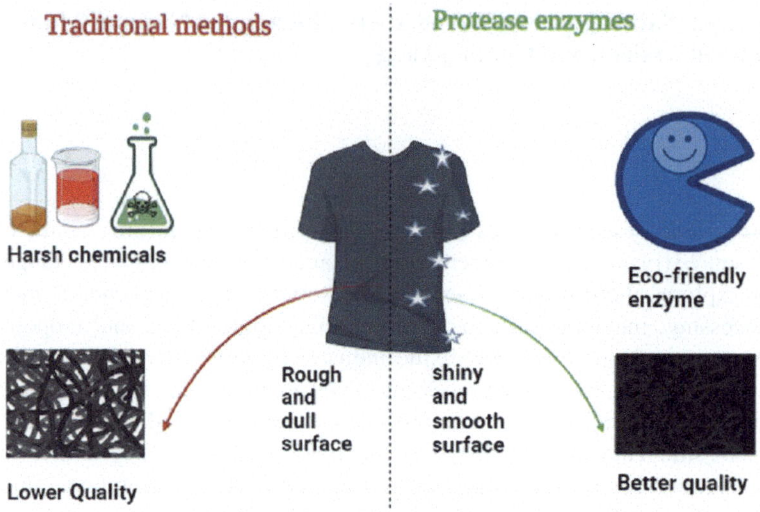

Keywords Proteases · Types · Textile industry · Applications · Impact on food and environment sustainability

1 Introduction

Enzymes are biological catalysts, sometimes referred to as bio-catalysts that quicken biological processes in living things. Additionally, they can be taken out of cells and utilized as catalysts for a variety of significant commercial activities [33]. Chemical reactions and other processes are frequently replaced by enzymes in industrial settings. However, the majority of businesses are still unable to use these enzymes due to their incapacity to function in harsh environments. Extremozymes' special adaptations enable them to function as bio-catalysts in industrial processes, even in adverse environments. Compared to chemical catalysis, they provide prospects for the advancement of enzyme technology and the development of very appealing, economical, sustainable, and environmentally friendly methods [39].

Given the market need, research into novel microbes is necessary to address this issue, as they are the primary source of all commercially significant proteases, which have a vast range of industrial uses. Proteases are employed in the textile industry to increase the lustre and softness of raw silk by removing the stiff and dull gum layer of sericin [28, 80].

The word for the enzymes that hydrolyze the peptide bonds in proteins to break them down is protease, also known as proteolytic enzyme. Regardless of the protein, proteases can identify the carboxylic side of an amino acid that forms a peptide bond even in the absence of substrate specificity [18]. Oxidoreductase and hydrolase are the two main categories of enzymes used in the pre-treatment and finishing processes of cotton. Laccase, catalase, peroxidase, and ligninase are the components of oxidoreductase. These enzymes work on finishing wool and fabric, bio-bleaching, bleach termination, and dye decolorization, among other things. On the other hand, hydrolases, which include cellulase/polishing, pectinase, cutinase, amylase, and lipase/esterase, are involved in a variety of activities, such as denim finishing, wool anti-felting, cotton softening, desizing, and wool finishing [14].

The use of enzymes in textile processing has significantly improved both the environment and the quality of the finished product. Before the fabric is coloured, it should be free of both artificial and natural contaminants. Several chemicals used in the preparing process, including as soda ash, caustic soda, hydrogen peroxide, hydrochloric acid, detergent, and auxiliaries, might have negative environmental effects [46, 67]. They are comprised of complex multi-dimensional molecules made up of polypeptide chains. Individual proteins with a relative molecular mass of 13,000 catalyze a single process, while multi-enzyme complexes with molecular mass of several millions catalyzes numerous reactions. Moreover, enzymes may require cofactors, which are tiny nonprotein molecules that aid in catalysis. Since enzymes are fragile molecules, they undergo denaturation that might be due to extreme change in temperature, light, alkalinity, acidity, ionizing radiation, and other biological factors, and thus become inactive [19].

Enzyme applications to textile wet processing have grown greatly in significance because of their energy-saving, stereo-specific, non-toxic, and environmentally friendly qualities. In textile wet processing systems, bioprocessing natural textiles like cotton, flax, jute, silk, and wool has shown to be incredibly successful because it has established improved functional qualities for the fibres in addition to other crucial benefits [11, 21]. The utilisation of enzymes in the textile processing industry is recognised due to their capacity to substitute harsh chemicals and conserve energy and water. Unfortunately, their uses were frequently hindered by their high cost and lack of long-term stability under storage and process condition [48].

Proteases represent a significant portion of the industrial enzyme business, accounting for 40% of worldwide enzyme sales and over 60% of the total enzyme market. Proteases are made up of a group of enzymes called amidases, peptidases, and proteinases in addition to a single enzyme [12, 50]. In these industries, enzymes are utilised to produce ecologically friendly fibre processing methods and enhance the quality of the final product [16]. Because proteases can catalyse the hydrolysis of peptide bonds in wool scales, they are frequently employed to alter the surface of wool to avoid wool felting [22].

Proteases are categorised according to the type of process they catalyse, the structure and makeup of their active site, and other factors. Their primary categorization into endoproteases and exoproteases is determined by the substrate's catalytic site [8, 28]. Proteases are useful in dehairing for the purpose of leather manufacture.

Since, the beginning of human civilization the conventional method of dehairing involves the use of lime and sodium sulphide as the lack of technology. But currently it is possible to replace chemical dehairing with enzyme-based dehairing processes using proteases. Procedures avoiding the use of lime and sulphide are being developed because of their environmental benefits [6, 7, 9, 28].

Proteases are a significant class of enzymes with a wide range of commercial applications that function as proteinases, peptidases, and amidases. They are significant industrial bio-catalysts [50].

2 Uses of Protease in Textile Industry

Proteases had been broadly used inside the fabric industry for numerous packages, inclusive of silk degumming, biopolishing of wool, and enzymatic remedy of wool with changed proteases. Silk degumming is the method of getting rid of sericin, a proteinaceous cementing layer surrounding silk fibroin filament, to acquire stepped forward lustre and softness. Historically, harsh chemical substances have been used in this procedure, which can reduce the energy of silk, shorten its shelf lifestyles, and negatively impact the surroundings.

Enzymes consisting of trypsin, papain, and bacterial proteases had been used as alternatives to these chemical compounds for silk degumming, as they can cleave peptide bonds of sericin without unfavourable the fibroin. Recombinant alkaline serine protease from *B. lehensis* has additionally been pronounced to enhance the softness and shine of silk fibres. Biopolishing of wool involves the usage of alkaline proteases to fabricate decrease-proof wool by means of hydrolysing the overlapping scales on wool fibres, which can factor toward the fibre pointers.

However, protease treatment can harm wool by means of causing immoderate lack of electricity and decreasing anti-felting capacity, which may also result in additional damage to the fibre indoors at some stage in wool processing. Consequently, precise proteases whose movement is restrained to the cuticle scale of wool fibres are preferred. Enzymatic treatment of wool with modified proteases can adjust the surface of wool and silk fibres to provide new and unique finishes.

Studies has been conducted at the utility of proteases to save you wool felting, that is a chief difficulty in wool processing. The bio-commercial viewpoints of microbial alkaline proteases had been reviewed, and it has been observed that they are ubiquitous, being observed in a wide diversity of resources together with vegetation, animals, and microorganisms. Proteases assault proteins via modes, yielding one of a kind products: exopeptidases act at the ends of the peptide chain, while endopeptidases or proteinases attack the interior of the peptide chain, yielding smaller polypeptides and peptides.

The predominant enzymes inside the industrial enzyme marketplace are those for processing cotton and cellulosic textiles, followed via enzymes for processing leather and fur. The enzyme marketplace for the remedy of silk and wool is minor,

Table 1 Types of proteases

Types of proteases	pH range	Uses of proteases	Classification	Sources
Alkaline	7–12	Majorly used in Detergent industry, Leather industry, Food industry	Serine proteases, chymotrypsins and the subtilisins	*Bacillus* sp., Deep-sea fungi, *Penicillium roquefortii* and *Aspergillus* flavipes, *Aspergillus* sp.
Acidic	3–4	In Food industry for the clotting of milk, proteolysis of wheat gluten	Aspartic endoproteases pepsins and pepsin-like enzymes, rennet and rennet-like enzymes	*Rhizomucor miehei*, *Mucor miehei*, *Rhizopus rhizopodiformis*, *Aspergillus* sp.
Neutral	5–7	Used for Meat tenderization, in bread production	Thermolysin, metalloprotease	*Bacillus* sp, *Aspergillus* sp.

accounting for 371% and 747%, respectively, of the global marketplace forecast in 2002 ($ million) (Table 1).

3 Different Types of Proteases Used in Textile Industry

There are several exclusive varieties of proteases used in the fabric industry, along with alkaline proteases, acid proteases, and impartial proteases. Every kind of protease has its personal particular residences and is used for different applications.

Alkaline proteases are the maximum normally used proteases within the textile industry. They are used for desizing, bioscouring, and biopolishing of cotton fabrics. Alkaline proteases also are used for removing sericin, a proteinaceous cementing layer surrounding silk fibroin filaments, to reap advanced lustre and softness in silk degumming. Alkaline proteases are produced through diverse microorganisms, together with Bacillus species, and are energetic at a pH range of 8–12.

Acid proteases are used for wool completing and for the removal of peroxidases. They're active at a pH range of 2–6. Acid proteases are used to modify the floor of wool and silk fibres to offer new and precise finishes. Research has been conducted on the application of acid proteases to prevent wool felting, which is a chief issue in wool processing [5].

Impartial proteases are used for decolouration of dyestuff and for the removal of protein-based totally stains. They are energetic at a pH range of 6–8. Neutral proteases are used to soften fabric, making them more cozy to put on and less difficult to manner.

4 Different Manufacturers of Proteases

There are several manufacturers of proteases for the textile enterprise, which includes Novozymes, Genencor, and DSM. Those businesses provide a number proteases for different packages, inclusive of stain removal, material softening, and fiber amendment. They are also targeted on developing sustainable answers for the textile enterprise, along with enzymes that can reduce water and electricity consumption in fabric processing.

5 Effect of Proteases on Quality of Textiles

Proteases are used in fabric processing to remove protein-based stains and to melt fabrics. They are particularly efficient, unique, and environmentally friendly, making them a sustainable desire for textile processing.

Proteases have numerous blessings over traditional methods for fabric processing. First, they require much less electricity and water than traditional strategies consisting of boiling or bleaching. Second, they are particularly particular, which means that they only break down proteins and do not damage the fibers or other additives of the fabric. Subsequently, they are biodegradable and environmentally friendly, which makes them a sustainable choice for fabric processing.

Alkaline proteases are the most normally used proteases inside the fabric industry. They're used for desizing, bioscouring, and biopolishing of cotton fabrics. Alkaline proteases are also used for getting rid of sericin, a proteinaceous cementing layer surrounding silk fibroin filaments, to acquire improved lustre and softness in silk degumming. Alkaline proteases are produced with the aid of various microorganisms, consisting of Bacillus species, and are active at a pH range of 8–12.

Acid proteases are used for wool completing and for the removal of peroxidases. They are active at a pH of 2–6. Acid proteases are used to adjust the surface of wool and silk fibres to offer new and unique finishes. Studies has been performed on the application of acid proteases to save you wool felting, that is a chief trouble in wool processing.

Neutral proteases are used for decolouration of dyestuff and for the elimination of protein-based stains. They are active at a pH range of 6–8. Impartial proteases are used to soften fabric, making them more relaxed to put on and less difficult to technique.

6 Kinds of Proteases

In biotic systems, proteases are widely distributed and have a variety of functions in the biochemical, physiological, and regulatory functions of cells and organisms [78]. They are obtained from different sources like microbes, plants and animals. The complex networks that encircle proteases, which also impact the temporal and spatial localization of their activity, are made up of peptides, receptors, substrates, activators, and inhibitors of small molecules [43]. Microbial proteases play a vital role for a number of physiological functions and may find use as therapeutic agents in the food, dairy, detergent, and leather sectors [64]. These enzymes are derived from microbial sources and are useful for a variety of tasks because they can effectively break down proteins. Microbial proteases are employed in the food industry to substitute chemical agents, increase flavor, and improve food quality—all of which contribute to environmentally beneficial practices. Furthermore, microbial proteases exhibit therapeutic potential in the medical domain and are utilized in various applications, including thrombolytic therapy, anticancer treatments, and wound healing.

An enzyme that hydrolyzes peptide bonds is called a protease. Endopeptidases and exopeptidases are two types of proteases that cleave the internal and terminal peptide bonds in substrates, respectively. Aminopeptidases and carboxypeptidases are two additional subclasses of exopeptidases [31]. Proteases can be divided into three groups according to pH: alkaline (pH 8.0–13.0), neutral (pH 6.0–8.0), and acidic (pH 2.0–6.0). For example, certain proteases require a pH of about 8 to be active, but others require a pH of about 6.9 or 4.5–5.5. In a similar vein, some proteases are most active at temperatures between 20 and 80 °C, while others prefer temperatures between 60 and 70 °C. Proteases vary in terms of their catalytic activity, substrate selectivity, pH, temperature, specificity of their active sites, and stability profiles [36].

With a 92% success rate in correctly identifying proteases and their kinds, ProtIdent is a web service that combines functional domain and evolutionary information [17]. Because it offers a dependable and effective method of identifying proteases and their types, this tool is beneficial to scientists and researchers who are researching proteases and their functions. The server makes use of a two-layer predictor: the first layer determines whether a query protein is a protease or not, and if it is, the second layer specifies which of the six functional categories of protease it is.

Based on their cellular location, substrate selectivity, and catalytic mechanism, proteases are categorized. Following the action mechanism, proteases can be divided into various kinds that are: aspartic proteases, metalloproteases, and threonine, cysteine, serine, and glutamic acid proteases [17]. These are:

Serine proteases: In this class, trypsin, chymotrypsin and elastase proteases are included. For the catalyzation of peptide bond hydrolysis, a serine residue can be used in their active site.

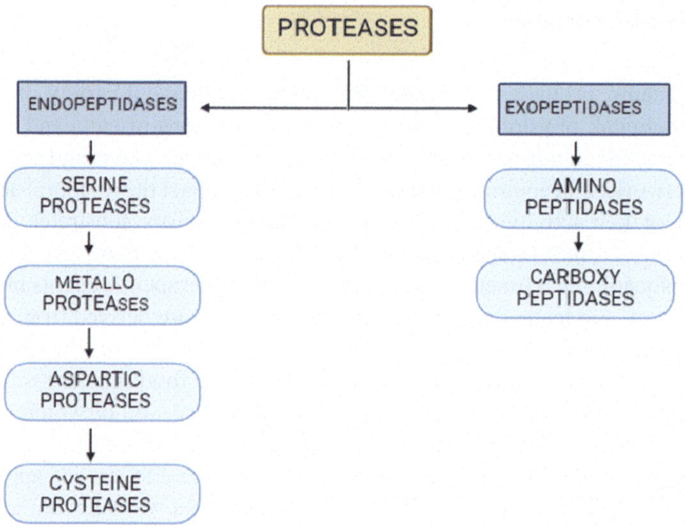

Fig. 1 Kinds of proteases

Cysteine proteases: In this class, papain, calpain, and lysosomal cathepsins are included. For the catalyzation of peptide bond hydrolysis, a cysteine residue can be used in their active site.

Aspartic Proteases: Pepsin and rennin fall into this category that can use aspartic residue in their active site for catalyzing peptide bond hydrolysis.

Metalloproteases: Thermolysin and carboxypeptidase A fall into this class that can utilize a metal ion in their active site for catalysis.

Glutamic proteases: These proteases, which include enzymes having a glutamic acid residue in their active site, work best at a pH of 2.0.

Threonine proteases: A family of proteolytic enzymes has an active site that contains a threonine residue. These residues are essential active sites in several of the proteasome's catalytic subunits (Fig. 1).

7 Classification of Proteases

Based on homology, catalytic type, and specificity, proteases, also known as peptidases, are categorized. Proteolytic enzymes are classified into two classes based on their specificity: endopeptidase, which target internal peptide bonds, and exopeptidases, which cleave one or more free amino acids from the N- or C-terminus.

7.1 Exopeptidase

The application of exopeptidases, which are enzymes that liberate terminal amino acid residues from peptides, is an additional intriguing debittering technique that lessens hydrolysate bitterness without lowering yields or necessitating the addition of substances with taste-active properties. Two classes of exopeptidases that cleave amino acid residues from the N- and C-terminals of peptides, respectively, are aminopeptidases and carboxypeptidases. The hydrolysate's bitterness can be reduced by exopeptidase treatment, which releases hydrophobic amino acids from one or both of the terminals. Hydrolysis with exopeptidases, however, may cleave off amino acid residues relevant to ACE-inhibitory activity, even while it is successful for debittering [15].

To prevent cleavage at the N or C termini of peptide substrates, exopeptidases employ terminal anchoring contacts [51]. Exopeptidases, also known as aminopeptidases or carboxypeptidases, function at or close to the end of polypeptide chains.

Aminopeptidases are exopeptidases that release an amino acid from their N-terminus. Dipeptidylpeptidase and tripeptidylpeptidase, respectively, are the names given to those that release a dipeptide or a tripeptide. Peptidyl-dipeptidase is the name given to the dipeptide that is released by carboxypeptidases, which can also release a single amino acid. The presence of amino acid substituents in the active site of carboxypeptidases allows them to be further classified into three groups: serine, metallo-, and cysteine carboxypeptidases [25] (Fig. 2).

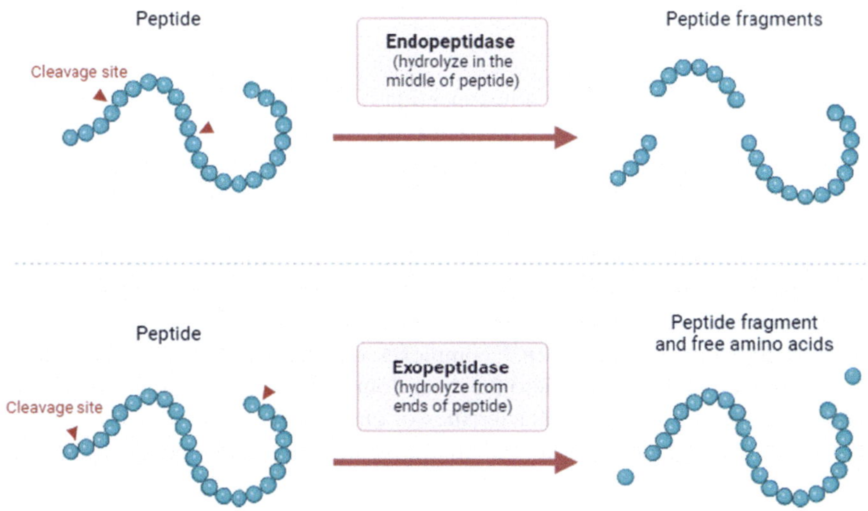

Fig. 2 Endopeptidase and exopeptidase

7.2 Aminopeptidases

Aminopeptidases (APN) contain carbohydrates and is extensively glycosylated and it makes up at least 20% of the proteins mass. An abundant enzyme, APN can be found in a wide variety of tissues, organs, and cells. It is highly expressed in the liver, small intestine mucosal cells, and the brush border membranes of the kidney. It is found in the apical region of polarized epithelial cells. Additionally, it is dispersed throughout the brain, spinal cord, and other nerve tissues and tripeptide, dipeptide, and single amino acid residues are released by aminopeptidases when they are oriented towards the free N-terminal of the polypeptide chain. Following identification, efforts are made to eliminate any potential N-terminal methionine found in proteins that express themselves heterologously but are absent from many mature, naturally occurring proteins. Many different microbiological strains, including bacteria and fungus, have aminopeptidases. Although one report indicated that aminopeptidases derived from *Aspergillus oryzae* are external enzymes, aminopeptidases often function as intracellular enzymes [50].

Aminopeptidase N is a metalloprotease that is widely distributed and a member of the M1 family. Since it depends on zinc, it is the same as CD13. The N-terminal of the APN, a type II integral membrane protein that is an ectoenzyme on the cell membrane, connects it to the cells. It has a molecular weight of cells. APN is discovered to be present on astrocytes and pericytes in addition to endothelial cells and synaptic membranes [41].

7.3 Carboxypeptidase

A zinc metalloprotease called carboxypeptidase N (CPN) (Enzyme Commission Number (EC) cleaves the basic amino acids arginine and lysine from the carboxy-terminus of proteins and peptides that are physiologically active. About 30 μg/ml of CPN are discharged into the bloodstream and are constitutively produced in the liver. It is a 280 Mr tetramer structurally made up of two big subunits (CPN2), which have 83 Mr each, and two tiny subunits (CPN1), which have 52 Mr each. The big subunits of CPN work to stop the protein from degrading or filtering from the bloodstream, whereas the small subunits contain the enzymatic active sites [45].

Model research employing Zn^{2+} complexes of different macrocyclic triamine and tetraamine derivatives (cyclen) has been found to be helpful in clarifying and comprehending the inherent characteristics of zinc ion recognition of substrate or inhibitor at the active centers of carbonic anhydrase and carboxypeptidases [34] (Fig. 3).

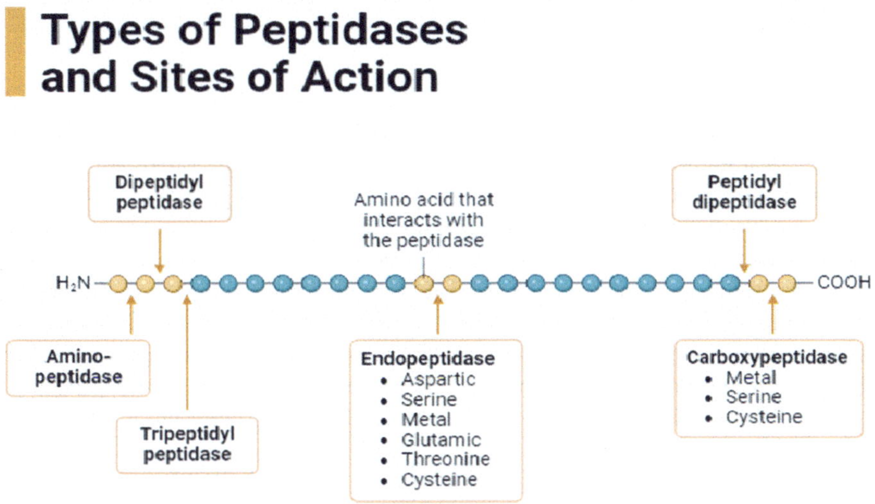

Fig. 3 Site of action of different types of peptidases

7.4 *Endopeptidase*

Unlike exopeptidases, endopeptidases are more difficult to classify and are instead categorized according to the size of the peptides that they cleave. Enzymes known as endopeptidases target peptides and proteins of any size. Endopeptidases disrupt the substrate's peptide bond at a distance [10].

Endopeptidases often affect a polypeptide's inner zone. Additionally, endoproteases are categorized according to the functional group that is present at the site and the specificity of their side chains [69]. Classification of endopeptidase on the basis of catalytic center is as follows:

- Serine protease.
- Aspartic acid protease.
- Cysteine proteases.
- Metalloprotease.
- Glutamic proteases.

Classification of endopeptidase on the basis of optimum temperature and pH range is as follows:

- Alkaline proteases: show maximum activity at pH range of 8–12.
- Acidic proteases: optimum activity at between pH 2–6
- Neutral proteases: optimum activity at pH 7 [27].

7.5 Serine Proteases

Serine proteases are one of the most extensively studied families of enzymes and approximately one-third of all proteases are serine. It is named because of the presence of a uniquely reactive nucleophilic Ser residue at the active site [30]. It is an important group of enzymes responsible for elastin degradation that occurs in a serious lung disease called emphysema. The majority of serine proteases for instance pancreatic enzyme like trypsin, chymotrypsin, and elastase, are considered to have evolved from common ancestors. However, they have gained their own specificity because of genetic mutations transferred by their evolutionary ancestors. The central portion of serine protease consists of three invariant residues, the Asp-His-Ser "charge relay system" or "catalytic traid". These residues catalyze the hydrolysis of peptide bonds within proteins and peptides by the following rate-limiting step:

$$\mathrm{Im} + \mathrm{R'—OH} + \mathrm{X—C(=0)R—Im} + \mathrm{—H} + \mathrm{R'OC(0-)(R)X}$$

where Im and R′—OH designate His and Ser [76]. The Asp-His-Sertraid is found in four different structural contexts, which shows that this catalytic machinery evolved on four different occasions [30]. Serine protease is an endoprotease that catalyzes the hydrolysis of bonds in the middle of polypeptide chains. Although serine protease is found in all kingdoms, including many viral genomes, there is a difference in its distribution in each clan across species. For example, clan PA protease in eukaryotes (rarely in prokaryotic and plant genome) [5, 23]. It performs many physiological and cellular functions like cellular and humoral immunity, digestion, fertilization, fibrinolysis, tissue remodeling, embryonic development, and protein processing [35].

7.6 Cysteine Protease

Cysteine protease is a protein found in all living organisms and is involved in the hydrolysis of amide, ester, peptide, thiol ester, and thiol ester bonds. It has a molecular mass of about 21–30 kDa and is further divided into exopeptidases (cathepsin X, carboxypeptidase B), which cleaves the peptide bond proximal to the carboxy or amino terminal of the substrate, and endopeptidases (bromelain, ficiain, cathepsins, papain), which cleave peptide bonds distant from the N or C terminal. The majority of cysteine proteases are evolutionarily related to papin and are synthesized in the active form. They are activated by proteolytic cleavage of the N-terminal [26, 80]. Papain and cathepsins are important because they belong to the most abundant family of cysteine proteases. Papain was found in 1937 *in Carica papaya* and was the first cysteine ever found. Cystiene has Cys-His-Asn residues at its active site. The histidine residue enhances nucleophilicity by acting as a proton donor. The nucleophilic cystine residue then attacks the carbon of the reactive peptide bond, forming a tetrahedral thioester and releasing an amine fragment. This intermediate is then stabilized by the formation of a hydrogen bond. The thioester bond is then hydrolyzed to form carboxylic acid from the remaining substrate fragment [77]. In the catalytic process, the formation of a thiolate/imidazolium ion pair (Cys-S-/His-Im+) is an important step, which occurs by the transfer of protons between Cys-25 and His-159. In all cysteine proteases, proteolytic activity is mediated by catalytic Cys and His, however, papain plays this role by Asn [55].

7.7 Aspartic Protease

Aspartic protease is a small group of proteolytic enzymes of the pepsin family that share the same catalytic apparatus. They are found in many organisms like humans, plants, and in retroviruses. The stomachs of mammals, lysosomes, fungi, and yeast are known to be the best sources of aspartic proteases [8, 63]. They usually function properly in acidic solutions, which limit the activity of aspartic proteases to only acidic environments. Therefore, it is less abundant than other protease groups, i.e., serine. In mammals, aspartic protease is synthesized as a zymogen (inactive), which is then converted to an active protease [73].

Certain microorganisms secrete aspartic protease as a virulence agent to break down proteins or cell membranes and enter the host cell. It is important in industries because it is used in the production of cheese because of its milk-clotting ability. It also has great interest as it can also be helpful in curing many human diseases [70].

7.8 Metalloprotease

Metalloproteases are a group of enzymes that facilitate the breaking of peptide bonds within proteins by including a catalytic metal ion at their active site. The most frequent metal ion used by metalloproteases is zinc (Zn^{2+}); however, active sites also contain other transition metals, including Co^{2+} and Mn^{2+}. Some physiological and pathological processes in which these enzymes are involved include the control of protein breakdown, cell proliferation, differentiation, extracellular matrix remodeling, vascularization, and cell migration [13].

Metalloproteases are subdivided into two groups: metalloendopeptidases and metalloexopeptidases. Peptide bonds within proteins are broken by metalloendopeptidases, whereas those at the extremities of proteins are broken by metalloexopeptidases. More precisely defined target sites and conserved catalytic residues specific to a given type of metalloprotease are nested within these divisions. *Bacillus thermoproteolyticus* secretes thermolysin, a Zn^{2+}-endopeptidase, an example of a metalloprotease. Thermolysin effectively breaks down the internal peptide bonds between leucine, isoleucine, and phenylalanine on the N-terminal side of large hydrophobic amino acids. Aspartame, a low-calorie sweetener, is prepared with this enzyme, which is also used commercially as a nonspecific protease in peptide sequencing.

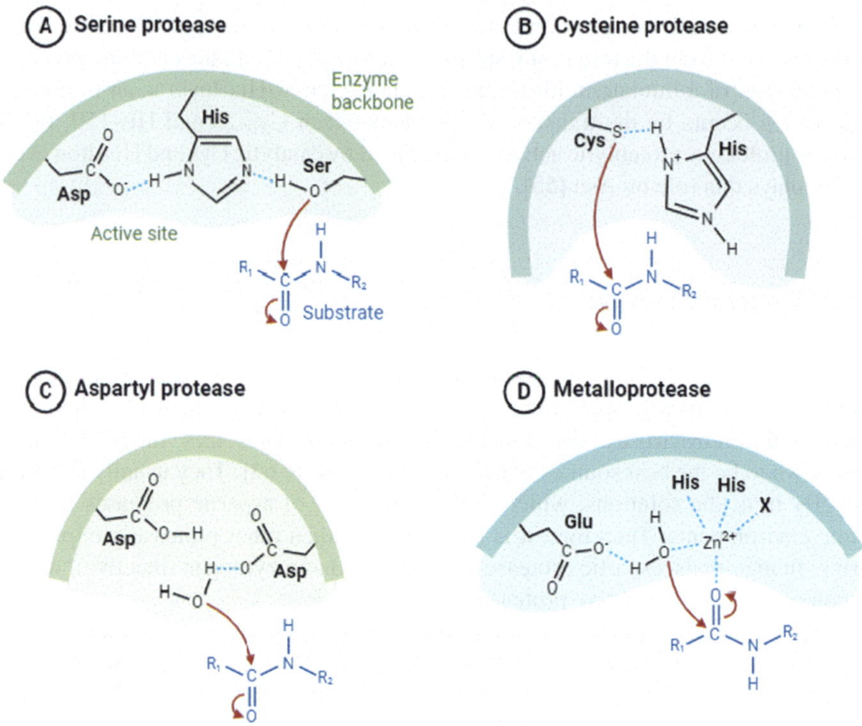

Fig. 4 Types of endopeptidases

Matrix metalloproteinases (MMPs), a class of enzymes that cleave protein substrates by a conserved process requiring activation of an active Zn^{2+} ion, are another example of a metalloprotease. Numerous physiological and pathological processes such as protein breakdown control, cell proliferation, differentiation, extracellular matrix remodeling, vascularization, and cell migration are influenced by MMPs [37] (Fig. 4).

8 Classification Based on Sources

Proteases are one of the main elements of different life forms of Earth including plants, animals, fungi, bacteria and viruses. The classification of proteases on the base of sources from which they are obtained is as follows.

8.1 Plant Proteases

Papain, keratinase, bromelain, actinidin and ficin are common proteases which are derived from plant sources [12, 28]. Papain has a long history and it is obtained from the latex of papaya fruit which is grown in the central and western regions of subtropical Africa, as well as India. Bromelain protease is an enzyme which is derived from the roots of pineapple plant when they are harvested. It is available commercially in powdered form. Actinidin is a cysteine or sulfhydryl protease which is obtained from the kiwi fruit also known as gooseberry. Further, Ficin is also an important and commonly known plant based cysteine protease. It is extracted from *Ficus carica* commonly known as Fig-Plant [4]. Ficin is used for meat tenderization because it increases the solubility of muscle proteins [11, 44]. Cysteine as proteolytic enzyme has been identified and characterized by leaves of maize in Pakistan [3].

8.2 Animal Proteases

The most commonly known proteases of animal origin are pancreatic trypsin, pepsin, rennin and chymotrypsin [28]. Trypsin is an intestinal enzyme and it is responsible for hydrolyzing protein content of food [50, 67]. Chymotrypsin protease is present in the extract collected from the pancreas of animals. Chymotrypsin is an important animal-based protease. The pure form of chymotrypsin is costly and it is primarily used in different medical and analytical applications. Pepsin is an acidic protease of animal origin and it is found in the stomachs of almost all vertebrates. Pro-rennin is present in the stomach of all nursing mammals. It is an inactive precursor of an important pepsin-like protease called rennin. Pepsin is responsible for converting pro-rennin into active rennin [1]. Calpain is an important enzyme firstly purified from chicken's skeletal muscles and it is highly utilized for the degradation of myofibrillar proteins [4].

8.3 Bacterial Proteases

Bacterial proteases are considered to be most important among all proteases because of their large number of industrial applications. They make up one of the biggest categories of industrial enzymes. They are used in detergent, leather, food, pharmaceutical and textile industries as well as in silver recovery and bioremediation procedures. In leather industry and laundry detergents, these enzymes are used for purpose of dehairing and to remove protein based stains from clothes, respectively. Most of the commercially available bacterial proteases are obtained from Bacillus and Pseudomonas bacteria. They are both acidic and alkaline in nature. The examples of protease releasing bacteria includes *Pseudomonas auroginosa* PD100, *Bacillus subtilis*, *Bacillus cereus*, *Bacillus isolate* C45 [28]. Research showed that *Bacillus subtilis* contains two major kinds of proteases, neutral protease and subtilisin [24].

8.4 Fungal Proteases

Fungal proteases have a broad substrate specificity and are active over a large pH range ranging from pH 4 to 11). For example, *Aspergillus oryzae* is a fungus which is a producer of three different proteases: acid, neutral and alkaline. But as compared to bacterial enzymes, they are less heat-tolerant and have a slower response rate [28]. Most of the proteases obtained from fungi are extracelluluar acid proteases. These acid proteases obtained from fungi are called as aspergilla opepsins. The important fungal species from which protease are extracted are *Aspergillus niger*, *Aspergillus oryzae*, *Aspergillus awamori* and *Aspergillus saitoi* [59].

8.5 Viral Proteases

Many different viruses have been found to have proteases. These proteases are unrelated to the complexity of their capsids, lipid envelopes, or genomic composition. Viral proteases not only cleave peptide bonds but also uncover numerous novel proteolysis mechanisms. It is discovered that both enclosed and non-enveloped viruses contain proteases. The proteases found in different viruses are specifically members of three families: aspartyl protease which is found in HIV-1, cysteine proteases which is found in adenoviruses and serine proteases which are found in hepatitis C virus (HCV), Flavivirus and herpesviruses [62]. Viral proteases have received more attention because of their well-organized roles in the generation of viral proteins which result in fatal diseases like cancer and AIDS [81].

8.6 Alkaline Proteases

Alkaline proteases are an essential class of enzymes for both industry and physiology. They are mostly utilized as detergent additives. In the hydrolysis of proteins, they have a specialized catalytic function [28]. Microorganisms that are alkaliphilic or neutralophilic both can produce alkaline proteases. However, alkaliphilic microorganism are abundant in nature and can be found in nearly every habitat regardless of the alkalinity level [54]. Most of the commercially important alkaline proteases are derived from genus Bacillus and Streptomyces. Alkaline proteases are active at alkaline pH which ranges from 9 to 11. The producers of alkaline proteases are distributed in soil, water and other highly alkaline conditions. Alkaline protease segregation has been reported from a range of sources including detergent pollution, dried fish, sand soil and slaughterhouses. Because of their distinct activity and ability to maintain a steady alkaline pH, alkaline proteases are used in a variety of pharmaceutical, food and other related industries. Despite the fact that several companies create alkaline proteases due to market demand and economic considerations, only strains exhibiting higher yields and hyperactivity will be approved in the current biotechnological advancement. *Bacillus* sp. yields two important kinds of alkaline proteases that are used as industrial enzymes: subtilisin Carlsberg and subtilisin novo. Serine proteases are also alkaline in nature and are used in detergent and leather industry [59].

8.7 Acidic Proteases

Acidic proteases, also known as acid proteolytic enzymes, are a type of protease that features optimally at acidic pH stages, normally underneath five. These enzymes are vital for diverse organic procedures, such as protein digestion, milk clotting, and cheese ripening. Within the context of food processing, acidic proteases have more than one applications. For instance, inside the brewing industry, acidic fungal proteases from *Aspergillus usamii* are applied to enhance the purposeful properties of wort, making sure a proper fermentation process [81].

These enzymes contribute to the discharge of sufficient peptides and amino acid degrees, which are essential for the growth of yeast all through fermentation. Moreover, acidic proteases are employed in the dairy industry to boost up cheese ripening, alter functional residences, and decrease allergenic properties of milk merchandise. They are also used to hydrolyze particular peptide bonds, generating paracasein and macropeptides, which play a vast role in cheese texture and flavor improvement.

Regarding protein digestion, acidic proteases, together with pepsin, are important for the initial breakdown of proteins in the stomach. Pepsin has a most proteolytic interest at pH 2, with an most appropriate pH range of two-5, and is irreversibly inactivated at pHs above 7.5. It acts preferentially on κ-casein on the casein micelles, contributing to the coagulation of the casein fraction of milk proteins below acidic conditions, at the same time as the whey protein fraction remains soluble.

The production of acidic proteases with the aid of microorganisms has been significantly studied due to their ability for business applications. The optimization of media composition and boom conditions is essential for the manufacturing of acidic proteases. It was determined that wheat bran dealt with acid/mineral salt answer produced the best protease interest (107.4 U/mL) compared to rice bran dealt with the identical mineral answer [47].

Acidic proteases have been pronounced to have potential applications in numerous industries. For instance, acidic proteases from *Bacillus halodurans* had been shown to have capability applications within the meals industry for the hydrolysis of gluten proteins, which may improve the digestibility of wheat-based products. Acidic proteases from *Aspergillus niger* had been used in the leather-based enterprise for the dehairing and bating techniques, that can enhance the satisfactory of leather products [71].

The impact of incubation time on the production of acid protease becomes also evaluated, and the highest protease hobby (95.2 U/mL) was recorded at 120 h. Similarly, an increase in the incubation time reduced the enzyme hobby. The effect of incubation temperature on acid protease production became additionally studied, and the highest protease activity (106.1 U/mL) was acquired at 30 °C. The effect of inoculum size on acid protease manufacturing turned into also evaluated, and the very best protease hobby (78.2 U/mL) changed into recorded from the isolate Z1BL1, whereas the lowest protease interest (31.7 U/mL) turned into noticed from the isolate Z1Y2 under stable-state fermentation [68].

8.7.1 Different Types of Acidic Proteases

Acidic proteases are categorized into three families, namely, pepsin (A1), retropepsin (A2), and enzymes from pararetroviruses (A3). These acidic proteases are pretty conserved, suggesting that they may have advanced from a common ancestor. The sequences of *E. coli* and *Cyprinus carpio* appear to be homologous to some extent; however, they do now not have common active-web site residues and therefore do no longer have a commonplace ancestor. Those, in turn, show no vast homology to the alternative seven alkaline proteases.

Acidic proteases can be categorised into two major classes based at the kind of amino acid used of their active site: aspartic proteases and glutamic proteases. Aspartic proteases use a catalytic aspartic acid in their energetic web site, while glutamic proteases use a catalytic glutamic acid (much less common). Aspartic proteases are further classified into businesses based totally on their catalytic residues: Aspartic proteases and Glutamic proteases. Aspartic proteases are the most common type of acidic protease and are extensively used in numerous industrial and applied sectors due to their extensive type of physiological characteristics, consisting of hydrolysis at pH extremes or elevated temperatures.

8.8 Neutral Proteases and Their Different Types

Neutral proteases are enzymes that catalyze the hydrolysis of peptide bonds in proteins under impartial pH conditions. The one-of-a-kind styles of impartial proteases encompass cathepsin G, elastase, and azurocidin. these proteases are enzymes that function below neutral pH conditions and are worried in numerous biological procedures and industrial packages. Cathepsin G is a chymotrypsin-like enzyme, at the same time as elastase and azurocidin also are sizable individuals of this group. Impartial proteases are usually determined within the genus Bacillus, with a pH variety normally among five to eight, and they are acknowledged for generating much less bitterness compared to other protease sorts [65].

8.8.1 Mechanisms of Neutral Proteases and Their Significance

Neutral proteases play essential function in various organic procedures and commercial programs. Experimental proof for enzymatic mechanisms is often constrained, main to discussions and debates about the suitable interpretation of facts. Molecular dynamics (MD) and quantum mechanical/molecular mechanical (QM/MM) studies have furnished insights into the mechanisms of neutral proteases, mainly in latest years. Those studies have targeted on most important protease instructions like aspartic, metallo-, cysteine, serine, and threonine proteases, elucidating their catalytic sports and structural functions.

The catalytic ability of neutral proteases, specifically in natural synthesis and commercial approaches, is increasingly identified for their excessive selectivity, catalytic performance, and moderate operational situations. Enzyme-primarily based procedures offer blessings consisting of minimum environmental effect, simpler separation, and reduced use of risky chemical substances. Lipases, a sort of impartial protease, are extensively applied within the production of optically active compounds, highlighting the importance of those enzymes in artificial chemistry. Despite the benefits of the usage of enzymes, challenges exist, inclusive of the need for further experimental and computational efforts to increase the knowledge of neutral proteases, particularly in biological structures and commercial packages [25] (Table 2).

9 Application of Proteases in Textile Wet Processing

Textiles undergo a number of treatments during the wet processing process to improve their qualities and attributes. Proteases are one of the many agents used in this procedure that have acquired importance due to their many uses. This chapter examines the function of proteases in several textile wet processing stages, emphasizing how they affect fabric characteristics and the environmentally friendly elements of enzymatic treatments (see Fig. 5).

9.1 Desizing

In order to prevent yarn breakage and damage, desizing is the process of removing adhesive compounds (sizes) from fabric threads that were applied during the weaving process. Prior to wet-processing techniques including dying, printing, and finishing cotton fibers, desizing, souring, and bleaching are all necessary [57]. Alkalis, acids, and oxidizing agents are just a few of the chemicals that are commonly used to remove particles of this size from textiles. But these substances also lessen the cloth's weight and strength in addition to eliminating non-cellulosic particles. Chemical desizing also harmed fabric surfaces and contaminated the surrounding area. Thus,

Table 2 A summary of some commercially important proteases

Product trade name	Microbial source	Applications	Supplier	Activity	Characteristics
Alcalase	*Bacillus licheniformis*	Food, detergent, and pharmaceutical industries, for protein hydrolysis,	Novozymes, DuPont (now part of Dow), and Amano	pH (pH 7–10) and temperature (40–70 °C), Activity at ≥0.75 Anson unit/ml	Resistant to inhibition by common inhibitors such as EDTA and urea selective hydrolysis of protein substrates
Savinase	*Bacillus licheniformis*, alkalophilic bacillus sp	Detergent, food, pharmaceuticals, and leather processing, stain removal, meat tenderization, pharmaceutical synthesis, and leather softening	Novozymes, DuPont, and DSM, among others	High proteolytic activity, Optimum temperatures ranging from 50 to 60 °C and pH range from 7.5 to 10 Activity: 12 KNPU-S/g	Suvinase is compatible with a range of detergent formulations broad pH range it remains active at high temperatures
Esparase	*Bacillus* species, *Aspergillus* species,	Detergent, food, pharmaceutical, and leather industries	Suppliers include Novozymes, Roche, and Sigma-Aldrich	Active at 55 °C, and pH range from 8 to 12.5 Activity: 8 KNPU-E/g	Active at low temperature and high pH, storage temperature is 0–10 °C
Biofeed proteases	From *Bacillus, aspergillus*, trichoderma, streptomyces species	In animal nutrition and in agriculture/or in feed Increase feed efficiency	Major are Novo nordisk, Denmark, *Bacillus licheniformis*, Sino biological, novozymes	Optimal pH range: 9.2 Optimal temperature range: 15 to 80 °C Specific activity: ≥2.80 Anson U/g	Enzyme activity number (EAN): 1–10 Available in powder, liquid, or granular form Storage at 25 °C Non-toxic and non-pathogenic

(continued)

Table 2 (continued)

Product trade name	Microbial source	Applications	Supplier	Activity	Characteristics
Durazym	Not explicitly stated yet but a variant of savinase	Animal feed, Detergent industry, Biotechnology	Novozymes, Denmark	Activity: ≥8.39 Anson U/mg, pH range 7–12, temperature: 20–80 °C	Density: 800 g/ml, size varies from 18–90 KDa, present in granules or liquid form
Neutrase	Source is *Bacillus subtilis*	Baking, brewing, and protein of animals and vegetable origin	Novozymes, Denmark, Univar solution	Activity: ≥0.8 Anson Units/mg, pH of about 8 and temperature: 20–80 °C	Stored at 2–8 °C In liquid state occurs
Novozymes3403	*B. Licheniformis*	Denaturation or for cleaning purpose	Novozymes, Denmark, Univar solution	Activity: ≥500 units/mg pH 8 Temperature 20–90 °C	Store at −8 °C Occurs on liquid state
Novozyme4551	*B. Licheniformis*	In food processing, leather	Novozymes, Denmark	Activity: ≥500 units/mg pH 6.9 Temperature 20–80 °C	Storage at 2–8 °C State powder form
Protease	Bacteria, Fungi, Archea, Pseudomonas, Staphylococcus, and Aeromonas being prominent producers	Food processing, in breakdown of protein rich waste	Solvay enzymes GmbH, Germany	Activity: ≥2.4 units/g pH between 6.5 and 8.5 Temperature of 60 °C	Storage at 2–8 °C In aqueous type solution

(continued)

Table 2 (continued)

Product trade name	Microbial source	Applications	Supplier	Activity	Characteristics
Proleather	Alkalophilic *Bacillus* sp.	Food industry	Amano pharmaceuticals ltd	Activity: ≥3.5 units/mg pH of about 4.5 and 5.5 Temperature 70 °C	Store at 2–10 °C State liquid
Protease P	*Aspergillus* sp.	Not known as such	Amano pharmaceutical Ltd	Activity: ≥0.6 units/g pH 4–7.5 Temperature <60 °C	Storage at 3–5 °C State liquid

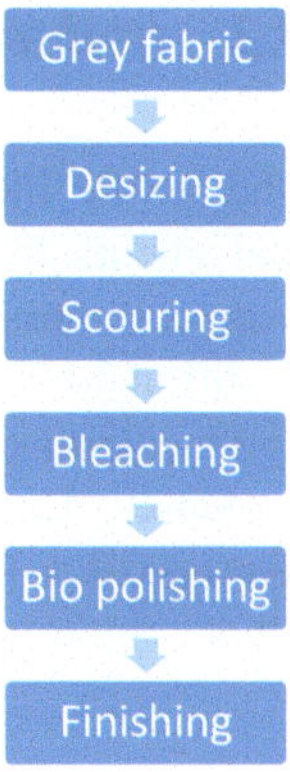

Fig. 5 Flowchart of textile wet processing

the preferred method these days is biological desizing [57]. In textile processing, proteases enzymes that degrade proteins can be utilized for desizing. They facilitate their removal from the fabric by hydrolyzing the protein-based sizing agents. In addition to ensuring effective size removal, this enzymatic desizing helps to lessen the environmental impact of conventional desizing techniques.

9.2 Bioscouring

An essential pre-treatment step in the dispensing of cotton and cotton mix products is scouring. Scouring is mostly done to remove the non-cellulosic elements from cotton fiber that cause the fiber to become impermeable. In order to remove hydrophobic substances from the main wall (such as protein, pectin, and organic acids) and cuticle (such as waxes and lipids), scouring is frequently done in a heated aqueous sodium hydroxide solution. When it comes to textile wet processing, enzymes have opened up a whole new field of research and the possibility of using eco-friendly materials to address the issue of extremely dangerous chemicals contaminating the environment. The removal of contaminants from fibers or fabrics, including pectins, waxes, and natural oils, is a critical step in the wet processing of textiles. This process is known as bioscouring. Enzymes known as proteases are frequently employed in bioscouring procedures to degrade contaminants based on proteins [42]. Alcalase is mostly used in textile processing for the scouring step, which is essential for getting fibers ready for other treatments like dyeing and finishing.

9.3 Biobleaching

The textile industry uses a procedure called "biobleaching," which uses biological agents like enzymes to increase the brightness of fabrics and remove leftover colorants. Biobleaching provides a more environmentally friendly option to traditional bleaching, which uses harsh chemicals like chlorine [52].

One kind of enzyme that is frequently utilized in biobleaching procedures is proteases. They assist in dissolving and removing protein-based stains and residues from textiles. Furthermore, proteases have the ability to help eliminate colorants and other contaminants, which adds to the overall brightening effect of biobleaching [58].

9.4 Biopolishing

The goal of the finishing technique known as "biopolishing" is to make textiles appear softer and smoother on the surface by eliminating fuzz and projecting fibers. Prominent fibers on the fabric surface are often broken down and removed during the biopolishing process using proteases, such as neutral or alkaline proteases. This procedure contributes to a smoother, softer appearance [74] (Table 3).

10 Economic Benefits of Wetting Agents

Since processing costs are frequently higher than manufacturing and material costs, they are viewed as one of the most important factors in determining a product's retail price, making economic processing one of the primary areas of focus. Because wastewater produced when chemicals are used needs to be treated before it can be released into the environment, the procedure's cost is increased. Approximately 2.5 billion tons of wastewater are released annually from China alone, which endangers the maintainability of the fabric sector and degrades natural water sources [61]. The primary purpose of wetting agents is to control the repellency of water in soils, however, they also have other advantages, such as improving water holding capacity and minimizing the physiological moisture stress on plants. Wetting agents decrease the capacity of surface overspill by increasing the rate of infiltration [49].

The use of wetting agents can increase the cross-sectional area of the top layer of soil, which results in the expansion of wheat emergence. Plant densities increase when banded wetting agents are used at high rates. Grain yield can be increased using banded wetting agents, and their correlation is positive [20]. The maintenance of playable surfaces requires a range of resources, including labor, water, fertilizer, and pesticides. Golf courses are currently under pressure to reduce inputs. In 2013, the most standard method of water management used by sports grounds was the application of soil wetting agents. Surfactants known as wetting agents are commonly

Table 3 Proteases used in desizing, bioscouring, biobleaching and other steps

Protease	Source	Work	References
Pectinase	Pectinase is a natural enzyme that is crucial for industry. It is produced by fruits	Pectinase has applications in bioscouring, biobleaching, and desizing processes. It also plays a role in biopolishing, which is the process of eliminating imperfections and protruding fibers from fabric surfaces	Kochhar et al. [38]
Alcalase	It is a serine endopeptidase enzyme that is produced by *Bacillus licheniformis*	The textile industry makes use of it. It is employed to enhance textile performance as well as break down desizing materials based on starch	Ikbal et al. [32]
Papain	It is a natural enzyme that is derived from papaya	Papain is used in a variety of textile and leather industry processes, including the removal of dust from wool and silk prior to dyeing, dehairing, baiting, and tanning	Contesini et al. [18]
Bromelain	It is a proteolytic enzyme that is present in pineapple	It may be used in textile processing for things like desizing and biobleaching. Additionally, it has potential uses in textile processing, including dyeing and finishing	Radhakrishnan and Radhakrishnan [56]
Savinase	It is a commercial alkaline protease that is produced by *Bacillus amyloliquefaciens*. It is well known for its substrate specificity	Its possible uses in the textile industry, especially in processing like shrink-resist finishing and enzyme-based treatments, have been thoroughly investigated	Wang et al. [79]
Subtilisin	It is initially produced by bacillus subtilis. It is serine protease that can also be derived from *Bacillus amyloliquefaciens*	It is utilized in the textile industry for a wide range of processes, including desizing and biobleaching. Synthetic fiber modification is another application for it	Singh et al. [66]
Esperase	It is a protease enzyme that is derived from *Bacillus subtilis*	It can be applied to the desizing process in the production of textiles. It works well to dissolve the starch-based desizing agents found in textiles	Soares et al. [60]
Xylanase	It is a protease enzyme that is obtained from Aspergillus and Trichoderma fungus	It is crucial to the bioprocessing of textiles. It is applied to textile fibers to extract xylan. It is important to the process of biobleaching wood pulp	Lau and Grosse [40]

showered on golf tees and greens. Soil wetting agents, like other surfactants, lower water surface tension, easing the even diffusion of water in permeable media like a turfgrass root zones [72].

Moreover, the fabrication of biological products, especially bio-catalysts, is overpriced and requires compatibility testing [2]. In addition, the enzymatic procedure is accompanied by a neutralizing phase, which increases the cost of the process [29]. Cost-effective production can be obtained by joining various wetting techniques or by reducing the number of process stages, which is often done in textile industries using conventional chemicals. However, the enzymatic methodology does not produce this, as enzyme reactions demand ideal environmental conditions [53]. The production of plastic has increased globally in recent years. By froth flotation, plastic is separated and recycled for future use. Plastics can be separately separated on the basis of their surface features using froth flotation. For this purpose, a number of wetting agents have been used, including methylcellulose, tannic acid, and lignin sulfonate [75]. Therefore, wetting agents are economically beneficial in reducing soil pollution, increasing crop yield, water conservation and water retention.

11 Conclusions

Enzyme-driven manufacturing procedures are the best substitutes for laborious, costly, and detrimental to the environment conventional technologies. In the textile business, proteases are essential for wool biopolishing, silk degumming, and fabric softening. When compared to conventional chemical procedures, they have benefits including energy efficiency, specificity in action, and ecological sustainability. Prominent manufacturers such as Novozymes and Genencor are spearheading sustainable solutions that boost eco-friendliness and efficiency in textile processing. Enzymes known for their diversity and significance in biological processes, proteases can be found in a variety of sources, including bacteria, plants, and animals. They also have different optimal pH and temperature levels. Based on their catalytic processes, they are divided into proteases with various active site residues and activities, including serine, cysteine, aspartic, metallo-, glutamic, and threonine-type proteases. Trypsin and chymotrypsin are examples of serine proteases, which include the catalytic triad Asp-His-Ser and are essential for a number of physiological functions. Metalloproteases use metal ions for catalytic activity and are involved in processes like cell proliferation and matrix remodeling, whereas cysteine proteases, like papain, and aspartic proteases hydrolyze peptide bonds in acidic environments, respectively. Enzymes that break peptide bonds in proteins are known as proteases, or peptidases. They fall into two groups: exopeptidases, which cleave amino acids from terminal locations, and endopeptidases, which target interior bonds. Endopeptidases target peptides that vary in length depending on catalytic centers and environmental conditions, whereas exopeptidases, such as aminopeptidases and carboxypeptidases, release amino acids from the N- or C-terminus, respectively.

Acidic proteases are essential to many industries, including dairy and food processing, because they work best in acidic environments. On the other hand, because to their great selectivity and catalytic effectiveness, neutral proteases function best in environments with a neutral pH and are useful in both industrial and biological processes. In textile wet processing, proteases are essential because they improve fabric quality and encourage eco-friendly methods. Their ability to effectively break down undesirable molecules and enhance fabric qualities makes them useful for operations including desizing, bioscouring, biobleaching, and biopolishing. In general, proteases provide a chemical-free, environmentally friendly substitute for conventional textile treatment techniques. By minimizing the expense of processing, particularly in wastewater treatment—a vital aspect of environmental sustainability—wetting agents provide financial advantages. They raise crop yields and water conservation in agriculture by improving the condition of the soil. They also optimize plastic and textile production processes, which lowers total prices and their negative environmental effects. Proteases are predicted to become more important in the textile sector in the future, especially in environmentally conscious processes like enzymatic fabric finishing and dye removal. Their environmentally friendly practices and uniqueness present viable substitutes for conventional chemical processes, opening the door to more sustainable and productive textile manufacturing techniques.

References

1. Adinarayana K, Ellaiah P (2002) Response surface optimization of the critical medium components for the production of alkaline protease by a newly isolated Bacillus sp. J Pharm Pharm Sci 5(3):272–278
2. Afreen S (2011) Developing a new combination and proportion of chemicals for the production of laundry detergent at low cost in context of Bangladesh. J Chem Eng 26:50–53
3. Ahmad R et al (2013) Bacterial expression, purification and partial characterization of new recombinant cysteine protease from maize leaves: post-transcriptional changes under ozone stress. Pak J Bot 45(S1):441–446
4. Arshad MS et al (2016) Plant and bacterial proteases: a key towards improving meat tenderization, a mini review. Cogent Food Agric 2(1):1261780
5. Arshad M, Adil M, Sikandar A, Hussain T (2014) Exploitation of meat industry by-products for biodiesel production: Pakistan's perspective. Pak J Life Soc Sci 12:120–125
6. Arshad M (ed) (24 Nov 2017) Perspectives on water usage for biofuels production: aquatic contamination and climate change. Springer
7. Arshad M (ed) (13 Nov 2020) Sustainable ethanol and climate change: sustainability assessment for ethanol distilleries. Springer Nature
8. Arshad M, Ansari AR, Qadir R, Tahir MH, Nadeem A, Mehmood T, Alhumade H, Khan N (2022) Green electricity generation from biogas of cattle manure: an assessment of potential and feasibility in Pakistan. Front Energy Res 25(10):911485
9. Arshad M (ed) (28 Apr 2023) Climate changes mitigation and sustainable bioenergy harvest through animal waste: sustainable environmental implications of animal waste. Springer Nature
10. Ashraf M, Hussain N, Baqar Z, Kumar A, Ferreira LFR, Iqbal HM (2023) Bioprospecting microbial proteases in various industries/sectors. In: Microbial biomolecules. Elsevier, pp 301–324

11. Bano I, Ghauri MA, Arshad M, Yasin T, Younus M (2017) Bioactivity of variant molecular weight chitosan against drug-resistant bacteria isolated from human wounds. Microb Drug Resist 23(8):958–965
12. Bano I, Arshad M (2018) Climatic changes impact on water availability. Perspect Water Usage Biofuels Prod: Aquat Contam Clim Chang 2018:39–54
13. Chang C, Werb Z (2001) The many faces of metalloproteases: cell growth, invasion, angiogenesis and metastasis. Trends Cell Biol
14. Chen J, Wang Q, Hua Z, Du G (2007) Research and application of biotechnology in textile industries in China. Enzyme Microb Technol 40(7):1651–1655
15. Cheung LK, Aluko RE, Cliff MA, Li-Chan EC (2015). Effects of exopeptidase treatment on antihypertensive activity and taste attributes of enzymatic whey protein hydrolysates. J Funct Foods 13:262–275
16. Choi J-M, Han S-S, Kim H-S (2015) Industrial applications of enzyme biocatalysis: current status and future aspects. Biotechnol Adv 33(7):1443–1454
17. Chou K-C, Shen H (2008) ProtIdent: a web server for identifying proteases and their types by fusing functional domain and sequential evolution information. Biochem Biophys Res Commun 376(2):321–325
18. Contesini FJ, Melo RRD, Sato HH (2018) An overview of Bacillus proteases: from production to application. CritAl Rev Biotechnol 38(3):321–334
19. Cosme F, Inês A, Vilela A (2023) Microbial and commercial enzymes applied in the beverage production process. Fermentation 9(4):385
20. Crabtree WL, Henderson CWL (1999) Furrows, press wheels and wetting agents improve crop emergence and yield on water repellent soils. Plant Soil 214(1):1–8
21. Dash AK, Sahoo SK (2021) Role of enzymes in textile processing. Bioprospecting of enzymes in industry, healthcare and sustainable environment, pp 395–410
22. Demirkan E, Kut D, Sevgi T, Dogan M, Baygin E (2019) Investigation of effects of protease enzyme produced by Bacillus subtilis 168 E6-5 and commercial enzyme on physical properties of woolen fabric. J Text Inst
23. Di Cera E (2009) Serine proteases. IUBMB Life 61(5):510–515
24. Food and D Administration (2001) Secondary direct food additives permitted in food for human consumption. Fed Reg 66(123):33829–33830
25. Gimenes NC, Silveira E, Tambourgi EB (2021) An overview of proteases: production, downstream processes and industrial applications. Sep Purif Rev 50(3):223–243
26. Grzonka Z, Kasprzykowski F, Wiczk W (2007) Cysteine proteases. Industrial enzymes: structure, function and applications, pp 181–195
27. Gurumallesh P, Alagu K, Ramakrishnan B, Muthusamy S (2019) A systematic reconsideration on proteases. Int J Biol Macromol 128:254–267
28. Hamza TA (2017) Bacterial protease enzyme: safe and good alternative for industrial and commercial use. Int J Chem Biomol Sci 3(1):1–10
29. Hebeish A, Kamel M, Helmy H, El Hawary N (2013) Science-based options for application of cellulase biotreatment and reactive dyeing to cotton fabrics. Life Sci J 10(4)
30. Hedstrom L (2002) Serine protease mechanism and specificity. Chem Rev 102(12):4501–4524
31. Hooper NM (2002) Proteases: a primer. Essays Biochem 38:1–8
32. Ikbal MS, Tisha FA, Asheque A, Hasnat E, Uddin MA (2024) Eco-friendly biopolishing of cotton fabric through wasted sugarcane bagasse-derived enzymes. Heliyon 10(4):e26346
33. Kakkar P, Wadhwa N (2022) Extremozymes used in textile industry. J Text Inst 113(9):2007–2015
34. Kettritz R (2016) Neutral serine proteases of neutrophils. Immunol Rev 273(1):232–248
35. Kimura E (2001) Model studies for molecular recognition of carbonic anhydrase and carboxypeptidase. Acc Chem Res 34(2):171–179
36. Klein T, Bischoff R (2011) Physiology and pathophysiology of matrix metalloproteases
37. Krem MM, Di Cera E (2001) Molecular markers of serine protease evolution. EMBO J 20(12):3036–3045

38. Kumar V, Dangi AK, Shukla P (2018) Engineering thermostable microbial xylanases toward its industrial applications. Mol Biotechnol 60:226–235
39. Kumar R, Vats R (2010) Protease production by Bacillus subtilis immobilized on different matrices. New York Sci J 3(7):20–24
40. Lau PC, Grosse S (2013) The fundamentals and fun of biocatalysis. In: Marine enzymes for biocatalysis. Elsevier, pp 17–69
41. Li Q, Yi L, Marek P, Iverson BL (2013) Commercial proteases: present and future. FEBS Lett 587(8):1155–1163
42. Luan Y, Xu W (2007) The structure and main functions of aminopeptidase N. Curr Med Chem 14(6):639–647
43. Madhu A, Chakraborty JN (2017) Developments in application of enzymes for textile processing. J Clean Prod 145:114–133
44. Mannhold R, Kubinyi H, Folkers G (2011) Aspartic acid proteases as therapeutic targets. Wiley
45. Maróstica MR, Pastore GM (2010) Some nutritional, technological and environmental advances in the use of enzymes in meat products. Enzym Res 2010
46. Matthews KW, Mueller-Ortiz SL, Wetsel RA (2004) Carboxypeptidase N: a pleiotropic regulator of inflammation. Mol Immunol 40(11):785–793
47. Mitra S, Vis E, Kumar R, Plumb R, Fam M (2006) Wetting agent and cultural practices increase infiltration and reduce runoff losses of irrigation water. Biologia 61(19):S353–S357
48. Mojsov K (2014) Industrial enzymes in textile processing and the healty environment: a review. Tekstilna industrija 61(1):12–16
49. Naveed M, Nadeem F, Mehmood T, Bilal M, Anwar Z, Amjad F (2021) Protease—a versatile and ecofriendly biocatalyst with multi-industrial applications: an updated review. Catal Lett 151:307–323
50. Naqvi N, Liu K, Graham RM, Husain A (2005) Molecular basis of exopeptidase activity in the C-terminal domain of human angiotensin I-converting enzyme: insights into the origins of its exopeptidase activity. J Biol Chem 280(8):6669–6675
51. Owusu-Fordjour EY, Yang X (2023) Bioleaching of rare earth elements challenges and opportunities: a critical review. J Environ Chem Eng, 110413
52. Queiroga AC, Pintado ME, Malcata FX (2012) Search for novel proteolytic enzymes aimed at textile and agro-industrial applications: an overview of current and novel approaches. Biocatal Biotransform 30(1):154–169
53. Radhakrishnan S, Radhakrishnan S (2019) Application of biotechnology in the processing of textile fabrics
54. Räisänen R, Nousiainen P, Hynninen PH (2001) Emodin and dermocybin natural anthraquinones as high-temperature disperse dyes for polyester and polyamide. Text Res J 71(10):922–927
55. Rao MB, Tanksale AM, Ghatge MS, Deshpande VV (1998) Molecular and biotechnological aspects of microbial proteases. Microbiol Mol Biol Rev 62:597–635
56. Rathod MG, Pathak AP (2018) Efficient decolorization of textile dyes by alkaline protease producing bacterial consortia
57. Razzaq A, Shamsi S, Ali A, Ali Q, Sajjad M, Malik A, Ashraf M (2019) Microbial proteases applications. Front Bioeng Biotechnol 7:110
58. Rehman A, Saeed A, Asad W, Khan I, Hayat A, Rehman MU, Shah TA, Sitotaw B, Dawoud TM, Bourhia M (2023) Eco-friendly textile desizing with indigenously produced amylase from Bacillus cereus AS2. Sci Rep 13(1):11991
59. Rzychon M, Chmiel D, Steć-Niemczyk J (2004) Modes of inhibition of cysteine proteases. Acta Biochimica Polonica 51(4)
60. Saxena S, Raja A, Arputharaj A (2017) Challenges in sustainable wet processing of textiles. In: Textiles and clothing sustainability: sustainable textile chemical processes, pp 43–79
61. Schwab R (2020) Improving water conservation on golf courses through wetting agents and unmown roughs University of Minnesota
62. Sharif N, Munir N, Hasnain M, Naz S, Arshad M (2021) Environmental impacts of ethanol production system. Sustain Ethanol Clim Chang: Sustain Assess Ethanol Distill 2021:205–223

63. Sharma A, Gupta SP (2017). Fundamentals of viruses and their proteases. Viral proteases and their inhibitors. Elsevier, pp 1–24
64. Singh S, Bajaj BK (2017) Potential application spectrum of microbial proteases for clean and green industrial production. Energy Ecol Environ 2:370–386
65. Singh R, Kumar M, Mittal A, Mehta PK (2016) Microbial enzymes: industrial progress in 21st century. 3 Biotech 6(2):174
66. Soares JC, Moreira PR, Queiroga AC, Morgado J, Malcata FX, Pintado ME (2011) Application of immobilized enzyme technologies for the textile industry: a review. Biocatal Biotransform 29(6):223–237
67. Sumantha A, Larroche C, Pandey A (2006) Microbiology and industrial biotechnology of food-grade proteases: a perspective. Food Technol Biotechnol 44(2):211
68. Suzuki K, Imajoh S, Emori Y, Kawasaki H, Minami Y, Ohno S (1988) Regulation of activity of calcium activated neutral protease. Adv Enzyme Regul 27:135–151
69. Szecsi PB (1992) The aspartic proteases. Scand J Clin Lab Invest 52(sup210):5–22
70. Tang J (ed) (2013) Acid proteases: structure, function, and biology, vol 95. Springer Science & Business Media
71. Tang J, Wong RN (1987) Evolution in the structure and function of aspartic proteases. J Cell Biochem 33(1):53–63
72. Theron LW, Divol B (2014) Microbial aspartic proteases: current and potential applications in industry. Appl Microbiol Biotechnol 98:8853–8868
73. Uddin MG (2015) Effects of biopolishing on the quality of cotton fabrics using acid and neutral cellulases. Text Cloth Sustain 1:1–10
74. Verma S, Dixit R, Pandey KC (2016) Cysteine proteases: modes of activation and future prospects as pharmacological targets. Front Pharmacol 7:193290
75. Wang CQ, Wang H, Gu GH, Fu JG, Lin Q, Liu YN (2015) Interfacial interactions between plastic particles in plastics flotation
76. Wang L, Yao J, Niu J, Liu J, Li B, Feng M (2018) Eco-friendly and highly efficient enzyme-based wool shrinkproofing finishing by multiple padding techniques. Polymers (Basel) 10(11)
77. Ward O (2011) Proteases. Compr Biotechnol, 571
78. Warshel A, Naray-Szabo G, Sussman F, Hwang J (1989) How do serine proteases really work? Biochemistry 28(9):3629–3637
79. Waste management, vol 46, pp 56–61
80. Yin LJ, Hsu TH, Jiang ST (2013) Characterization of acidic protease from Aspergillus Niger BCRC 32720. J Agric Food Chem 61(3):662–666
81. Zephyr J et al (2021) Viral proteases: structure, mechanism and inhibition. Enzym 50:301–333

Turning Waste Into Power: Bioenergy from Textile Effluent Treatment

Iram Liaqat, Abdul Rehman, Guojing Yang, Sikander Ali, and Muhammad Arshad

Abstract The textile industry heavily relies on dyes for coloration purposes. However, the effluent discharged from these processes contains both dyes and heavy metals. Traditional methods of treating wastewater from the textile industry, such as coagulation, membrane processes, electrolysis, and ion exchange, present economic and efficiency challenges. Furthermore, some of these approaches generate harmful sludge, diminishing their environmental sustainability. These microbial cells leverage bacterial activity to change their chemical power from textile industrial effluents substrates into electrical power. Through electron transfer to electrodes, bypassing conventional electron acceptors, microbial cells demonstrate recent technical feasibility. This breakthrough has sparked significant excitement, among the researchers highlighting MFCs as a potential source of sustainable "Bioenergy" by harnessing textile industrial effluents for power generation. Microalgae, efficient microorganisms, offer dual benefits in wastewater treatment: biomass recovery and value-added product generation (biofuels), while curbing greenhouse gases. Their photosynthesis-driven growth reduces CO_2 emissions and aids in heavy metal and antibiotic removal from textile industrial effluents, highlighting their versatile role in sustainable solutions. The present review is aimed to comprehensively study the generation of bioenergy from effluents treatment in textile industry through the utilization of MFCs (microbial fuel cells) and microalgae, two extensively employed methods within the textile sector for harnessing bioenergy from the effluents produced by textile

I. Liaqat (✉) · A. Rehman
Department of Zoology, Government College University, Lahore, Pakistan
e-mail: dr.iramliaqat@gcu.du.pk; iramliaq@hotmail.com

G. Yang
Hainan Medical University, Laboratory of Tropical Environment and Health, Haikou, Hainan, China

S. Ali
Institute of Industrial Biotechnology, Government College University, Lahore, Pakistan

M. Arshad
Jhang-Campus, University of Veterinary and Animal Sciences Lahore, Lahore, Pakistan

M. Arshad (ed.), *Enzymes in Textile Processing: A Climate Changes Mitigation Approach*, SDGs and Textiles, https://doi.org/10.1007/978-981-97-8058-7_12

industrial processes. This study seeks to contribute to the advancement of bioenergy production from textile effluents by exploring these techniques and their application in treating textile effluents.

Graphical Abstract

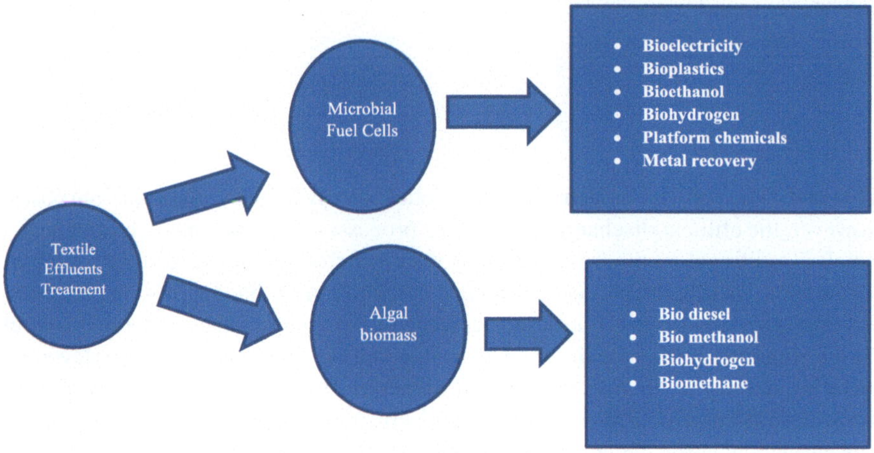

Keywords Effluents · Textile industry · Waste water treatment · Microbial fuel cells · Bioenergy · Microalgae

1 Introduction

Industrialization has a central part in propelling a nation's financial development than reshaping his societal norms. Large- and medium-scale industries, crucial drivers of progress, play a significant role in fostering economic growth, irrespective of a nation's developmental status. Within these sectors, the textile industry emerges as a key player, owing toward this aptitude to generate a wide array of occupation opportunities than its complicated ties by various financial domains, together internationally then nearby. It stands out as the fastest-growing and greatest promising subdivision in this context. Yarn, angora, artificial fibres, artificial colours, substances, and procedure aquatic remain fundamentally rare resources that greatly enhance the economic importance of the textile industry. The fabric sector is broadly considered into two main procedures based on fiber production: dry processing and wet fabric processes [17]. During dry processing, the primary waste generated is in solid form, whereas in the wet processing steps, liquid waste takes precedence.

Wet fabric processing involves a diverse array of activities such as dyeing, finishing, and printing. These processes, occurring throughout the fiber creation stage, require the use of drinking water, diverse chemicals, and dyes. Consequently, wastewater is produced; the composition of which varies according to the specific processes applied. This discharged wastewater serves as a notable source of pollution, carrying dyes and diverse pollutants due to the extensive use of pure water and also carries dangerous contaminants like dyes, chromium, sodium hydroxide (NaOH), starch, acids, and additional pollutants. As a result, textile factories are known to consume a significant volume of clean water and chemicals, surpassing many other industries. Most of their discharged effluents are heavily polluted. On average, the textile industry utilizes around two hundred liters of clean water per kilogram of product [19].

Research has shown that the dyeing and finishing segment within the textile industry is accountable for producing a substantial portion, approximately 17–20%, of global industrial wastewater. This wastewater, saturated with pigments, adversely impacts aquatic flora, disrupting their photosynthesis process. Specifically, the effluents from printing and dyeing operations consist of a high chemical or dye concentration, making their decomposition a challenging task.

Presently, a majority of dyes employed in the textile industry consist of complex perfumed mixes and heterocyclic constructions. These intricate and stable dye structures make them highly resistant to decomposition.

Biological remediation, also known as bioremediation, stands out as a promising technique owing to its cost-effectiveness, environmental friendliness, then on-going efficiency. This urges bacteria towards clean environment using their own microbial processes. Various microbes, such as bacteria, mushrooms, have been harnessed for bioremediation purposes [24, 25] (Abinandan et al. 2015). This primarily happens by main binary strategies.

Exploring unconventional possessions for reducing hefty metallic levels in the atmosphere takes led to the detection of economically viable then harmless materials for wastewater action. Organic resources have arose as eco-friendly and economical replacements. Microalgae, owing to their aptitude to withstand contaminated liquid, have gained prominence as bio sorbents. Despite extensive research on bioremediation techniques involving numerous microbes, microalgae-based bioremediation expertise has garnered significant consideration due to its compensations. These can effectively eliminate contaminants and generate value added products (Karla et al. 2021) [1]. This organic method holds immense promise for the lasting treatment of business wastewater and has the potential to promote the expansion of more maintainable and globular financial practices in world.

Basically, the idea to utilizing lime algae as a raw material for biofuel manufacture has gained significant traction owing to escalating oil charges, the swift reduction of usual oil assets, and, greatest outstandingly, the detrimental impact of relic petroleum usage on world-wide heating. As per the 2020 Creation Oil View Report by the Organization of the Fuel Exporting Republics, oil is predictable to continue the predominant petroleum in the upcoming. World-wide request is projected to surge around the globe. However, OPEC member countries are rapidly depleting their current oil reserves. It is evident that fossil fuel resources will become increasingly scarce.

Renewable energy resources, which emit minimal or not at all CO_2, can significantly fulfil a substantial helping of the biosphere's vigour requirements. Soybeans, sugarcane, and herbal lubricants are utilized for producing first-generation biofuels. Anxieties around balancing future nourishment manufacture with energy demands have removed attention towards second-generation biofuels. Fuel sources like lawn grasses, timber, atrophy, switch grass, and organic waste yield additional net vigour per acre compared to corn and sugarcane. But the intricate thermal and biological functions required for biofuel production after these sources often result in low produces besides tall values, among other feedstock's, offer superior environmental and economic advantages. They can meet growing demands though minimalizing ecological impact. Algae consume previously established their utility in producing biological products [4, 26].

The quest for other oils, especially in the transportation sector, has led to heightened interest in microalgae-derived biofuels. These biofuels are viewed as a potential solution to the challenges faced in both production and consumption. Extensive research has been conducted on the welfares of microalgae basis along with manufacture worth enhanced bio products.

Polluted water is increasingly viewed as a valuable resource for pure water, vigour, and plant inseminating nutrients. However, current polluted water action know-hows face several limits in attaining these change goals. Aimed at instance, conservative aerobic triggered mud (CAAS) action schemes are known for their high energy consumption, generation of huge amounts of residuals, then inability to recover possible capitals from wastewater. Cutting-edge the US, it was reported that around 0.5 kWh/m3 of energy was needed to use for wastewater treatment [28].

Aerobic digestion (AD) is a key skill that utilizes polluted water as an vigour foundation, changing organic content addicted to CH4, air, which can then be converted into power; studies have shown that AD is not highly efficient to capture the desirable energy. Polluted water fails to meet stringent controlling standards, necessitating technically demanding steps. Moreover, methane is a potent conservatory gas with a worldwide heating possible about 25 eras that of CO_2, necessitating strict control to prevent its release into the atmosphere [9].

Additionally, water reuse practices are becoming more common, especially in arid regions. However, this approach typically demands more energy for treatment, primarily due to increased water quality requirements for reuse.

Review provides a deep thorough analysis of the bioenergy production derived from effluent treatment in the textile industry. It explores a diverse array of technologies, including biodegradation strategies utilizing mushrooms, microorganisms, and bacteriological fuel cells. Additionally, it covers chemical approaches like photocatalytic corrosion and treatments involving ozone and the Fenton's process. Furthermore, the review delves into various physicochemical techniques. These methods are not only assessed for their effectiveness in waste treatment but also for their potential in generating bioenergy during the treatment of textile industry effluents [12].

2 Microbial Fuel Cells (MFCs)

Microbial fuel cells are emerged as the most promising solution, especially in remote areas, where they address energy and environmental concerns. Loaded capabilities, and serving as an in-situ control source for treatment and polluted water treatment, MFCs offer unique compensations. A general structure of fuel cell is shown in Fig. 12.1. These include the direct change of substrate energy into power, generating fewer extra triggered mud compared to methods like anaerobic digestion (AD) and conservative aerobic triggered slush (CAAS). Additionally, MFCs prove insensitive to operational environments, even at low temperatures, eliminating the need for gas treatment and energy input for aeration. Their versatility shines in areas lacking adequate electrical infrastructures, making them a sustainable solution for wastewater treatment in resource-constrained regions [11].

While the current utilization of microbial fuel cells (MFCs) is primarily limited to laboratory settings, their potential for industrial applications has been convincingly demonstrated. Several advancements contribute to their viability. Firstly, the discovery of networks involving nanowires, cytochromes, and within has deepened the understanding of electron transmission machines. Secondly, ongoing developments in reactor design and architecture have resulted in innovative solutions. Thirdly, there has been a substantial increase in reactor volume, scaling up to ten litters, and operational periods have remained protracted toward span annually. Quarterly, diverse addition expedites integration of cells into polluted water treatment applications. Lastly, the scope of polluted water types investigated in MFCs has broadened, encompassing a range from humanoid feces to heavy metals. These advancements collectively underscore the promising industrial prospects of MFC technology [22].

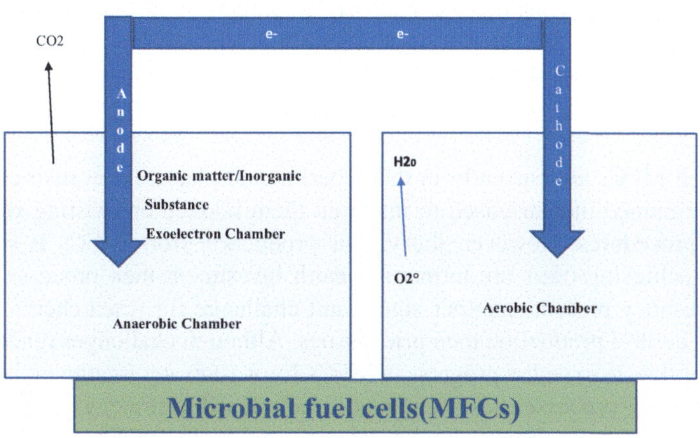

Fig. 12.1 Microbial fuel cells

2.1 General Features of the MFC

In a typical microscopic fuel cell (MFC), the simultaneous metabolism of substrates and conversation of electrons with a conductor by electrochemically active microbes is crucial. In the terminal, anaerobic breathing bacteria professionally break down carbon-based matter into CO_2. These transport anode through a bacteria-driven outside system. Power is produced, ensuring electro-neutrality as they move between electrodes. Conventionally, MFCs are predominantly biological on positive side, and microorganisms remains inside the cell membrane. The air–cathode typically comprises a catalyst layer, electrode, and separator. Notably, the separator's role in MFCs has become increasingly critical, impacting internal resistance and overall MFC performance. Preceding efforts mainly focused catalyst enhance discount reaction rates, these materials are expensive when applied in MFCs, primarily because MFC control and current thicknesses are relatively lower than conservative petroleum cells. In contrast, bio-cathodes present an appealing alternative to abiotic catalysts, utilizing bacteria as biocatalysts on inexpensive carbon cathodes. Despite their superiority, there is a scarcity of studies delving into the metabolic mechanisms of bio-cathodes. Consequently, fundamental research on bio-cathodes is urgently needed [20].

According to reports, the project and structure of bacteriological fuel cells (MFCs) play a pivotal role in determining their applicability in wastewater treatment. Researchers like Logan and Regan have emphasized that attaining high power thicknesses in MFCs is significantly influenced by system architecture rather than composition. A study consistently exposed by cells varied immunisation outperform folks with unadulterated attitude immunisation in both control density and contaminants elimination competence. The diverse organic components originate in many waste biomass foundations necessitate a varied bacteriological communal for effective degradation. A well-designed system that can be easily scaled up is cost-effective, high-performing, and can be seamlessly integrated with existing wastewater treatment facilities, is crucial for promoting the commercial viability of MFCs [7, 13].

Although MFCs are currently in the experimental phase, innovative enterprises consume remained industrialised to integrate them hooked on existing wastewater treatment procedures. However, the vigour production from MFCs is still insufficient for achieving costs are terms of wealth investment then process. Ensuring lasting constancy presents another significant challenge for researchers, alongside addressing control production then price issues. Although challenges rendered cells less competitive, to expedite progress in MFCs for wastewater treatment, integrating MFCs with other processes has been considered a feasible strategy [2, 3, 5, 15].

In the realm of wastewater treatment, microbial fuel cells (MFCs) exhibit three key features: energy efficiency, minimal sludge production, and reduced energy consumption. Numerous studies have delved into their capabilities in treating diverse types of wastewater. These investigations have revealed MFCs' ability to effectively remove multiple contaminants, including biological waste, heavy metals, polyalcohol's, petroleum products, dyes, phenols, furans, quinolones, and pyridine derivatives. However, for practical implementation, it is imperative to assess the performance of MFCs using real wastewater scenarios [8].

2.2 *Textile Industrial Waste*

Firstly, beet-sugar plants yield abundant wastewater sources rich in carbohydrates, making them ideal substrates for microorganisms in MFCs. Secondly, wastewater from starch processing is notable for its high starch and protein content, rendering it a valuable energy resource for MFC applications. Thirdly, large quantities of brewery wastewater are generated during cooling, scarification, fermentation, and washing processes, posing environmental challenges. Despite its high chemical oxygen demand (COD), researchers have successfully utilized brewery wastewater, ironic in carbon-based matter, as an appropriate substance form microbes within cells. Fourthly, the dairy industry, a prominent food sector, produces wastewater as it processes raw exploit into various products. Such dairy wastewater serves as an effectual analyse in MFCs [23].

Industrial wastewaters, particularly from colour engineering and the fabric manufacturing, often comprise significant amounts of artificial biochemical dyes. While some efforts have been made to explore MFCs' decolourization abilities for specific dyes, few studies have tackled real dye wastewater scenarios. Addressing this challenge without the need for luxurious resources for simultaneous dye decolourization and green energy generation, [21] developed a Gritty Triggered Biological Bacteriological Petroleum Cell (GACB-MFC). Their findings revealed impressive results: approximately 73.0% and 77.0% of color removal occurred at the terminals, respectively. Moreover, elimination was achieved in both terminals. Venomousness capacities indicated that Polluted water significantly less toxic, even anode effluent's toxicity markedly decreased compared to the unique dye wastewater of process. The GACB-MFC system generated route power of 0.33 V. Notably, the scheme's pH mechanically adjusted, though maintaining a stable pH of 8.0 at both the terminals–a key advantage of the GACB-MFC system.

Generally, numerous types of Polluted waters have been examined in Bacteriological Fuel Cells (MFCs) for both wastewater action and power generation purposes. Over the past insufficient years, considerable advancements have been made in enhancing the control density and efficiency of Polluted water action using MFCs. These studies have showcased the exceptional treatment capabilities of MFCs, specifically the elimination specific contaminants. Notably, studies have revealed that MFCs can achieve remarkable treatment effects, with COD removal efficiency surpassing ninety dropping to less than 20 mg/L. However, given the various constraints outlined in wastewater discharge standards, a comprehensive evaluation is essential to further develop this skill as a viable wastewater treatment solution in future research [12, 24, 25].

Metabolic Mechanism

2.3 *Metabolism in Anode Chamber*

Microorganisms in bacteriological fuel cells (MFCs) can transfer electrons to the anode conductor through four possible mechanisms: They relied on adding exogenous mediators to generate power, but their toxicity and instability limited their MFC applications. However, recent breakthroughs have shown that naturally existing bacterial communities can generate electricity without the need for exogenous mediators. Certain electrochemically active microorganisms can straight transmission electrons resulting in organic matter breakdown to the anode. Notable examples of these mediator-free., *Rhodoferax spheroides* remains unclear [21].

Despite advances in "photo MFCs," which incorporate photosynthetic microorganisms, the precise mechanism of their terminals is not fully understood. The cells equip MFCs to handle diverse wastewaters, some of which are challenging for established biological treatment methods. Simple organic compounds like monosaccharides and volatile fatty acids are easily degraded, while complex substances like cellulose require the breakdown matter. In this context, syntrophic interactions are crucial for breaking them. Recent reports suggest that MFCs are particularly suitable polluted waters due to the high proportion of the cells that are challenging for electrochemically active microorganisms to utilize [21].

Metabolism in Cathode

2.4 *Abiotic Cathodes*

Popular cathodes processes often require the acceleration of oxidation-reduction reactions on electrode surfaces. Platinum, renowned for its exceptional catalytic abilities, has long been considered the ideal. However, its widespread use is severely hampered by exorbitant costs and toxicity concerns. To address these challenges, researchers have turned to non-platinum (non-Pt) catalysts for oxygen reduction reactions in MFCs. Various non-Pt catalysts have been explored for non-living.

2.5 *Biocathodes*

Biocathodes container can be categorized hooked on aerobic and anaerobic types based on the catholic chamber circumstances. In aerobic biocathodes, oxygen helps as the fatal electron acceptor, while hard peacekeepers. On the other hand, biocathodes have a notable advantage in preventing through the alternate membrane, thus preventing electron loss to oxygen. The superiority of biocathodes over abiotic cathodes is evident for two main reasons. Firstly, biocathodes enhance MFC sustainability and can be employed for various beneficial purposes. Secondly, bacteriological breakdown bio terminal harnessed to harvest valuable goods, eliminate undesirable compounds, and even treat wastewater, including processes such as denitrification, heavy metal removal, and sulphate reduction [18].

Apart from studies focusing on the metabolic mechanisms of biocathodes, research has delved into the subsequently extensively examined cutting-edge cathodes owing toward ability to supply O_2, capture CO_2, produce appreciated then polish polluted water effluents. The activity from microbes likely involves enhancing production. Researchers consume emphasized the importance of best light strengths for MFCs with photosynthetic cathodes, as lower light intensity can generate more electricity in these setups.

3 Reactors Used in Wastewater Treatment

3.1 Two-Compartment MFC

Traditionally, bi-chamber microbial fuel cells consist an positive terminal, negative cavity, conversation crust, while, due to their intricate structures, aeration needs, unsustainable oxidants negative cavity, do not directly contribute to reducing energy costs. Though these MFCs are still far from profitable viability, their designs have significantly progressed for important wastewater action. For instance, [14] introduced a salty bond MFC to explore control production, indicating the potential of MFCs for generating energy. Industrialised an up-flow bacterial petroleum cell (UMFC) by incorporating an inner cathode to ensure unceasing conduct and enhance presentation. The UMFC joint the benefits of the UASB organisation through the necessities of a dual-chamber MFC. Moreover, the ion conversation film, which incurs the highest cost and influences the long-term steadiness of an MFC, has been a focal point of research. Jang et al. utilized a membrane-less MFC to diminish construction costs; although these UMFCs and membrane-less MFCs excel in contaminant elimination, they often exhibit minimum energy.

3.2 Air Cathode MFC

In a single-compartment MFC, the air cathode serves as a fundamental structure, comprising five essential components: the proton conversation film, substance coating, cathode, dispersal coating, and protective layer. Liuo and Loga designed an MFC with both terminals. This assembly was created by directly membrane exchange. Another innovation introduced by Liu et al. involved a solitary tubular Plexiglas chamber including eight graphite rods arranged concentrically around a central cathode. The air-porous cathode was incorporated. In a different approach, a tubular MFC system was developed, with an outer cathode and an inward anode fashioned from graphite grains. Designed a positioned between the anode and cathode. In this SCMFC system, the anode was positioned above the runny layer, while the cathode's top surface was unprotected to the air.

3.3 Benthic MFC

Benthic BMFCs or residue bacteriological firewood cells (SMFCs) are organisations designed to produce power by harnessing the natural latent alterations amongst anoxic sediment besides toxic salt-water. In BMFCs, microbes are busy to oxidize substrates, relocating electrons to a terminal rooted in or resting on the sediment. These electrons are then transferred to a cathode postponed in the superimposing seawater, where melted oxygen is reduced to water. Over recent years, various configurations of BMFCs have been planned and enhanced. Initial designs often utilized a postponed cathode in the water. However, these enterprises were found to be fragile, resulting in extremely low power generation. Nielsen et al. introduced a chamber-based BMFC with a deferred and semi-enclosed anode, meaningfully reducing the system footprint and increasing power output to around 3v.

3.4 Stacked MFC

The voltage production of a typical microbial fuel cell (MFC) rarely surpasses the theoretical voltages, even if internal losses are disregarded. Comparatively, stacking MFCs in series or parallel configurations is considered an effective method to increase both powers and flows. Loaded microbial fuel cells (SMFCs) are categorized into dual types: sequences and similar connections. For example, created a SMFC consisting of six individual MFCs. Both anode and cathode electrodes were constructed using graphite grains, connected to the external circuit by a graphite rod. This study showed that stacking MFCs in either series or parallel configuration significantly increased power densities. However, an over potential phenomenon was observed, partially limiting the current density increase in series connections. Types of microbial fuel cells are shown in Fig. 12.2.

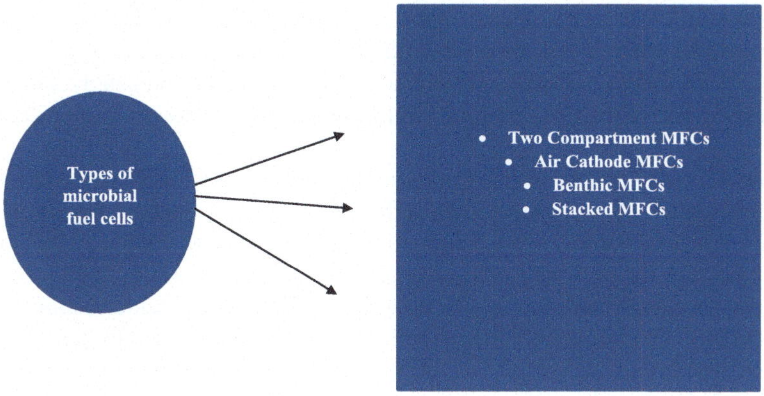

Fig. 12.2 Types of microbial fuel cells

4 Integration Reactors

4.1 Integration with Anaerobic Digestion

Traditional anaerobic polluted water treatment processes grounded on bioconversion have been widely used for treating textile industrial wastewater due to their cost-effectiveness. However, certain challenges such as low temperatures and low carbon-based attentions have been encountered in anaerobic treatment, leading to the need for waste "polishing" or post-treatment steps.

It is important to note that anaerobic treatment and microbial fuel cells (MFCs) are distinct technologies for taking energy possible in wastewater. MFCs have lower liveliness change competence and higher wealth speculation compared to anaerobic digestion (AD). Nevertheless, the effluent quality from MFCs surpasses that after anaerobic digestion. Generally, MFCs are well-suited for handling intermediate- and low-strength wastewaters with relatively simple compositions, while AD excels in dealing with high-strength wastewaters. MFCs can effectively degrade volatile fatty acids like acetate and butyrate, even at low concentrations, whereas these acids can accumulate and affect the treatment process in anaerobic digestion. This has gained interest in polluted water treatment [16].

Researchers like Pham et al. have critically compared the procedures of AD and MFCs, demonstrating the potential for integrating these technologies in wastewater treatment despite their differences. Alderman et al. utilized both technologies separately for treating in height- and low-strength wastewaters. In these treatment processes, significant quantities of extra slush were produced, which could be converted to biogas using conventional anaerobic digesters. Additionally, a novel system, the Up-flow Anaerobic Mud Blanket Device–Microbial Fuel Cell–Biological Sparkling Filter (UASB-MFC-BAF) was developed to address high-strength molasses wastewater. In this UASB-MFC-BAF process, a thoroughgoing power concentration of 1410.2 mW/m^2 and a current density of 4947.9 mA/m^2 were achieved when high-strength treacle wastewater (with a chemical oxygen demand (COD) of 127,500 mg/L) was used as influent. The final results indicated removal efficiencies of 53.2% for COD, 52.7% for sulphate, and 41.1% for color after treatment [27].

Another prevalent approach involves integrating microbial fuel cells (MFCs) into anaerobic reactors. A straightforward method of integration is incorporating an MFC into the up-flow anaerobic slush comprehensive. For instance, enhanced the up-flow membrane-less bacterial fuel cell by incorporating toothless biological aerated filters at the reactor's top to treat palm oil mill effluent. These enhancements significantly impacted NH3-N removal, with the immobilized biological aerated filters playing a pivotal role in further NH3-N and COD removal. The resulting wastes exhibited COD and NH3-N levels below 350 and 8 mg/L, respectively, with removal efficiencies exceeding 96.5% and 93.6%.

Devised a scheme by uniting an anaerobic fluidized bed (AFB) with an MFC for liquor distillery wastewater treatment. This AFB-MFC system featured an anode chamber, a cathode chamber, and an external resistor, with the anode and cathode compartments during the stabilization period. Simultaneously, the system achieved a maximum power density of 124.03 mW/m^2 under various operational settings and an external resistance of 120 Ω.

Integrated an MFC into the anaerobic–anoxic–toxic (A2/O) procedure, a widely used wastewater treatment method known for simultaneous removal. This integrated system not only generated electricity but also addressed issues related to substrate competition and high energy consumption. During stable operation, the MFC-A2/O reactor demonstrated an average increase of 15.9% in COD removal efficiency, 9.3% in TN removal, and 1.4% in TP removal compared to control.

4.2 Integration with Ecological Treatment Technologies

Unique integration of bacteriological fuel cells (MFCs) with ecological action technologies involves incorporating MFCs into bog conduct systems. Investigators have observed that florae secrete organics, known as rhizodeposits, into soils and sediments, constituting around 40.0% of plants' photosynthetic activity. These rhizodeposits, made finished the photosynthetic action of rice paddy turfs, exhibit bioelectricity manufacture cell setup. This integration offers specific applications combining MFCs and ecological treatment technologies, primarily focusing on sediment microbial fuel cells (SMFCs).

Mohan et al. (2015) intended a miniature fluctuating macrophyte-based bionetwork by integrating environmental action skills with sediment-type microbial fuel cells (SMFCs) for bioelectricity cohort from domestic manure and agitated distillery wastewater. Effective bioelectricity generation was observed in two fuel cell assemblies (A1 and A2), correlating with increased organic load. Power generation (A1:

211.14 mA/m^2; A2: 224.93 mA/m^2) and removal efficiencies of COD (86.67%) and VFA (72.32%) were notably high.

A miniature environmentally engineered flooded was developed macrophyte-based combined system embedded with sediment-type petroleum cells for manure and agitated distillery wastewater treatment. This scheme efficiently removed colour, nitrates, and turbidity. The education future mixing biohydrogen manufacture to enhance acid-rich waste treatment or serve as a tertiary action unit for improving purposes.

A flat subsurface movement built swamp could purpose as an MFC for wastewater treatment under suitable weather conditions. The MFC operated effectively at low organic loading rates but ceased functioning under high organic loading rates.

A constructed wetland-microbial fuel cell (CW-MFC) was developed to treat swine wastewater in batch mode or continuous, upward flow mode. The CW-MFC system efficiently achieved wastewater treatment objectives, particularly when combined with cathode aeration, leading to high power generation.

A study was conducted seeing bio-cathode materials and reactor configurations to assess the performance of an upright subsurface flow created wetland-microbial fuel cell. Consequences specified that a granular galvanised carbon-stainless steel mesh bio-cathode was most apt for CW-MFC submissions due to its large external area, vessel water engagement, and developed supremacy yield.

One innovative integration of microbial fuel cells (MFCs) with environmental conduct machineries involves incorporating phototrophic microbes hooked on the MFC system. These microorganisms compensate for the MFC's volume shortage in nutrient removal and catalyse power manufacture. Zhang et al. (2016) introduced an effective bacteria-algae coupled sediment-type photo microbial fuel cell capable of efficiently removing carbon, nitrogen, and phosphorus from wastewater in a single step. The system generated achieved impressive removal efficiencies, with organic carbon, nitrogen, and phosphorus removal rates reaching 99.6%, 87.6%, and 69.8%, respectively. Mass balance analysis revealed that algae mechanism for nitrogen (75.0%) and phosphorus (93.0%) removal, with contributions from nitrification and denitrification (22.0%) for nitrogen removal.

Diverse bacterial taxa in the biofilm on the cathode conductor or in the catholyte postponement played functional roles in pollutant degradation and nutrient cycling, demonstrating that algal biofilm formation enhanced oxygen transfer to the electrode, decreasing Ohmic and charge transfer resistance. However, the addition of bicarbonate for algae growth reduced cathode catalytic activity, indicating the need for alternative carbon sources or different catalyst materials to sustain current generation. Gouveia et al. explored the impact of light on bioelectricity production and microalgae biomass in a photosynthetic alga-microbial fuel cell. Utilizing *Chlorella vulgaris*, the PAMFC system verified the potential to generate valuable microalgae biomass, enhancing the economic possibility of these bio electrochemical systems.

Associated to conservative wastewater conduct units, ecological behaviour know-hows offer characteristic compensations, including minimal or no chemical usage, absence of foul odours, ease of operation, and cost-effectiveness. They are perceived as ecologically complex, environmentally approachable, and automatically straightforward. Though, there is a need to enhance power generation and reduce costs, primarily focusing on MFC development. Further investigation is obligatory to understand the synergy mechanisms within the system, and optimization of factors such as plant selection, construction methods, and operational modes is essential. Considering organic loading limitations, transforming this technology into a tertiary conduct unit for improving purposes could be a viable option.

4.3 Integration with Membrane Bioreactor

Many studies aimed to enhance power cohort in microbial fuel cells (MFCs) by optimizing their construction, electrode and skin resources, challenge, though, is meeting rigorous seepage superiority standards. Film knowledge has remained widely employed for pollutant separation or valuable chemical product recovery due to its high efficacy.

The incorporation of film parting processes with MFCs, moreover finished consecutive connecting or straight connexion, has shown promise in various studies. For instance. Previously, an innovative osmotic bacteriological fuel cell utilizing a onward osmosis membrane as a sieve was developed. This design improved proton transport and allowed simultaneous water extraction, wastewater treatment, and power group. Wang et al. introduced a bio electrochemical film reactor (BEMR) using a stainless strengthen network with biofilm formation as together finished, while the BEMR achieved impressive COD and NH4 + -N subtraction productivities, optimizing its configuration could enhance electricity recovery further.

Another novel approach involved an electrochemical film bioreactor using non-woven material as the separator and filter. This system effectively removed contaminants, achieving high COD removal efficiencies and moderate NH4-N and TN removal efficiencies. Li et al. shown to control membrane fouling in a fluidized bed membrane bio electrochemical reactor (MBER). Though the MBER demonstrated electricity generation and contaminant removal, improved performance was observed when linked to a regular MFC. Types of integration reactor are shown in Fig. 12.3.

Ren et al. scrutinised a two-stage shared behaviour progression, involving bacteriological petroleum cell crust bioreactor. These integrated systems highlight the potential of combining membrane technology with MFCs for efficient polluted water treatment and energy group.

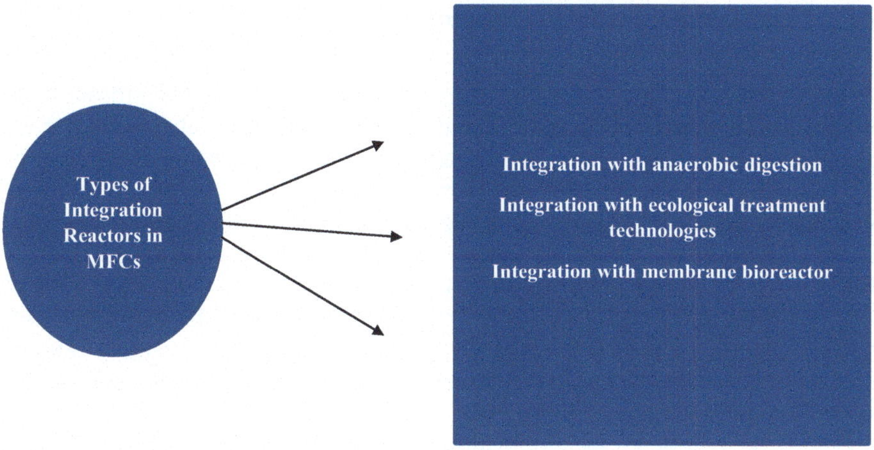

Fig. 12.3 Types of integration reactors in microbial fuel cells

5 Microalgae Polluted Water Treatment

Businesses release substantial contamination into water streams, leading to wastewater covering a range of pollutants like reserves, unstable organic mixes, oils, heavy metals, and pesticides, depending on the industry type. Moreover, industrial and sewer wastewater is rich in nutrients, with the build-up of contaminants like nitrogen and phosphorus adversely affecting both river and brine ecologies. To ensure safe discharge into the setting and for human consumption, certain contaminants must be removed from wastewater. Applying these nutrients is crucial for microalgae to function. Cultivating microalgae in wastewater proves beneficial and frugally feasible, especially considering the environmental issues arising from large volumes of wastewater containing harmful chemicals. Extensive research has been dedicated to wastewater treatment, leading to the exploration of various wastewater recycling methods. Water pollution poses challenges such as water scarcity for essential household and industrial services, underscoring the importance of wastewater treatment to enhance water quality and alleviate water scarcity. Recent studies have highlighted microalgae bioremediation as a promising approach for industrial wastewater treatment. The growth of microalgae relies heavily produce useful products. Essential micronutrients, vitamins, trace minerals, and macronutrients like nitrogen and phosphorus are crucial for optimal microalgae growth. Successful outcomes often requires state-of-the-art infrastructure.

5.1 Textile Wastewater Treatment

Polluted water stemming after the fabric manufacturing poses significant environmental challenges due to its multitude of pollutants, particularly dyes. If not adequately treated, these pollutants can lead to aesthetic degradation, the accumulation of toxins in aquatic ecosystems. Harnessing the power of microalgae for treatment has emerged as a viable substitute to traditional methods. Bioenergy from textile effluents treatments is also shown in Fig. 12.4. During microalgae growth, dyes are degraded through bioconversion/biodegradation or bio sorption processes. Microalgae treatment not only eliminates from cloth wastewater but also curtails the various negative environmental impacts associated with environments. Additionally, in the advantage of generating valuable biomass, convertible into bio products, biofuels, and bioenergy, fuels, etc.

Studies have explored the elimination of wastewater using microalgae, such as Spirulina platensis, achieving removal rates of 94–99% at pH 2 and apartment disease, comparable to activated carbon dye removal rates (94–98%) (Cardoso et al. 2012). In another research, a fed-batch reactor utilizing demonstrated substantial removals,

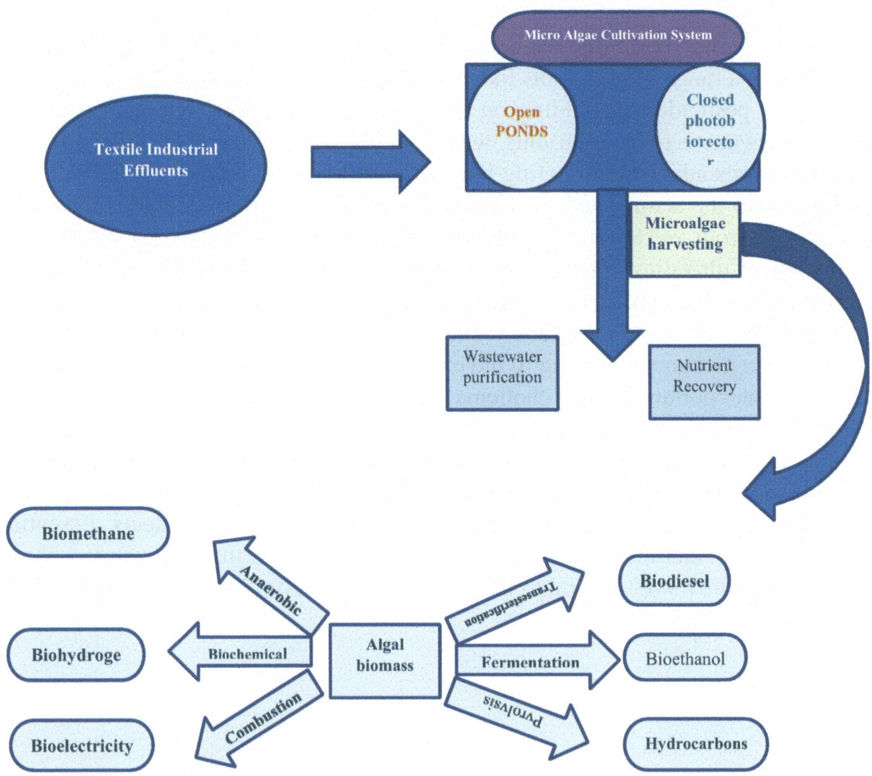

Fig. 12.4 Bioenergy from algal biomass

with total nitrogen at 70% and total phosphorus at 95%. Reduced by 68–72%. This study suggests the viability of integrating bacteriological algal agriculture with cloth wastewater treatment (Kumar et al. 2018).

Microalgae bioremediation of polluted water can occur through the cells. The immobilization method offers advantages such as space efficiency and ease of harvest. For instance, immobilized *Desmodesmus* sp. exhibited exceptional subtraction rates accomplishment (Al-Fatwas and Abdullah 2016). Studies also observed substantial decolourizations and nitrogen elimination rates (80% and 71%, respectively) since material wastewater using powerless (Adam et al. 2015).

In the context of wastewater-based microalgae growth, understanding the wastewater's characteristics is pivotal to reducing biomass production costs. Knowledge of nutrient content, especially the ratio essential for microalgae growth, is crucial. Additionally, the wastewater's entire biological weight, indicated by COD content, must be comprehensively considered, given, present in industrial wastewater (Abdel-Rauf et al. 2012).

5.2 Biofuel Production from Algal Biomass

Bioenergy, sourced from natural or biological origins, is a vital renewable energy form often considered sustainable and eco-friendly. Scientists worldwide have intensified their research in this domain, recognizing bioenergy's potential to combat world-wide heating and deliver unsoiled vigour in the short and medium term. The call for biofuels as remnant petroleum substitutes has surged due to the current fossil fuel consumption surpassing production levels. By 2030, the United States is anticipated to produce a substantial biomass volume, ranging from 991 to 1147 million dry tons annually, yet this would only meet a quarter of the nation's liveliness requirements (Khoie and Yee 2015). In the past decade, algae have gained significant attention as a biofuel source, effectively reducing CO_2 emissions and maintaining an equilibrium compared to fossil fuels. Biodiesel derived from microalgae reduces CO_2 emissions significantly, offering a 78.5% reduction associated to petroleum-based diesel (Van Gerpen 2005). Algal biofuels, categorized into four generations based on feedstock, production feasibility, and technological advancement, hold distinct advantages. They are readily available, are eco-friendly, benefit the economy, and promote (Ahmad et al. 2016). These biofuels include biodiesel, bioethanol, biogas, and biohydrogen, generated through various conversion processes from algal biomass (Hussain et al. 2021). The production of fourth-generation biofuels involves like microalgae, yeast, and fungi. Efforts are ongoing to enhance microalgae-based biofuel production (Chowdhury and Loganathan 2019; Schenk et al. 2008). Biodiesel is typically produced through the chemical conversion or transesterification of algal oil, while algal biomass combustion generates electricity. Biochemical conversion technologies employ anaerobic digestion and fermentation. Thermochemical conversion methods, such as pyrolysis, gasification, and liquefaction, yield bio-oil, charcoal, and syngas, respectively. Despite their promise, biofuels pose challenges related to policy, technology, and feedstock selection.

5.3 *Biodiesel Production*

Biodiesel, a maintainable fuel derived after renewable and surplus lipids, stands as a viable alternative to firewood. The lipid composition of algal biomass significantly impacts the quality and yield of biodiesel, making it a possible substitute for conventional diesel (Milano et al. 2016). Similar to petroleum-derived diesel, biodiesel from algal sources is sulphur-free and emits fewer particulate matter and greenhouse gases. It is a concern due to its subpar presentation at cold temperatures. Milano et al. (2022) explored biodiesel production from microalgae strains rich in oleic acid. Microalgae species, along with the oleic acid-rich *Saccharomyces cerevisiae*, are instrumental in enhancing biodiesel's oxidative stability.

Biodiesel primarily originates after microalgae through transesterification, a procedure where fats, particularly triacylglycerides, react with alkali or acid, yielding biodiesel and glycerol (Sirajunnisa and Surendhiran 2016). The choice of microalgae strain significantly influences. Quality parameters like cetane number (CN), viscosity, biodiesel quality and its engine performance. Biological is an emerging method to enhance biodiesel quality. Crude oil extracted from algae possesses higher viscidness than diesel oil, making it unsuitable for direct engine use. To reduce viscosity and enhance fluidity, a chemical procedure called transesterification is necessary. This process involves esterification, where fatty acids interact with alcohol to form esters. In transesterification, esters and alcohol react, leading to the creation of greasy acid-alcohol esters by replacing the alkoxy group. Large and branched triglyceride molecules are broken down into shorter, straight-chain esters, increasing biodiesel usability. Untried studies on Chlorella species demonstrated their high fatty acid content, producing significant biodiesel quantities in Malaysia.

5.4 *Bioethanol Production*

Bioethanol, a highly fruitful biofuel derived from microalgae biomass (third-generation), offers numerous advantages, notably its minimal requirement for arable land, coupled with its ability to absorb CO_2 from the atmosphere, thereby benefiting the environment. Microalgae possess cell walls rich in lipids and carbohydrates like cellulose, man nans, sulphated glycan's, xylenes, and starch. These intricate mechanisms are enzymatically or chemically abridged to simple sugars, which are then converted into bioethanol under anaerobic conditions (Chaudhary et al. 2014). The production process of bioethanol from microalgae involves various stages such as algal biomass selection and cultivation, pre-treatment, liquefaction, scarification, anaerobic fermentation, and distillation for bioethanol sanitisation.

Unadulterated ethanol serves as an efficient alternative to gasoline, possessing a advanced octane number and heat of evaporation than petrol. It contains the same vigour gratified as 66% petrol by volume (Singh et al. 2011). When compared to traditional crops like maize, soybeans, or sugarcane, microalgae are regarded as more effective feedstocks for bioethanol production [26]. Bioethanol yield from microalgae is nearly twice that of sugarcane and five times that of maize. Numerous microalgae species have shown promise for, exhibiting a Furthermore, studies have explored enzyme-catalyzed hydrolysis for marine algae *P. cruemtum* and freshwater *P. cruemtum*, demonstrating their potential as bioethanol contenders, especially in simultaneous saccharification and fermentation (SSF) processes, yielding promising bioethanol outcomes.

5.5 Biomethane Production

Microalgae have the potential to produce biogas, a valuable resource that can be used to produce power, fuel cells. Their biomass, rich in polysaccharides and lipids and containing minimal, is an outstanding feedstock for biogas manufacture through anaerobic (Harun et al. 2011a). Biogas, the outcome of anaerobic digestion, comprises 55–75% methane (CH4) and 25–45% CO_2, making it a potent energy source. Microalgae are particularly advantageous for biogas production due to their higher polysaccharide and lipid concentrations, absence of lignin, lower cellulose content, ease of harvest, reduced land requirements, rapid conversion to biogas, and faster growth compared to lignocellulosic biomass. Moreover, the solid waste generated from anaerobic digestion serves as a valuable soil supplement.

Seaweeds, in particular, hold significant promise for biogas cohort, with various algae classes like, Spirulina, and Ulva utilized in biogas production processes. The biogas production process involves hydrolysis, acetogenesis, and methanogenesis steps. Challenges such as the risk of eutrophication and the generation of hazardous chemicals complicate the process, along with factors like low C:N ratio leading to ammonia production, which is harmful to anaerobic species. Additionally, sodium ions can suppress microbial activity, necessitating the use of salt-tolerant microorganisms for efficient biogas production from microalgae biomass.

Studies have shown that microalgal cultures grown under specific conditions can significantly enhance biogas production. For instance, certain conditions. In another study, *Arthrospira platensis* produced 4 g VS/L/day biogas when co-cultivated with carbon-rich substrates at different C:N ratios oil palm empty fruit bunches (OPEFB) resulted in a specific biogas production rate of 0.128–0.129 m^3/kg COD/day, showcasing the potential of microalgae in biogas production. Furthermore, co-cultivation of marine algae *Nannochloropsis oculata* and palm oil mill effluent (POME) produced higher yield, highlighting the effectiveness of algae in biomethane production.

5.6 Biohydrogen Production

Biohydrogen, with its tall fattening worth and clean corrosion possessions, has gained significant consideration as a possible green get-up-and-go foundation. Unlike biological, the biological is under ambient conditions. Recently, microalgae have emerged as a promising avenue for biohydrogen production, offering an eco-friendly approach to sustainable energy generation.

The manufacture of biohydrogen after microalgae falls hooked on binary categories: processes. In light-dependent processes, microalgae and cyanobacteria bio-photolyse in water, while photosynthetic bacteria primarily conduct photo-fermentation. In light-independent processes, dark fermentation occurs, where anaerobic bacteria ferment organic molecules. This Biohydrogen production is environmentally friendly, emitting no greenhouse gases and producing water as a by-product. Biohydrogen stands out as an efficient energy source due to its high available.

The enzyme hydrogenase, initially found in microalgae, plays a crucial role in splitting aquatic to yield hydrogen under shadowy anaerobic fermentation circumstances (Cuellar-Bermudez et al. 2015). Though, a limitation of this method is the restricted response while outstanding to oxygen group, which rapidly deactivates the hydrogenase enzyme. Advancements in technology are necessary to enhance the oxygen broad-mindedness of the hydrogenase enzyme (Chye et al. 2018; Cuellar-Bermudez et al. 2015).

Academics consume discovered innovative techniques to effectively generate Biohydrogen from microalgae. For instance, using carbohydrates from *Chlorella vulgaris* biomass pre-treated with 1% H2SO4, fermentation with. Another study achieved production from *Chlorella sp.* (~11.6 mL/gives) finished shady by integrating an algae-bacteria system, offering a sustainable and cost-effective approach. This shady fermentation method foodstuffs by-products like commercialized easily. Since it doesn't require a light source or aeration, it significantly reduces additional costs, making it an economically viable and environmentally friendly solution for Biohydrogen production.

6 SDGs (Sustainable Development Goals)

In the contemporary landscape of business operations, the significance of sustainability and lime source cable organisation cannot be overstated (Li et al. 2019; Kalanick and Playback 2020; Li and Wu 2020). The Combined Countries' ambitious maintainable expansion agenda for 2030, which has been in effect since 2016, encompasses 17 supportable expansion goalmouths (SDGs). These goals encapsulate the new benchmarks for monetary, social, and conservational progress, encompassing objectives such as eradicating poverty, fostering monetary development, and safeguarding the environment. The SDGs call upon nations, corporations, non-governmental organizations, and the public worldwide to contribute collectively toward these goals.

In the realm of private enterprise, it has been demonstrated that sustainable business practices can unlock economic opportunities of up to US$12 trillion (source). SDGs are also shown in Fig. 12.5. Notably, major fashion conglomerates have taken proactive measures to align their strategies with the United Nations' SDGs. For instance, Inditex has integrated the SDGs into its sustainable supply chain management efforts, diligently working toward each specific goal. Inditex enforces a stringent Code of Conduct and conducts meticulous compliance evaluations through audits, ensuring the well-being and good health of its workforce. Similarly, H&M has pledged to make substantial contributions to the sustainable development goals, as outlined in its "Programme 2030." By 2030, it has committed to utilizing only second-hand or sustainably obtained resources cutting-edge hers foodstuffs, thereby ensuring answerable manufacture practices. These initiatives represent significant steps by private enterprises to align their operations with the global pursuit of sustainable development objectives.

The style stock cable, unique to the fabric then clothing industry, stands out for its labour-intensive engineering, lengthy worth hawser, and relatively high conservational impact in comparison to supply chains in other sectors. Economically and socially, global production capitalizes on little work costs in emerging nations like. However, incidents such as the Rana Plaza collapse in Bangladesh have shed light on the serious social challenges within the textile and apparel industry, prompting heightened public awareness about labour rights, worker safety, fair wages, and more.

From an environmental standpoint, the industry inflicts significant harm on the environment. Enormous quantities of resources like cotton, linen, wool, as well as energy (such as power) and aquatic are expended. Particularly, in the dyeing procedure, wastewater laden with toxic chemicals is generated. In cases where proper treatment is lacking, this wastewater severely contaminates rivers and streams. In some developing towns hosting apparel factories, "colourful rivers and streams" have become a grim reality due to the absence of clean technology investments for wastewater treatment.

According to World Bank statistics, textile dyeing accounts for marine contamination from manufacturing ingesting, contributing to the creation of 72 toxic chemicals. Furthermore, the rise of fast fashion and rapidly changing trends has led to frequent consumer purchases, shorter lifespans of garments, and a surge in post-consumer clothing left-over. Consequently, the textile and clothing industry faces a significant challenge in meeting the new purposes of economic, community, and conservational development.

Presently, concerns surrounding the sustainability of fast fashion have garnered significant attention. The rapid turnover in fast fashion directly leads to substantial waste generation, exerting immense pressure on the environment. Under mounting external pressure, fashion brands have begun adopting sustainable measures.

One prominent sustainable practice involves the widespread adoption of "organic strand." For instance, H&M embraced sustainable sourcing of cotton, constituting

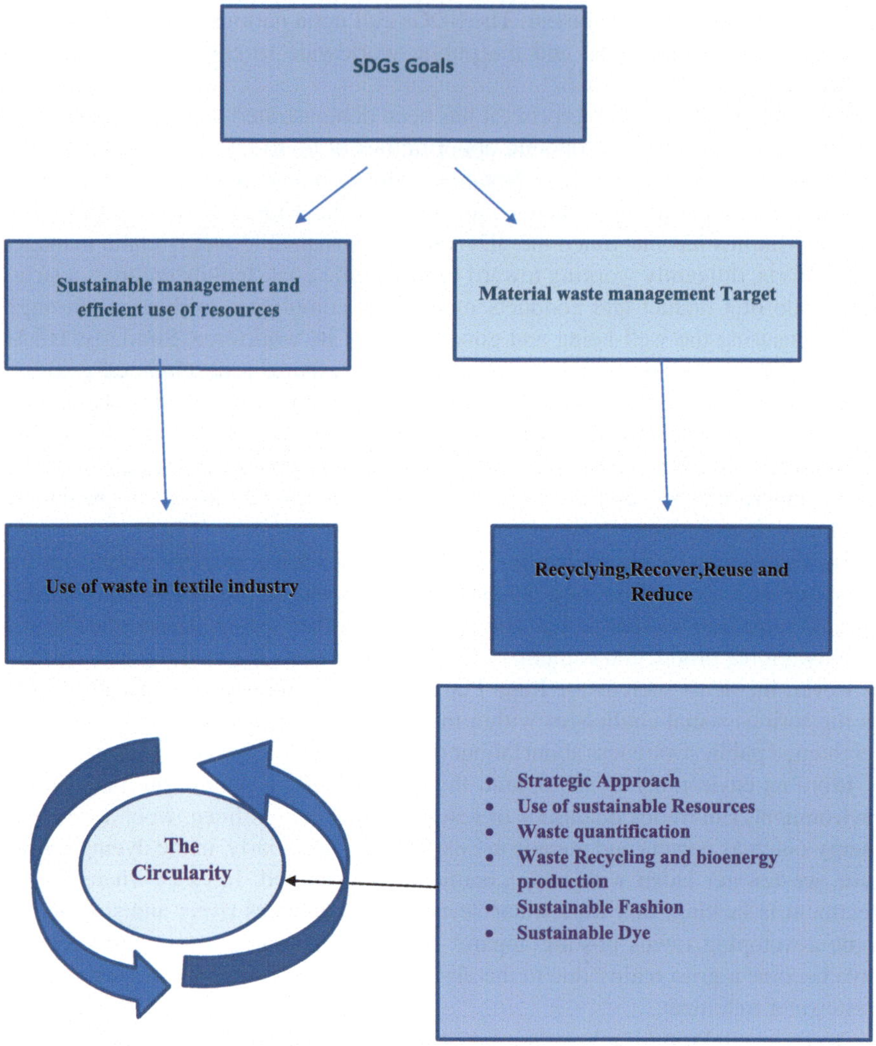

Fig. 12.5 SDGs (sustainable development goals)

43% of its entire strand ingesting in 2018. The company aims for all its brands to utilize maintainable uncooked constituents besides second-hand constituents. To address aquatic contamination stemming from the denim colouring procedure, Levi's introduced a "Scorched Colouring Procedure" in 2016, significantly reducing aquatic ingesting and pollution. In 2018, Levi's Eureka Lab pioneered a new laser technology, making distressed and washed jeans more eco-efficient. In a move to protect animal rights, luxury brand Gucci announced its decision to cease using genuine leather and fur in 2017.

Maintainable approach advocates for the use of artificial leather in handbags and shoes. This type of artificial leather not only boasts high functionality but also competes with genuine leather. In efforts to safeguard the oceans, sustainable fashion companies like ECOALF collect and reprocess plastic waste, such as PET bottles, transforming them into fashionable clothes and shoes. In a bid to close the loop, H&M has initiated a used clothing assortment strategy. Uniqlo collects post-consumer products for contribution, while Nike recycles old shoes to create new ones. These practices underscore the promise of major, deep-rooted global fashion products to sustainability.

6.1 Sustainable Fashion Supply Chain Management

In the textile and apparel (TA) sector, sustainability issues rooted in the “3BL” framework (covering economic, social, and environmental aspects) are of paramount importance. This section delves into research papers exploring these concerns from the perspective of Fashion Supply Chain Management (FSCM). Numerous facets are examined, such as info revelation in sustainability and more. Existing literature addresses these issues from both strategic and operational management viewpoints.

After a planned organisation standpoint, Chi (2011) examines the role of CSR in ornamental sustainability in global way supply manacles. Daija et al. (2016) explore principled standards and supportable development in the Romanian garment sector, proposing ethical measures for sustainability. Shi et al. (2017) assess the economic and environmental impacts of sustainable investments, revealing mutual benefits for manufacturers and retailers. Gurney and Yu (2012) analyse world-wide associations in clothing supply networks, emphasizing the facilitative role of strong relationships in achieving sustainability goals.

From an manoeuvres organisation (OM) standpoint, Na and Na (2015) study sustainability efforts in Korean textile and fashion companies, focusing on environmental materials, product reuse, and eco-marketing. Atitlan (2007) examines eco-labelling in European countries, highlighting its importance human health and environmental issues in the TA sector. Shen (2014) investigates the structure, using H&M as a case study, indicating progress but with room for improvement in environmental considerations. Li et al. (2017) analyse water usage in China’s TA industry, identifying large industry scale as a significant driver for increased water consumption. Additionally, Li and Wu (2017) explore the impact, finding a reduction in profits due to environmental management system adoption.

The literature on “maintainable style supply hawser organization” offers valuable understandings, informing practices to enhance sustainability across economic, social, and environmental dimensions.

6.2 Sustainable Design

Sustainable design in apparel emphasizes the use of biodegradable supplies, patterns, and systems, contrasting with conventional approaches. Emerging concepts for the environment further enrich maintainable enterprise methodologies.

Jin Gam et al. (2011) focus on disassembling in men's jacket design, emphasizing the simplicity of fabric choices to enhance product administration. Alitalia et al. (2015) identify key factors for prolonging clothes' use, categorizing them into behaviour levels. Ecolabel impact on sustainable apparel design, suggested the need for designer guidelines to enhance sustainability efforts. De Angelis et al. (2017) study customers' discernments of avocado extravagance approach, advising luxury brands to maintain familiar product features, ensuring brand consistency in new green product development.

In summary, the literature on "sustainable design" offers valuable strategies grounded in the 3BL framework, providing a foundation for sustainable environment.

6.3 Sustainable Dyeing

The dyeing process within the chain poses significant pollution challenges, particularly in terms of water pollution, with potential consequences for workers' health and surrounding communities when toxic chemicals are not properly treated. Key factors influencing sustainability in the dyeing section encompass choices related to dye selection (e.g., natural or chemical), dyeing methods and operational management.

It was explored maintainable pattern techniques aimed at reducing the environmental impact of dyeing processes. They apply eucalyptus tree leaf and bark extracts to local fabrics, finding that screen printing enables local fabrics to achieve superior durability in daily use. This model contributes to the expansion of maintainable pattern practices.

It was examined eco-dyeing by means of normal hemp fiber, experimenting with different hemp fabrics and natural dye pigments derived from sources. Their research reveals that natural dyes offer excellent coloration performance, suggesting wider adoption in the TA subdivision.

Also, the conservational recital of two colouring approaches, besides conservative colouring was assessed. They discover that spin-dyeing devours less liveliness, water, and generates fewer carbon productions. They recommend the implementation of spin-dyeing in the garb industry to reduce conservational impression.

The environmental colouring methods compared to traditional dyeing processes was scrutinized. They demonstrated that approaches such as "enzymatic amalgamation" and "plasma pre-treatment" promote sustainable colouring while reducing raw material, get-up-and-go, and water consumption.

In conclusion, the “maintainable colouring” prose offers valuable understandings for sustainable performs within the 3BL framework, addressing critical environmental challenges in the TA supply chain.

6.4 Sustainable Sourcing

Effective maintainable obtaining in the upstream source cable is crucial for both apparel manufacturers and retailers, encompassing activities such as raw material preparation and supplier selection. OEM manufacturers must ensure a steady supply of fabrics and accessories, while fashion retailers are responsible for selecting reliable suppliers, placing orders, managing lead times, and maintaining quality assurance. Collaborating with eco-friendly contractors is a planned tactic for maintainable obtaining.

Choi (2013b) investigates contractor collection in the manner source chain considering environmental carbon taxation regulations. The study discloses that carbon production levy schemes influence fashion retailers’ supplier choices, with non-linear quadratic carbon taxation forms promoting indigenous contractor collection.

Administration intervention is recommended to equilibrium sustainability and proceeds. Amin oust and Sagafinia (2017) tackle hoard handcuffs, proposing a “fuzzy implication system” aimed at contractor assessment besides position based on indices.

6.5 Sustainable Production

Manufacture in the fabric and attire (TA) source cable raises concerns related to labour rights, resource and liveliness feasting, leftover cohort, and more. Numerous studies delve into sustainable operations within the production process, seeking to mitigate environmental impact and improve overall efficiency.

Pineda-Henson and Cuiaba (2004) evaluate green output in sustainable engineering, importance factors like aquatic practise, get-up-and-go consumption, and land eco toxicity.

In summary, the “maintainable manufacture” works offers valuable insights for maintainable performs founded on the principles of the triple bottom line (3BL).

7 Conclusion and Future Perspective

The utilization of bioenergy from effluent treatment in the textile industry holds significant promise as a sustainable and environmentally friendly approach. This research has delved into the various methods and technologies employed in harnessing bioenergy from textile wastewater treatment processes. By exploring different biological treatment methods such as anaerobic digestion, microbial fuel cells, and algae-based systems, this study has highlighted the potential for generating bioenergy while simultaneously addressing the issue of textile wastewater pollution.

One of the key results of this investigation is the effectiveness of anaerobic digestion in converting organic pollutants present in textile effluents a valuable foundation of renewable liveliness. Additionally, the integration of microbial fuel cells into wastewater treatment systems has shown promising results in generating electricity from the microbial metabolism of organic matter. Furthermore, the use of algae-based systems, particularly microalgae and cyanobacteria, has demonstrated the ability to absorb pollutants and produce bioenergy through processes like photosynthesis and biohydrogen production.

The environmental implications of this research are profound. By implementing these bioenergy generation methods, textile industries can meaningfully decrease their carbon footmark and confidence on non-renewable vigour foundations. Moreover, the treatment of textile effluents using these bioenergy technologies can mitigate water pollution, protecting aquatic ecosystems and human health. It is essential for the textile industry to embrace these sustainable practices and invest in research and development to optimize the efficiency of bioenergy generation from effluent treatment. Co-operative labours among industries, investigators, and politicians are crucial to generating a supportive environment for the adoption of these eco-friendly technologies.

The research on bioenergy from effluent treatment in the textile industry signifies a crucial step toward achieving a more sustainable future. By harnessing the potential of bioenergy while addressing wastewater pollution, the textile industry can production a noteworthy role in the changeover to a greener and additional maintainable liveliness countryside. Continued research, technological advancements, and industry-wide adoption are necessary to fully unlock the economic, environmental, and social benefits offered by bioenergy from textile effluent treatment.

References

1. Arshad M, Adil M, Sikandar A (2014) Hussain T (2014) Exploitation of meat industry by-products for biodiesel production: Pakistan's perspective. Pakistan J Life Soc Sci 12:120–125
2. Arshad M (ed) (2017) Perspectives on water usage for biofuels production: aquatic contamination and climate change. Springer
3. Arshad M (ed) (2020) Sustainable ethanol and climate change: sustainability assessment for ethanol distilleries. Springer Nature
4. Arshad M, Ansari AR, Qadir R, Tahir MH, Nadeem A, Mehmood T, Alhumade H, Khan N (2022) Green electricity generation from biogas of cattle manure: an assessment of potential and feasibility in Pakistan. Front Energy Res 25(10):911485
5. Arshad M (ed) (2023) Climate changes mitigation and sustainable bioenergy harvest through animal waste: sustainable environmental implications of animal waste. Springer Nature; Arun J, Gopinath KP, Sivaramakrishnan R, SundarRajan P, Malolan R, Pugazhendhi A (2021) Technical

insights into the production of green fuel from CO_2 sequestered algal biomass: a conceptual review on green energy. Sci Total Environ 755:142636
6. Bano I, Ghauri MA, Arshad M, Yasin T, Younus M (2017) Bioactivity of variant molecular weight chitosan against drug-resistant bacteria isolated from human wounds. Microb Drug Resist 1;23(8):958–65
7. Bano I, Arshad M (2018) Climatic changes impact on water availability. Perspect Water Usage Biofuels Prod: Aquat Contam Clim Chang 39−54
8. Bhatia L, Sarangi PK, Nanda S (2020) Current advancements in microbial fuel cell technologies. Biorefinery Altern Resour: Target Green Fuels Platf Chem 477–494
9. Bhatia SK, Mehariya S, Bhatia RK, Kumar M, Pugazhendhi A, Awasthi MK, Yang YH (2021) Wastewater based microalgal biorefinery for bioenergy production: progress and challenges. Sci Total Environ 751:141599
10. Dębowski M, Zieliński M, Grala A, Duduks M (2013) Algae biomass as an alternative substrate in biogas production technologies. Renew Sustain Energy Rev 27:596–604
11. Gadow SI, Li YY (2020) Development of an integrated anaerobic/aerobic bioreactor for biodegradation of recalcitrant azo dye and bioenergy recovery: HRT effects and functional resilience. Bioresour Technol Rep 9:100388
12. Kadam SK, Watharkar AD, Chandanshive VV, Khandare RV, Jeon BH, Jadhav JP, Govindwar SP (2018) Co-planted floating phyto-bed along with microbial fuel cell for enhanced textile effluent treatment. J Clean Prod 203:788–798
13. Luo J, Tian W, Jin H, Yang J, Li J, Wang Y, Zhou M (2022) Recent advances in microbial fuel cells: a review on the identification technology, molecular tool and improvement strategy of electricigens. Curr Opin Electrochem 101187
14. Min H, Kim M, Lee SU, Kim H, Kim G, Choi K, Seok SI (2019) Efficient, stable solar cells by using inherent bandgap of α-phase formamidinium lead iodide. Science 366(6466):749–753
15. Nawaz A, Hafeez A, Abbas SZ, Haq IU, Mukhtar H, Rafatullah M (2020) A state of the art review on electron transfer mechanisms, characteristics, applications and recent advancements in microbial fuel cells technology. Green Chem Lett Rev 13(4):365–381
16. Pandit S, Savla N, Jung SP (2020) Recent advancements in scaling up microbial fuel cells. In: Integrated microbial fuel cells for wastewater treatment. Butterworth-Heinemann, pp 349–368
17. Prabakar D, Manimudi VT, Mathimani T, Kumar G, Rene ER, Pugazhendhi A (2018) Pre-treatment technologies for industrial effluents: critical review on bioenergy production and environmental concerns. J Environ Manage 218:165–180
18. Sharif N, Munir N, Hasnain M, Naz S, Arshad M (2021) Environmental impacts of ethanol production system. Sustain Ethanol Clim Chang: Sustain Assess Ethanol Distill 205−223
19. Singh V, Mishra A, Srivastava P (2023) Textile and domestic effluent treatment via co-cultivation of diplosphaera mucosa VSPA and scenedesmus obliquus. Biomass Bioenerg 172:106756
20. Suresh R, Rajendran S, Kumar PS, Dutta K, Vo DVN (2022) Current advances in microbial fuel cell technology toward removal of organic contaminants–a review. Chemosphere 287:132186
21. Vegamoor VK, Kalathil D, Rathinam S, Darbha S (2019) Reducing time headway in homogeneous CACC vehicle platoons in the presence of packet drops. In: 2019 18th European control conference (ECC). IEEE, pp 3159–3164
22. Wang J, Ren K, Zhu Y, Huang J, Liu S (2022) A review of recent advances in microbial fuel cells: preparation, operation, and application. Biotech 11(4):44
23. Wang Z, Mei X, Ma J, Wu Z (2012) Recent advances in microbial fuel cells integrated with sludge treatment. Chem Eng Technol 35(10):1733–1743
24. Wu XT, Ansari AR, Pang XX, Li HZ, Zhang ZW, Luo Y, Arshad M, Song H (2019) Visfatin plays a significant role in alleviating lipopolysaccharide-induced apoptosis and autophagy through PI3K/AKT signaling pathway during acute lung injury in mice. Arch Immunol Ther Exp 1(67):249–261
25. Wu D, Zheng X, Kalathil D, Xie L (2019) Nested reinforcement learning based control for protective relays in power distribution systems. In: 2019 IEEE 58th conference on decision and control (CDC). IEEE, pp 1925–1930

26. Yap CX, Henders AK, Alvares GA, Wood DL, Krause L, Tyson GW, Gratten J (2021) Autism-related dietary preferences mediate autism-gut microbiome associations. Cell 184(24):5916–5931
27. Yaqoob AA, Mohamad Ibrahim MN, Rafatullah M, Chua YS, Ahmad A, Umar K (2020) Recent advances in anodes for microbial fuel cells: an overview. Materials 13(9):2078
28. Zhou M, Wang H, Hassett DJ, Gu T (2013) Recent advances in microbial fuel cells (MFCs) and microbial electrolysis cells (MECs) for wastewater treatment, bioenergy and bioproducts. J Chem Technol Biotechnol 88(4):508–518

The Circular Economy Concept in the Textile Industry

Taj Ud Din, Nasib Zaman, Muhamad Nazir Uddin, Wajid Khan, Nabila Qayum, Sumayya Qayum, Sajid Iqbal, Iram Liaqat, Aqsa Mumtaz, Sher Zaman Safi, Arfa Tahir, and Muhammad Arshad

Abstract The circular economy in textiles requires a transformative lifecycle, starting with design and emphasizing durability, long life, and reuse or recycling. Changes in consumer, public, industry, and authority preferences, such as product sharing, renting, or buying as a service, are necessary to increase utilization while maintaining economic viability. The textile sector has a significant environmental impact due to its linear and intricate nature, including the cultivation of natural fibers, spinning and sizing fabrics, fine-tuning, printing, and dyeing of finished products. The textile manufacturing process in Europe has a significant impact on agricultural land and the alteration of natural land, especially in relation to natural fibers like cotton. The textile industry accounts for 30–50% of greenhouse gas emissions from the supply chain, positioning textiles as the sixth most environmentally detrimental category among home products in the European Union. The establishment of a textile circular economy requires inventive corporate strategies, environmental practices, and legislative measures throughout the entire value chain. Circular economy principles encompass several stages of a product's life cycle, including design, production, marketing, consumption, waste management, reusing, and recycling. To achieve a sustainable transition, substantial modifications are required throughout all phases of the product's life cycle. The efficacy of circular business models heavily relies on the

T. U. Din · N. Zaman · M. N. Uddin · W. Khan · N. Qayum · S. Iqbal
Centre for Biotechnology and Microbiology, University of Swat, Swat, Pakistan

S. Qayum
Centre for Plant Sciences and Biodiversity, University of Swat, Swat, Pakistan

I. Liaqat
Microbiology Lab, Department of Zoology, Government College University, Lahore, Pakistan

A. Mumtaz · M. Arshad (✉)
Biochemistry Section, Jhang-campus, University of Veterinary and Animal Sciences Lahore, Lahore, Pakistan
e-mail: Muhammad.arshad@uvas.edu.pk

S. Z. Safi
Faculty of Medicine. Bioscience and Nursing, MAHSA University, Jenjarom, Selangor, Malaysia

A. Tahir
Department of Environmental Sciences, Lahore College of Women University, Lahore, Pakistan

M. Arshad (ed.), *Enzymes in Textile Processing: A Climate Changes Mitigation Approach*, SDGs and Textiles, https://doi.org/10.1007/978-981-97-8058-7_13

capabilities of the supply chain and network infrastructure. The integration of circular economy ideas into a company's long-term strategy and implementation necessitates a fundamental transformation inside the organization. Digital technologies, such as IT devices, robots, and mobile applications, have the potential to support a transition towards circular economy, improving product design, consumer engagement, and service offerings. Future studies should prioritize the use of digital technologies in the textile and garment industry, specifically in the domain of circular economy (CE), such as smart mirrors, 3D printing, and AI.

Graphic Abstract

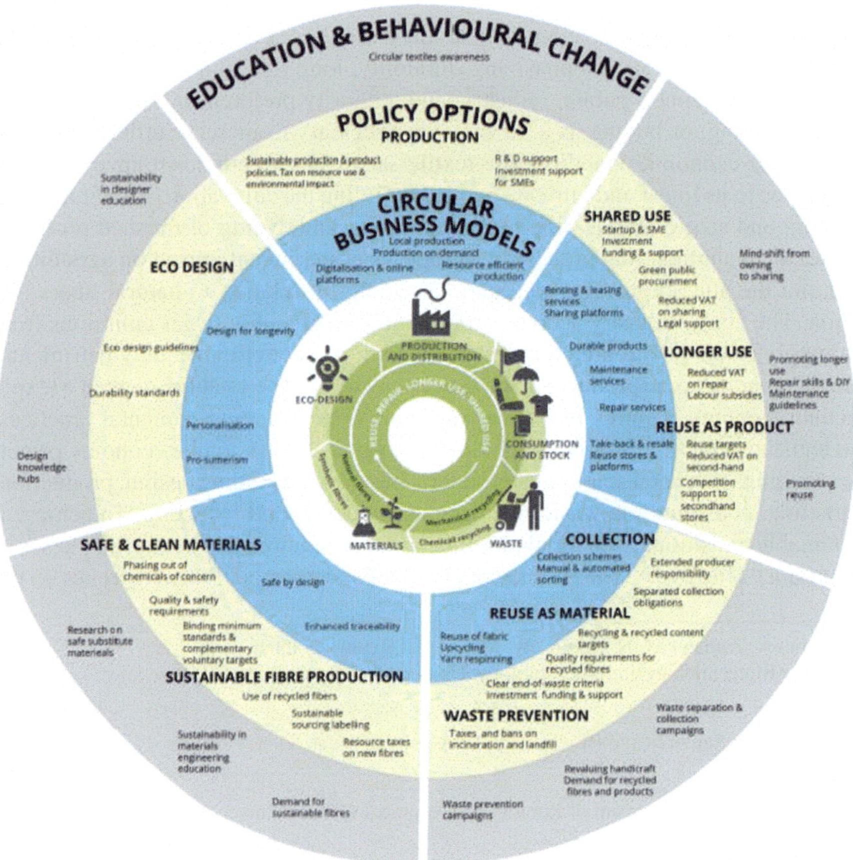

Source: EEA and ETC/WMGE, adapted from EEA (2016)

Keywords Textile industry · Circular economy · Climate change · Sustainable development goals

1 Introduction

The textile industry, a cornerstone of global commerce, has long faced challenges related to sustainability and environmental impact. In recent years, there has been a paradigm shift in addressing these concerns through the adoption of a circular economy model. Unlike the traditional linear approach of 'take, make, and dispose,' a circular economy in textiles seeks to create a closed-loop system where resources are continuously reused, recycled, and repurposed. At its core, the circular economy in the textile industry aims to minimize waste, reduce environmental footprint, and promote the efficient use of resources throughout the entire product life cycle. This holistic approach encompasses everything from the extraction of raw materials to production, consumption, and eventual disposal.

The traditional textile production model often results in significant environmental consequences, including depletion of natural resources, pollution, and the accumulation of non-biodegradable waste. In contrast, the circular economy concept encourages innovative strategies to mitigate these issues by designing products for durability, reusability, and recyclability. The dynamics of global economies have long been entwined with the intricate dance of production, consumption, and financial interactions. Whether enveloped in the folds of Feudalism, Socialism, Communism, or Capitalism, the essence of an economy is sculpted by the interplay of population, resource allocation, wealth distribution, and cultural tapestries. Money, wealth, production, and consumption conduct this economic symphony, yet the challenge that echoes across the globe is the metamorphosis of these systems into sustainable entities [48].

As nature whispers its wisdom, we discern the inherent limits guiding each species, ensuring harmony within the constraints of their ecosystems. The principles governing sustainability and feasibility emerge, and when mirrored in the blueprint of a new economy, the outcome is not just economic gains but a legacy of lasting benefits [32, 42].

2 The Circular Symphony: Rhythms of Sustainability

Governments worldwide now seek economies that shun the 'take-make-dispose' paradigm, opting instead for the circular dance of sustainable design, remanufacturing, reuse, and recycling. Embedded in the laws of thermodynamics, the second law warns us of the increasing entropy when energy transforms or matter moves without restraint. The essence lies in the quality loss of materials and energy when exploited limitlessly, exemplified by the contrast between a kilo of gold fashioned into an ornament and the same gold distributed across mobile phone microchips [29, 55].

In 1989, environmental economists David W. Pearce and R. Kerry Turner unfurled the circular economy as the antidote to the environmental wasteland of linear models.

Their vision, rooted in non-linear living systems, espoused waste and pollution reduction, maximal product and material utilization, and the rejuvenation of natural systems [55].

3 The Dual Phases: Biological and Technological Ballet of Circularity

The circular economy elegantly pirouettes through two phases—the biological and the technological. In the biological realm, materials like cotton or wood reenter the parental system through composting or anaerobic digestion, breathing life into the soil and enhancing renewable resources. On the technological stage, scientific prowess orchestrates the recovery and refurbishment of components/materials. The 6R's—Recycle, Rethink, Refuse, Reuse, Repair and Reduce—waltz together, curbing resource consumption and mitigating waste accumulation [29, 48].

4 Conceptual Crescendos: From Functional Service to Blue Economy

4.1 Functional Service Economy

The economic chronicle unfolds from Stone Age bulk goods transactions to the Industrial Economy's smart goods era. The Performance Economy ushers in a shift, internalizing risks, fostering job creation, and optimizing resource utilization [13, 26].

4.2 Cradle to Cradle Design Philosophy

Nature bequeaths resources designed for disassembly and reuse, differentiating between biological and technical nutrients. This philosophy illuminates the path where every element in nature transforms into a resource for another, advocating for design innovation and creativity [45, 48].

4.3 *Biomimicry*

Nature, a repository of ingenious solutions, becomes the muse for design. The humpback whale's flipper inspires wind turbine blades, unveiling nature's blueprint for sustainable solutions [41].

4.4 *Industrial Ecology*

An intricate dance of material and energy flows, environmental impact, and sociopolitical influences, Industrial Ecology classifies ecosystems into linear, partially cyclic, and cyclic. It envisages a future steeped in Natural Capitalism, marrying economics, ecology, and innovation [14, 48].

4.5 *Blue Economy Systems*

Cascading systems intertwine production cycles, turning one system's waste into the next's resource. The Blue Economy manifesto pledges '100 innovations for 100 million jobs in the next 10 years,' underlining the transformative power inherent in nature's wisdom [24, 43, 48].

5 The Global Tapestry of Textile Industries Weaving Through Transformations

5.1 *Globalization's Warp and Weft: Navigating Textile Growth*

The globalization tapestry unfurls, intertwining production and financial threads across borders. The wealth of nations now weaves through economic agents in far-off lands, transcending national boundaries, policies, and political constraints. Transnational firms, financial juggernauts, and cutting-edge technologies thread through the fabric of a global economy, reshaping the traditional landscape [10, 57].

5.2 *The Rise of Asian Tigers: Textile Transformations in East Asia*

A crescendo echoes through the textile industry's symphony, with East Asia emerging as a vibrant hub. The 'Four Tigers'—Hong Kong, Korea, Singapore, and Taiwan—dance to the tunes of remarkable growth, fueled by heavy investments, infrastructure allurements, and compulsory education. Their textile tale intertwines with the global narrative, weathering financial storms and standing resilient in the face of economic crises [27].

5.3 *Georgia's Textile Journey: A Tale of Cotton and Innovation*

Georgia, nestled at the crossroads of Europe and Asia, spins its textile narrative. From the initial twirls of silk production to the cotton-centric crescendo post-cotton gin invention, Georgia's looms hummed with innovation. Through the trials of war and the boll weevil's shadow, the textile symphony persevered, crafting bedspreads, carpets, and tire fabrics for the evolving automobile industry [2].

5.4 *Threads of Heritage: India's Rich Textile Tapestry*

India, an ancient weaver of textile stories, intertwines its narrative with the threads of history. From the flourishing textile centers during 1000–1500 AD to the upheavals brought by mechanical inventions in the eighteenth century, India's looms wove through the fabric of time. The British trade invasion, the decline of ancient industries, and the surge of the cotton mill marked epochs in India's textile saga [22].

5.5 *Dragon's Silk: China's Textile Tale*

China's textile odyssey begins in the late 1800s, with cotton yarn imported from British India weaving into the Northern Chinese tapestry. The benefits of a vast market, raw materials, and cheap labor beckoned foreign enterprises. World War I's hush ushered in a rapid transformation, with China and Japan sprinting to meet global demands. The Sino-Japanese War painted a turbulent canvas, yet China's resurgence post-World War II shaped a new textile narrative, fueled by resilience and industry [15].

5.6 The Looms of Innovation: England's Textile Symphony

England's textile saga, a tale of transformations, unfolds across the looms of history. From the seventeenth-century influx of Indian cotton fabrics to the mechanization marvels of the eighteenth century, England's looms echoed with progress. The flying shuttle, water frame, Spinning Jenny, Spinning Mule, and power loom orchestrated a symphony of mechanization, casting England as the world's textile dominator. The textile industry in modern England, valued at GBP 9 billion, continues to thread its story, supported by innovation and sustainable growth [48].

5.7 Textile Industry of America

In days of yore, the American textile industry burst onto the scene in a rebellion against British control and tariffs on goods from England. People snubbed British wares, opting instead to craft their own local products dubbed 'home spun.' This movement commenced with fabrics and apparel, expanding to provide uniforms and blankets for Revolutionary War soldiers in the Continental Army [48]. In England, technological innovations paved the way for faster textile production, but the Brits kept these breakthroughs shrouded in secrecy under the threat of treason. However, in 1789, Samuel Slater, leveraging his design skills and experience from England, replicated textile mill construction in New England, USA, kickstarting the American Industrial Revolution. Samuel Slater etched his name as the 'Founder of the American Industrial Revolution' [44].

Fast forward to 1814, the first textile factory emerged in Lowell, Massachusetts, setting the stage for other industries to follow suit. Massachusetts blossomed into the textile and economic nucleus. World War I and II prompted many textile industries to pivot towards war needs, marking the USA's shift towards self-sufficiency. Post-war, economic growth persisted, and by the late 1990s, textile manufacturing morphed into marketing as industries migrated overseas for cheap labor and materials [48]. North and South Carolina, former textile hubs, now beckon Chinese companies due to the allure of lucrative labor, materials, and energy. The textile industry's resurgence looms on California's horizon. According to the US Bureau of Labor Statistics, in 2015, the US textile industry employed 579,300, contributing around $100 billion to the nation's GDP. Exports, including fiber, yarns, fabrics, made-ups, and apparel, totaled $21.6 billion, clinching the fourth spot in global exports [48].

By 2016, the National Council of Textile Organizations (NATO) reported a $74.4 billion milestone in textile and apparel shipments, marking an 11% uptick since 2009. The US textile supply chain employed approximately 565,000 in 2016, with capital expenditure hitting $2 billion in 2015. Amidst this, the textile industry stands as a stalwart pillar, supporting families and societies through myriad stages of growth. The statistics and analyses abound, exploring the catalysts for this growth—favorable raw material environments, cost-effective labor, expansive markets, technological leaps, and a skilled workforce. Yet, amidst this narrative, the impact of the textile industry on society and the environment, essential pillars of sustainability, begs exploration.

6 Issues with Linear Models

As our global population skyrockets, nearing 7.7 billion today and forecasted to reach 8 billion by 2025, the strain on our natural resources intensifies [46]. The linear economy, a mode entrenched in 'take-make-dispose,' has driven a voracious consumption cycle. From the Industrial Revolution through World Wars and beyond, this linear approach propelled progress but at an exorbitant cost to our environment. Today, as we stand on the precipice of an exponentially growing world economy, the linear model's flaws become glaring.

6.1 Population Explosion and Resource Scarcity

The United Nations predicts a population surge—8 billion by 2025, 9 billion by 2040, 9.6 billion by 2050, and a staggering 11.2 billion by 2100. This relentless growth places immense pressure on resources, set to be utilized at 160% of their capacity [25]. The linear economy's mantra of 'take-make-dispose' fueled the Industrial Revolution, but its aftermath reveals the toll overutilized resources with no plans for rejuvenation. Our once abundant resources, pictured lush and bountiful, now wear the scars of overuse.

6.2 The Linear Economy's Predicament

The linear economy, encapsulated in the 'take-make-dispose' philosophy, perpetuates a continuous production flow, maximizing products with minimal consideration for waste. Its legacy, built on massive resource extraction, has left a trail of environmental degradation and resource depletion. In the wake of these issues, the world began exploring alternatives—first with the 'Reuse Economy,' emphasizing the secondary use of products, and later transitioning to the 'Circular Economy' [38].

7 From Linear to Circular Economy

In contrast to the linear model's extractive approach, the circular economy aims for sustainability through design. Products are meticulously crafted to maintain a pure raw material stream throughout the value chain, promoting one-time investments, easy retrieval, and multiple uses. Waste diminishes, and if new raw material is required, it is sourced sustainably to avoid environmental damage. The circular economy introduces 'eco-effectivity,' minimizing negative impacts through process changes, system innovations, and entropy reduction [23].

8 Designing for the Future

The role of design development becomes paramount in classifying products under these economic models. Take, for instance, Apple's sleek yet glued iPad, demanding battery replacement, versus Google's Nexus 7, slightly thicker but designed for repair and extended lifespan. This minute difference in design reverberates in the consumer's mindset, creating a substantial impact on sustainability perceptions and influencing purchasing decisions [53].

9 Implications of the Linear Model

9.1 Risks in Supply

The linear economy's insatiable appetite for material and energy during the 1900–2000 era witnessed a 20-fold GDP rise. However, this came at the expense of material richness and consumption. With a fixed amount of planetary materials and uncertain replenishment mechanisms, industries dependent on materials face risks and higher prices, impacting profit margins [38].

9.2 Volatility in Prices

The linear model's challenges have ushered in an era of 'resource stress,' where unpredictable material prices, environmental degradation, and supply disruptions are commonplace. Since 2006, commodity prices have fluctuated, introducing risks in material supply investments [58].

9.3 Materials of Importance

Critical materials like metals, electronics, and computer components are vital for production. Their scarcity, coupled with price fluctuations and uncertain availability, poses challenges for industries dependent on these materials [34].

9.4 Global Connectivity

Global trade has interconnected products and their dependencies on resources like oil, leading to intricate linkages. Regional price changes can reverberate globally due to financial market integration and efficient transportation systems [6, 19].

9.5 Increases in Demand for Material

With the world population nearing 8 billion by 2025 and a rising GDP, the demand for material and energy to manufacture products skyrockets. Estimates suggest that consuming resources at half the US rate would deplete 'strategic materials' within 40 years [31].

9.6 Degradation of Ecosystems

The linear economy's disregard for the environment has generated vast amounts of pre and post-consumer waste, burdening ecosystems. Plastics production alone reached 78 million tons in 2013, with only a small fraction entering the recycling loop, putting immense strain on the environment [61].

9.7 Decreased Product Lifetime

Consumer trends toward rapid product turnover shorten product lifespans. The craving for the latest features and designs results in manufacturers producing goods with shorter lifespans. This impacts raw material procurement and elevates waste percentages, adding to the linear model's detrimental effects [21].

In the realm of industrial production and economics, the pivotal areas of concern revolve around environment, climate change, and sustainability. Laws and regulations are in a constant state of flux, requiring industries, be it manufacturing or retail, to be answerable for the waste generated at any stage of a product's life cycle. A study conducted across six Indian cities illuminated the role of rag pickers, who, in recycling three million tons of waste, earned INR 24,500 per ton and reduced carbon dioxide emissions by 721 kg annually [34]. While traditionally, the responsibility for waste management in India has fallen on the government, recent legal shifts place the onus on manufacturers, retailers, or even consumers who have utilized a product. Waste audits become indispensable for controlling waste during production and use through sustainable design and development [60].

The global concern for environmental degradation and its impact on humanity is shared by consumers, politicians, and governments. A U.S. study delved into consumer accountability for carbon emissions, revealing the influence of carbon footprint labels on purchasing decisions. These labels, detailing CO_2 emissions throughout a product's life cycle, empower consumers to make environmentally conscious choices. When consumers opt for sustainable products, their decisions, based on ecological footprints, hold sway over a brand's reputation. The imminent shift from a linear to a circular economy and sustainable business practices is on the horizon as policymakers and nations worldwide recognize the need for change [49].

The textile industry, driven by an insatiable demand for clothing, stands as a formidable player in global economies. Regrettably, it also claims the title of the second-largest polluting industry. From fiber production to the final product, the industry's myriad processes contribute to air, water, and solid waste pollution. The Chinese Textile Industry, for instance, has discharged a staggering two and a half billion tons of wastewater, laden with toxic components, into the environment [4–6]. Air pollution arises from various sources, including solvents, spills, and wastewater treatment, as well as emissions from ovens, boilers, and storage tanks. The colossal consumption of energy and chemicals in fiber production, coupled with the use of approximately one million tons of chemical dyes annually, exacerbates the industry's environmental impact [56].

Fashion's dominance in the apparel industry, fueled by consumer desire for the latest trends, has given rise to fast fashion and a throwaway culture. A study reveals a 36% reduction in the number of times a garment is worn globally compared to 15 years ago. This culture, prevalent in both Germany and China, contributes to an annual cost of USD 460 billion for disposable clothes, with 60% of German and Chinese consumers owning more clothes than necessary [5, 8, 28]. The textile industry's environmental toll is staggering, with greenhouse gas emissions in 2015 reaching 1.2 billion tons of carbon dioxide equivalents, surpassing the emissions of all international flights and naval ships combined [39]. Moreover, textile industries are responsible for 20% of water pollution due to the discharge of wastewater from dyeing and finishing processes.

The urgency to transition from a linear to a circular economy is evident, necessitating immediate attention and action. Closed-loop manufacturing, encompassing reuse and recycling, emerges as a viable solution. A circular economy aims to eliminate waste, and while achieving zero waste is challenging, various strategies such as reuse, recycling, renting clothes, and producing completely or partially biodegradable products play crucial roles [4, 7].

Closed-loop manufacturing, however, comes with its set of challenges and opportunities. Sustainable product design, certification of non-use of hazardous chemicals, and effective collaboration between consumers, charities, manufacturers, retailers, governments, and society are essential for successful closed-loop processes. Transparency in manufacturing and reporting is a powerful tool employed by fashion retailers like H&M to enhance their customer base. Several initiatives and projects, such as the Common Thread Initiative, Fabric of Change Initiative, and Textile Waste Diversion, are making strides toward closing the loop and fostering a circular economy. While the challenges are manifold, addressing criteria like sustainable product design, collection systems, and economic considerations can pave the way for a successful transition to a circular economy [20].

In the pursuit of a circular economy, the textile industry faces the challenge of managing its extensive global supply chain, dealing with diverse products, and overcoming technical hurdles in recycling. The shift to high-value textile-to-textile recycling is crucial for minimizing the environmental impact. However, this transition brings its own set of challenges, including the difficulty of recycling cotton fibers compared to polyester, the labor-intensive nature of recycling processes, and the need for optimization of blend ratios for recycled content.

The economics of recycling, in terms of infrastructure, personnel, and capital, pose a significant challenge. As the demand for recycled materials increases, questions arise about the cost-effectiveness of recycled products and the compromise in quality accepted by consumers. The shift to a circular economy requires a shift in mindset, focusing on sustainability, innovation, and resource efficiency. It demands transformative changes in manufacturing processes, where innovation and resource efficiency take center stage [16].

The challenges notwithstanding, the transition to a circular economy presents vast opportunities. Planning, sustainable product design, effective collaboration, and consumer willingness to participate in closed-loop processes are essential components of a successful circular economy. As the world strives for a more sustainable future, the circular economy stands as a beacon of hope, offering a model where economic development and environmental preservation can coexist harmoniously [40].

10 The True Potential of High-Value Textile Recycling—Case Studies

10.1 Closed Loop Blankets from Dutch Workwear

In the USA and UK, approximately 75–85% of used textiles, and in the Netherlands, around 30 Kt/year of uniforms, find their way to landfills for incineration. Reputed agencies and government organizations prefer incineration to protect brand reputation and prevent uniform misuse. However, this waste could be a goldmine for fiber extraction through mechanical or chemical methods [48].

Taking the case of Dutch Defense uniforms, which have an average composition of 50:50 cotton/polyester, BigaGroep sorted them by color and type. These garments were then cut into smaller pieces and sent to recover in Spain. After cleaning to remove non-recyclable parts like buttons and zippers, the garment pieces underwent mechanical fiber extraction using steel wires. The resulting 3/1 Nm recycled yarn, comprising 80% recycled fiber and 20% virgin polyester fiber, was used as weft yarns to create blankets for humanitarian aid.

The final Recover recycled yarn boasted a blend composition of 40:40:20 recycled cotton fiber/recycled PET fiber/virgin PET fiber. Environmental assessments revealed an 87% reduction in water consumption (Virgin 26–34.5 m^2 Recycled), a 42% reduction in energy consumption (Virgin 115–67 GJ Recycled), and a 33% decrease in greenhouse gas emissions (Virgin 4.66–3.13 tons Recycled) compared to 'all virgin polyester' use [30].

10.2 Closed Loop Denim from G-Star RAW Denim

G-Star RAW's collaboration with Wireland Textiles and Recover in 2016 led to a lifecycle assessment study on closed-loop recycled denim. Recycled denim fabric from returned garments commanded a 12.5% premium compared to virgin equivalents. G-Star explored three approaches: incineration, downcycling, and high-value recycling.

In the high-value recycling approach, used denim was transported to Wireland Textiles, sorted, bundled, cleaned by Recover, and spun into yarn with 30% recycled fiber. This yarn constituted 40% of the fabric weight, and the resulting fabric had 12% recycled content. Environmental impacts indicated a 9.8% reduction in water consumption (9.017–81.210 m^3), a 4.2% decrease in energy consumption (455–436 GJ), and a 3.8% drop in carbon emissions (33.8–32.5 tons) compared to virgin denim fabric [48].

10.3 Closed Loop Recover Yarns from Post-consumer Waste

Dutch charity Sympany sorts post-consumer garments into re-wearable and recyclable categories. Recover in Spain then separate non-recyclable parts, like zippers and buttons, through a cleaning process. The remaining textiles are transformed into recycled fiber, carded, spun, and plied into 20/2 Nm yarn for new fabric production [52].

Three colored yarns and one white cream yarn were produced through this high-value recycling process. Each yarn had a unique composition, with the white cream yarn showing significant environmental benefits: a 62% reduction in water consumption (286–108 m^3), a 33% decrease in energy consumption (88.7–59 GJ), and an 18% drop in greenhouse gas emissions (3.84–3.16 tons) due to the avoidance of water-intensive dyeing and the use of post-consumer waste [18].

11 Roadmap for a Circular Economy

The roadmap for a circular economy in textiles is divided into two segments. The first segment focuses on sustainability measures in the production cycle, from raw material to product development, and distribution to disposal. The second segment outlines steps, systems, and strategies for stakeholders to transition towards a circular economy.

Regardless of the suggestions, the implementation of new technologies and research is paramount to making these concepts a reality. Transparency in assessments, evaluations, and reports, accessible without cost, fosters new ideas and facilitates implementation at both field and market levels. These efforts emphasize the imperative need to achieve a circular economy, allowing nature to replenish resources for future generations [36].

12 Impact of the Circular Economy on the Environment

Because of the linearity and complexity of the textile value chain, the manufacturing and consumption of textiles in Europe have a substantial impact on the environment. Material resource depletion, water pollution, land use changes, chemical toxicity, and climate change are all issues that the textile industry exacerbates. Everything from growing natural fibers to spinning and sizing fabrics to finishing, printing, and dying final items is part of the production step. Then there's shipping textiles all over the world, which generates emissions and packaging waste [1].

Washing, drying, and ironing are the most impactful aspects of textile maintenance throughout the use phase. On average, households in the European Union do 6.2 washes every week, which is quite high. In addition to the considerable amount of textiles that are ultimately burned or disposed of in landfills, the transportation and processing of these items also contribute to the effects of their collection and management [50].

Both renewable and non-renewable resources are vital to the textile industry's operations. These include natural fibers, oil, processing and dyeing chemicals, and fertilizers and insecticides used to grow cotton. All of the apparel, footwear, and home textiles bought by EU households in 2017 required almost 675 million metric tons of primary raw materials. The majority of this material use occurred outside of Europe, which suggests that the agricultural and fiber production processes, which are material and energy-demanding, are primarily located outside of Europe in the textile supply chain [50].

More than 90% of the water used in telecom production is consumed in regions other than Europe. Producing cotton in areas where water is scarce, in addition to meeting the needs of other crops, can worsen this problem. On a worldwide scale, one kilogram of cotton uses just over 10,000 L of water [37].

Agricultural land and natural land change are primarily affected by the textile production phase in Europe. This is especially true for natural fibers like cotton, which have a substantial influence on land use. Similar to the food-related and land-intensive product sectors of beverages, restaurants, and hotels, European households consumed textiles in 2017. Despite this, the agricultural portion of the textile supply chain is mostly located outside of Europe, therefore only 7% of the land use for textile products consumed by European families occurs in Europe [17].

One fiber crop, cotton, accounts for over 2.4% of the world's arable land area, or more than 31 million hectares, and has a significant influence on textile land use. By interfering with food crop cultivation, this land use can reduce local food availability, which in turn can cause hunger and contribute to food shortfalls. As an example, in India, where cotton is grown on 8% of the fertile land, it is responsible for 9% of the country's malnutrition [47].

Silk and wool from sheep, cashmere, angora goats, alpaca, etc., are examples of animal-based fibers and fabrics that produce a large amount of land and methane emissions during production, leaving a substantial carbon and land footprint. Reduced need for arable land as a result of recycling and fair trade efforts is encouraging [35].

Producing textiles was the leading cause of greenhouse gas emissions in 2015, with 1.2 billion tons CO_2-eq, surpassing even the total emissions from international flights and shipping. The majority of textiles' climatic impact happens during manufacture, with a further 44% occurring during use and 5% during transportation. In terms of production-phase climate change, acrylic is the most impactful fiber type, followed by nylon and polyester; silk, on the other hand, has the least influence [33].

Approximately 30–50% of the greenhouse gas emissions from the textile supply chain are attributed to this industry, rather than the transportation, food, or housing sectors. More than three-quarters of the emissions from the manufacturing of textiles that end up in European Union homes end up in other parts of the globe. This makes textiles the sixth most polluting category of EU household products when it comes to the effect they have on global warming [51].

The usage of dyes is a significant contributor to the high chemical content of textile production processes in Europe. Approximately three kilograms of chemicals are needed to produce one kilogram of cotton t-shirts. During the finishing steps, numerous chemicals are added to the garment to impart specific properties, such as repelling water and dirt for outdoor wear, stiffening or anti-wrinkle agents, and biocides to prevent mold or bacteria-induced odors during storage and transportation.

The textile industry employs almost 3,500 different compounds, with 750 of those being considered harmful to humans and 440 to the environment. There are 240 of these compounds that pose a significant risk to human health and 120 that pose a threat to the environment. Dyeing and finishing textiles is a major contributor to water pollution, accounting for almost 20% of all pollutants. Workers and communities who rely on polluted water sources for activities like swimming, fishing, and drinking may be at risk from the discharge of dirty wastewater.

Workers' health is negatively impacted by many of the chemical substances used in textile production, and consumers' health is also potentially impacted by some of these chemicals that remain in the final textile goods, whether intentionally or unintentionally. Some textile auxiliaries, including dispersion dyes, finishing chemicals, water repellents, flame retardants, formaldehyde, biocides, and textile scents, can elicit allergic contact dermatitis (ACD) (see Fig. 1).

To show conformity with REACH, make monitoring and enforcement easier, and educate consumers about the materials and chemicals used in textile products, the European Commission is now considering new labeling regulations. It is challenging to maintain track of all the chemicals that have been utilized along the lengthy, complicated, and worldwide textile supply chains. Softeners like phthalates are part

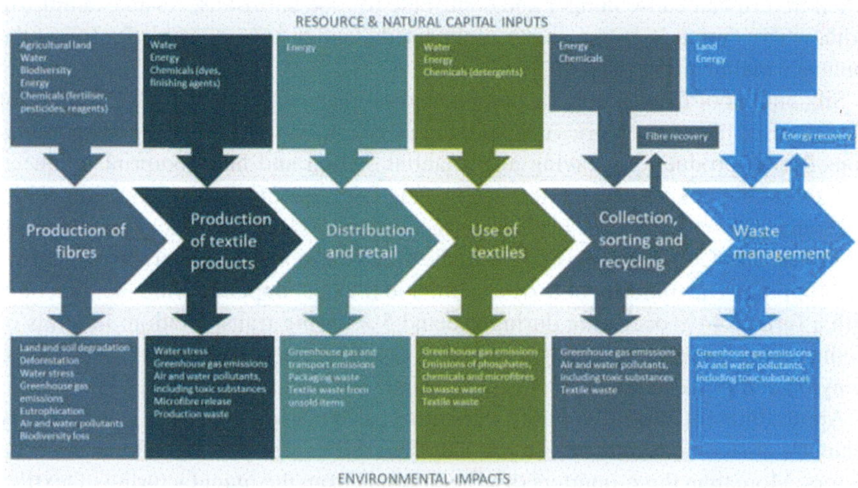

Source: EEA and ETC/WMGE, based on EEA (2014)

Fig. 1 Environmental impacts of textile industries

of the polyvinyl chloride (PVC) plastics used in T-shirt patterns, shoes, and purses. These compounds mimic hormones and can disrupt fertility in humans.

Fluorinated compounds, including the highly persistent and poisonous perfluoroalkyl and polyfluoroalkyl substances (PFAS), have recently received a lot of focus. Regulation 2018/1513 limits the use of chemicals identified as carcinogenic, mutagenic, or hazardous for reproduction (CRMs) in textiles, apparel, and footwear, and an EU plan is being developed to reduce the use of PFAS. A proposal to limit the marketing of textile, leather, hide, and fur products containing skin-sensitizing chemicals has been put up by France and Sweden.

Chemicals released into home wastewater after washing clothes are another source of water and sludge contamination. There appears to be a gradual release of processing chemicals during the use phase, as the quantities of these compounds gradually drop throughout washing, according to the research. Another major contributor to chemical runoff into home wastewater is the usage of detergents.

There is a loss of economic value and environmental difficulties caused by the present take-make-dispose strategy in the fashion business. One garbage truck's worth of textiles is landfilled or burned every second. In addition to using a fifth of the world's carbon budget by 2050, the sector is responsible for the annual discharge of half a million tons of plastic microfibers into the ocean from washing clothes—the equivalent of more than 50 billion plastic bottles. The environmental and social implications are prompting efforts to create a more sustainable and circular system of production and consumption.

As a priority product category with a high potential for circularity, the European Commission has chosen textiles, garments, and fabrics. Sustainable resource usage is the central focus of the New Circular Economy Action Plan put out by the newly inaugurated President of the European Commission, Ursula von der Leyen. This plan would particularly help industries like textiles and construction, which are both resource-intensive and have a significant impact. Separate textile collection and suitable end-of-life treatment are two goals that all Member States are enjoined to achieve by 2025, according to the 2018 Circular Economy Package and WFD.

In a perfect world, everyone would be able to get high-performance textiles that are suitable for a broad range of uses, as well as reasonably priced, stylish apparel that reflects their style. Business, society, and the environment should all reap the benefits of a sustainable, circular textile system that is inherently restorative and regenerative. All along the textile value chain, major structural shifts are required to realize a sustainable and regenerative system.

A textile circular economy cannot be realized without first innovating the business strategy. By reusing items and materials, circular business models seek to reduce the output of waste and the input of primary resources while still creating profitable company solutions. For the sake of clean and safe material cycles, textile value chains should be open and traceable.

Finally, new ways of making clothes, different ways of doing business, more sustainable habits, and enabling legislative actions along the whole value chain are all needed for a textile circular system. Europe can take a step towards a greener, more equitable future by reducing the negative effects of textile production on society and the environment.

To promote sustainable practices and decrease waste, circular business models are becoming increasingly popular in Europe. Innovation policies, financial incentives, legislation, quality standards, trade policies, and voluntary efforts are the five pillars that the European Sustainable Business Federation (Ecopreneur) has highlighted as essential to hasten the shift to a textile and clothing system that is based on reusing and recycling materials. Sustainable business models, increased innovation, and textile production and consumption regulation are all within reach with the help of the European Union's Circular Economy Action Plan.

The development of more sustainable production techniques, the encouragement of more sustainable consumption, and the increase in the reuse and recycling of textiles should accompany regulations of textile production and consumption to achieve a more circular and sustainable system. Nevertheless, these lofty goals are not without their obstacles; for example, some players in the value chain may see a drop in sales and earnings if they reduce their textile consumption through repair and reuse. Consumer prices could rise as a result of the internalization of the costs associated with environmental impacts like emissions and resource use, as well as the adoption of more sustainable manufacturing methods and fair working conditions.

Everyone involved in the textile value chain needs to alter their habits of thought and action if the industry is going to make a concerted effort to shift towards circularity in both production and consumption. In this process, raising awareness and educating people are crucial. Choosing the right materials for yarns and textiles is the first step in creating textile goods. The environmental and climatic implications of various fibers vary since they are made from various resources. There are compounds of concern that are present in both virgin and recycled fiber and fabric production processes.

The production strategies for sustainable fibers, which aim to increase circularity and sustainability, will differ among fiber kinds. Improving agricultural practices to lessen demands on land and water resources should be prioritized when dealing with natural fibers like cotton. Improved water and agrochemical efficiency, the use of less harmful pesticides, and the introduction of soil-conserving farming practices including composting, crop rotation, and reduced tillage would all constitute sustainable cotton production. Although many clothing businesses now offer organic cotton as a way to meet the need for more sustainable clothes, it does not mean that dangerous chemicals are not used in the manufacturing or dyeing processes.

The majority of the negative effects on the environment from synthetic fibers like polyester stem from their energy consumption and the fact that they are made from fossil fuels. Reduced energy and resource use is good for the environment and the economy when made from recycled fibers. One way that technological advancements can help cut expenses and waste is by creating materials that are long-lasting, safe, and recyclable. Another way is by combining used and repaired materials. Those who are interested in purchasing products made of sustainably derived materials, despite the potential price premium, can get trustworthy and transparent information through eco-labeling, which allows consumers to distinguish between products according to their social and environmental implications.

Numerous textile labels, developed by governmental agencies, private businesses, and trade groups, attest to the increasing interest in the circular economy across Europe. Nevertheless, numerous certification schemes and labels employing diverse methodologies have proliferated due to the absence of a universal consensus on sustainability and circularity indicators. The European Union's eco-label, which looks like a flower, is an optional label that covers standards for things like water usage, reduced emissions to water and air, and limited use of harmful substances. It has the potential to significantly improve the environmental effect and circularity of EU-purchased textiles.

When it comes to chemical usage, the Nordic Swan eco-label is notoriously strict. The most recent update addresses the role of textiles in the circular economy by examining re-design, fiber composition, take-back systems, and product lifetime extension. Nevertheless, even in Nordic nations, the proportion of textiles with eco-labels to overall textile sales is still quite small. An independent and trustworthy source needs to control and use simple, standardized data on social, environmental, and climate product performance across brands and products if consumers are going to be able to make conscious choices and increase demand for sustainable fibers.

Refunded virgin payments (RVP) schemes could be a good way to encourage businesses to use recycled fibers, especially since virgin textile fibers are not very expensive right now. For RVP to be successful, it is crucial to establish fair fees and ensure that all participants in the value chain are transparent when reporting the amount of virgin fibers used.

The issue of micro-plastic discharge to water bodies as a result of synthetic fiber textile production, washing, and waste management has recently garnered considerable attention. Possible solutions include more efficient filters in home washing machines and wastewater treatment facilities, smarter textile construction, prewashing, and vacuum exhaustion on production sites. To further support the circular economy in Europe, more study and information of textile micro-plastics is required.

A lot of study is going on in the textile sector right now to find ways to lessen the amount of chemicals used and find safer alternatives. Water and energy can be saved by printing fabrics with pigments instead of dyes. Several eco-friendly alternatives have been created that attach to fiber more easily [11]. There are efforts to find biological dyes that can replace chemical ones, but the industry is struggling to afford these more sustainable options because of the razor-thin profit margins. Because of their gentle working conditions and excellent specificity, enzymes are frequently utilized commercially as chemical substitutes.

An attitude shift among product designers is needed to accomplish the transition to the notion of safe by design. Including environmental awareness in education and training is crucial. To eliminate potentially harmful substances from manufacturing processes and final goods, it is crucial that businesses and government agencies work together to establish quality and safety standards. Legally binding standards for textile products based on measurable criteria surrounding durability, farming methods, labor conditions, recycled content, usage of chemicals, microfiber leaching, durability and recyclability can assist achieve this goal.

Companies can now monitor the movement of materials and finished goods from the beginning of the fiber production process all the way to the consumer's hands thanks to improved traceability in textile supply chains. This assures quality and compliance, mitigates fraud, and engages each player in the product value chain to bear direct accountability for their procedures and activities. Technical solutions, such block-chain technology, and stakeholder participation are necessary for setting up a traceability system.

Design is vital in the development of a circular value chain for textiles. To create a circular value chain, it is vital to choose materials and design products in a way that promotes a long life. In order to make textile items last longer for consumers, eco-design principles including long-lasting style, design for disassembly, and design for durability are crucial.

Sustainable, long-lasting textiles made without harmful chemicals are the goal of the textile industry, and public authorities can play a significant role in guiding this trend. Energy efficiency in energy-consuming devices is the primary emphasis of the EU Eco-design Directive, but material choices and circularity considerations have received less attention thus far. Eco-design standards for furniture and textiles that prioritize reparability and recycling have been proposed by Bauer et al. (2018b). Durable fasteners, readily available spare parts, colorfastness, fabric resilience, promotion of recycled materials to counter virgin manufacture, and reduced water and energy use during maintenance are all objectives of these standards, which aim to enhance the active lifespan of textile products.

13 Circular Economy and Sustainable Development Goals in Textile Industry

A common understanding of "sustainability" is growth "that meets the needs of the present without compromising the ability of future generations to meet their own needs" provided that this growth does not negatively impact the environment. This concept is closely related to the circular economy. Even in the business world, the idea has been on the rise in recent years. An organization must take into account all of the contexts in which it operates if it wants to be genuinely sustainable in the long run, according to the triple bottom line theory put forward by Elkington. Social, ecological, and economic sustainability—sometimes called "people, planet, and profit" are all a part of this [12]. In order to guarantee the company's long-term survival and competitiveness, as well as the delivery of benefits, the economic dimension is concerned with the efficient use of tangible and intangible resources. The ecological dimension, on the other hand, is concerned with the availability, use, and treatment of natural resources. Lastly, the social dimension is concerned with human well-being, society, inter-societal relations, and fairness. Accordingly, sustainability is an all-encompassing concept that needs a firm foundation in corporate strategy and a matching business model as well as connections to company culture, procedures, and operations. Simultaneously, the ideal way to evaluate sustainability is through the organization's value chain, which includes both the upstream and downstream operations of the supply chain and the complete supply network. The United Nations Sustainable Development Goals (SDGs) have been a resource for many businesses as they have crafted sustainability plans and initiatives. The SDGs, officially known as the "2030 Agenda for Sustainable Development", aim to "shift the world on to a sustainable and resilient path" and were put into force in January 2016. Regarding the industrial sector, PriceWaterhouseCoopers noted that "sustainability is moving from the corporate side-lines into the mainstream" due to the advent of the SDGs, and the European Commission stated that "the 2030 Agenda into grates in a balanced manner the three dimensions of sustainable development—economic, social and environmental" (see Table 1).

Table 1 Sustainable development goals (SDGs)

Number	Sustainable development goal
1	End poverty in all its forms everywhere
2	End hunger, achieve food security and improved nutrition and promote sustainable agriculture
3	Ensure healthy lives and promote well-being tor all at all ages
4	Ensure inclusive and equitable quality education and promote lifelong learning opportunities Tor all
5	Achieve gender equality and empower all women and girls
6	Ensure availability and sustainable management of water and sanitation for all
7	Ensure access to affordable, reliable, sustainable and modern energy for all
8	Promote sustained, inclusive and sustainable economic growth, full and productive employment and decent work for all
9	Build resilient infrastructure, promote inclusive and sustainable industrialization and foster innovation
10	Reduce inequality within and among countries
11	Make cities and human settlements inclusive, safe, resilient and sustainable
12	Ensure sustainable consumption and production patterns
13	Take urgent action to combat climate change and its impacts
14	Conserve and sustainably use the oceans, seas and marine resources for sustainable development
15	Protect, restore and promote sustainable use of terrestrial ecosystems, sustainably manage forests, combat desertification, and halt and reverse land degradation and halt biodiversity loss
16	Promote peaceful and inclusive societies for sustainable development, provide access to justice for all and build effective, accountable and inclusive institutions at all levels
17	Strengthen the means of implementation and revitalize the global partnership for sustainable development

More and more, those who use a broad definition of sustainability see the CE as a critical tool to start moving towards a more sustainable future. Because it provides ideas and methods for how businesses might apply sustainability, it is interesting and relevant to academics and practitioners alike. An elementary definition of a circular economy is "one which balances economic development with environmental and resource protection". According to the Ellen McArthur Foundation, a CE is "restorative and regenerative by design, and aims to keep products, components, and materials at their highest utility and value at all times". An "economy where the value of products, materials and resources is maintained in the economy for as long as possible, and the generation of waste minimized" is how the European Commission has described the CE. Design, manufacturing, marketing, consumption, waste management, reusing, and recycling are all part of a product's life cycle that the CE concept covers. Not only will production and product design need to undergo significant adjustments, but so must consumer spending habits, if we are to make a sustainable shift to a CE. Waste management provides a chance to put as much trash as possible back into productive use via waste prevention, material reuse, and recycling. As a result, businesses should spend more time thinking about how to handle their products as they reach the end of their useful life. Building upon the CE principles, also known as R-principles, new and creative business models, products, and processes are needed to achieve this goal. These will generate sustainable value for the organization and its stakeholders. According to Geissdoerfer et al., one of the most important ways for organizations to implement CE is through business model innovation.

Only via collaborative, coordinated efforts inside supply chain networks can a sustainable CE be established in today's economy, defined by the worldwide division of labor and outsourcing. Important success criteria for innovation with a circular focus include solid relationships and effective cooperation. The ability of the supply chain and the network infrastructure are crucial for the success of circular business models, say Geissdoerfer et al., while the comprehensive incorporation of circularity into supply chain management is what really makes them work, say Montag et al. In order to make the transition from a linear to a circular supply chain, they offer a model of supply chain maturity that outlines the adjustments needed at the strategic, tactical, and operational levels. Incorporating CE principles into a company's long-term strategy and all the efforts to put it into action demands, first and foremost, a paradigm shift within the company. The writers argue that modifications are necessary at every stage of the product's life cycle—from production to usage, disposal, and recycling—in order to transform the linear supply chain into a circular one. This is done at the tactical level. The authors show that most of the eight SCOR processes need to be changed to close the product and material loops by referring to the Supply Chain Operations Reference (SCOR) model that has been developed for circular supply chains at the process level. Obviously, the organization and its leadership are the first two tiers. According to the model, in order to put CE principles into practice at the operational level, a company's strategy, business model, and product/service portfolio must be aligned with the CE. Companies make changes to their internal operations and the supply chain as a result of their strategic and tactical decisions. Thus, in order to apply the CE, it is necessary to restructure the procedures involved in the supply chain. Therefore, getting to a corporate level of circularity is all about the supply chain. Joint, coordinated strategies and practices within supply chain networks are essential for implementing a sustainable CE in today's economy, which is marked by a global division of labor and outsourcing. Key success elements for circular-oriented innovation, according to Brown et al., include well-founded partnerships and functioning teamwork. Montag et al., cite the comprehensive incorporation of circularity into supply chain management as the catalyst for circular business models, while Geissdoerfer et al. state that network infrastructure and supply chain capabilities are crucial in enabling such models [54]. In order to make the transition from a linear to a circular supply chain, they offer a model of supply chain maturity that outlines the adjustments needed at the strategic, tactical, and operational levels. Incorporating CE principles into a company's long-term strategy and all the efforts to put it into action demands, first and foremost, a paradigm shift within the company. The writers argue that modifications are necessary at every stage of the product's life cycle—from production to usage, disposal, and recycling—in order to transform the linear supply chain into a circular one. This is done at the tactical level. The authors show that most of the eight SCOR processes need to be changed to close the product and material loops by referring to the Supply Chain Operations Reference (SCOR) model that has been developed for circular supply chains at the process level. Obviously, the organization and its leadership are the first two tiers. According to the model, in order to put CE principles into practice at the operational level, a company's strategy, business model, and product/service portfolio must all be in line with the CE. Companies make changes to their internal operations and the supply chain as a result of their strategic and tactical decisions. Thus, in order to apply the CE, it is necessary to restructure the procedures involved in the supply chain. Therefore, getting to a corporate level of circularity is all about the supply chain. Using the SCOR model modified for CSCs, Montag and Pettau present "a theoretical and conceptual approach for measuring supply chain performance in the circular economy era" within this framework. According to the authors, the model takes into account economic, environmental, social, and cyclical factors when measuring the effectiveness of the supply chain. To achieve "thus reconciling the goals of sustainability and circularity", the framework incorporates not just the three conventional sustainability dimensions—economic, environmental, and social—but also the novel circular viewpoint of performance. Also, their study differentiated between two potentially competing viewpoints: the environmental and the circular. Environmental sustainability seeks to lessen the impact on Earth's ecosystem by decreasing waste and other negative outputs, like CO_2 emissions, while the CE strives to maximize the amount of time and value that products, components, and materials remain in circulation through strategies like recycling and reusing. The framework "enables a clarification for the complex relationship between circularity and sustainability" by providing a horizontally integrated composition of performance measures that analyze the CSC's performance from an economic, environmental, social, and circular perspective. It is based on SCOR processes that have been adapted for circularity (see Fig. 2).

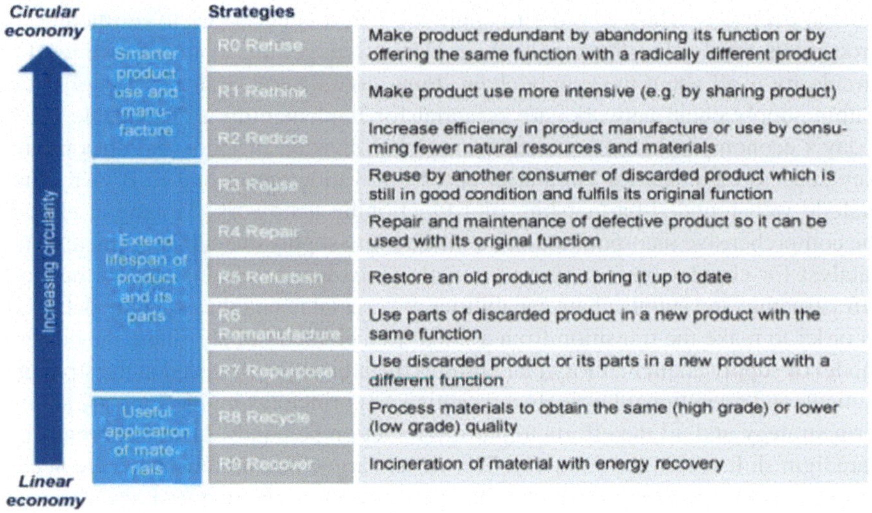

Fig. 2 The 9Rs principles

Numerous ideas, principles, and models pertinent to this research have been discussed in recent literature, and a multi-dimensional approach incorporating different tactics and measurements is necessary for the CE's implementation in the T&C business. To achieve a comprehensive rollout, it is important to incorporate CE principles early on, especially in the product design phase. Some possible steps in this direction include using recyclable or mono-materials, making things easy to disassemble and repair, making the most efficient use of materials, and making things last longer. A new business model can be necessary if the level of change is significant enough. One example is the Product/Service System (PSS) models that Abbate et al. found in the Italian clothing sector. These models involve leasing or renting instead of selling. There is a focus on methods that improve worker health and safety, decrease material consumption, and lessen the production of harmful by-products and waste, such as wastewater and emissions, in relation to the production processes [9]. According to other Italian case studies, in order for supply chain orchestration to work in a circular supply chain, it is necessary to have intense and long-term relationships with suppliers, as well as to digitalize and integrate production processes vertically. In order to keep products and materials in the supply chain loop, it is essential to create and optimize reverse logistics processes. This includes measures like increasing the capacity of collection and sorting facilities, developing and improving textile-to-textile recycling technologies, and investing in the education and awareness of individuals, industry stakeholders, and governments. There has been an uptick in interest in digital technology and circular economy (CE) ideas in recent years, from both the corporate and academic worlds. Several recent articles offer varying viewpoints on the connection between these two ideas. Many of the first digitalization models and frameworks centered on how new technology will affect existing processes and organizational structures, perhaps leading to the creation of whole new types of businesses. Lang recently demonstrated the interconnections among digitalization, the CE, and sustainability by identifying product transformation as one of four critical pillars and sustainability as a driver of digital transformation. Electronic instruments, automated systems, and technological gadgets that enable the processing, transmission, and storage of extremely massive data sets are known as digital technologies. Some people use the acronyms SMAC (Social Media, Mobile, Analytics/Big Data, and Cloud) and BRAID (Blockchain, Robotics, Artificial Intelligence/Knowledge Work Automation, the Internet of Things, and Digital Fabrication) to broadly describe these technologies. Several writers have highlighted the unique importance of these technologies in advancing and bolstering the CE. Internet of Things (IoT) devices, robots, and mobile apps—when combined with analytical tools and optimization software—are emphasized by Frost and Sullivan. In a study conducted by Salminen et al., it was found that the Internet of Things (IoT) allows for better management and analysis of data coming from various sources, which in turn "enhances services provision and the co-evolution of the circular economy". Reuter argued that the IoT can promote CE in the metallurgy industry. Similarly, Bressanelli et al. investigated the role of analytics, big data, and the Internet of Things in facilitating a CE transition. The eight key functionalities that were deemed important during the transition were as follows: enhancing product design, attracting target customers, monitoring and tracking product activity, providing technical support, optimizing product usage, upgrading the product, improving renovation and end-of-life activities, and providing preventive and predictive maintenance. Online sensors can monitor the whereabouts, state, and accessibility of supply chain assets, as pointed out by Owen-Jackson. These communications can be kept secure by directly exchanging information through decentralized, secure methods, such as blockchain. It is possible to optimize resources, lengthen lifecycles, and aid in the regeneration of natural resources through the combined efforts of these advances. Electronic media "helps support dematerialization by reducing our reliance on physical resources e.g., retailers selling e-books directly to consumers or digital video and audio being delivered online rather than through physical media," he said, adding that digital technology has made this possible. Pagoropoulos et al. pointed out that this necessitates a shift in business models towards a PSS model for manufacturing enterprises. "Meeting the same client demands with less environmental impact" means moving away from selling only things and instead selling the utility, which is achieved through a combination of products and services. The authors discovered that digital technologies frequently support the shift to a PSS business model, which has many social, environmental, and economic advantages. Using interview and document data from four innovative Northern-European enterprises, Ranta et al. performed a multiple case study of CE business strategies made possible by digital technologies. "Four important categories of digitally-enabled business model innovation for CE that differ in the degree to which they enhance resource flows, value creation, and capture" according to the authors. Incorporating visibility and intelligence into products, according to Antikainen et al., digitalization can be considered as a facilitator of the CE. This means that assets can be better informed about their location, condition, and availability. In order to examine these ideas across supply chains, several scholars have gone outside the context of specific companies. To illustrate the point, Del Giudice et al. examined how CE practices affected the performance of enterprises in a circular supply chain. After conducting an online survey with 378 Italian company managers who had already implemented CE principles, the authors discovered that in order to design and promote CE initiatives, businesses need to create value-added relationships by utilizing big data. This will encourage management and employees to work together more effectively. Displays the proposed framework by Bressanelli et al., which "shows the linkages between digital technologies, circular strategies and practices, and sustainability performance". This framework takes a different approach. According to the authors, digitalization allows for a comprehensive redesign of products, business models, and value chains [3]. This redesign impacts every phase of a product's life cycle, leading to reduced energy and material consumption, product reuse, component remanufacturing, and material recycling. As a result, "the achievement of enhanced sustainability performance in terms of environmental, economic, and social benefits" is improved.

According to other writers, the use of digital technologies to bolster the CE is quite restricted. Using data from 600 manufacturing companies, Neligan investigated digitalization's potential to facilitate a sustainable CE in Germany. He discovered that, although "the opportunities offered by digital networking for increasing material efficiency are only used to a limited extent", traditional efficiency-raising measures that optimize manufacturing processes are still widely used in the manufacturing sector. Interviews with top IT executives in eight European organizations by Wynn and Jones revealed that while all of the organizations were actively working toward a CE, digital technology implementation was not directly tied to any of these efforts. All of these organizations did, however, make use of digital technologies; doing so had broader advantages in terms of efficiency improvements and cost savings, which aided sustainability goals in general, rather just for this particular target or in support of the CE. Both digitalization and sustainability have been treated independently up until now, as pointed out by Cagno et al. While evaluating the connection between digital technologies and the CE, they used McKinsey's ReSOLVE methodology. Their observation about the absence of a comprehensive study of the connection between the two ideas prompted them to call attention to the importance of conducting empirical investigations into decision-making procedures and particular CE practices. A digital performance viewpoint, showing how digital technology affects CSC performance, presents "future research opportunities," as pointed out by Mon tag and Pettau.

Few studies have examined the application of digital technologies within the framework of the CE in the T&C business. "The mainstream recycling technologies and structural solutions that could eliminate barriers to the introduction of a global closed loop in the textiles industry are still few", according to Koszewska, who studied industry examples in order to identify and evaluate challenges faced by the industry in adapting to the CE model. She did find digitalization as one of the "new trends (that) are emerging towards the circular economy". The digital product passport will greatly help achieve the objectives of the EU Strategy for circular textiles, according to

Centobelli et al. Nevertheless, this can only be achieved by sharing data at each point in the supply chain and by maintaining transparent traceability from raw materials to recycling. Transparency regarding material composition and process traceability is one of the fundamental conditions for a circular and sustainable textile supply chain, according to Saccani et al., who conducted a multiple case study. After reviewing a number of options, Alves et al. concluded that the best way to adopt CE in the T&C business and achieve the required transparency along the supply chain would be to use a mix of blockchain and Internet of Things (IoT) technology. Through a series of case studies, Bürklin and Wynants uncovered several potential applications of digital technologies in the T&C sector. For example, smart mirrors could increase consumer happiness by guiding them more accurately through the buying process, while 3D printing, virtual prototyping, and 3D body scanning could all contribute to better fit and enable demand-driven production. In their discussion of "the potential for the adoption of circular economy principles enabled by digitalization, including innovative resale and sharing models", Pal and Jayarathne offer some suggestions for the use of various digital technologies in the T&C sector, such as 3D modeling, AI, and Big Data. "A number of barriers that currently prevent widespread digitalization of the CSCs, which are related to lack of available technology, economic viability, strategic development and digital knowhow" is what they conclude, nevertheless. Therefore, there is a lack of research on how digital technologies promote the CE across supply chains, specifically in the T&C industry but also more generally. A few studies have touched on digitalization in relation to the CE, but this is by no means exhaustive.

14 Conclusion

The circular economy in textiles demands a transformative lifecycle, with design as its initial phase—emphasizing durability, long life, and reuse or recycling. Achieving a circular economy necessitates significant collaboration across the supply chain, with education, research, finance, and regulation all geared towards a mindset focused on material input and output. Transformation requires altering consumer, public, industry, and authority preferences, with concepts like product sharing, renting, or buying as a service gaining traction. As seen in the rental of designer handbags and traditional attire in India, these shifts increase utilization while maintaining economic viability [59]. The textile sector has a substantial environmental impact as a result of its linear and intricate nature. It encompasses multiple facets, including the cultivation of natural fibers, the process of spinning and sizing fabrics, the fine-tuning, printing, and dyeing of the finished products. The sector necessitates the utilization of both renewable and non-renewable resources, with a majority of material consumption taking place in regions outside of Europe. The textile manufacturing process in Europe has a significant impact on agricultural land and the alteration of natural land, especially in relation to natural fibers such as cotton. The textile business accounts for around 30–50% of greenhouse gas emissions originating from the supply chain, positioning textiles as the sixth most environmentally detrimental category among home products in the European Union. The priority product category with high potential for circularity has been identified by the European Commission as textiles, garments, and fabrics. The establishment of a textile circular economy necessitates the implementation of inventive corporate strategies, environmental practices, and legislative measures throughout the entire value chain. The implementation of the circular economy (CE) provides enterprises with a vital mechanism to transition towards a more sustainable trajectory. The objective is to achieve a harmonious equilibrium between economic progress and the preservation of the environment and resources, with the ultimate goal of maximizing the utility and worth of products, components, and materials consistently. The notion of Circular Economy (CE) encompasses several stages of a product's life cycle, including design, production, marketing, consumption, waste management, reusing, and recycling. In order to achieve a sustainable transition towards a circular economy (CE), substantial modifications are required throughout all phases of the product's life cycle, encompassing manufacture, utilization, disposal, and recycling. In order to adopt circular economy principles, enterprises should prioritize waste prevention, material reuse, and recycling. The efficacy of circular business models heavily relies on the capabilities of the supply chain and network infrastructure. The integration of circular economy (CE) ideas into a company's long-term strategy and implementation necessitates a fundamental transformation inside the organization. Montag and Pettau propose a conceptual framework that aims to assess the performance of supply chains within the context of the circular economy. This framework encompasses various elements of sustainability, including economic, environmental, and social aspects. Both environmental sustainability and circular economy (CE) principles are distinguished from each other. The successful implementation of circular economy (CE) ideas within the textile and garment sector necessitates a comprehensive strategy including various dimensions. These dimensions encompass the early integration of CE principles, product design, and the vertical integration of production processes. Digital technologies, including Internet of Things (IoT) devices, robots, and mobile applications, have the potential to support a transition towards circular economy (CE), thereby improving product design, consumer engagement, and service offering. Nevertheless, digital technologies possess restricted capacity to facilitate clean energy in the T&C industry. Future study should prioritize the utilization of digital technologies in the T&C industry, namely in the domain of CE. This includes investigating the implementation of smart mirrors, 3D printing, and AI.

References

1. Abad-Segura E, Fuente ABDl, González-Zamar M-D, Belmonte-Ureña LJ (2020) Effects of circular economy policies on the environment and sustainable growth: worldwide research. Sustainability 12(14):5792

2. Andrews MG (1987) The men and the mills: a history of the southern textile industry. Mercer University Press
3. Arshad M, Adil M, Sikandar A, Hussain T (2014) Exploitation of meat industry by-products for biodiesel production: Pakistan's perspective. Pak J Life Soc Sci 12:120–125
4. Arshad M (ed) (24 Nov 2017) Perspectives on water usage for biofuels production: aquatic contamination and climate change. Springer
5. Arshad M (ed) (13 Nov 2020) Sustainable ethanol and climate change: sustainability assessment for ethanol distilleries. Springer Nature
6. Arshad M (ed) (28 Apr 2023) Climate changes mitigation and sustainable bioenergy harvest through animal waste: sustainable environmental implications of animal waste. Springer Nature
7. Arshad M, Bano I, Younus M, Khan A, Rahman A (2018) Health concerns associated with biofuel production. Perspectives on water usage for biofuels production: aquatic contamination and climate change, pp 97–105
8. Arshad M, Hussain T, Chaudhry N, Sadia H, Aslam B, Tahir U, Abbas M, Qureshi N, Nazir A, Rajoka MI, Iqba M (1 Jan 2019) Enhancing profitability of ethanol fermentation through gamma ray mutagenesis of saccharomyces cerevisiae. Pol J Environ Stud 28(1)
9. Arshad M, Ansari AR, Qadir R, Tahir MH, Nadeem A, Mehmood T, Alhumade H, Khan N (2022) Green electricity generation from biogas of cattle manure: an assessment of potential and feasibility in Pakistan. Front Energy Res 25(10):911485
10. Bairoch P, Kozul-Wright R (1998) Globalization myths: some historical reflections on integration, industrialization and growth in the world economy. In: Transnational corporations and the global economy. Springer, pp 37–68
11. Bano I, Ghauri MA, Arshad M, Yasin T, Younus M (2017) Bioactivity of variant molecular weight chitosan against drug-resistant bacteria isolated from human wounds. Microb Drug Resist 23(8):958–965
12. Bano I, Arshad M (2018) Climatic changes impact on water availability. Perspect Water Usage Biofuels Prod: Aquat Contam Clim Chang 2018:39–54
13. Benyus JM (1997) Biomimicry: Innovation inspired by nature. Morrow New York
14. Boons F, McMeekin A (2019) An introduction: mapping the field (s) of sustainable innovation. In: Handbook of sustainable innovation. Edward Elgar Publishing, pp 1–25
15. Broggi CB (2016) Trade and technology networks in the Chinese textile industry. Springer
16. Charnley F, Cherrington R, Mueller F, Jain A, Nelson C, Wendland S, Ventosa S (2024) Retaining product value in post-consumer textiles: how to scale a closed-loop system. Resour Conserv Recycl 205:107542
17. Chen B, Han M, Peng K, Zhou S, Shao L, Wu X, Wei W, Liu S, Li Z, Li J (2018) Global land-water nexus: agricultural land and freshwater use embodied in worldwide supply chains. Sci Total Environ 613:931–943
18. Chioatto E, Sospiro P (2023) Transition from waste management to circular economy: the European Union roadmap. Environ Dev Sustain 25(1):249–276
19. CICC Research, CICC Global Institute xinran. liu@ cicc. com. cn (2024) Textiles and apparel—expanding production bases overseas while moving up the industry chain. In: The reshaping of china's industry chains. Springer, pp 331–350
20. Coggio W, Hevus I, Webster C, Schrock A, Thompson B, Ulrich K, Dzadek N (2021) Advancing adhesives: bio-based succinic acid polyester polyols. Adhes Sealants Ind Mag (June 1, 2015):36e42
21. Commission E (2011) Communication from the commission to the European parliament, the council, the European economic and social committee and the committee of the regions youth opportunities initiative. European Commission Brussels, Belgium
22. Dantwala ML, Vakil CN (1937) Marketing of raw cotton in India
23. De Angelis R, Howard M, Miemczyk J (2018) Supply chain management and the circular economy: towards the circular supply chain. Prod Plan Control 29(6):425–437
24. Eggermont M (2022) A bio-inspired design and space challenges cornerstone project. In: Biomimicry for aerospace. Elsevier, pp 41–62
25. Federico G, Tena Junguito A (2023) How many people on earth? World population 1800–1938

26. Fish FE, Weber PW, Murray MM, Howle LE (2011). The tubercles on humpback whales' flippers: application of bio-inspired technology. Oxford University Press
27. Gordon D (1988) The global economy: new edifice or crumbling foundations? New Left Rev 168(1):24–65
28. Haeggblom J, Budde I (2021) Circular design as a key driver for sustainability in fashion and textiles. Sustain Text Fash Value Chain: Driv Concepts Theor Solut, 35–45
29. Hamilton T (2008) Whale-inspired wind turbines. MIT Technol Rev
30. Hasanbeigi A, Price L (2015) A technical review of emerging technologies for energy and water efficiency and pollution reduction in the textile industry. J Clean Prod 95:30–44
31. IEA I (2015) Carbon capture and storage, the solution for deep emissions reductions. International Energy Agency Publications Paris
32. Jawahir I, Dillon O, Rouch K, Joshi KJ, Venkatachalam A, Jaafar IH (2006) Total life-cycle considerations in product design for sustainability: a framework for comprehensive evaluation. In: Proceedings of the 10th international research/expert conference, Barcelona, Spain
33. Kirchain R, Olivetti E, Miller TR, Greene S (2015) Sustainable apparel materials. Materials Systems Laboratory, Massachusetts Institute of Technology, Cambridge, pp 1–34
34. Kumar S, Smith SR, Fowler G, Velis C, Kumar SJ, Arya S, Rena, Kumar R, Cheeseman C (2017) Challenges and opportunities associated with waste management in India. R Soc Open Sci 4(3):160764
35. Liikanen, J. (2020). Production of hydroxy carboxylic acids from waste cellulosic material.
36. de Sousa L, Jabbour AB, Jabbour CJC, Godinho Filho M, Roubaud D (2018) Industry 4.0 and the circular economy: a proposed research agenda and original roadmap for sustainable operations. Ann Oper Res 270:273–286
37. Lukman RK, Glavič P, Carpenter A, Virtič P (2016) Sustainable consumption and production–Research, experience, and development–the Europe we want. J Clean Prod 138:139–147
38. MacArthur E (2013) Towards the circular economy. J Ind Ecol 2(1):23–44
39. MacArthur E (2017) A new textiles economy: redesigning fashion's future. http://www.ellenmacarthurfoundation.org/publications>. Acesso em 25(08), 2021
40. Maldini I, Duncker L, Bregman L, Piltz G, Duscha L, Cunningham G, Vooges M, Grevinga T, Tap R, van Balgooi F (2017) Measuring the Dutch clothing mountain: data for sustainability-oriented studies and actions in the apparel sector
41. Manahan SE (2006) Environmental science and technology: a sustainable approach to green science and technology. CRC Press
42. Manavalan E, Jayakrishna K (2019) An analysis on sustainable supply chain for circular economy. Procedia Manuf 33:477–484
43. Miret C (2022) How does the circular economy interact with urban food system transformation? In: Social and cultural aspects of the circular economy. Routledge, pp 64–85
44. Mohajan H (2021) Third industrial revolution brings global development
45. Oguntona OA, Aigbavboa CO (2019) Biomimicry approaches for innovative sustainable solutions in the construction industry. In: Innovative production and construction: transforming construction through emerging technologies. World Scientific, pp 335–352
46. Organization WH (2023) Trends in maternal mortality 2000 to 2020: estimates by WHO, UNICEF, UNFPA, World Bank Group and UNDESA/Population Division: executive summary
47. Pan J, Chu C, Zhao X, Cui Y, Voituriez T (2008) Global cotton and textile product chains. International Institute for Sustainable Development. Manitoba, Canada: Suisse
48. Radhakrishnan S (2021) Circular economy in textiles and fashion. Circular economy: assessment and case studies, pp 163–202
49. Rhoades C (2014). There is more to closed-loop textile recycling than technological innovation. The Guardian Sustainable Business Forum. The Guardian Online, London
50. Romano P, Vinelli A (2001) Quality management in a supply chain perspective: strategic and operative choices in a textile-apparel network. Int J Oper Prod Manag 21(4):446–460
51. Roy Choudhury A (2014) Environmental impacts of the textile industry and its assessment through life cycle assessment. Roadmap to sustainable textiles and clothing: environmental and social aspects of textiles and clothing supply chain, 1–39

52. Schnitzer H, Braunegg S (2021) 20th European roundtable on sustainable consumption and production
53. Semplonius NN (2021) Reuse, repair, remanufacture and recycle, rethinking the circular economic model. A case study on business park Haskerveen, Heerenveen (The Netherlands)
54. Sharif N, Munir N, Hasnain M, Naz S, Arshad M (2021) Environmental impacts of ethanol production system. Sustain Ethanol Clim Chang: Sustain Assess Ethanol Distill 2021:205–223
55. Stahel WR, Stahel WR (2010) Selling performance. Perform Econ, 86–178
56. Szokan N (2016) The fashion industry tries to take responsibility for its pollution. The Washington Post
57. Tokareva O, Jacobsen M, Buehler M, Wong J, Kaplan DL (2014) Structure–function–property–design interplay in biopolymers: spider silk. Acta Biomater 10(4):1612–1626
58. Turner A (2018) Black plastics: linear and circular economies, hazardous additives and marine pollution. Environ Int 117:308–318
59. Wu XT, Ansari AR, Pang XX, Li HZ, Zhang ZW, Luo Y, Arshad M, Song H (2019) Visfatin plays a significant role in alleviating lipopolysaccharide-induced apoptosis and autophagy through PI3K/AKT signaling pathway during acute lung injury in mice. Arch Immunol Ther Exp 1(67):249–261
60. Zhang Y, Sun M, Yang R, Li X, Zhang L, Li M (2021) Decoupling water environment pressures from economic growth in the Yangtze River Economic Belt, China. Ecol Indic 122:107314
61. Zhou Q, Huang Z, Liu J, Zhao Y, Jung JC-Y, Zhang J, Xu S (2021) A closed-loop regeneration of $LiNi_{0.6}Co_{0.2}Mn_{0.2}O_2$ and graphite from spent batteries via efficient lithium supplementation and structural remodelling. Sustain Energy Fuels 5(19):4981–4991

Effluent Treatment in Textile Industry to Achieve SDGs

Zahra Abid, Zirwa Sarwar, Neelma Munir, Sher Zaman Safi, and Muhammad Arshad

Abstract Textile effluents are notorious for their intense coloration, high biological and chemical oxygen demands, and the presence of toxic compounds and pose severe health risks to communities living near industrial areas. Consumption of large amounts of water in the textile sector generates huge ratio of wastes including life-threatening chemicals like heavy metals and dyes organic compounds. The treatment of wastewater from textile sectors and various linked industries is particularly branched into three prime stages that include primary, secondary, and tertiary treatment. The textile effluent industry is continuously evolving as a response to growing environmental concerns, regulatory pressures, and technological advancements. Several future trends and advancements are likely to shape the landscape of textile effluent treatment and management such as circular economy practices. Various strategies are utilized to reduce pollution when dealing with dye in textile effluent, e.g, filtration, adsorption, and biodegradation utilizing specific microbes are examples of these approaches. It consists of a succession of stages that use different methods at each stage to remove contaminants and pollutants. Textile mills can use effective treatment strategies to lessen their environmental impact and potentially reuse treated water by following prescribed standards and taking into account the individual properties of the effluent. The industry will increasingly adopt eco-friendly dyes, chemicals, and materials that generate fewer pollutants in effluents, aligning with the principles of sustainable chemistry. Technological advancements, regulatory pressures, and changing consumer preferences will drive the industry toward more responsible and efficient management of effluents, contributing to a cleaner and more sustainable environment.

Z. Abid · Z. Sarwar · N. Munir (✉)
Department of Biotechnology, Lahore College for Women University, Lahore, Pakistan
e-mail: neelma.munir@yahoo.com

S. Z. Safi
Faculty of Medicine. Bioscience and Nursing, MAHSA University, Selangor, Malaysia

M. Arshad
Biochemistry Section, Jhang-Campus, University of Veterinary and Animal Science, Lahore, Pakistan

M. Arshad (ed.), *Enzymes in Textile Processing: A Climate Changes Mitigation Approach*, SDGs and Textiles, https://doi.org/10.1007/978-981-97-8058-7_14

Graphical Abstract

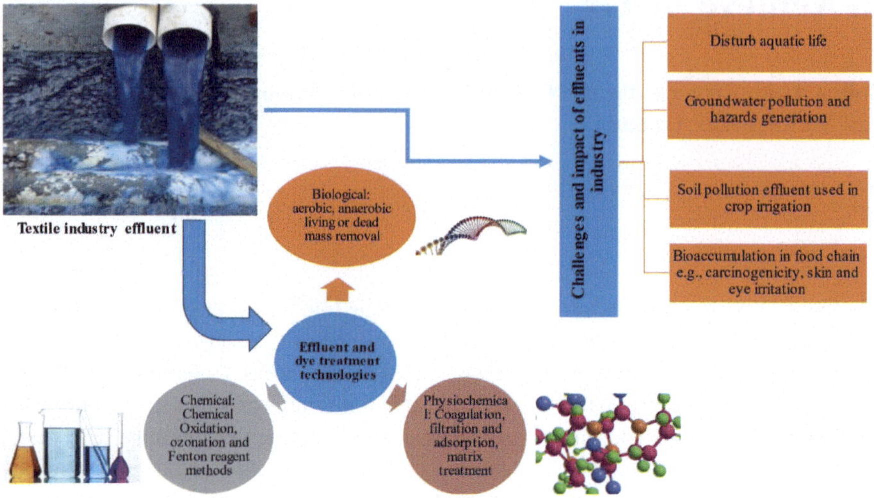

Keywords Industrial effluents · Treatment · Heavy metals · Wastewater · Sustainable

1 Introduction to Effluent Treatment in the Textile Industry

Effluent treatment in the textile industry has become an indispensable process to mitigate the environmental impact caused by the rapid growth of this sector. Textiles, stemming from the Latin word "texere," meaning to weave, have a rich history. From their origins as a basic necessity for clothing, the textile industry has evolved into a global force that allows individuals to express their fashion statements and cultural identity [1].

In the face of climate change, the textile industry's exponential growth raises serious concerns about its environmental footprint, particularly regarding the generation of effluents [1]. Effluent treatment processes are vital not only for mitigating pollution but also for aligning with Sustainable Development Goals (SDGs), especially those related to clean water and sanitation (SDG 6), responsible consumption and production of SDGs and climate action [6, 8].

Effluent treatment plays a critical role in achieving SDGs by reducing water pollution from textile manufacturing processes, thereby safeguarding aquatic ecosystems and ensuring access to clean water (SDG 6). Moreover, by implementing efficient treatment methods, the textile industry can minimize resource consumption, promote recycling, and adopt sustainable practices, contributing to responsible consumption and production SDGs [5, 7].

Furthermore, effluent treatment in the textile industry directly addresses climate change concerns SDG by mitigating greenhouse gas emissions associated with pollution. By adopting eco-friendly treatment technologies and reducing energy consumption during effluent treatment processes, the industry can significantly reduce its carbon footprint and contribute to global efforts to combat climate change.

Effluent treatment in the textile industry is not only crucial for mitigating environmental impacts but also for advancing progress towards Sustainable Development Goals, particularly those related to clean water and sanitation, responsible consumption and production, and climate action [13, 14].

There should be a tool to deal with the pollution caused by textile industry as it is growing faster day by day [49]. As Clothing is a basic need of mankind, this industry has grown globally and it has provided a chance to the people to make innovations according to their fashion sense and cultural back ground. The industry of textile is considered to be top most necessity as well as important industry as it accounts for 14 percent of total input of country. Apart from the benefits, this excessive growth has led to some serious environmental issues—one of them is release of effluents [1].

Textile effluents can be defined as the wastes that are produced during different processes in the industry like pretreatment, dyeing, printing, and finishing. As in all the above-mentioned processes large amount of water and energy is used and it generates a lot of wastes. These wastes include life-threatening chemicals like heavy metals and dyes organic compounds [12]. High color density, biological and chemical oxygen needs, as well as the presence of harmful compounds, make these effluents very dangerous if they are not treated. The effect of these effluents includes contamination of rivers lakes, etc, disturbance of aquatic life and harmful impact on the organisms who live near the industry [23].

In order to save the environment from harms of these effluents, "effluent treatment" is regarded as one of the integral process in any textile industry. Effluent treatment includes removing toxic chemicals or converting them into the less harmful form with the help of several methods and technology so that surroundings can be saved. Considering the diversity and toxicity of the pollutants, industry must wisely use correct physical and chemical techniques for gaining good results [54].

Treatment of wastes is a challenging task that evolves by time to achieve the goal of protection of environment from so many heavy metals dyes and chemicals that are excreted by industry. Researchers and industries are working passionately to find out better methodology to deal the effluents [17]. Some very reliable treatment processes include oxidation processes (AOPs), membrane filtration, biological treatment phytoremediation, adsorption, etc. In addition, industries are trying to produce eco-friendly and biodegradable dyes for better results. Effluent treatments protects the globe and also makes the industry more stable and easy to work in [56]. People are also supporting greener textiles that are made from very environment friendly textile industries [23].

To combat the environmental consequences of textile effluents, effluent treatment has emerged as an essential industrial procedure. In order to get rid of pollutants, latest technologies and high throughput processes aimed at removing or degrading pollutants from the wastewater, ensuring that it meets regulatory standards and poses minimal harm to the environment. Given the diverse range of pollutants present in textile effluents, an integrated approach utilizing physical, chemical, biological, and emerging technologies has become imperative for achieving effective and sustainable results [54].

The textile industry's effluent treatment is a complex and evolving task, driven by heavy metals, and various organic compounds that includes synthetic dyes. A lot of products are made with heavy metals like chromium (Cr), lead (Pb), zinc (Zn), arsenic (As), iron (Fe), sulfur (S), iron (Fe), mercury (Hg), copper (Cu), magnesium (Mg) and cadmium (Cd) colorants used in textile industry. It is to be known that the amount of heavy metals are being used more than the normal usage that lead to great rise in the textile industry of heavy metals due to which heavy metals have undoubtedly become important, resulting in increased metal flux in aquatic environments. Removal of these metallic materials is now necessary to maintain the ecosystem's balance. The activation of carbon has proved to be one of the very applicable and advance form of technique to inactive harmful and toxic heavy metals. But the expense of the heavy metal removal has become a major challenge due to activated carbon [55]. There are numerous studies to remove heavy metal ions utilizing various materials from wastewater, especially those that are dangerous to humans, and different adsorbents have been used. Researchers and industries are actively exploring eco-friendly alternatives and innovative techniques to enhance treatment efficiency. Membrane filtration methods, technology advanced in oxidation process and various biological techniques, adsorption, and phytoremediation are among the promising approaches being explored. Additionally, efforts are underway to develop and adopt biodegradable and eco-friendly dyes to minimize the environmental impact of dyeing processes [66, 72, 74].

Effluent treatment not only serves to protect the environment but is also crucial for the industry's long-term sustainability [34]. Governments, regulatory bodies, and industry stakeholders play a pivotal role in promoting and enforcing effluent treatment practices. Support for research and innovation in greener textile manufacturing processes is also essential to foster a more environmentally responsible textile industry [68].

By embracing eco-friendly practices and responsible wastewater management, the textile industry can continue to flourish while minimizing its ecological footprint [32]. Effluent treatment ensures that the industry's growth and economic prosperity do not come at the expense of environmental degradation and public health risks [41]. A concerted effort towards sustainable practices and the adoption of cleaner technologies will pave the way for a more vibrant and ecologically conscious textile industry, contributing to a cleaner and healthier world for generations to come [54].

The textile industry is a crucial sector that has been an important key that lead to growth of economy as well as employment worldwide [41]. However, the rapid expansion of this industry has come with significant environmental consequences, particularly concerning the generation of effluents [50]. Textile effluents refer to the wastewater produced during various manufacturing processes, including dyeing, printing, and finishing of fabrics [63]. These effluents are laden with a plethora of chemicals, dyes, and other pollutants, making them highly detrimental to the environment and human health [9]. Textile effluents are notorious for their intense coloration, high biological and chemical oxygen demands, and the presence of toxic compounds. If left untreated, these effluents can contaminate water bodies, disrupt aquatic ecosystems, and pose severe health risks to communities living near industrial areas [38]. To address these environmental challenges, effluent treatment in the textile industry has become a paramount concern [25]. Effluent treatment involves the implementation of various technologies and processes aimed at removing or degrading pollutants from wastewater, ensuring that the discharged water meets regulatory standards and poses minimal harm to the environment [25]. The treatment of textile effluents is a complex task due to the diverse range of pollutants present, including synthetic dyes, heavy metals, and organic compounds [39]. Therefore, an integrated approach utilizing physical, chemical, biological, and emerging technologies is necessary to achieve effective and sustainable results [1].

Consumption of large amounts of water in the textile sector generates huge ratio of effluent having every kind of contaminants and pollutants coming out from dyes and clothing industry [39]. Organic substances offer biochemical oxygen demand BOD as well as chemical oxygen demand COD. Organics substances that are dissolved, halogenated organic compounds, surfactants, variety of dyes and heavy metals, sulphur and sulphates compounds, effluents that are suspended or you can say the particles and other contaminants may be present in discharged wastewaters depending on the textile finishing techniques [63]. If the nutritional deficiencies (nitrogen and phosphorus) are supplemented, biological substances can be eliminated from wastewater. Textile wastewater treatment requirements vary due to the amount of dyes or textile effluents being suspended and advanced and latest techniques used in textile manufacturers [63]. The effluent quality is decided for each application due to the wide variety of finishing procedures and industries. Wastewater containing discharged effluent are often manufactured in (WWTPs) local effluent or textile treatments plants managed by textile enterprises, or in rare circumstances in municipal WWTPs. There are certain processing steps that are involved in the treatment in such a way the effluent is taken step by step and then released into the water that is receiving it and this has gained exponential success in the recent years. The most basic wastewater treatment procedures for textile production and finishing include mechanical sewage pre-treatment, storing and homogenizing, the flotation process, pH neutralization, precipitation of chemicals, biological (aerobic and anoxic treatment), along with process automation [47].

One of the industries with the largest percentage of development, the clothing and textiles sector significantly contributes to the country's economy. This sector of the economy also consumes huge amount of water which, as a result, consists of a lot of suspended particles, effluents and wastewater with a lot of toxins. A significant cause of water pollution is the leakage of colours into the environment during the dyeing and finishing processes of textile fibres. All of the necessary steps for treating wastewater are explained in Fig. 1. Physical, biological, or chemical processes used for individual wastewater treatment are typically highly expensive and result in large

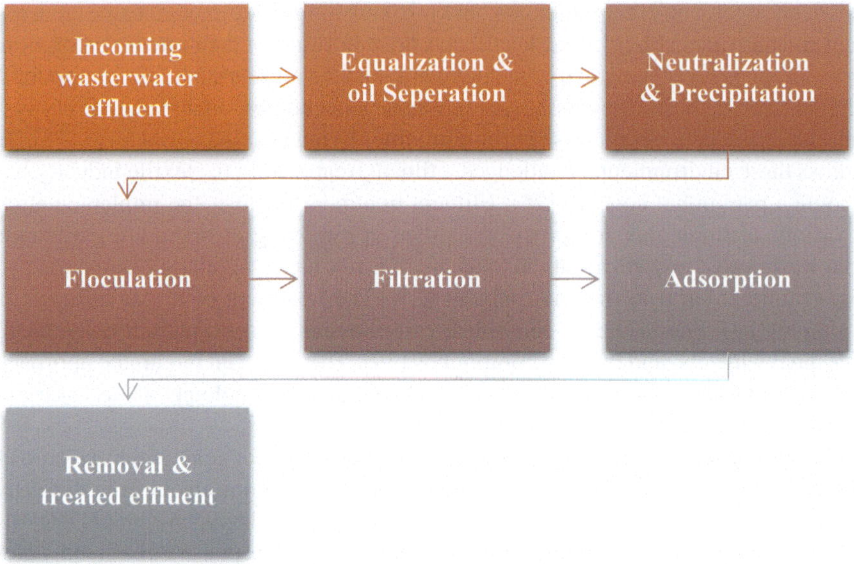

Fig. 1 Process of effluent treatment

amounts of sludge. Alternative treatment methods that cover the whole wastewater treatment cycle are therefore necessary [54].

2 Challenges and Impacts of Textile Effluents

Textile mills emit a significant volume of effluent, which contains harmful and hazardous compounds that harm the environment. Aquatic macrophytes and algae are poisoned by industrial wastewater [11]. Hazardous wastes can quickly spread throughout the environment, contaminating new areas and causing further health risks when they are released into the air, water, or land. Along with the issue of colour fading during the dyeing process, the textile sector also generates enormous amounts of effluent [11]. These effluents are complex combinations of numerous contaminants, from connected pesticides and heavy metals to the original colours lost during the dyeing process, and if not properly controlled, these pollutants can seriously harm water supplies [39]. Textile wastewaters from various phases of textile processing include massive levels of contaminants that are extremely damaging to all the surroundings and its environment if discharged when not being treated properly [11].

The challenges posed by textile effluents are multi-dimensional and have far-reaching consequences on both the environment and human welfare such as textile effluents contain a vast array of printing and dyes that are synthetic and used in these industries [76]. These dyes are often recalcitrant, meaning them are resistant to degradation, leading to the discharge of brightly colored effluents into water bodies. This impairs light penetration, hindering photosynthetic activity in aquatic plants and disrupting the ecological balance [76].

In toxic compounds, a sophisticated chemical dance occurs during the manufacture of fabrics that beautify our lives in the brilliant and colorful realm of the textile industry. A symphony of chemicals brings fabrics to life, from the mesmerizing colors of dyes to the sparkling brilliance of metallic finishes. However, as this beautiful show spreads over the world, a darker undertone throws a shadow on the atmosphere. Heavy metals, once gleaming alchemical marvels, today play a malevolent function. Cadmium, with its enticing luster, and mercury, once dubbed "liquid silver of dreams," join forces with lead and chromium to form a poisonous foursome that endangers aquatic life [73]. These metals, which are imprisoned in textiles through various procedures, escape into wastewater and weave their way into bodies of water, where their enchantment gives rise to ecological disaster [35]. They accumulate, persistent and uncompromising, disrupting the equilibrium of aquatic ecosystems and jeopardizing aquatic creatures' survival. Though sometimes neglected, the textile industry's supporting cast of auxiliary chemicals plays an important role in this story. Surfactants, solvents, and emulsifiers entwine with the fabric, leaving traces of their presence as they help with dyeing, printing, and finishing processes. While they may appear to be insignificant on their own, their combined impact is significant. As they enter the waterways, these chemicals work together to contaminate natural ecosystems, stoking the flames of environmental degradation [35].

The economy with global concern include textile industry as an important part for growth and strong pillar in the industry, creating jobs and contributing to the manufacturing of diverse fabrics and materials. The environmental impact of textile effluents, on the other hand, cannot be neglected [16]. High levels of biochemical oxygen demand (BOD) and chemical oxygen demand (COD) in these effluents indicate that they contain significant amounts of organic materials [16]. They pose a serious hazard to aquatic ecosystems when discharged into areas of water when not being treated effectively and properly [16]. Microorganisms begin the process of organic matter decomposition as textile effluents are discharged into bodies of water [76]. This causes a dramatic drop in the dissolved oxygen levels in the water. Aquatic creatures such as fish, insects, and plants rely substantially on dissolved oxygen for survival and growth [26]. Reduced oxygen availability generates a suffocating environment in which marine life struggles to breathe, reproduce and thrive [76].

There are some dyes that are synthetic; they are not affected by any kind of biological degradation. As a result, those dyes stay in the air for long time, continuing to exert their harmful effects [39]. The results are frequently disastrous, with large fish deaths and the formation of dead zones where marine life cannot live [39]. To solve this serious issue, the textile industry must implement sustainable practices and invest in effective wastewater treatment technology that can decrease the amount of BOD and COD as well as other contaminants in their effluents. Strict environmental rules and the promotion of proper wastewater treatment will aid in safeguarding our

Fig. 2 Standards of biological oxygen demand

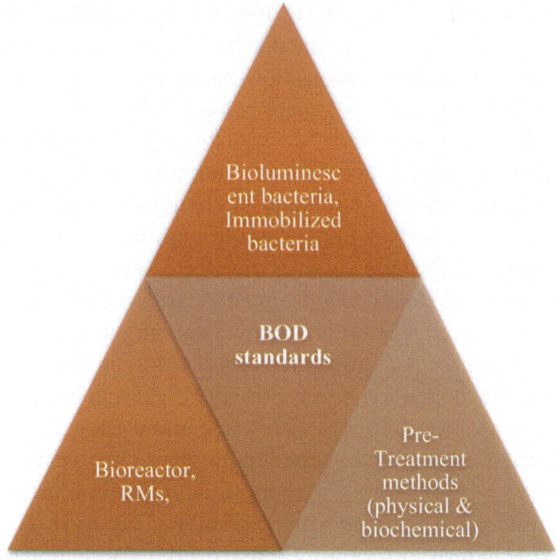

water supplies and maintaining the delicate balance of aquatic ecosystems. Figure 2 highlights the main standards of biological oxygen demand. The textile sector can help to make the environment cleaner and healthier for present and future generations by acting ethically [51].

The indiscriminate discharge of untreated textile effluents pollutes water bodies and harms the livelihoods of communities dependent on these water sources for drinking, irrigation, and fishing [26]. The ecological and socio-economic impacts can be severe and long-lasting. To address these challenges and mitigate the impacts of textile effluents, effective and sustainable effluent treatment technologies are imperative [51].

3 Effluent Treatment Technologies

The treatment of wastewater from textile sectors and various linked industries is particularly branched into three prime stages that include: primary, secondary, and tertiary treatment (Fig. 3). Large particles and trash are removed from the water during initial treatment. Microorganisms then break down organic matter in the secondary treatment to decrease pollution. Finally, modern procedures are utilized in tertiary treatment to further purify the water and remove any leftover impurities. Various strategies are utilized to reduce pollution when dealing with dye in textile effluent. Filtration, adsorption, and biodegradation utilizing specific microbes are examples of these approaches [36]. A BOD5/COD5 test is performed prior to treatment to determine whether the wastewater is biodegradable. The reuse of treated water is also

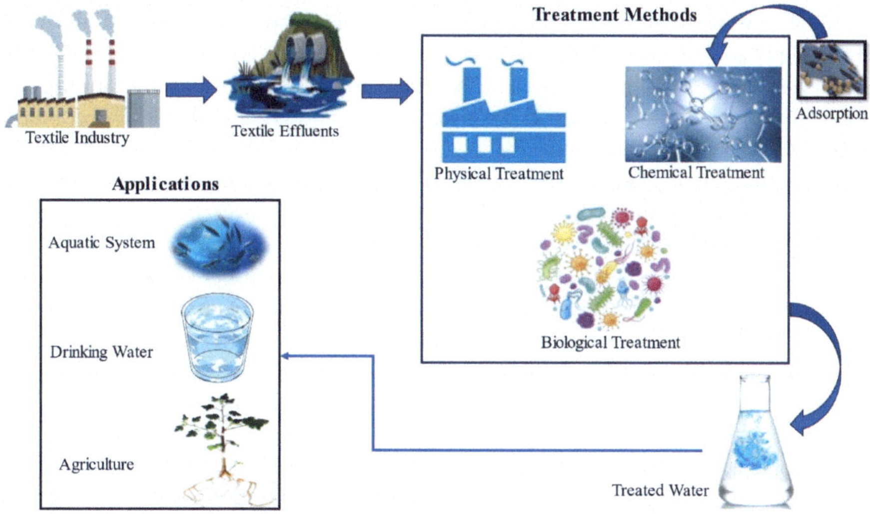

Fig. 3 Treatment methods required to remove pollutants from the textile industrial effluents

evaluated, and specific parameters such as COD, color, pH, turbidity must be met to assure its safety. Figure 2 shows the treatment methodologies required to remove pollutants from the industrial effluents. The wastewater treatment procedure tries to make it cleaner and safer for the environment. It consists of a succession of stages that use different methods at each stage to remove contaminants and pollutants. Textile mills can use effective treatment strategies to lessen their environmental impact and potentially reuse treated water by following prescribed standards and taking into account the individual properties of the effluent [36].

A network of water treatment facilities known as effluent treatment combine and treat all wastewater gathered from water-using operations. Before being released, process industry effluent must be cleansed. The variables that affect variations in removal efficiency include concentrations of effluent contaminants, coagulant dose, mixing and flocculation stirring speed, durations, and settling time. A range of methods and techniques are used in the textile industry for effluent treatment to remove or degrade pollutants from wastewater [11].

3.1 Preliminary Treatment

Eliminating pollutants that are floatable and settleable, such as floating particles, biological waste, too much grease and oil, and grit (both physical and chemical), is the first step of the treatment of textile waste. During the initial treatment, textile effluent is screened using bars and fine screens for coarse suspended particles such yarn, lint, fabric fragments, fibres, and rags. Untreated wastewater undergoes preliminary treatment to remove granules and suspended particles, especially oils and lubricant solutions [11].

Rags, sections of fabric, particles, threaded processes, and dust are among the entangled things that are removed during the screening process. Prior to secondary treatment, suspended fibres must be removed; otherwise, the secondary treatment would be impacted, leading to clogged trickling filters or carbon beads. One of the main benefits of microtiter-based screening methods over more conventional screening methods using Petri plates or test tubes is the capacity to swiftly, effectively, and economically screen a large number of strains under a variety of conditions [11].

Textile wastewater needs to be pre-treated in order to ensure an ongoing supply of bodily fluids throughout the equalisation and the neutralisation process. The majority of the time, a fixed flexible container is used to control the wastewater. To prevent the cotton fur and slurry from sinking to the bottom of the container, air is used to mix them together. Before moving on to the next stage of treatment, sulfuric acid or hydrochloric acid is used to neutralise the alkalinity of the wastewater [11].

However, with the assistance of charged particles certain artificial chemical and toxic compounds are mixed to the fluid and various liquid material In order to provoke the manufacturing of organic or inorganic compounds that can perform coagulation and eventually coagulate in water bodies. The coagulants are inorganic or organic cationic coagulants because colloids frequently have negative charges. Coagulation brought on by organic polymers results in the mixing of sediments that can be extracted. AI and FeC13 are two widely used substances. Due to their propensity to generate in solution multi-charged polynuclear complexes with improved adsorption capabilities, aluminum and iron coagulants are useful [11].

3.2 *Physical Treatment Methods*

Typically utilized in sanitation as a foot-treatment or as a predecessor to biological procedures, the physical and chemical operations attempt to minimize synthetic dyes pollution or erase substances that may disrupt the biological process [31]. Any approach that uses facilities to separate or mix components through physical processes not having any addition of chemicals or toxins for example decantation process and filtration process to filter effluent can be used to remove pollutants [16]. Physical treatment techniques entail physically removing contaminants from wastewater without significantly altering its chemical makeup [36].

Adsorption is one of the key processes currently being investigated. It is effective for small volumes of wastewater, but it is a slow and costly approach [31]. Charcoal is one of the most adsorbent and actively adsorbent material, which can be derived from a range of sources for instance woody material, coconut, peat and shells of walnut, anthracite, bituminous coal. In this process, a solid substance interacts with a moving medium, causing the transfer of mass of the fluid or gaseous state to the outermost layer of the material [15]. This is a result of the unbalanced forces of attraction on the solid's surface [31]. According to research, the adsorption efficiency is influenced by the pH level, humidity, polarisation of the subject matter, pressure, and surface area. Adsorption using activated carbon and diverse substances like peat moss fly ashes and lignite combined, chipped wood, ceramics, and wheat husk; filtration through membranes, ion exchange, radiation treatment, and coagulation are physical methods for removing pollutants from wastewater. Physical or biological (chemisorption) adsorption depends on the forces that hold the molecules of both the adsorbent and adsorbate together. As a result, data must be collected in order to characterize the adsorption performance of distinct adsorbent types [31]. The variables already mentioned in the context can be investigated as parameters that can influence the process [36].

Membranes are one of the most well-known and widely utilized filter media on the market. This is due to their ability to separate chemicals that conventional filters cannot [37]. This separation process can clarify, concentrate, and, most importantly, constantly separate the dye from the wastewater, as well as treat huge amounts of wastewater, but the costs are high and maintenance is challenging because cleaning the filter is difficult elements [3]. Their method is based on successfully isolating color molecules of suitable size from wastewater. "Membranes are physical barriers capable of controlling the passage or retention of different substances, depending on the treatability problem to be solved," a study found. Membranes may separate particles as well as fractionate molecules of different weights due to their variable pore sizes [37]. Numerous inorganic and organic substances, including polymer compounds, porcelain, carbide-based silicon, and steel made of stainless steel, can be used to create membranes [36]. There are certain aspects that are included in membrane filtration.

Reverse Osmosis (RO) is considered to be highly effective purifying water treatment method that separates dissolved salts, minerals, and pollutants from water using microscopic pores with semi-permeable membrane [2]. The RO membrane allows only water molecules to flow through while successfully preventing bigger ions and particles by using the reverse osmosis concept. Water is driven through the membrane under pressure, moving from a more concentrated solution to a lower concentration and leaving contaminants in a reject stream. This procedure produces filtered water with much lower amounts of pollutants [2]. Reverse osmosis is the process that is very much linked with various other industries of textiles, including saltwater desalination, ultrapure water production for industrial uses, and domestic water purification systems [36].

Ultrafiltration membranes are essential instruments in concentration, dialysis, and purification operations, particularly when dealing with macromolecules such as proteins [28]. Their adaptability and efficiency have made them indispensable in a wide range of industries. That is one of the most widely used treatment method for disposing and then reuse. Water of exceedingly high quality, often exceeding drinking water regulations, is now achievable with ultrafiltration membranes [28]. These specialized membranes successfully remove suspended particles, bacteria, viruses, and bigger organic components from treated water, ensuring that it passes high quality standards. Ultrafiltration improves water conservation efforts and supports sustainable practices in industrial processes by separating and concentrating important chemicals while allowing smaller molecules such as water to pass through. It not only helps with efficient resource management, but it also helps with environmental protection by promoting the recycling and reuse of water resources, lessening the demand on freshwater supplies [36].

Nanofiltration: This membrane is narrower than ultrafiltration but thicker than reverse osmosis. The pore diameter is around 0.001 m. The separation mechanism includes solubility and diffusion in addition to filtration. The passage of solvent through the membrane is influenced by osmotic pressure [36]. Below is the Table 1 showing mostly used physical treatment methods for the removal of pollutants.

Table 1 Physical treatment methods to remove pollutants

Primary treatment method	Description	Chemicals used	References
Screening	Removal of coarse suspended particles such as yarns, lint, cotton, fabrics, fibers, and rags	N/A	[44]
Sedimentation	Settling of fine floating and suspend able matter (e.g, clay or silts) in tanks due to gravity force	N/A	[44]
Skimming	Removal of scum (e.g, greasy and oily material) suspended during sedimentation process	N/A	[44]
Equalization	Eliminating shock loads by controlling it and maintain the flow velocity of effluent	N/A	[44]
Neutralization	Adjustment of pH to a uniform range (typically 5 to 9) with the help of reducing surplus acidic and alkaline composition	Sulfuric acid, hydrochloric acid, caustic soda, ammonia, etc	[44]
Chemical Coagulation	Addition of chemicals to coagulate and remove colloidal particles from the effluent	Alum, poly aluminum chloride, ferric chloride, etc	[44]
Mechanical Flocculation	Physical process enabling small particles to agglomerate and settle through mixing with paddles	N/A	[44]
Chemically Enhanced Primary Treatment (CEPT)	Use of chitosan coagulants and alum sludge to lower the cost of sludge disposal	Alum sludge, chitosan coagulants	[44]

3.3 Chemical Treatment Methods

Chemical treatment methods involve the use of chemicals to alter the chemical composition of pollutants, making them easier to remove or degrade. Common chemical treatment methods include:

In coagulation and flocculation, the process involves adding chemicals known as coagulants and flocculants to the wastewater. Coagulants neutralize the electric charge of particles, while flocculants promote the clumping together of particles, forming larger flocks that settle more rapidly.

In oxidation, various oxidizing agents, such as ozonation, Fenton's reagent and hydrogen peroxide are used to break down organic pollutants and dyes. Oxidation helps convert complex and recalcitrant compounds into simpler, more biodegradable substances. Table 2 shows techniques of chemically treated matter for such efficient and productive treatment methods that can lead to sustainable system.

Certain chemicals are added to the wastewater to induce the formation of insoluble precipitates with pollutants, facilitating their separation from the water. To improve the elimination of particular pollutants, the acid level of the effluent can be adjusted. For instance, raising pH can enhance precipitation's ability to get rid of contaminants such as heavy metals [36].

3.4 Biological Treatment Methods

Biological treatment is a popular and successful approach for treating various types of wastewater and organic contaminants [62]. It uses naturally occurring microbes to break down and eliminate toxins from wastewater, converting it into harmless byproducts non-toxic matter or particulate like fluid, CO2 and biological substance. Aerobic, anaerobic, and anoxic processes are the three basic types of biological treatment. Aerobic treatment uses oxygen to encourage the growth of aerobic microbes that devour organic matter, whereas anaerobic treatment uses particular anaerobic microbes to degrade and decrease rate of toxicity in organic components in absence of oxygen. Anoxic treatment is generating a low-oxygen environment to encourage the decrease of nitrates and other nitrogen molecules. Depending on the method, each has its own set of advantages and applications [11].

Activated sludge is a popular biological treatment process that includes mixing wastewater with a microbial culture (activated sludge) in aeration tanks [11]. The activated sludge bacteria actively devour the organic contaminants present in the wastewater as their food supply, resulting in the breakdown of complex organic molecules into simpler, less hazardous chemicals [62]. Following the treatment, the biological treatment methods have evolved as a highly effective and widely used approach in the wastewater treatment sector for breaking down organic pollutants. These methods rely on the exceptional capacity of microorganisms, using fungal and bacterial organisms to break down and remove pollutants from sewerage [62]. The

Table 2 Name of algal species involved in effluent treatment

Treatment method	Description	Examples of reactors	References
Aerobic Treatment	Breakdown of organic content in wastewater using microorganisms (bacteria, algae, fungi) in the presence of oxygen	Rotating Biological Contactor (RBC): Uses rotating discs for high interfacial area and bonding among organic ones and the microbes	[42]
		Membrane Bioreactor and filtration combines effluent or sludge for efficient treatment	[42]
		Large volumes of wastewater can be handled using oxygenated fluid bed biological reactors with a shorter stay duration	[42]
		Circulating Fluidized Bed Bioreactor (CFBBR): Operates with two fluidization regimes, offering high sludge concentration	[42]
Anaerobic Treatment	Anaerobic microorganisms' degradation of organic pollutants in the absence of oxygen	Up-flow Anaerobic Sludge Blanket (UASB) Reactor: High organic load handling capacity, low sludge formation rate	[42]
		Anaerobic Baffled Reactor (ABR) Divides processes into compartments using baffles, efficient dyeing effluent treatment	[27]
		Sulfidogenic ABR: Employs sulfidogenic conditions for effective COD and dye removal in dyeing effluents	[27]

treatment method is appropriate for organic chemical-containing wastewater, such as that found in the textile industry. While biological treatment approaches have various advantages, including economic effectiveness and environmental friendliness, they do have certain downsides. One prevalent concern is the production of odours by microbes during their metabolic processes. Furthermore, if trickling filters are not properly maintained, the filter material might get severely blocked, reducing treatment efficacy. Furthermore, the effluent from biological treatment may require additional treatment in some circumstances to fulfil severe regulatory or discharge standards. Despite these disadvantages, the advantages of biological therapy options surpass the negatives in most cases. They offer an effective and long-term solution for treating textile dyeing wastewater and other organic-rich industrial effluents. Biological treatment technologies contribute greatly to pollution reduction and waste minimization by utilizing the intrinsic capabilities of microorganisms. [11]. Two most basic techniques are applied—one is aerobic treatment and the other one is anaerobic treatment [23].

Aerobic treatments have a considerable impact on the textile industry because they provide a long-term and effective option for treating wastewater generated during various textile processes. This can be well understood by the fact that chemicals that are used in dyeing, printing, and finishing activities and also the organic components present in various dyes, pigmented stuff of cloths, fabrics that are discharged from textile industry. If these pollutants are released into bodies of water untreated, they can cause major environmental problems. Organic pollutants, on the other hand, can be easily broken down into innocuous byproducts like as carbon dioxide, water, and biomass through usage of aerobic treatment techniques for example activated sludge systems, peat, moss or aerobic bioreactors. As a result, aerobic treatments efficiently lower the BOD and COD of textile wastewater, ensuring regulatory compliance and conserving the environment ensuring environmental standards are followed and the surrounding ecosystems are protected. Furthermore, by implementing aerobic treatment methods, textile manufacturers may demonstrate their dedication to sustainable practices by reducing their environmental impact and supporting a cleaner and greener approach to textile production [23].

Secondly, anaerobic treatments are critical in eliminating effluent from the textile industry, solving the issues provided by high-strength organic wastewater created during various textile processes. In anaerobic treatment, bond is formed between the organic material and the microbes such as yeast and bacteria that will eventually break the toxicity and acidity found in textile sludge in non-oxygenated environment. These bacteria transform organic materials into biogas, which mostly consists of methane and carbon dioxide, via a series of metabolic processes. The biogas produced during anaerobic digestion is a useful source of renewable energy that may be used to power a variety of industrial activities, reducing dependency on traditional fossil fuels. This not only aids the textile industry's sustainability, but it also helps to reduce greenhouse gas emissions. Furthermore, the residual anaerobic sludge or biomass after treatment includes significant nutrients and can be used as a soil conditioner or fertilizer, effectively finishing the resource recovery loop. Textile businesses may successfully treat their wastewater, reduce their environmental impact, and encourage a more circular and environmentally responsible approach to wastewater management by adding anaerobic treatments. [23].

Because of its multiple advantages, biological treatment technologies are crucial in the field of wastewater management. Notably, these approaches are not only effective but also cost-effective, as they require less infrastructure and energy than more elaborate physical or chemical treatments. Furthermore, their ability to produce less sludge provides a considerable environmental benefit, reducing waste quantities and lowering disposal costs. Microorganisms including yeast, fungi, bacteria, green algae, and digestive enzymes are used in agriculture to digest and eliminate contaminants from wastewater ads to the sustainability of biological treatments. These bacteria have an intrinsic ability to degrade organic materials, making the effluent less hazardous to the environment. However, it is critical to recognize that biological treatment

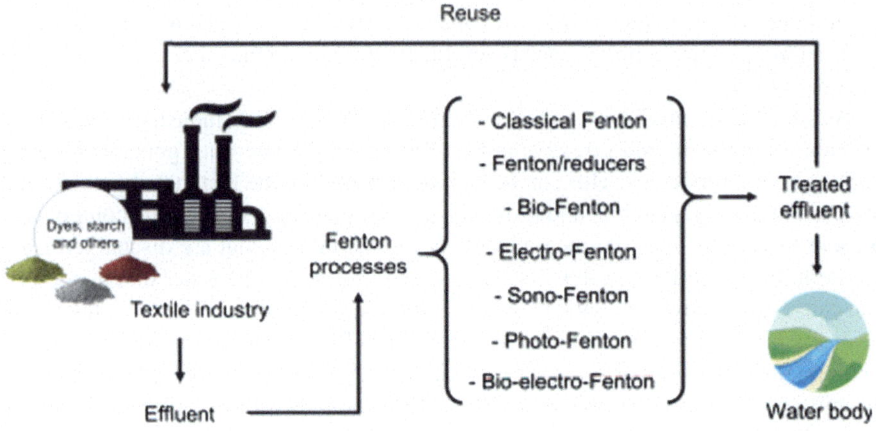

Fig. 4 Biological methods for effluent treatment

technologies may be limited in their ability to remove certain types of pollutants or handle fluctuations in wastewater composition. Continuous research and development are focused on new and innovative technologies to optimize their efficiency and solve these problems. Figure 4 illustrates the biological method of effluent treatment. Biological treatment methods will continue to play an important part in developing a sustainable and responsible approach to wastewater treatment and conserving the health of our ecosystems as the textile industry and other sectors strive to embrace more eco-conscious practices [23].

A. Aerobic Processes Involved:

Active sludge is used in the silt process—one of the biological remediation techniques that the clothing sector uses the most frequently. To decompose organic materials and remove pollutants from wastewater, aerobic microorganisms are implemented. In an aeration tank, wastewater is mixed with a culture of activated sludge including bacteria and other microorganisms. As a food supply, the bacteria devour the organic molecules in the wastewater, resulting in the breakdown of dyes and other contaminants. Before being discharged, the treated effluent is separated from the activated sludge in a settling tank [23].

SBR is a biological treatment technique that treats textile wastewater in batches. It entails filling a reactor with textile effluent and allowing microorganisms to breakdown contaminants throughout an aerobic and anoxic phase sequence. Aeration, settling, and decanting are all alternated during the process. SBR systems are adaptable, simple to run, and capable of achieving significant levels of pollution removal. [23]

Modern wastewater treatment systems called membrane bioreactors (MBRs) combine biological treatment with highly technical membrane filtration. Bacteria similarly to conventional biological treatments decompose and digest biological material and eliminate pollutants from wastewater. MBR, on the other hand, distinguishes itself by incorporating a membrane filtration system that functions as a physical barrier, separating the treated effluent from the active biomass. The tiny pores of the membrane selectively block suspended particles and bacteria, allowing only high-quality treated water through. This outstanding purification results in effluent that exceeds severe water quality regulations, making it suitable for a wide range of reuse applications. MBR has several advantages, including compact design, greater process control, lower sludge output, and resilience to influent changes.

Trickling filters are used in the methods for biological removal of pollutants so this can be used in order to have effective effluent treatment methods that may have an adverse effect. Biological treatment technologies have evolved as a highly effective and commonly utilized approach for breaking down organic contaminants in the wastewater treatment industry. These approaches rely on microorganisms' extraordinary ability to breakdown and eliminate pollutants from wastewater, such as bacteria and fungi. The treatment procedure is ideal for wastewater containing organic chemicals, such as those found in industries [23].

B. Anaerobic Procceses

In the absence of oxygen, anaerobic treatment is used to break down molecules of organic matter in textile effluent. Anaerobic microorganisms break down complex organic matter, producing biogas (methane and carbon dioxide) as a by-product. Prior to exposing the wastewater to additional aerobic treatment, this procedure can be employed as a pretreatment step to remove the majority of organic debris [23]

Algae-based treatment involves using seaweed to remove pollutants and fertilisers from textile wastewater. For their growth, algae use nutrients like nitrogen and phosphorus, which are frequently present in excess in textile wastewater. They absorb contaminants during the process and transform them into biomass. The algae species used in wastewater treatment are listed in Table 3. Algal treatment can be combined with other biological treatment methods to remove pollutants more effectively [23].

3.5 Enzyme Treatment

In the realm of textiles, where vibrant colors and intricate designs rule, enzymes step in as secret agents of change. Picture them as tiny superheroes that break down tough dyes in wastewater.

Dyes can be stubborn, but enzymes have the power to dismantle them into smaller, harmless pieces. They work like a tag team with other helpful microbes to make sure all the dye bits are dealt with.

Enzymes are like puzzle pieces that fit perfectly with certain dyes, making them super effective. They do their job under gentle conditions, unlike harsh chemicals. Plus, they are kind to the environment.

After the enzyme action, the treated water goes through a final cleanup to ensure its safe. This teamwork of enzymes and other treatments makes the water cleaner

Table 3 Name of algal species involved in effluent treatment

Algae genus	Effluent removal capabilities	References
Chlorella sp.	Nitrogen and Phosphorus Removal	Palmer (1974), Erganshev and Tajiev (1986)
Ankistrodesmus sp.	Nitrogen and Phosphorus Removal	Palmer (1974)
Scenedesmus sp.	Nitrogen and Phosphorus Removal	Palmer (1974), Erganshev and Tajiev (1986)
Euglena sp.	Organic Matter, Nitrogen, and Phosphorus	Palmer (1974), Erganshev and Tajiev (1986)
Chlamydomonas	Organic Matter, Nitrogen, and Phosphorus	Palmer (1974), Erganshev and Tajiev (1986)
Oscillatoria sp.	Organic Matter, Nitrogen, and Phosphorus	Palmer (1969), Erganshev and Tajiev (1986)
Micractinium sp.	Organic Matter, Nitrogen, and Phosphorus	Palmer (1969)
Golenkinia sp.	Organic Matter, Nitrogen, and Phosphorus	Palmer (1969)

and ready for reuse. Enzymes might be small, but they are mighty allies in the textile world. They take part in a special dance that transforms wastewater, leaving cleaner water behind as a thank-you bow [23]. Table 4 contains the enzymes that are responsible for effluent removal effectively.

Table 4 Name of enzymes involved in removal of contaminants

Enzyme	Substrate	Function	References
Amylases	Starches, Carbohydrates	Break down starches and carbohydrates	Palmer (1969)
Cellulases	Cellulose	Degrade cellulose, found in cellulosic fibers	
Proteases	Proteins	Break down proteins from dyeing and finishing	
Lipases	Fats and Oils	Degrade fats and oils in wastewater	
Pectinases	Pectin	Break down pectin from fruit and vegetable dyes	
Laccases	Lignin	Involved in the degradation of lignin	

4 Emerging and Innovative Technologies

Environmental concerns plays crucial role in choosing the dyes, clothing material, fragments of textile used material that are utilized in textile industry throughout the world. According to the World Bank, the World Bank estimates that 20% of the world's water pollution is caused by the textile industry, indicating that these operations produce a significant volume of wastewater. The concentrations of suspended solids (SS), phosphorus, pigments, sodium chloride, organ-pesticides, biodegradable organic waste, and metals that are found in these effluents are often high. The textile sector is looking for sustainable wastewater treatment techniques that will assist them reduce their water footprint and operational costs due to rising water shortages and environmental restrictions. Sustainable wastewater treatment may be the best choice for the textile industry given the current challenges. Therefore, it is crucial to talk about and promote awareness-raising strategies that help to lessen the current difficulties with textile wastewater. As a result, this chapter will discuss a variety of environmentally friendly wastewater treatment techniques, including membrane biological reactors (MBR), electrocoagulation (EC), ultrasonic treatment, an advanced oxidation process (AOP), ozonation, and granular activated carbon (GAC) (Periyasamy et al, 2018). Emerging and innovative technologies in wastewater treatment are continually being developed to address the challenges posed by industrial wastewater, including that from the textile industry. These technologies aim to improve treatment efficiency, reduce environmental impacts, and explore more sustainable approaches [70].

Some of the notable emerging and innovative technologies include:

MBRs separate treated water from biomass by combining biological treatment with membrane filtration. When compared to traditional biological treatment procedures, this results in higher effluent quality, less sludge generation, and a smaller footprint.

AOPs use strong oxidants like ozone, hydrogen peroxide, or UV light to break down complex chemical molecules that are resistant to conventional treatment procedures. AOPs are efficient in treating stubborn pollutants such as dyes and other harmful textile industry chemicals [19].

To clean wastewater, constructed wetlands imitate natural wetland ecosystems. These systems use physical, biological, and chemical methods to remove contaminants from aquatic plants, soil, and microorganisms. They are environmentally beneficial, save energy, and give extra benefits such as habitat construction and biodiversity support [19].

Nanotechnology is being researched in order to provide new materials and procedures for more efficient pollutant removal. Nanomaterials can be utilized as adsorbents or catalysts, improving pollutant degradation and overall treatment efficiency [70].

BES integrates biological treatment with electrochemical processes to promote microbial degradation and energy generation. These systems hold promise for sustainable wastewater treatment and resource recovery [19].

FO is a membrane-based method that uses the osmotic pressure gradient to draw water through a semi-permeable membrane while leaving impurities behind. It has potential applications in wastewater concentration and resource recovery [19]. Hybrid treatment systems combine multiple treatment methods that include chemical procedures, physical procedures and biological processes to achieve higher efficiency in pollutant removal. Biological nanomaterial's naturally occurring substances 596 produced by microorganisms definitely will eliminate the pigmented material, colors, 597 pigments from the sector of textile [19].

Innovative and latest technologies are playing role in effluent treatment that can lead to more sustainable and environmentally friendly practices. Ongoing research and implementation efforts are crucial to unlocking the full potential of these technologies in addressing the challenges of wastewater treatment and ensuring the protection of water resources and the environment. As the need for more efficient and sustainable effluent treatment grows, researchers and engineers have been exploring emerging and innovative technologies to address the challenges posed by textile effluents.

Effluent treatment technologies must be carefully selected and integrated into the textile industry's manufacturing processes to ensure optimal performance and environmental compliance. Additionally, a comprehensive monitoring and compliance framework is necessary to track the efficiency of the treatment systems and maintain [1].

From fibre cultivation to product life cycle analysis, the textile industry is undergoing a significant change towards sustainability. The "Shift to Sustainability," a movement, places a strong emphasis on environmentally friendly procedures throughout. The production methods, raw material farming, and product disposal are all being modernized for social and environmental responsibility. High standards are ensured by ISO certifications throughout a product's life cycle. Considering how much used clothing there is in landfills, textile waste is still a problem. Used clothing is gathered by charities like Goodwill, which then gives it to people in need. This method embodies sustainability in action by converting one person's garbage into another's priceless resource [54]

Growing consumer awareness has been a major factor in the development of greener products, such as recycled clothing made from textile waste. Environmental friendliness has also become a focus of advertising and marketing. Recycled products are not rejected by consumers. According to a new study, buying recycled textile products that are safe for the skin and the environment is a good idea. The study also shows that men are more likely than women to experience this [54].

5 Integration of Sustainable Development Goals

The United Nations adopted the 17 interrelated Sustainable Development Goals (SDGs), commonly referred to as the Global Goals, in 2015 to address numerous global concerns and promote sustainable development by 2030 [46]. Integrating the SDGs involves incorporating these goals into policies, strategies, and actions at various levels from the local to global, to achieve a more equitable, prosperous, and environmentally responsible future [33]. Goals requires a comprehensive and coordinated approach involving governments, businesses, communities, and individuals. It is about transforming our world to ensure a more just, equitable, and sustainable future for all [20].

SDGs referring to Sustainable Development Goals in effluent treatment within the textile industry is crucial for achieving environmentally responsible and socially sustainable practices [21, 22]. The textile industry is known for its significant water consumption and pollution, making it important to align its activities with the SDGs to minimize negative impact [52]. The following SDGs can be integrated into effluent treatment in the textile industry:

5.1 Goal 6: Clean Water and Sanitation

The Sustainable Development Goals (SDGs) of the United Nations are designed to ensure that everyone has access to clean water and sanitation and that these resources are managed sustainably. A fundamental human right that is necessary for maintaining environmental sustainability, public health, and general well-being is access to clean water and proper sanitation [10]. The following are the key purposes of Goal 6.

- Implement advanced effluent treatment technologies to ensure the proper treatment of wastewater before discharge.
- Reduce water usage through efficient processes and recycling initiatives.
- Monitor water quality regularly and adhere to local and international standards.

It is of great concern that whatever is the location or the status of people whether they are living in urban areas or in local town places should have reach to clean drinking water and basic sanitation facilities (H. Delanka-Pedige et al, 2021). Goal 6 emphasizes the importance of maintaining water quality and reducing pollution [67]. Efforts are directed towards minimizing the release of hazardous chemicals, untreated wastewater, and other pollutants into water bodies.

Efficient and sustainable use of water resources is encouraged. This involves promoting water-saving technologies, responsible water management practices, and reducing water waste in various sectors [24]. The goal recognizes the interconnectedness of water and ecosystems. Protecting and restoring water-related ecosystems, such as rivers, lakes, wetlands, and watersheds, is crucial for maintaining the balance of natural systems. Goal 6 underscores the value and worth of filtered and sterilized water and sanitation facilities for the mankind, environment that will lead to balanced development. Its integration into the textile industry's effluent treatment practices is vital for minimizing negative impacts on water resources, fostering innovation, and contributing to a more sustainable and responsible industry [75]. By aligning with this goal, the textile industry can achieve the global targets for clean water and sanitation by 2030.

6 Relevance to Effluent Treatment in the Textile Industry

The textile industry is the major consumption area of water and fluid and can generate substantial wastewater containing pollutants from dyeing, finishing, and other processes [60]. Table 5 represents the SDG goals and their objectives relevance to effluent treatment in the textile industry. Integrating Goal 6 into effluent treatment in the textile industry is crucial for several following reasons:

Wastewater Treatment: Effluent treatment processes must adhere to stringent standards to ensure that the wastewater discharged is properly treated and does not contaminate water bodies or harm aquatic life.

Table 5 SDG goals and their objectives relevance to effluent treatment in the textile industry

SDGs Goals	Objectives	References
Goal 9: Industry, Innovation, and Infrastructure	• Invest in innovative and sustainable technologies for effluent treatment to minimize environmental impacts • Promote research and development to identify new methods for reducing pollutants in wastewater	[40]
Goal 12: Responsible Consumption and Production	• Adopt cleaner production practices to minimize the generation of hazardous substances in the production process • Encourage circular economy principles, such as recycling and reusing water and materials	[59]
Goal 13: Climate Action	• Implement energy-efficient technologies for effluent treatment to reduce greenhouse gas emissions • Consider renewable energy sources for powering treatment facilities	[29]
Goal 14: Life Below Water	• Ensure that effluent discharge does not harm aquatic ecosystems, including rivers, lakes, and oceans • Implement measures to prevent the release of harmful chemicals and microplastics into water bodies	Molony et al. (2022)
Goal 15: Life on Land	• Prevent soil and land pollution by treating effluents effectively before they are released into the environment • Adopt sustainable land management practices to protect ecosystems	[64]
Goal 17: goals and its partnership	• Collaborate with industry associations, governments, NGOs, and other stakeholders to share best practices and knowledge • Engage in multi-stakeholder partnerships to enhance efforts in achieving sustainable effluent treatment	[53

Resource Efficiency: Applying water-efficient technologies and practices in the textile industry's processes can reduce overall water consumption and minimize the environmental impact associated with water use [58].

Pollution Prevention: Effluent treatment should focus on removing pollutants, such as chemicals and dyes, from wastewater before discharge. This aligns with Goal 6's emphasis on controlling water pollution.

Innovation and Technology: Integrating Goal 6 encourages the textile industry to invest in innovative and sustainable technologies for effluent treatment, enhancing the industry's contribution to clean water and sanitation.

Collaboration and Partnerships: Partnerships between the textile industry, regulatory bodies, and civil society can help ensure compliance with water quality standards and promote sustainable water use practices [43].

6.1 Social and Labor Aspects (Not Explicit SDGs, But Important):

Ensure the health and safety of workers involved in effluent treatment processes. Provide training and capacity building for employees to handle effluent treatment systems effectively. Comply with local and national regulations regarding effluent treatment and discharge. Advocate for stricter regulations and support the development of effective enforcement mechanisms [18]. Publish annual sustainability reports that detail efforts and progress in effluent treatment and environmental performance. Communicate with stakeholders about the company's commitment to sustainable practices [65]. By integrating these SDGs into effluent treatment practices, the textile industry can contribute to a more sustainable and responsible production process, reduce environmental impacts, and support the broader global agenda for sustainable development.

7 Mitigating the Impacts of Textile Effluents on Marine Ecosystems

Mitigating the impacts of textile effluents on marine ecosystems is a critical step in promoting environmental sustainability and safeguarding the health of our oceans (Mdlalose & Chimuka, 2022). Textile industry effluents, which contain toxicity due to dyes, harmful chemicals, lead, cadmium that are heavy metals and plastics can pose serious threats to marine life and ecosystems. Implementation of advanced wastewater treatment technologies that effectively remove or neutralize harmful substances from textile effluents before discharge into water bodies [4]. These technologies may include biological treatment, membrane filtration, and advanced oxidation processes.

Adopt closed-loop systems within textile manufacturing processes to minimize consumption of water and important fluids. Adopting water recycling and reuse practices to reduce the volume of effluents released into the environment [71]. Promote the use of eco-friendly and non-toxic dyes and chemicals in textile production to minimize the generation of hazardous substances in effluents. Regular monitoring and stringent enforcement can help ensure that effluents meet specified quality standards [61]. Foster collaboration between textile manufacturers, environmental organizations, and government agencies to develop best practices and industry standards for sustainable effluent management [69]. Raise awareness among consumers about the environmental impacts of textile production and encourage responsible purchasing decisions. Educate consumers about the importance of supporting sustainable and eco-friendly textile products [30].

8 Conclusion

The integration of effluent treatment practices in the textile industry with the Sustainable Development Goals (SDGs) refers to a balanced and highly concerned developmentally goals areas for a responsible future. The textile industry, known for its significant water consumption and pollution, holds immense potential to align its actions with the broader global agenda of achieving economic prosperity, social equity, and environmental stewardship. By focusing on the specific SDGs that intersect with effluent treatment, such as Goal 6 (Clean Water and Sanitation), the textile industry can steer its efforts towards transformative change. Ultimately, the successful integration of effluent treatment practices with the SDGs is a testament to the industry's dedication to a sustainable legacy, where fabrics of progress are woven with threads of responsibility, compassion, and resilience. Adopting advanced treatment technologies, exploring sustainable materials, and promoting transparency in supply chains can drive positive change. Additionally, educating consumers about the environmental consequences of their choices empowers them to support eco-friendly products.

9 Future Trends and Advancements in the Textile Effluent Industry

The textile effluent industry is continuously evolving as a response to growing environmental concerns, regulatory pressures, and technological advancements. Several future trends and advancements are likely to shape the landscape of textile effluent treatment and management such as circular economy practices. Textile manufacturers are expected to embrace circular economy principles, focusing on reducing waste, reusing materials, and recycling water and resources. Advanced treatment technologies will lead to effective treatment methodologies. Emerging techniques, such as electrocoagulation, membrane filtration, and biological treatments, will gain prominence for their effectiveness in removing contaminants.

The consumption of nanotechnology holds promise in treating textile effluents, particularly in removing micropollutants and microplastics. Nanomaterials can be designed to selectively capture pollutants, enhancing the efficiency of treatment processes. Bioremediation techniques involving microbes such as bacteria, algae and yeast effluents will gain attention. Bioaugmentation, where beneficial microorganisms are introduced to enhance the degradation of pollutants, will also advance. Small-scale, decentralized treatment systems will become more prevalent, especially in regions with limited infrastructure. Modular and mobile treatment units will offer flexible solutions for managing effluents.

Governments and international bodies will continue to tighten regulations on wastewater discharge, forcing textile manufacturers to invest in better treatment technologies and practices. The industry will increasingly adopt eco-friendly dyes, chemicals, and materials that generate fewer pollutants in effluents, aligning with the principles of sustainable chemistry. Technological advancements, regulatory pressures, and changing consumer preferences will drive the industry toward more responsible and efficient management of effluents, contributing to a cleaner and more sustainable environment.

References

1. Adane T, Adugna AT, Alemayehu E (2021) Textile industry effluent treatment techniques. J Chem 2021:1–14
2. Ahuchaogu AA, Chukwu OJ, Obike A, Igara CE, Nnorom IC, Echeme JBO (2018) Reverse osmosis technology, its applications and nano-enabled membrane. Int J Adv Res Chem Sci 5(2):20–26
3. Aragonés-Beltrán P, Mendoza-Roca J, Bes-Piá A, García-Melón M, Parra-Ruiz E (2009) Application of multicriteria decision analysis to jar-test results for chemicals selection in the physical–chemical treatment of textile wastewater. J Hazard Mater 164(1):288–295
4. Arias AH, Alfonso MB, Girones L, Piccolo MC, Marcovecchio JEJEP (2022) Synthetic microfibers and tyre wear particles pollution in aquatic systems: relevance and mitigation strategies 295:118607
5. Arshad M (2010) Bioethanol: asustainable and environment friendly solution for Pakistan. A Scientific J COMSATS–Science Vision 2011:16–7
6. Arshad M, Adil M, Sikandar A, Hussain T (2014) Exploitation of meat industry by-products for biodiesel production: Pakistan's perspective. Pakistan J Life Soc Sci 12:120–5
7. Arshad M (2017) Clean and sustainable energy technologies. InClean energy for sustainable development Jan 1 (pp. 73–89). Academic Press
8. Arshad M, Ansari AR, Qadir R, Tahir MH, Nadeem A, Mehmood T, Alhumade H, Khan N (2022) Green electricity generation from biogas of cattle manure: an assessment of potential and feasibility in Pakistan. Front Energy Res 25(10):911485
9. Arundel J (1995) Sewage and industrial effluent treatment: a practical guide. Blackwell Science Ltd.

10. Avellán, T, Nagabhatla, N, Jalan, I, & Liao, D. (2022). Integrating circularity to achieve sustainability: Examples of various wastewater treatment systems. In Circular economy and sustainability (pp. 21–35). Elsevier.
11. Azanaw A, Birlie B, Teshome B, Jemberie M (2022) Textile effluent treatment methods and eco-friendly resolution of textile wastewater. In: Case studies in chemical and environmental engineering, p 100230
12. Babu BR, Parande A, Raghu S, Kumar TP (1995) Textile technology. Technology 11:110–122
13. Bano I, Ghauri MA, Arshad M, Yasin T, Younus M (2017) Bioactivity of variant molecular weight chitosan against drug-resistant bacteria isolated from human wounds. Microbial Drug Resist 1;23(8):958–65
14. Bano I, Arshad M (2018) Climatic changes impact on water availability. Perspectives on Water Usage for Biofuels Production: aquatic Contamination and Climate Change (pp. 39–54)
15. Barredo-Damas S, Iborra-Clar M, Bes-Pia A, Alcaina-Miranda M, Mendoza-Roca J, Iborra-Clar A (2005) Study of preozonation influence on the physical-chemical treatment of textile wastewater. Desalination 182(1–3):267–274
16. Baumgartner RJ, Rauter R (2017) Strategic perspectives of corporate sustainability management to develop a sustainable organization. J Clean Prod 140:81–92
17. Bencheikh I, Azoulay K, Mabrouki J, El Hajjaji S, Moufti A, Labjar N (2021) The use and the performance of chemically treated artichoke leaves for textile industrial effluents treatment. Chem Data Collect 31:100597
18. Bengtsson M, Alfredsson E, Cohen M, Lorek S, Schroeder PJSS (2018) Transforming systems of consumption and production for achieving the sustainable development goals: Moving Beyond Effic 13:1533–1547
19. Bera SP, Godhaniya M, Kothari C (2022) Emerging and advanced membrane technology for wastewater treatment: a review. J Basic Microbiol 62(3–4):245–259
20. Chen TL, Kim H, Pan SY, Tseng PC, Lin YP, Chiang PC (2020). Implementation of green chemistry principles in circular economy system towards sustainable development goals: challenges and perspectives 716:136998
21. Delanka-Pedige HM, Munasinghe-Arachchige SP, Abeysiriwardana-Arachchige IS, Nirmalakhandan N, Ecology W (2021). Wastewater infrastructure for sustainable cities: assessment based on UN sustainable development goals (SDGs) 28(3):203–209
22. Delanka-Pedige HM, Munasinghe-Arachchige SP, Abeysiriwardana-Arachchige IS, Nirmalakhandan N (2021). Evaluating wastewater treatment infrastructure systems based on UN Sustainable Development Goals and targets 298:126795
23. Deng D, Lamssali M, Aryal N, Ofori-Boadu A, Jha MK, Samuel RE (2020) Textiles wastewater treatment technology: a review. Water Environ Res 92(10):1805–1810
24. Diep L, Martins FP, Campos LC, Hofmann P, Tomei J, Lakhanpaul M, Parikh PJSD (2021) Linkages between sanitation and the sustainable development goals: a case study of Brazil. 29(2):339–352
25. Eskandarinezhad S, Khosravi R, Amarzadeh M, Mondal P, Magalhães Filho FJC (2021) Application of different Nanocatalysts in industrial effluent treatment: a review. J Compos Compounds 3(6):43–56
26. Fletcher K (2013) Sustainable fashion and textiles: design journeys. Routledge
27. Ghaly A, Ananthashankar R, Alhattab M, Ramakrishnan VV (2014) Production, characterization and treatment of textile effluents: a critical review. J Chem Eng Process Technol 5(1):1–19
28. Ghosh R (2003) Protein bioseparation using ultrafiltration: theory, applications and new developments. World Scientific
29. Gibb N (2016) Getting climate ready: a guide for schools on climate action and the whole-school approach. UNESCO Publishing
30. Goh P, Kang H, Ismail A, Khor W, Quen L, Higgins DJC (2022) Nanomaterials for microplastic remediation from aquatic environment: Why nano matters? 299:134418
31. Gosavi VD, Sharma S (2014) A general review on various treatment methods for textile wastewater. J Environ Sci, Comput Sci Engin Technol 3(1):29–39

32. Gray NF (2004) Biology of wastewater treatment, vol 4. World Scientific
33. Guo H, Hackmann H, Gong KJBED (2021) Big data in support of the Sustainable Development Goals: a celebration of the establishment of the International Research Center of Big Data for Sustainable Development Goals (CBAS). In: vol 5: Taylor & Francis, pp 259–262
34. Holkar CR, Jadhav AJ, Pinjari DV, Mahamuni NM, Pandit AB (2016) A critical review on textile wastewater treatments: possible approaches. J Environ Manage 182:351–366
35. Islam MM, Mahmud K, Faruk O, Billah M (2011) Textile dyeing industries in Bangladesh for sustainable development. Int J Environ Sci Develop 2(6):428
36. Jahan N, Tahmid M, Shoronika AZ, Fariha A, Roy H, Pervez MN, Cai Y, Naddeo V, Islam MS (2022) A Comprehensive review on the sustainable treatment of textile wastewater: zero liquid discharge and resource recovery perspectives. Sustainability 14(22):15398
37. Kiran S, Adeel S, Nosheen S, Hassan A, Usman M, Rafique MA (2017) Recent trends in textile effluent treatments: a review. advanced materials for wastewater treatment, pp 29–49
38. Kuo W-CJ, Smith R (1997) Effluent treatment system design. Chem Eng Sci 52(23):4273–4290
39. Kuo W-C, Smith R (1998) Designing for the interactions between water-use and effluent treatment. Chem Eng Res Des 76(3):287–301
40. Lekan A, Aigbavboa C, Babatunde O, Olabosipo F, Christiana A (2022) Disruptive technological innovations in construction field and fourth industrial revolution intervention in the achievement of the sustainable development goal 9. 22(14), 2647–2658
41. Li W-W, Yu H-Q, He Z (2014) Towards sustainable wastewater treatment by using microbial fuel cells-centered technologies. Energy Environ Sci 7(3):911–924
42. Lin SH, Chen ML (1997) Treatment of textile wastewater by chemical methods for reuse. Water Res 31(4):868–876
43. Luthra S, Sharma M, Kumar A, Joshi S, Collins E, Mangla SJTRPEL, Review T (2022) Overcoming barriers to cross-sector collaboration in circular supply chain management: a multi-method approach 157:102582
44. Madhav S, Ahamad A, Singh P, Mishra PK (2018) A review of textile industry: wet processing, environmental impacts, and effluent treatment methods. Environ Qual Manage 27(3):31–41
45. Mdlalose L, Chimuka LJ (2022) Mitigation approaches to prevent microplastics effects in the aquatic environment: exploration of microbeads from personal care and cosmetic products 16(5):84
46. Mesquita RFd, Klein B, Xavier A, Matos FRN (2017) Mining and the sustainable development goals: a systematic literature review
47. Moga IC, Covaliu IC, Matache MG (2018) Advanced wastewater treatment stage for textile industry. Industria Textila 69(6):478
48. Molony BW, Ford AT, Sequeira AM, Borja A, Zivian AM, Robinson C, Lønborg C, Escobar-Briones EG, Di Lorenzo E, Andersen JH, Müller MN (2022) Sustainable development goal 14-life below water: towards a sustainable ocean 8:829610
49. Naresh B, Jaydip J, Prabhat B, Rajkumar P (2013) Recent biological technologies for textile effluent treatment. Int. Res J Biol Sci 2(6):77–82
50. Neto F (2003) A new approach to sustainable tourism development: moving beyond environmental protection. Natural resources forum
51. Niinimäki K (2015) Ethical foundations in sustainable fashion. Textiles Clothing Sustain 1:1–11
52. Obaideen K, Shehata N, Sayed ET, Abdelkareem MA, Mahmoud MS, Olabi AG (2022). The role of wastewater treatment in achieving sustainable development goals (SDGs) and sustainability guideline 7:100112
53. Oliveira-Duarte L, Reis DA, Fleury AL, Vasques RA, Fonseca Filho H, Koria M, Baruque-Ramos J (2021). Innovation Ecosystem framework directed to Sustainable Development Goal# 17 partnerships implementation 29(5):1018–1036
54. Pang YL, Abdullah AZ (2013) Current status of textile industry wastewater management and research progress in Malaysia: a review. Clean-Soil, Air, Water 41(8):751–764
55. Panigrahi T, Santhoskumar A (2020) Adsorption process for reducing heavy metals in textile industrial effluent with low cost adsorbents. Prog Chem Biochem Res 3(2):135–139

56. Peralta-Zamora P, Esposito E, Pelegrini R, Groto R, Reyes J, Durán N (1998) Effluent treatment of pulp and paper, and textile industries using immobilised horseradish peroxidase. Environ Technol 19(1):55–63
57. Periyasamy AP, Ramamoorthy SK, Rwawiire S, Zhao Y (2018) Sustainable wastewater treatment methods for textile industry. In: Sustainable innovations in apparel production, pp 21–87
58. Prasad MNV (2023) Resource recovery from urban flood, municipal and industrial wastewaters in the context remediation technologies and circular economy. In: Water in circular economy. Springer, pp 103–120
59. Programme UND (2016) Goal 12: responsible consumption and production
60. Ramasamy R, Aragaw TA, Balasaraswathi Subramanian RJES, Research P (2022) Wastewater treatment plant effluent and microfiber pollution: Focus on industry-specific wastewater. 29(34):51211–51233
61. Roy P, Mohanty AK, Misra MJESA (2022) Microplastics in ecosystems: their implications and mitigation pathways 1(1):9–29
62. Sarayu K, Sandhya S (2012) Current technologies for biological treatment of textile wastewater–a review. Appl Biochem Biotechnol 167:645–661
63. Sathya K, Nagarajan K, Malar CG, G, Rajalakshmi, S, Raja Lakshmi P (2022) A comprehensive review on comparison among effluent treatment methods and modern methods of treatment of industrial wastewater effluent from different sources. Appl Water Sci 12(4):70
64. Sayer J, Sheil D, Galloway G, Riggs RA, Mewett G, MacDicken KG, Arts B, Boedhihartono AK, Langston J, Edwards DP (2019) SDG 15 Life on land–the central role of forests in sustainable development. In Sustainable development goals: their impacts on forest and people. Cambridge University Press, pp 482–509
65. Schröder P, Lemille A, Desmond P (2020) Making the circular economy work for human development. Resourc Conservat Recycl 156:104686
66. Sharif N, Munir N, Hasnain M, Naz S, Arshad M (2021) Environmental impacts of ethanol production system. Sustainable Ethanol and Climate Change: Sustainability Assessment for Ethanol Distilleries. 205–23
67. Shehata N, Egirani D, Olabi AG, Inayat A, Abdelkareem MA, Chae KJ, Sayed ET (2023) Membrane-based water and wastewater treatment technologies: Issues, current trends, challenges, and role in achieving sustainable development goals, and circular economy 320:137993
68. Sungur Ş, Gülmez F (2015) Determination of metal contents of various fibers used in textile industry by MP-AES. J Spectroscopy
69. Upadhyay D, Tripathi A, Joshi KD, Jindal MB, Shukla K (2023) Chemical dye-induced alteration in hematological, biochemical, and enzymatic parameters of gymnocorymbus ternetzi exposed to synthetic food colour wastewater
70. Vandevivere PC, Bianchi R, Verstraete W (1998) Treatment and reuse of wastewater from the textile wet-processing industry: Review of emerging technologies. J Chem Technol Biotechnol: Int Res Process, Environm Clean Technol 72(4):289–302
71. Weis JS, De Falco FJM (2022) Microfibers: environmental problems and textile solutions 1(4):626–639
72. Wu XT, Ansari AR, Pang XX, Li HZ, Zhang ZW, Luo Y, Arshad M, Song H (2019) Visfatin plays a significant role in alleviating lipopolysaccharide-induced apoptosis and autophagy through PI3K/AKT signaling pathway during acute lung injury in mice. Arch immunol ther exp 1(67):249–61
73. Yang C-L, McGarrahan J (2005) Electrochemical coagulation for textile effluent decolorization. J Hazard Mater 127(1–3):40–47
74. Zaini MS, Arshad M, Syed-Hassan SS (2023) Adsorption isotherm and kinetic study of methane on palm kernel shell-derived activated carbon. J Bioresour Bioprod 1;8(1):66–77
75. Zhong J, Li XJL (2022) Interlinkages among county-level construction indicators and related sustainable development goals in China 11(11):2008
76. Şahinkaya S (2013) COD and color removal from synthetic textile wastewater by ultrasound assisted electro-Fenton oxidation process. J Ind Eng Chem 19(2):601–605

Microbial Enzymes in Bioremediation of Water Polluted by Textile Industry Effluents

Mohammad Ejaz, Mehmoona Sharif, Sher Zaman Safi, Sabir Nawaz, Sheryar Jamil, Muhammad Ali Syed, and Waqar Ahmed

Abstract The process of textile production is purported to account for ~20% of the worldwide contamination of clean water, which is attributed to the dyeing chemical treatments, and finishing of textile products. Emerging contaminants found in the natural environment have been linked to various immediate and long-term harmful impacts on living organisms, ultimately causing the degradation of ecosystems and presenting substantial hazards. The conventional techniques employed for the elimination of pollutants have demonstrated inefficiency, as they often result in the production of additional harmful substances. Microbial enzymes are crucial in bioremediation processes as they enable the enzymatic reactions that break down contaminants. The recognition of various microbial enzymes by pollutant substrates under optimal conditions including pH, temperature, concentration, and contact time can facilitate their efficient transformation into less harmful byproducts. The present chapter aims to investigate the importance of enzymatic bioremediation and underscore its benefits in mitigating environmental pollution, particularly in the context of textile industrial effluents that pose a threat to the purity of water resources.

M. Ejaz (✉) · S. Nawaz · S. Jamil
Department of Microbiology, Government Postgraduate College Mandian, Abbottabad, Pakistan
e-mail: mohammadejaz113@gmail.com

M. Sharif · S. Nawaz
Department of Microbiology, Quaid-I-Azam University, Islamabad, Pakistan

S. Z. Safi
Faculty of Medicine, Bioscience and Nursing, MAHSA University, Jenjarom, Selangor, Malaysia

M. A. Syed · W. Ahmed
Department of Microbiology, The University of Haripur, Haripur, Pakistan

M. Arshad (ed.), *Enzymes in Textile Processing: A Climate Changes Mitigation Approach*, SDGs and Textiles, https://doi.org/10.1007/978-981-97-8058-7_15

Graphical Abstract

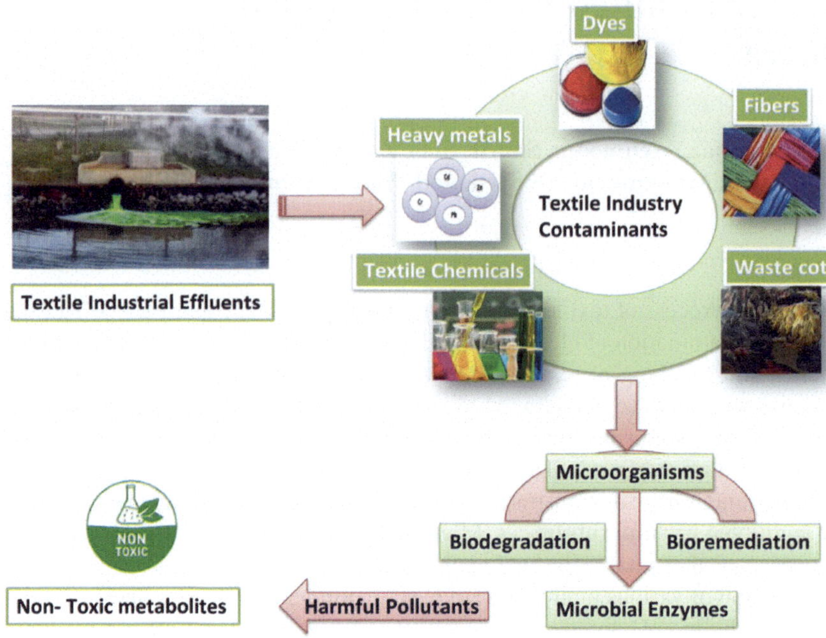

Keywords Microbial enzymes · Bioremediation · Microorganism · Textile effluents · Biodegradation

1 Introduction

The Earth is today facing substantial and far-reaching repercussions due to human activities, including the use of advanced agricultural methods, industrialization, the rapid expansion of the human population, and unhealthy rivalry for power. Industrialization plays a crucial function in promoting economic progress and enabling the modernization of society. In both developed and developing countries, many big and medium-scale industries significantly contribute to the economic development of the nation [5–7].

The textile industry has a significant portion of the market share, primarily attributed to its capacity to generate diverse employment prospects due to global demand for textile-based products and its extensive integration with many economic sectors on a global and local scale. The textile production industry, being one of the most ancient and intricately sophisticated industries, showcases a remarkable level of technological complexity. The proliferation of textile products has correspondingly escalated in response to the burgeoning demand from the populace. It is estimated that the size of the global textile market was recorded at USD 1,695.13 billion in the year 2022. The revenue is expected to grow at a compound annual growth rate (CAGR) of 7.6% from 2023 to 2030 [89]. The rapid industrialization and ever-increasing demand for textiles led to the use of activities that are responsible for the generation of pollutants such as phenols, polyaromatic hydrocarbons, azo dyes, polychlorinated compounds, pesticides, heavy metals, and other similar substances at levels that have never been observed before. These chemical compounds exhibit resistance to the natural process of biodegradation and persist in the environment over a prolonged duration, thereby posing acute and chronic harm to the biotic constituents of ecosystems [33]. The presence of these pollutants has resulted in the manifestation of teratogenic, carcinogenic, mutagenic, and toxic properties, both in humans and other organisms, thereby exerting a comprehensive impact on various ecosystems across the entire planet [69].

Regular exposure to effluents from textile dyes carries a possible harm to human health when considering the larger picture. Research has shown that this type of exposure can directly or indirectly contribute to many health problems. These encompass immunosuppression, the emergence of autoimmune disorders, complications in the respiratory and circulatory systems, impairment of the central nervous system and neurobehavioral functions, allergic responses, diverse types of cancer, as well as symptoms such as emesis, excessive diarrheal episodes, tissue harm, ocular and dermatological disorders, eye infections, and respiratory difficulties. These are just a few examples of the potential consequences associated with such textile effluents [98].

The utilization of microbial enzymes in the process of bioremediation for water contaminated by effluents from the textile industry aligns with the objectives outlined by the United Nations Sustainable Development Goals (SDGs). This novel methodology encompasses various essential SDGs, notably emphasizing "Clean Water and Sanitation," as it strives to rehabilitate and sustain the integrity of water resources impacted by discharges from the textile industry. Moreover, microbial enzymes' technological advancement is in accordance with SDG 12, which pertains to responsible consumption and production. It achieves this by endorsing sustainable industrial methodologies and mitigating the ecological impact associated with the textile sector. Furthermore, it provides support for SDG 15, which pertains to terrestrial ecosystems, and SDG 14, which focuses on marine ecosystems. Bioremediation plays a crucial role in the restoration of these ecosystems and serves as a protective measure against the deleterious impacts of textile pollutants on aquatic organisms. The utilization of microbial enzymes in water bioremediation serves as a prime illustration of how cutting-edge scientific methodologies can effectively contribute to the progression of our worldwide sustainability objectives.

Historically, various traditional methods such as precipitation, oxidation, flocculation, coagulation, adsorption, and flotation utilized for the eradication of contaminants have proven to be inefficacious, as they frequently lead to the generation of secondary pollutants [53]. The utilization of microorganisms or microbial enzymes has superseded traditional methods in the treatment of wastewater from the textile industry due to their ability to efficiently eliminate contaminants without the production of additional pollutants. The global recognition and adoption of bioremediation techniques have been steadily growing, primarily attributed to their remarkable public acceptance, cost-effectiveness in comparison to traditional remediation methods, abundant supply of enzymes, and minimal ecological footprint [33]. The investigation of enzymes for bioremediation has garnered significant attention owing to their remarkable capacity to operate within broader pH and temperature spectrums, even in the presence of contaminants and elevated saline concentrations. Enzymatic bioremediation represents a highly advantageous and cutting-edge approach within the realm of ecological and economical strategies. The procedure involves investigating the inherent traits of microorganisms or genetically engineered organisms that possess the ability to generate particular enzymes to metabolize or catalyze the pollutants. This results in the conversion of the toxic form into a non-toxic form, and occasionally, the creation of novel products [94]. Within the realm of bioremediation processes, a selection of enzymes that play a crucial role include azoreductases, peroxidases, laccases, dehalogenases, and hydrolases.

In this chapter, an in-depth exploration is undertaken to elucidate the intricate mechanisms and functions of microbial enzymes employed in the process of bioremediation for the purification of textile industrial effluents. The chapter additionally elucidates the adverse impacts of textile industrial effluents on the environment, fauna, and human well-being, while also exploring the potential of employing microorganisms for bioremediation purposes in wastewater treatment.

2 Textile Industrial Effluents

Industrialization is the main determinant of a nation's economic progress and the advancement of living standards. A country's economic performance heavily relies on the presence of several large and medium-sized firms, both in developed and emerging nations [1, 73]. One of the industries with the highest water consumption is the textile production process, which is expected to use thousands of cubic meters of water each day. Among the main input raw materials in the textile sectors are cotton, wool, synthetic fibers, synthetic dyes (direct dyes, reactive dyes, vat dyes, acid–base dyes, etc.), chemicals, and process water [103]. The de-sizing, bleaching, dyeing, printing, and washing processes used to produce fabrics result in the emission of effluents that can be extremely colored as well as resistant to biodegradation [34, 65]. Moreover, the biological oxygen demand (BOD), chemical oxygen demand (COD), salt, total dissolved solids (TDS), total suspended solids (TSS), and total nitrogen (TN) in the textile effluent have all been at very high levels, followed by higher acidity–salinity and elevated temperature conditions. Due to the presence of dyes and other harmful chemicals that are difficult to degrade, wastewater from the dyeing and printing sector is more contaminated. Solid wastes are primarily produced during dry processing, whereas liquid wastes are primarily produced during wet processing phases [1, 49].

The polluting effects of the coloring operations of both printing and dyeing are problematic for the textile industry. The majority of textile dyes on the market are difficult to discolor due to their intricate remnant structures. Additionally, some of the dyes used in the dyeing process are untreatedly released directly into the textile wastewater pipelines [104]. The operations rely on potable water, the utilization of diverse chemicals and colors, the release of wastewater during wet processing stages, and the process of fiber creation. The process of wet processing unit operations has become the main cause of water pollution due to the escalated use of clean water, dyes, and chemicals. This contaminated water is laden with colors and other harmful substances. The wastewater released by the textile industry contains toxic substances such as colors, sodium hydroxide, chromium, acids, starches, etc. [1, 70]. As a consequence, as compared to other industries, textile factories are anticipated to rely on more potable chemical and water supplies, and nearly all the industry's effluents remain heavily contaminated. For each kilogram of the final product, the textile industry utilizes 200 L of potable water throughout the processing [112]. According to the World Bank estimate, the portion of industrial wastewater generated by dyeing and finishing is close to 17–20 [17, 35]. Currently, the majority of dyes are heterocyclic complexes with both polar color-display moieties and aromatic compounds. Dye molecules have a more intricate and stable structure and are challenging to decompose [17, 112]. Table 15.1 describes the chemical composition, properties, and applications of different dyes used in the textile industry.

2.1 Impact of Textile Industry Pollutants on the Environment and Organisms

The chemical processing of textiles, particularly dyeing, uses high concentrations of chemical intermediates and dyes that can cause various health and environmental problems. Removing contaminants from wastewater is crucial for meeting the standards, laws, and regulations that safeguard the environment and human health [8]. This ensures that the effluent may be discharged without posing any hazard [70, 104]. Water bodies can readily be shielded from sunlight by textile effluent and dyes, which impedes the development of aquatic life. Textile effluent contains highly hazardous, carcinogenic, and mutagenic synthetic dyes [23, 31, 92]. The textile industry's usage of synthetic colors results in significant water contamination due to their inadequate adhesion to the fabric. As a result, the environment and public health are significantly harmed by the ongoing untreated outflow of wastewater from several textile industries. Aquatic plants experience significant photosynthetic repercussions as a result of the widespread discharge of coloring substances into freshwater bodies due to widespread dumping. Significant amounts of chemicals or dyes, along with their challenging degradation, are present in the wastewater effluent resulting from printing and dyeing unit activities [35]. Textile dyes may also intoxicate aquatic organisms, which could subsequently infiltrate the food chain, and contaminate aquatic environments. Textile dyes have a negative impact on bodies of water, degrading their aesthetic quality. They do this by increasing biochemical and COD levels, hindering

Table 15.1 Chemical composition, properties, and applications of different dyes used in the textile industry

Types of dyes	Chemical composition	Properties	Application
Direct dyes	Phthalocyanines, poly azo compounds, certain stilbenes, and oxazines	Anionic, water-soluble dyes	Paper, rayon, leather, cotton, and a small quantity of nylon
Basic dyes	Cyanine, hemicyanine, thiazine, acridine, oxazine, triaryl methane, and diazahemicyanine	Water-soluble	Medicine, cation dyeable polyethylene terephthalate, polyacrylonitrile, modified polyesters, modified nylons and paper
Acid dyes	Xanthene, triphenylmethane, anthraquinones, nitroso, nitro, and azine as well as azo (including premetallized)	Water-soluble	Paper, inkjet printing, leather, modified acrylics, silk, wool, nylon, and food
Reactive dyes	Chromophores such as azo, triarylmethane, anthraquinone, phthalocyanine, oxazine and formazan	Water-soluble	Nylon and wool, fibers, cellulosic fibers and cotton
Disperse dyes	Styryl, nitro, azo, anthraquinone, and benzodifuranone	Non-ionic, water-insoluble dyes	Nylon, acrylic fiber, cellulose, cellulose acetate, and polyester
Vat dyes	Cellulose, cotton fibers, rayon and wool	Water-insoluble	Anthraquinone (polycyclic quinones) and indigoids
Solvent dyes	Triarylmethane, phthalocyanine, anthraquinone and azo	Water-insoluble, solvent-soluble, nonpolar and some minute polar	Lubricants, gasoline, waxes, plastics and oils
Sulfur dyes	Organic compounds with nitro, amino groups, and Na-polypeptide and sulfur	Water-soluble	Rayon and cotton

photosynthesis, restricting plant growth, entering the food chain, causing recalcitrance and bioaccumulation, and potentially leading to toxicity and mutagenicity [3].

Textile dyes pose significant health risks and can contribute to environmental degradation and various illnesses in humans and animals. Similar to a microbiologist, the dyes have a tendency to persist in soil and sediments and can also make their way into public water supply systems. This is because they are resistant to breaking down in aerobic conditions, especially in traditional treatment facilities [100, 108]. Despite the majority's resistance to environmental degradation, azo-type compounds that produce hazardous aromatic compounds can be reduced or partially destroyed in the presence of anoxic sediments. Another potential is the creation of further mutagenic and carcinogenic chemicals by mixing dyes with intermediate synthetic compounds or their breakdown byproducts [50, 108]. The xenobiotic and refractory characteristics of the dyes ultimately affect the structure and functionality of ecosystems. Extended periods of exposure to complex metal dyes have significant impacts on both human health and aquatic biota. Due to their long-lasting nature and half-lives ranging from 2 to 13 years, the textile industry often utilizes this particular set of dyes. Their composition includes the metals nickel, copper, cobalt, and most notably chromium [64, 87].

The textile industry also discards enormous amounts of untreated or treated effluent including azo dyes and other organic contaminants. Furthermore, when azo dyes degrade, whether it is before or after being disposed of, the resulting treated effluents contain amino acids that are likely to be more dangerous than the original substances. On the other hand, untreated wastewater has a range of harmful impacts on aquatic ecosystems and organisms, thereby significantly compromising the well-being of aquatic plants and mammals [80]. Dye-containing textile wastewater's ecotoxicological effects on the environment and organisms are shown in Fig. 15.1.

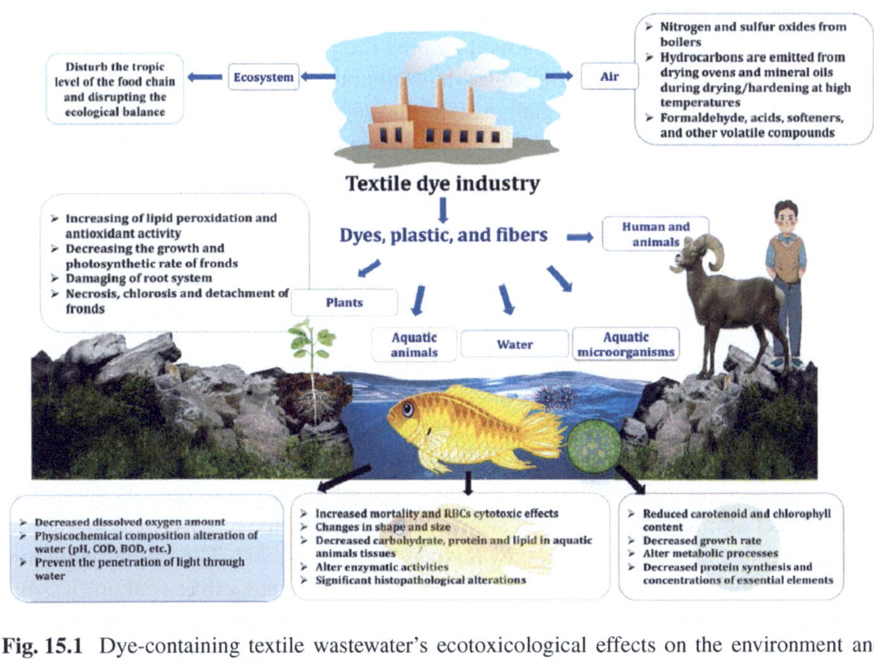

Fig. 15.1 Dye-containing textile wastewater's ecotoxicological effects on the environment and organisms (Adopted from [3])

3 Bioremediation Using Microorganisms

The application of indigenous microbial communities and their metabolic activities to mitigate pollution is commonly referred to as bioremediation. The bioremediation approach exhibits practicality and yields a favorable outcome. According to the study conducted by Ahmedna et al., microbial communities play a crucial role in the effective bioremediation of metal-contaminated soil and water [2]. Microbial entities possess the capacity to induce alterations in the chemical composition and motility of contaminated substances through processes involving the reduction, accumulation, mobilization, and immobilization of metallic elements. The utilization of biological methodologies for pollutant remediation offers three primary advantages: firstly, the implementation of biological processes directly at the contaminated site; secondly, the inherent environmentally friendly nature of bioprocess technologies, devoid of any secondary pollution; and thirdly, the enhanced cost-effectiveness of ex-situ approaches.

Microorganisms play a crucial role in the bioremediation of heavy metals, carcinogens, and textile dyes. Microorganisms possess diverse mechanisms by which they can engage with contaminants. By harnessing the contaminants as a vital source of sustenance and fortifying the organism against the pernicious repercussions, there exist two primary approaches through which the organism can effectively contend with contaminants. The microorganism exhibits responses to environmental contaminants through both direct and indirect mechanisms, including but not limited to biosorption and biotransformation [30].

By harnessing the metabolic capabilities of diverse microorganisms such as bacteria, yeasts, algae, and fungi, bioremediation systems have emerged as a prevalent approach for the remediation of industrial effluents. These systems exploit the capacity of these microorganisms to accumulate and metabolize a wide range of contaminants, facilitating the purification process. The phenomenon of dye biosorption or biodegradation by microorganisms has been elucidated through the observation and analysis of the chromatic properties exhibited by the cellular matrix [29, 46]. Given the diverse nature of strains, it is observed that they exhibit distinct modes of attack on the dye molecule, targeting different sites. Additionally, certain strains may utilize degraded products generated by one strain, which are subsequently subjected to further decomposition by another strain. Consequently, the process of degradation is enhanced through the utilization of mixed cultures [27, 77]. The implementation of pure microbial cultures in extensive wastewater treatment facilities encompassing a variety of pollutant dyes is hindered by their specificity towards particular dyes, thereby posing challenges in achieving effective decolorization [81]. The catabolic potential of various strains can be synergistically augmented within a consortium of mixed cultures, thereby leading to the proficient degradation of dyes [81].

4 Microbial Enzymes in Bioremediation of Textile Industrial Effluent

Bioremediation techniques have garnered global recognition owing to their broad public acceptance, economic efficiency in comparison to traditional remediation methods, ample availability of enzymes, and minimal ecological repercussions [18]. Enzymes have brought in considerable interest in the field of bioremediation research owing to their remarkable ability to operate effectively under diverse pH and temperature conditions, even within the presence of pollutants and fluctuating salt concentrations. Enzyme mediated bioremediation is an evolving technology that shows great potential in the field of environmental science. It is considered novel due to its innovative approach, and it is highly regarded for its environmentally friendly nature. The process entails examining the intrinsic characteristics of microorganisms or genetically modified organisms that possess the ability to produce particular enzymes. These enzymes exhibit catalytic activity or engage in metabolic processes to facilitate the transformation of pollutants. This transformation results in the conversion of pollutants from their detrimental state to a harmless form and occasionally leads to the generation of novel products [16]. Thus far, only a limited variety of bacterial species have shown the ability to produce highly efficient biodegrading enzymes, potentially elongating the bioremediation process involving microbial enzymes. Consequently, the use of genetically engineered organisms, capable of producing substantial quantities of enzymes under optimal conditions, has gained more widespread acceptance [16].

Aerobic microorganisms like *Pseudomonas, Mycobacterium, Alcaligenes, Sphingomonas, and Rhodococcus* produce enzymes that are used for the bioremediation of insecticides and organic compounds. However, oxygen independent bacteria are utilized for the bioremediation of chlorodiphenyl, decolorization of trichloroethene (TCE), and chloroform. The key enzymes that are utilized for the process of bioremediation include the cytochrome P450, laccases, proteases, lipases, hydrolases, dehalogenases, and dehydrogenases [16]. Furthermore, fungi have the capability to breakdown substances using enzymes for example azoreductases, lignin (LiP) peroxidases, manganese (MnP) peroxidases, as well as laccases. For instance, peroxidases and laccases found in wood decay fungi can effectively break down textile dyes [64].

Laccases, dehalogenases, and hydrolases are a specific class of enzymes that are utilized in bioremediation procedures. Laccases exhibit the potential to facilitate the decomposition of aromatic amines, polyphenols, and their derivatives. Dehalogenases, conversely, facilitate the degradation of a diverse array of halogenated compounds through the cleavage of C-X bonds (where X represents halogen atoms like Cl). In contrast, hydrolases utilize the hydrolysis process, which involves the incorporation of water molecules to cleave chemical bonds. This mechanism effectively reduces the toxicity of larger molecules by disassembling them into smaller constituents. These enzymes exhibit a vital role for the process of bioremediation by facilitating the degradation of various chemical bonds, including C–C, C–O, C–N, S–S, S–N, S–P, and C–P [16].

Enzymes engage with dyes present in textile effluent, resulting in the formation of precipitates that can be readily eliminated or chemically changed into simpler compounds [90]. The pace of dye deterioration under enzyme attack is affected by different elements such as the dye's chemical constitution, salt content, metal ion levels, pH, and wastewater temperature [106]. The degradation of the contaminants within the effluents of the textile industry using microbial enzymes presents a pragmatic and effective strategy aligning with the principles of green chemistry, offering various advantages including substrate specificity and selectivity [48].

A wide variety of enzymes have been employed in bioremediation as environmentally friendly biocatalysts, augmenting the reaction rate by reducing the activation energy of the molecules. Enzymes employed in bioremediation offer numerous advantages as compared to alternative traditional microbial decomposition [97]. The duration of retention durations can vary significantly ranging from 30 min to 72 h depending on factors such as the specific enzyme employed, the reaction conditions, and the nature of the contaminant [19]. For instance, Silva et al. observed that a reactive textile dye (Remazol Brilliant Blue R) was removed at a rate of 86% after being retained for 13 min [102]. Nazari et al. conducted a review on biological methods for emerging contaminants degradation and found that the retention durations ranged from 3 to 21 days [82]. Furthermore, it is important to note that the enzymes are not depleted during the reaction and the process proceeds without the formation of harmful byproducts. Moreover, a broader spectrum of temperature and ionic strength conditions can be utilized, surpassing the tolerated range of the corresponding microbial strains. Enzymes lack the ability to replicate that allows a more regulated process in terms of enzyme concentration and the availability of cofactors [97].

4.1 Enzyme-Mediated Bioremediation of Textile Effluents

The extraordinary adaptability and efficiency exhibited by enzyme-mediated bioremediation in the decomposition of persistent organic contaminants have garnered significant recognition, leading to its extensive implementation in diverse industries, biotechnological procedures, and environmental settings [14]. This approach involves extracting the extracellular and intracellular metabolites from distinct bacterial, fungal, algal, and plant taxa, which can subsequently be employed for the biosynthesis of the aforementioned enzymes [76].

The scientific literature extensively reports on the biosynthesis of various enzymes by dye-decolorizing microorganisms, encompassing azoreductase, riboflavin mononucleotide reductase, laccase, peroxidases, tyrosinase, aminophenazone N-demethylase, lignin peroxidases (LiPs), and alcohol oxidase [76]. Notably, in the domain of enzymatic catalysts, azoreductases, laccases, and lignin peroxidases occupy a pivotal role in the decolorization process associated with azo dyes [90]. Various enzymes of microbial origin are being broadly utilized in the bioremediation of textile effluents that contain harmful chemicals and products (Table 14.1).

4.1.1 Azoreductases

Azoreductases (EC 1.7.1.6) belong to a class of enzymatic reducers that employ a reductive cleavage mechanism to break down azo dyes into amines having no color. The redox process necessitates the involvement of an electron donor, such as FADH or NADH, as a substitute for a reducing equivalent with a low molecular weight [13]. The classification of this enzyme is contingent upon its utilization of either NADH, NADPH, or both as its coenzyme [105].

Azoreductases are widely regarded as the principal enzymatic agents responsible for the decomposition of azo compounds. This class of enzyme exists in two distinct states: one being membrane-bound, while the other resides within the confines of the cell. Both categories exhibit thermos resistance, hydrophilic properties, and generally demonstrate enhanced efficacy for the decomposition of azo dyes [49, 96].

Recent discoveries have unveiled the existence of azoreductase and naphthalene-degrading genes within the bacterial population inhabiting coastal regions, which play a major part in the decolorization of naphthalene-based dyes [120]. The bacterial culture thus exhibits the potential to deplete the coloration of a wide spectrum of dyes, encompassing amaranth, naphthalene, as well as both acidic and anthracene Dione dyes. This particular process is instrumental in the mitigation of ecotoxicity linked to azo dyes in textile effluents.

Yeasts have the potential to biologically decolorize the azo compounds due to the presence of azoreductases, enzymes catalyzing the reductive cleavage of azo groups (−N = N−). Candida zeylanoides and *Candida oleophila* are two varieties of such yeasts. The azo bond is cleaved and the necessary amines are generated by these strains, allowing them to decolorize. *Candida tropicalis, Debaryomyces hansenii,* and *Candida sorbosa* are just a few of the ascomycete yeast species that have been investigated in their capacity to decolorize azo dyes. Decolorizing azo and reactive high-exhaust textile dyes also can be carried out by *Galoctomyces geotrichum* MTCC 1360 [95]. Similarly, *Trichosporon beigelii* has exhibited the ability to fully degrade Navy blue HER [56]. Based on various scientific investigations, it has been observed that photosynthetic cyanobacteria and algae possess the remarkable capability to enzymatically degrade azo dyes through the utilization of azoreductases [107]. Several species of *Chlorella* and *Oscillatoria*, known for their potential to dissociate azo dyes, have the potential to enzymatically convert these dyes into the corresponding aromatic amines. Furthermore, it is worth highlighted that these microorganisms possess the ability to enzymatically transform aromatic amines into comparatively simpler organic compounds or carbon dioxide subsequently [95].

4.1.2 Laccases

Laccases, being copper-dependent enzymes, exhibit the capacity to facilitate the oxidation of diverse molecules harboring various functional groups, including aromatic amines and phenolic compounds. This particular catalytic process entails the generation of two water molecules concomitantly with the conduction of an electron from a solitary oxygen molecule [26]. These enzymes exhibit the capacity to oxidize nonphenolic compounds with lower solubility and high stability, thereby playing a role in the remediation of groundwater and underwater pollution. Fungal laccases exhibit significant utility in the remediation of various xenobiotics, encompassing textile dyes that are frequently encountered in aqueous environments and soil matrices. The observed efficacy can be ascribed to their substantial redox potential, which was evaluated in relation to laccases found in plants and bacteria [47]. Laccases exhibit remarkable biochemical stability across a broad range of conditions, encompassing fluctuations in pH, temperature, organic compounds, and salt concentrations [41]. As exemplified by the CotA protein derived from *Bacillus subtilis*, laccase frequently showcases remarkable stability even at high temperatures of 75 °C. This characteristic makes it highly pertinent in industrial applications as a thermostable enzyme. On the contrary, the temporal decay rate of recombinant CotA laccase derived from Escherichia coli is observed to be 2 h [26]. The remarkable enzymatic prowess of laccase in the degradation of xenobiotics and subsequent production of polymeric compounds renders it an invaluable resource in the realm of bioremediation methodologies. Significantly, the prevailing contaminant currently acknowledged, polycyclic aromatic hydrocarbons (PAHs), characterized by a benzene ring arranged in a linear fashion, exhibit a homogeneous dispersion across various natural habitats [118]. Laccase exhibits promising capabilities in the mitigation and detoxification of phenolic compounds and dyes that are commonly produced as byproducts within the textile industry. In a comprehensive investigation, it was observed that the recombinant CotA laccase derived from *Escherichia coli* exhibited the remarkable ability to effectively remove colour from synthetic textile effluents (STE). Recombinant CotA laccase exhibited efficient decolorization capabilities towards seven dyes possessing diverse structural compositions, both in their purified and unpurified states. The decoloring kinetics observed for both purified and unpurified CotA laccase exhibited enhanced rates when the sodium chloride-Tris–EDTA (STE) buffer was maintained at a neural pH [110]. Similarly, the investigation by Lu et al. unveiled that an alkaline laccase, originated from a genetically modified bacterial strain called *Bacillus licheniformis*, exhibited remarkable efficacy in the complete decomposition of synthetic dyes named reactive black and crimson within a timeframe of one hour. At a pH value of 9.0, the pure recombinant laccase exhibits a remarkable decolorization efficiency, surpassing 93% for the range of colors investigated within a time frame of 4 h [71].

The study revealed varying levels of laccases, lignin peroxidase (LiPs), NADH-DCIP reductase, azoreductase, polyphenol oxidase, and aminophenazone N-demethylase activity during the investigation of the enzymatic processes implicated in the microbial degradation of Methyl Red by *Saccharomyces cerevisiae* MTCC 463. As per scientific investigation, these particular compounds undergo subsequent degradation into aliphatic amines, potentially facilitated by the enzymes involved in oxidation such as lignin peroxidase and laccase [51]. In a study, *Pseudomonas putida* F6 was isolated from soil which yielded two laccases; one of these, designated CopA, was able to destroy five of the seven colors tested. Reactive Black

10B, Reactive Blue 19 (also known as Azovan Blue), Brom Cresol Purple, and Amido Black 10B [75]. Moreover, Bacillus sp.'s purified laccase transforms 92% of 4,4′ isopropylidene diphenol into 2-methoxy-4-ethylphenol [86]. Similarly, the laccase known as Lac15, which originates from the metagenome of aquatic microorganisms, demonstrated equivalent results by successfully decolorizing 50 mM azo dyes, specifically reactive orange 13 and reactive dark-blue 194, in a one-hour timeframe [42]. Consequently, laccase exhibits significant promise in terms of economically viable wastewater treatment, targeting a broad spectrum of pollutants including phenols and non-phenolic compounds, polyhydroxyalkanoates, synthetic dyes, and other pertinent contaminants.

4.1.3 Peroxidases

Peroxidases (1.11.1.7) are a class of oxidoreductases that possess heme as a crucial component. These enzymes exhibit a pivotal function in oxidizing a wide variety of organic and inorganic compounds, while concurrently participating in the reduction of peroxides, such as hydrogen peroxide (H_2O_2) [28]. The peroxidase action entails the supply of electrons with the capacity to engage in binding interactions with supplementary substrates such as ferricyanides and ascorbic acid, ultimately leading to their degradation into innocuous constituents. These enzymes are derived from various origins, including botanical, zoological, and microbial sources. Peroxidases originating from various microorganisms, such as bacteria, cyanobacteria, fungi, yeast, *Coprinopsis cinerea, Phanerochaete chrysosporium, Pseudomonas sp.*, *Bacillus sphaericus*, and *Bacillus subtilis*, are utilized in pollutant degradation, dye decolorization in textiles, and the elimination of environmental contaminants [113]. The association of peroxidases with enzymes that possess the ability to reduce oxides is known to exert a substantial impact on the toxicity of dyes [76]. Peroxidases exhibit remarkable potential in the biodegradation of wastewater polluted with phenolic compounds, 2-methylphenol, and chlorine containing substances. The aqueous solution derived from the textile industry often exhibits a pronounced chromaticity owing to the existence of rhodamine dyes, which possess a notable resistance towards conventional bleaching methodologies. However, these dyes can be effectively degraded through the action of peroxidase.

The peroxidases also exhibit catalytic activity towards the oxidation of methoxybenzene (anisole), lignin dimers, phenolic compounds, alkanes, dyes, and phenylmethanols, even without the presence of manganese II. An additional variant of peroxidase, exhibiting the capacity for dye-decolorization, has been discovered within the *Agaricus* fungi. This particular peroxidase has been recognized as a catalyst for oxidizing phenols and dyes [72]. Peroxidases derived from diverse origins also demonstrate a degree of inherent non-specificity. The edible macroscopic fungus *Pleurotus ostreatus* exhibits the production of an extracellular peroxidase enzyme, which showcases the remarkable potential for catalyzing the oxidation of different color compounds, like remazol brilliant blue [99, 114, 117]. Various peroxidases obtained from bacteria have also been employed for decolorizing different synthetic dyes. The potential of *Brevibacterium casei* in the breakdown of an azo dye, named 2-naphthol orange and the reduction of chromium (VI) has been investigated in nutrient-limited conditions. In the process of Chromium (VI) reduction, *Brevibacterium casei*'s reduction enzyme effectively employs AO7 as an electron donor. Upon the combination of oxidized 2-naphthol orange and reduced chromium (VI), an intriguing purple intermediate was observed [83].

Lignin Peroxidases

Lignin peroxidases constitute a category of monomeric enzymes, exhibiting molecular weights ranging from 38 to 45 kDa having iron in the form of ferric. Due to their elevated redox ability, LiPscan facilitate the decomposition of phenols and nonphenolic compounds, methylation, and break down of aromatic rings present in dyes [111]. LiPs can oxidize nonphenolic structures because they possess high redox ability [36]. Different fungal species including *Phanerochaete chrysosporium, Boletus adustus, Trametes versicolor, Phlebia radiate,* and *Ganoderma* spp produce LiPs. Lignin peroxidase obtained from *Phanerochaete chrysosporium* and chloroperoxidase obtained from *Caldariomyces fumago* have the ability to convert pentachlorophenol (PCP) into totetrachloro-p-benzoquinone through the process of oxidative dehalogenation within the presence of H_2O_2 [11]. Lignin peroxidases also catalyze the in vitro reduction of trichloroethylene into chlorine containing radicals within the existence of tertiary alcohol, EDTA, oxalic acid, and H_2O_2 [116]. Moreover, LiP obtained from *Pleurotus pulmonarius* strain CPG6 was effectively purified and decolorized dyes like 2-anthracenesulfonic acid, Allura red 129, Brilliant blue, Navy blue 5R, Reactive blue 19, and triazine dye red A from textile effluents [10, 39]. A study carried out by Mahmood et al. showed that among the three ligninolytic enzymes, lignin peroxidases obtained from fungus causing brown rot, named *Piptoporus betulinus* strain IEBL-3, have high potential for the breakdown of dyes and effluents from textile industry [74].

Manganese Peroxidases

Manganese peroxidases (MnPs) are glycoproteins containing heme group and their molecular weight range in size from 32 to 62 kDa. MnP utilizes H_2O_2, just as other ligninolytic peroxidases. MnPs may oxidize Mn^{2+} to Mn^{3+} by utilizing H_2O_2. MnP can remove aromatic compounds containing polycyclic hydrocarbons and nitrogen group, azo dyes, and endocrine-disrupting substances like bisphenol A and alkylphenols [78]. In addition, MnP may oxidize nonphenolic compounds by using mediators such as lipid and thiyl radicals. MnPs can be produced by several fungal species named *Phanerochaete chrysosporium, Irpex lacteus, Coriolus versicolor, Dichomitus squalens,* and *Ganoderma* spp. [79]. Industrial dyes, polycyclic hydrocarbons with benzene ring, 2-hydroxychlorobenzene, and antimicrobials—all of these can be harmful to human health can all be broken down by MnPs. Moreover, MnP obtained from the strain named *Bjerkandera adusta* CX-9 known as MnP BA30 demonstrated strong stability in solvent and effective decolorization of dye [20]. Likewise, *Trametes pubescens* (bracket fungus) strain i8 generated manganese peroxidase TP55, having a molecular mass of 56 kDa, which may be utilized to create cleansers, remove colors from textiles, and break down lignin [88]. In 2021, Gomathy and his coworker reported that manganese peroxidase produced by the alkaliphilic bacterium *Pseudomonas putida* was capable of oxidizing mixed azo colors at a basic pH [40];).

4.1.4 Dehalogenases

Dehalogenases (EC 3.8.1.5) from microbial source (EC 3.8.1.5) has received a lot of interest due to its important role in the bioremediation of halogenated organic chemicals. Dehalogenase breakdown C-X bonds by using three mechanisms, such as hydrolysis, reduction, and oxygenolysis, that can perform dehalogenation by replacing the halogen (X) atom with a hydroxyl (OH) group sourced from water and a hydrogen obtained from H2, respectively [109].

Zu and his colleagues identified a strain of *Bacillus* named GZT obtained from waste recycling site sludge that has an outstanding capacity to de-brominate and mineralize a bromophenol [9, 85, 121]. Liang et al. also isolated the identical species in favorable circumstances and the gene sequence of TBP dehalogenase obtained through *Bacillus* spp. GZT was transferred into BL21 (DE3) competent *Escherichia coli* strain to validate tribromophenol dehalogenase activity. The isolated enzyme destroyed 2,4,6-tribromophenol within 120 min at a temperature of 35 °C and 200 rpm with an efficiency of 80%. H_2O_2, NADPH, Mn^{2+}, along with Mg^{2+} stimulated the activity of dehalogenase, whereas it is inhibited by EDTA, 1,1′-dimethyl-4,4′-bipyridinium dichloride, Ni^{2+}, Fe^{2+}, Cu^{2+}, and Ca^{2+} [67]. The *Rhizobium* spp. gene that encodes haloalkane dehydrogenase [22] and haloalcohol dehydrogenase produced by bacterium *Pseudomonas umsongensis* strain (YCIT1612) [115], expressed within *E. coli*, both contribute to bioremediation through the breakdown of haloalkane and epoxide integration reaction. Different bacterial species, including *Pseudomonas* species [68], *Ancylobacter aquaticus* strain named UV5 [59], *Pseudomonas umsongensis* strain such as YCIT1612 [115], and *Rhizobium* spp. [22], produced enzymes having the potential to breakdown diverse halogenated substrates. Fricker et al. utilized the pure culture of *Dehalococcoides mccartyi* strain named JNA to remove chlorine extremely hazardous widespread environmental contaminants 1-hydroxypentachlorobenzene and other chlorine containing phenolic compounds into 3, 5-dichloro-phenol (DCP) by using three methods [37].

Zhang and his colleagues discovered two 2-haloalkanoic acid dehalogenases (dehalogenase 1 and dehalogenase 2) from aquatic bacterium named *Pseudomonas perfectomarinus* strain DEH130. The dehalogenase activity experiment was performed on both enzymes, and dehalogenase 2 showed selective activity with substrate 2-chloropropionic acid whereas they were unable to successfully decontaminate dehalogenase. Dehalogenase 2 exhibited strong stereospecificity for halogen containing substrate compared to dehalogenase 1 [119]. Boyer et al. identified reductive dehalogenase specific to 2,4,6-trichlorophenolate obtained from the bacterium *Desulfitobacterium frappieri* strain PCP-1. This enzyme facilitates the dichlorination of pentachlorophenol (PCP) at the ortho, meta, and para locations into hydroxychlorobenzene. Harmful environmentally halogen containing chemicals are converted to non-toxic and decomposable forms by reductive dehalogenation [21].

4.1.5 Cytochrome P450 Monooxygenase

Cytochrome p450 monooxygenases (CYP) constitute a group of enzymes containing heme and catalyze a variety of processes including N-hydroxylation, N-dealkylation, oxidation, dehalogenation, and C-H bond hydroxylation. NADPH-cytochrome p450 reductase provides required electrons to CYP during its processes, with the enzyme obtaining electrons from atmospheric oxygen. Hence, the presence of reducing agents such as NADPH or FAD becomes necessary [63]. CYPs are multifunctional enzymes present in various bacteria, fungi, plants, and mammals. *Saccharomyces, Rhodococcus, Streptomyces, Escherichia, Salmonella, Dehalococcoides,* and *Bacillus* are some of the species utilize their CYPs for bioremediation purpose [61].

Research has been conducted in both protein engineering and non-engineering domains to explore cytochrome p450 monooxygenases for the bio-remediation of organic contaminants and hydrocarbons. Recent research on synthetic P450s and the use of microbial based P450s for biodegradation and hydroxylation of ecologically important substrates were conducted [61]. Within the realm of microbial based P450s, protein engineering research on the *Bacillus megaterium* strain P450, CYP102A1, confirmed its ability for oxidizing Polycyclic aromatic hydrocarbons [24]. Kumar and his coworker investigated modified version of CYP102A1, demonstrated improved activity in dealing with polycyclic aromatic hydrocarbons (PAHs), chlorinated biphenyls, and straight chain alkanes. These substances often employed in the biodegradation of hazardous chemicals, decontamination of alkanes in gaseous form, and terpenes [62]. P450s exhibited the potential, especially in the breakdown of persistent halogen containing compounds that resist dioxygenases. The mutants F87W, F98W, Y96F, and V247L of monooxygenase from *Pseudomonas putida* CYP101A1 (P450cam) displayed the potential to breakdown insoluble pentachlorobenzene, all without relying on surfactants or organic chemicals [52].

4.1.6 Hydrolases

Hydrolases utilize water to cleave chemical bonds, facilitating the transformation of macromolecules into micromolecules, thereby reducing toxicity. These enzymes aid in the breaking various bonds including C–C, C-N, C-O, C-P, C-N, S–S, and S–N [16]. Hydrolases also catalyze other related processes, such as condensation and alcoholysis reaction. The noteworthy benefits of this enzyme family include its rapid accessibility, absence of cofactor stereoselectivity, and tolerance to the introduction of water-soluble solvents. Classified as enzymes of group, hydrolases are additionally categorized based on the kind of bond hydrolyzed. Extracellularly producing hydrolases include amylases, cellulases, xylanases, proteases, esterases, DNases, and pullulanase offer a wide range of potential applications across diverse sectors including food, feed preservatives, biotechnology, and various chemical industries such as textile industry [93].

4.1.7 Cellulases

The microorganisms that produce cellulase present in soil are spread over various genera of a bacterial domain [38]. *Micromonospora, Pseudomonas, Acidothermus, Paenibacillus,* and *Streptomyces* exemplify distinctive and innovative species that hydrolyze cellulose, implying that the ecosystem holds promise as an intriguing source for discovering novel enzymes capable of breakdown polysaccharide and cellulose [84]. Various microorganisms have been studied for their capacity to degrade synthetic textile dyes, including azo dyes, benzhydrylbenzene dyes, and dyes. Research has delved into the biotechnological potential of microbial cells isolated from textile effluents. As reported by Khokhar et al. a fungal species named *Trichoderma viride* obtained from a textile wastewater exhibited cellulolytic activity under culture conditions fortified with carboxymethylcellulose [57]. Filamentous fungi isolated from textile industry effluent that are capable of producing hydrolytic enzymes including lipase, amylase, cellulase, and laccase have proven to be an effective bioremediation technique [32]. In 2021, Bernal et al. isolated twenty-four species of fungus, including *Aspergillus* spp., *Trichoderma* spp., and *Penicillium* spp., obtained through various materials such as textile effluents and agricultural waste. In the textile industry, *Trichoderma reesei* fungal cellulases are the most commonly used enzyme. With the exception of that, actinomycetes belonging to genera *Streptomyces* and *Thermobifida*, along with bacteria within the genera *Pseudomonas* and *Rhizomonas*, are a few of the sources of enzymes employed in the decoloring and degrading textile dyes [15].

4.1.8 Lipases

Lipase, an enzyme containing serine and belonging to the hydrolase family, is a widely recognized biocatalyst known for its role in the breakdown of triglycerides' ester bonds into fatty acids and glycerine. It plays a crucial role in degrading lipids produced by different microbes, animals, as well as plants thereby decreasing the organic content of polluted soil. The process of fat degradation take place at the interface between lipid and water, where lipolytic substances often reach an equilibrium between the monomeric, micellar, and emulsified phases [55]. Lipid degrading enzymes isolated from microorganisms find diverse commercial applications, particularly in the bioremediation targeting residual oil, textile wastewater, petroleum impurities, and soil improvement. They are also applicable in a variety of readily available raw materials because they require low energy, specific to a substrate, high stability, short processing time [4].

Lipid degrading enzymes can improve the breakdown of oily effluents discharged from diverse areas that contain oils, fats, and protein [12]. In light of growing concern over environmental pollutants, bio-decomposable polymers have been marketed as a promising solution to this problem. Specifically, lipase PL, a purified enzyme obtained by *Alcaligenes* species, catalyzes the break down process of poly(L-lactic acid) (PLA) polymers into oligomers, and subsequently into monomers [45]. Bernal et al. exhibited the biotechnological capability of filamentous fungi such as *Aspergillus sydowii* (ITF 30) from textile effluent for the production of lipase degradation enzymes and effectively discolorize a synthetic dye such as Remazol Brilliant Blue R [15].

4.1.9 Proteases

Protease is an enzyme that belongs to the hydrolase family and is responsible for catalyzing the peptide bonds in proteins. They can also be obtained from microbes such as *Amycolatopsis* sp., *Aspergillus* sp., and *Bacillus* sp [16]. Microbial proteases are highly significant due to their cost-effectiveness, high yield, and exceptional efficacy. These industries, including the leather industry, the food business, and wastewater treatment, extensively utilize them [60]. Microbial protease comprises 66% of the overall output of protease used in commercial settings. These enzymes are widely utilized and contribute to over 60% of total enzyme sales. Protease has the ability to break down α-ester bonds, poly(hydroxybutyrate) (PHB) depolymerase breaks ß-ester bonds, and lipase breaks γ-ω bonds. Therefore, protease can be employed in bioremediation to degrade polymers [44].

Textile wastewater frequently contains proteins derived from many sources, such as dyes, sizing agents, and fibers. Proteases degrade these proteins into smaller peptides and amino acids via hydrolysis, hence easing their elimination from the effluent. The enzymatic breakdown aids in decreasing the amount of organic matter and enhances the overall effectiveness of the treatment procedure. Proteases mainly focus on breaking down proteins but they can also play a role in breaking down certain dye molecules, particularly those found associated with proteins. Proteases can facilitate the removal of color from the effluent by indirectly assisting in the decolorization process when dyes are attached to proteins [66] (Table 15.2).

Table 15.2 Utilization of microbial enzymes for bioremediation of textile effluents and their mechanisms

Enzymes	Microorganisms	Substrate	Mechanism	References
Azoreductases	*Pseudomonas paucimobilis, Galoctomyces geotrichum* MTCC 1360, *Trichosporon beigelii*	Azo dyes, naphthalene-based dyes, nitr0-aromatic compounds	Reduction of substrate with NADH and FADH as a cofactor	[56, 91, 95]
Laccases	*Bacillus subtilis, Escherichia coli, Pseudomonas putida* F6, *Bacillus vallismortis* fmb103	o, p-phenol, aminophenol, polyphenol, polyamines, aromatic compounds, and lignin	Reduction of oxygen molecule, involving one electron's oxidation with a variety of aromatic chemicals	[26, 75, 101, 105]
Cytochrome P450 monooxygenases	*Bacillus megaterium* CYP102A1, *Pseudomonas putida* CYP101	Alkanes, steroids, fatty acids, polycyclic aromatic compounds	Reduction and oxidization of heme iron to carry out electron transfer processes	[24, 43, 52]
Lignin peroxidases	*Pleurotus pulmonarius* CPG6, *Piptoporus betulinus* IEBL-3, *Phanerochaete chrysosporium, Boletus adustus, Trametes versicolor*	Phenolic compounds containing halogens, aromatic compounds, and PAHs	Oxidation of compounds in the presence of cosubstrate such as H_2O_2	[36, 39, 74]
Manganese peroxidases	*Phanerochaete chrysosporium, Irpex lacteus, Trametes pubescens* i8, *Bjerkandera adusta* strain CX-9	Lignin, Phenolic compounds	Oxidation of phenolic compounds in the presence of mediator H_2O_2 and Mn^{2+}	[20, 36, 88]
Dehalogenases	*Pseudomonas umsongensis* YCIT1612, *Dehalococcoides mccartyi* strain JNA, *Bacillus sp.* GZT	Halogenated organic compounds	Substitution of halogen atom with a hydroxyl group, while the water molecule acts as a cofactor Introducing one or two atoms of molecular oxygen to the substrate	[37, 54, 115, 121]

(continued)

Table 15.2 (continued)

Enzymes	Microorganisms	Substrate	Mechanism	References
Cellulases	*Trichoderma viride, Aspergillus* spp., *Penicillium* spp., *Pseudomonas* and *Sphingomonas* spp.	Cellulose	Degradation of cellulose into oligo and monosaccharide	[15, 57, 58]
Lipases	*Aspergillus sydowii* ITF 30, *Alcaligenes* spp., *Acinetobacter* spp., *Mycobacterium* spp., and *Rhodococcus* spp.	Triglycerides, lipid/oil	Hydrolysis of mono, di, and triglycerides into fatty acid and glycerol	[15, 25]

5 Conclusion

Textile industries emit effluents with elevated amount of harmful and persistent chemicals that have detrimental impacts on both the environment and human health. To address and potentially eradicate the adverse effects associated with textile dyes, it is possible to employ biomasses derived from plants, bacteria, extremophiles, and fungi to induce discoloration, transformation, or mineralization of these colors. Although microbial species have been considered as potential agents for bioremediation in treating textile effluents, the effectiveness of their use in comparison to the utilization of microbial enzymes is less satisfactory. The enzymes that are widely used to treat textile effluents include azoreductases, laccases, hydrolases, peroxidases, lipases, and cellulases. Various classes of microorganisms possess the ability to synthesize such enzymes that exhibit specificity towards distinct pollutants, thereby rendering them remarkably proficient in the degradation of diverse effluents originating from the textile industry. Furthermore, these enzymes function optimally under gentle conditions, thereby minimizing the requirement for aggressive chemicals and energy-demanding procedures. This not only reduces the overall ecological footprint but also enhances the economic viability of bioremediation.

References

1. Adane T, Adugna AT, Alemayehu E (2021) Textile industry effluent treatment techniques. J Chem 1–14
2. Ahmedna M, Marshall WE, Husseiny AA, Rao RM, Goktepe I (2004) The use of nutshell carbons in drinking water filters for removal of trace metals. Water Res 38(4):1062–1068
3. Al-Tohamy R, Ali SS, Li F, Okasha KM, Mahmoud YA-G, Elsamahy T, Jiao H, Fu Y, Sun J (2022) A critical review on the treatment of dye-containing wastewater: ecotoxicological and health concerns of textile dyes and possible remediation approaches for environmental safety. Ecotoxicol Environ Saf 231:113160
4. Arora NK, Mishra J, Mishra V (2020) Microbial enzymes: roles and applications in industries. Springer
5. Arshad M (2010) Bioethanol: a sustainable and environment friendly solution for Pakistan. A Scientific J COMSATS–Science Vision 2011:16–7
6. Arshad M, Adil M, Sikandar A, Hussain T (2014) Exploitation of meat industry by-products for biodiesel production: Pakistan's perspective. Pakistan J Life Soc Sci 12:120–125
7. Arshad M (2017) Clean and sustainable energy technologies. In: Clean energy for sustainable development 1. Academic Press, pp 73–89
8. Arshad M, Ansari AR, Qadir R, Tahir MH, Nadeem A, Mehmood T, Alhumade H, Khan N (2022) Green electricity generation from biogas of cattle manure: an assessment of potential and feasibility in Pakistan. Front Energy Res 25(10):911485
9. Bano I, Ghauri MA, Arshad M, Yasin T, Younus M (2017) Bioactivity of variant molecular weight chitosan against drug-resistant bacteria isolated from human wounds. Microbial Drug Resistance 1;23(8):958–65
10. Bano I, Arshad M (2018) Climatic changes impact on water availability. Perspect Water Usage Biofuels Prod: Aquat Contam Clim Chang 2018:39–54
11. Bansal N, Kanwar SS (2013) Peroxidase (s) in environment protection. Sci World J

12. Basheer SM, Chellappan S, Beena P, Sukumaran RK, Elyas K, Chandrasekaran M (2011) Lipase from marine Aspergillus awamori BTMFW032: production, partial purification and application in oil effluent treatment. New Biotechnol 28(6):627–638
13. Bell J, Plumb JJ, Buckley CA, Stuckey DC (2000) Treatment and decolorization of dyes in an anaerobic baffled reactor. J Environ Eng 126(11):1026–1032
14. Bento RM, Almeida MR, Bharmoria P, Freire MG, Tavares AP (2020) Improvements in the enzymatic degradation of textile dyes using ionic-liquid-based surfactants. Sep Purif Technol 235:116191
15. Bernal SP, Lira M, Jean-Baptiste J, Garcia PE, Batista E, Ottoni JR, Passarini MR (2021) Biotechnological potential of microorganisms from textile effluent: isolation, enzymatic activity and dye discoloration. An Acad Bras Ciênc 93:e20191581
16. Bhandari S, Poudel DK, Marahatha R, Dawadi S, Khadayat K, Phuyal S, Shrestha S, Gaire S, Basnet K, Khadka U (2021) Microbial enzymes used in bioremediation. J Chem 2021:1–17
17. Bhatt P, Rani A (2012) Textile dyeing and printing industry: an environmental hazard. Asian Dyer 10:51–54
18. Bilal M, Asgher M, Parra-Saldivar R, Hu H, Wang W, Zhang X, Iqbal HM (2017) Immobilized ligninolytic enzymes: an innovative and environmental responsive technology to tackle dye-based industrial pollutants–a review. Sci Total Environ 576:646–659
19. Bittencourt GA, de Souza Vandenberghe LP, Martínez-Burgos WJ, Valladares-Diestra KK, de Mello AFM, Maske BL, Brar SK, Varjani S, de Melo Pereira GV, Soccol CR (2023) Emerging contaminants bioremediation by enzyme and nanozyme-based processes–a review. Iscience
20. Bouacem K, Rekik H, Jaouadi NZ, Zenati B, Kourdali S, El Hattab M, Badis A, Annane R, Bejar S, Hacene H (2018) Purification and characterization of two novel peroxidases from the dye-decolorizing fungus Bjerkandera adusta strain CX-9. Int J Biol Macromol 106:636–646
21. Boyer A, Pagé-BéLanger R, Saucier M, Villemur R, Lépine F, Juteau P, Beaudet R (2003) Purification, cloning and sequencing of an enzyme mediating the reductive dechlorination of 2, 4, 6-trichlorophenol from Desulfitobacterium frappieri PCP-1. Biochem J 373(1):297–303
22. Cairns SS, Cornish A, Cooper RA (1996) Cloning, sequencing and expression in Escherichia coli of two Rhizobium sp. genes encoding haloalkanoate dehalogenases of opposite stereospecificity. Eur J Biochem 235(3):744–749
23. Cao J, Sanganyado E, Liu W, Zhang W, Liu Y (2019) Decolorization and detoxification of Direct Blue 2B by indigenous bacterial consortium. J Environ Manage 242:229–237
24. Carmichael AB, Wong LL (2001) Protein engineering of Bacillus megaterium CYP102: the oxidation of polycyclic aromatic hydrocarbons. Eur J Biochem 268(10):3117–3125
25. Casas-Godoy L, Duquesne S, Bordes F, Sandoval G, Marty A (2012) Lipases: an overview. Lipases and phospholipases: methods and protocols:3–30
26. Chandra R, Chowdhary P (2015) Properties of bacterial laccases and their application in bioremediation of industrial wastes. Environ Sci Process Impacts 17(2):326–342
27. Chang J-S, Chou C, Chen S-Y (2001) Decolorization of azo dyes with immobilized Pseudomonas luteola. Process Biochem 36(8–9):757–763
28. Chanwun T, Muhamad N, Chirapongsatonkul N, Churngchow N (2013) Hevea brasiliensis cell suspension peroxidase: purification, characterization and application for dye decolorization. AMB Express 3:1–9
29. Chen K-C, Wu J-Y, Liou D-J, Hwang S-CJ (2003) Decolorization of the textile dyes by newly isolated bacterial strains. J Biotechnol 101(1):57–68
30. Chugh M, Kumar L, Bharadvaja N (2022) Fungal diversity in the bioremediation of toxic effluents. In: Development in wastewater treatment research and processes. Elsevier, pp 61–88
31. Croce R, Cinà F, Lombardo A, Crispeyn G, Cappelli CI, Vian M, Maiorana S, Benfenati E, Baderna D (2017) Aquatic toxicity of several textile dye formulations: acute and chronic assays with Daphnia magna and Raphidocelis subcapitata. Ecotoxicol Environ Saf 144:79–87
32. Dhiman K, Shirkot P (2015) Bioprospecting and molecular characterization of laccase producing bacteria from paper mills of Himachal Pradesh. Proc Natl Acad Sci, India Sect B: Biol Sci 85:1095–1103

33. Elekwachi CO, Andresen J, Hodgman TC (2014) Global use of bioremediation technologies for decontamination of ecosystems. J Bioremediation & Biodegredation 5(4):1
34. Eslami A, Moradi M, Ghanbari F, Mehdipour F (2013) Decolorization and COD removal from real textile wastewater by chemical and electrochemical Fenton processes: a comparative study. J Environ Health Sci Eng 11:1–8
35. Essadki A, Bennajah M, Gourich B, Vial C, Azzi M, Delmas H (2008) Electrocoagulation/ electroflotation in an external-loop airlift reactor—application to the decolorization of textile dye wastewater: a case study. Chem Eng Process 47(8):1211–1223
36. Falade AO, Nwodo UU, Iweriebor BC, Green E, Mabinya LV, Okoh AI (2017) Lignin peroxidase functionalities and prospective applications. MicrobiologyOpen 6(1):e00394
37. Fricker AD, LaRoe SL, Shea ME, Bedard DL (2014) Dehalococcoides mccartyi strain JNA dechlorinates multiple chlorinated phenols including pentachlorophenol and harbors at least 19 reductive dehalogenase homologous genes. Environ Sci Technol 48(24):14300–14308
38. Garcia IG, Simal-Gandara J, Gullo M (2022) Advances in food, bioproducts and natural byproducts for a sustainable future: from conventional to innovative processes, vol 12. MDPI
39. Giap VD, Duc HT, Huong PTM, Do TH, Do HN, Duy VD, Quynh DT (2022) Purification and characterization of lignin peroxidase from white-rot fungi Pleurotus pulmonarius CPG6 and its application in decolorization of synthetic textile dyes. J Gen Appl Microbiol 68(6):262–269
40. Gomathy S, Sridharan R, Kumar PS, Gayathri KV, Al Farraj DA, Chen T-W (2021) Application of alkaline MnP immobilized Luffa fibers in mixed azo dyes degradation. Environ Technol Innov 24:101964
41. Guan Z-B, Luo Q, Wang H-R, Chen Y, Liao X-R (2018) Bacterial laccases: promising biological green tools for industrial applications. Cell Mol Life Sci 75:3569–3592
42. Guan Z-B, Zhang N, Song C-M, Zhou W, Zhou L-X, Zhao H, Xu C-W, Cai Y-J, Liao X-R (2014) Molecular cloning, characterization, and dye-decolorizing ability of a temperature-and pH-stable laccase from Bacillus subtilis X1. Appl Biochem Biotechnol 172:1147–1157
43. Guengerich FP (2018) Mechanisms of cytochrome P450-catalyzed oxidations. ACS Catal 8(12):10964–10976
44. Haider TP, Völker C, Kramm J, Landfester K, Wurm FR (2019) Plastics of the future? The impact of biodegradable polymers on the environment and on society. Angew Chem Int Ed 58(1):50–62
45. Hoshino A, Isono Y (2002) Degradation of aliphatic polyester films by commercially available lipases with special reference to rapid and complete degradation of poly (L-lactide) film by lipase PL derived from Alcaligenes sp. Biodegradation 13(2):141–147
46. Hu T (1992) Sorption of reactive dyes by Aeromonas biomass. Water Sci Technol 26(1–2):357–366
47. Iark D, dos Reis Buzzo AJ, Garcia JAA, Côrrea VG, Helm CV, Corrêa RCG, Peralta RA, Moreira RdFPM, Bracht A, Peralta RM (2019) Enzymatic degradation and detoxification of azo dye Congo red by a new laccase from Oudemansiella canarii. Biores Technol 289:121655
48. Ihsanullah I, Jamal A, Ilyas M, Zubair M, Khan G, Atieh MA (2020) Bioremediation of dyes: current status and prospects. J Water Process Eng 38:101680
49. Imran M, Crowley DE, Khalid A, Hussain S, Mumtaz MW, Arshad M (2015) Microbial biotechnology for decolorization of textile wastewaters. Rev Environ Sci Bio/Technol 14:73–92
50. Ito T, Adachi Y, Yamanashi Y, Shimada Y (2016) Long–term natural remediation process in textile dye–polluted river sediment driven by bacterial community changes. Water Res 100:458–465
51. Jafari N, Soudi MR, Kasra-Kermanshahi R (2014) Biodegradation perspectives of azo dyes by yeasts. Microbiol 83:484–497
52. Jones JP, O'Hare EJ, Wong LL (2001) Oxidation of polychlorinated benzenes by genetically engineered CYP101 (cytochrome P450cam). Eur J Biochem 268(5):1460–1467
53. Judd S, Jefferson B (2003) Membranes for industrial wastewater recovery and re-use. Elsevier
54. Jugder B-E, Ertan H, Lee M, Manefield M, Marquis CP (2015) Reductive dehalogenases come of age in biological destruction of organohalides. Trends Biotechnol 33(10):595–610

55. Karigar CS, Rao SS (2011) Role of microbial enzymes in the bioremediation of pollutants: a review. Enzyme research 2011
56. Khan R, Bhawana P, Fulekar M (2013) Microbial decolorization and degradation of synthetic dyes: a review. Rev Environ Sci Bio/Technol 12:75–97
57. Khokhar I, Haider MS, Mushtaq S, Mukhtar I (2012) Isolation and screening of highly cellulolytic filamentous fungi. J Appl Sci Environ Manag 16(3)
58. Korsa G, Konwarh R, Masi C, Ayele A, Haile S (2023) Microbial cellulase production and its potential application for textile industries. Ann Microbiol 73(1):13
59. Kumar A, Pillay B, Olaniran AO (2016) L-2-haloacid dehalogenase from ancylobacter aquaticus UV5: sequence determination and structure prediction. Int J Biol Macromol 83:216–225
60. Kumar A, Sharma S (2019) Microbes and enzymes in soil health and bioremediation. Springer
61. Kumar S (2010) Engineering cytochrome P450 biocatalysts for biotechnology, medicine and bioremediation. Expert Opin Drug Metab Toxicol 6(2):115–131
62. Kumar S, Jin M, Weemhoff JL (2012) Cytochrome P450-mediated phytoremediation using transgenic plants: A need for engineered cytochrome P450 enzymes. J Pet & Environ Biotechnol 3(5)
63. Lamb SB, Lamb DC, Kelly SL, Stuckey DC (1998) Cytochrome P450 immobilisation as a route to bioremediation/biocatalysis. FEBS Lett 431(3):343–346
64. Lellis B, Fávaro-Polonio CZ, Pamphile JA, Polonio JC (2019) Effects of textile dyes on health and the environment and bioremediation potential of living organisms. Biotechnol Res Innov 3(2):275–290
65. Li M, Li K, Wang L, Zhang X (2020) Feasibility of concentrating textile wastewater using a hybrid forward osmosis-membrane distillation (FO-MD) process: performance and economic evaluation. Water Res 172:115488
66. Li Q, Yi L, Marek P, Iverson BL (2013) Commercial proteases: present and future. FEBS Lett 587(8):1155–1163
67. Liang Z, Li G, Mai B, Ma H, An T (2019) Application of a novel gene encoding bromophenol dehalogenase from Ochrobactrum sp. T in TBBPA degradation. Chemosphere 217:507–515
68. Liu JQ, Kurihara T, Hasan A, Nardi-Dei V, Koshikawa H, Esaki N, Soda K (1994) Purification and characterization of thermostable and nonthermostable 2-haloacid dehalogenases with different stereospecificities from Pseudomonas sp. strain YL. Appl Environ Microbiol 60(7):2389–2393
69. Liu L, Bilal M, Duan X, Iqbal HM (2019) Mitigation of environmental pollution by genetically engineered bacteria—current challenges and future perspectives. Sci Total Environ 667:444–454
70. Louhichi B, Gaied F, Mansouri K, Jeday MR (2022) Treatment of textile industry effluents by electro-coagulation and electro-fenton processes using solar energy: a comparative study. Chem Eng J 427:131735
71. Lu L, Wang T-N, Xu T-F, Wang J-Y, Wang C-L, Zhao M (2013) Cloning and expression of thermo-alkali-stable laccase of Bacillus licheniformis in Pichia pastoris and its characterization. Biores Technol 134:81–86
72. Lundell TK, Mäkelä MR, Hildén K (2010) Lignin-modifying enzymes in filamentous basidiomycetes–ecological, functional and phylogenetic review. J Basic Microbiol 50(1):5–20
73. Madhav S, Ahamad A, Singh P, Mishra PK (2018) A review of textile industry: wet processing, environmental impacts, and effluent treatment methods. Environ Qual Manage 27(3):31–41
74. Mahmood RT, Asad MJ, Asgher M, Khan FS, Muzammil K, Nasir N, Anwar P, Awais M (2022) First report on the bioremediation of textile industrial effluents by Piptoporus betulinus IEBL-3 by using response surface methodology. Appl Sci 12(3):1090
75. Mandic M, Djokic L, Nikolaivits E, Prodanovic R, O'Connor K, Jeremic S, Topakas E, Nikodinovic-Runic J (2019) Identification and characterization of new laccase biocatalysts from Pseudomonas species suitable for degradation of synthetic textile dyes. Catalysts 9(7):629

76. Mishra S, Maiti A (2019) Applicability of enzymes produced from different biotic species for biodegradation of textile dyes. Clean Technol Environ Policy 21:763–781
77. Mohana S, Shrivastava S, Divecha J, Madamwar D (2008) Response surface methodology for optimization of medium for decolorization of textile dye direct black 22 by a novel bacterial consortium. Biores Technol 99(3):562–569
78. Moon D-S, Song H-G (2012) Degradation of alkylphenols by white rot fungus Irpex lacteus and its manganese peroxidase. Appl Biochem Biotechnol 168:542–549
79. Mousavi SM, Hashemi SA, Iman Moezzi SM, Ravan N, Gholami A, Lai CW, Chiang W-H, Omidifar N, Yousefi K, Behbudi G (2021) Recent advances in enzymes for the bioremediation of pollutants. Biochemistry research international 2021
80. Mudhoo A, Ramasamy DL, Bhatnagar A, Usman M, Sillanpää M (2020) An analysis of the versatility and effectiveness of composts for sequestering heavy metal ions, dyes and xenobiotics from soils and aqueous milieus. Ecotoxicol Environ Saf 197:110587
81. Murugesan K, Kalaichelvan P (2003) Synthetic dye decolourization by white rot fungi
82. Nazari MT, Simon V, Machado BS, Crestani L, Marchezi G, Concolato G, Ferrari V, Colla LM, Piccin JS (2022) Rhodococcus: A promising genus of actinomycetes for the bioremediation of organic and inorganic contaminants. J Environ Manage 323:116220
83. Ng TW, Cai Q, Wong C-K, Chow AT, Wong P-K (2010) Simultaneous chromate reduction and azo dye decolourization by Brevibacterium casei: azo dye as electron donor for chromate reduction. J Hazard Mater 182(1–3):792–800
84. Poulsen JS, de Jonge N, Macêdo WV, Dalby FR, Feilberg A, Nielsen JL (2022) Characterisation of cellulose-degrading organisms in an anaerobic digester. Biores Technol 351:126933
85. Qasim M, Bukhari SA, Ghani MJ, Masoud MS, Huma T, Arshad M, Haque A, Ibrahim Z, Javed S, Rajoka MI (2016) Relationship of oxidative stress with elevated level of DNA damage and homocysteine in cardiovascular disease patients. Pak J Pharm Sci 2:29
86. Rajeswari M, Bhuvaneswari V (2016) Production of extracellular laccase from the newly isolated Bacillus sp. PK4. Afr J Biotechnol 15(34):1813–1826
87. Rawat D, Mishra V, Sharma RS (2016) Detoxification of azo dyes in the context of environmental processes. Chemosphere 155:591–605
88. Rekik H, Jaouadi NZ, Bouacem K, Zenati B, Kourdali S, Badis A, Annane R, Bouanane-Darenfed A, Bejar S, Jaouadi B (2019) Physical and enzymatic properties of a new manganese peroxidase from the white-rot fungus Trametes pubescens strain i8 for lignin biodegradation and textile-dyes biodecolorization. Int J Biol Macromol 125:514–525
89. Report GV (2022) Textile market size, share & trends analysis report by raw material (Wool, Chemical, Silk), by product (natural fibers, polyester), by application, by region, and segment forecasts. pp 2023−2030. https://www.grandviewresearch.com/industry-analysis/textile-market#:~:text=Report%20Overview,revenue%20from%202023%20to%202030
90. Routoula E, Patwardhan SV (2020) Degradation of anthraquinone dyes from effluents: a review focusing on enzymatic dye degradation with industrial potential. Environ Sci Technol 54(2):647–664
91. Roy DC, Biswas SK, Saha AK, Sikdar B, Rahman M, Roy AK, Prodhan ZH, Tang S-S (2018) Biodegradation of crystal violet dye by bacteria isolated from textile industry effluents. PeerJ 6:e5015
92. Samuchiwal S, Gola D, Malik A (2021) Decolourization of textile effluent using native microbial consortium enriched from textile industry effluent. J Hazard Mater 402:123835
93. Sánchez-Porro C, Martin S, Mellado E, Ventosa A (2003) Diversity of moderately halophilic bacteria producing extracellular hydrolytic enzymes. J Appl Microbiol 94(2):295–300
94. Sangwan S, Dukare A (2018) Microbe-mediated bioremediation: an eco-friendly sustainable approach for environmental clean-up. Adv Soil Microbiol: Recent Trends Futur Prospect: Vol 1: Soil-Microbe Interactl 45–163
95. Saratale RG, Saratale GD, Chang J-S, Govindwar SP (2011) Bacterial decolorization and degradation of azo dyes: a review. J Taiwan Inst Chem Eng 42(1):138–157

96. Sarkar S, Banerjee A, Chakraborty N, Soren K, Chakraborty P, Bandopadhyay R (2020) Structural-functional analyses of textile dye degrading azoreductase, laccase and peroxidase: a comparative in silico study. Electron J Biotechnol 43:48–54
97. Shah MP, Rodriguez-Couto S (2022) Development in wastewater treatment research and processes: microbial degradation of xenobiotics through bacterial and fungal approach. Elsevier
98. Sharif A, Nasreen Z, Bashir R, Kalsoom S (2020) 8. Microbial degradation of textile industry effluents: a review. Pure Appl Biol (PAB) 9(4):2361–2382
99. Sharif N, Munir N, Hasnain M, Naz S, Arshad M (2021) Environmental impacts of ethanol production system. In: Sustainable ethanol and climate change: sustainability assessment for ethanol distilleries. pp 205−223
100. Sharma A, Kumar V, Shahzad B, Ramakrishnan M, Singh Sidhu GP, Bali AS, Handa N, Kapoor D, Yadav P, Khanna K (2020) Photosynthetic response of plants under different abiotic stresses: a review. J Plant Growth Regul 39:509–531
101. Shekher R, Sehgal S, Kamthania M, Kumar A (2011) Laccase: microbial sources, production, purification, and potential biotechnological applications. Enzym Res
102. Silva MC, Torres JA, de Sá LRV, Chagas PMB, Ferreira-Leitão VS, Corrêa AD (2013) The use of soybean peroxidase in the decolourization of Remazol Brilliant Blue R and toxicological evaluation of its degradation products. J Mol Catal B Enzym 89:122–129
103. Singh RL, Singh PK, Singh RP (2015) Enzymatic decolorization and degradation of azo dyes–a review. Int Biodeterior Biodegradation 104:21–31
104. Stergiopoulos D, Dermentzis K, Giannakoudakis P, Sotiropoulos S (2014) Electrochemical decolorization and removal of indigo carmine textile dye from wastewater. Global NEST J 16(3):499–506
105. Sun J, Zheng M, Lu Z, Lu F, Zhang C (2017) Heterologous production of a temperature and pH-stable laccase from Bacillus vallismortis fmb-103 in Escherichia coli and its application. Process Biochem 55:77–84
106. Teerapatsakul C, Parra R, Keshavarz T, Chitradon L (2017) Repeated batch for dye degradation in an airlift bioreactor by laccase entrapped in copper alginate. Int Biodeterior Biodegradation 120:52–57
107. Vijayaraghavan K, Yun Y (2014) A study on biosorption potential of Aspergillus sp. of tannery effluent. Adv Biosci Biotechnol 266–291
108. Vikrant K, Giri BS, Raza N, Roy K, Kim K-H, Rai BN, Singh RS (2018) Recent advancements in bioremediation of dye: current status and challenges. Biores Technol 253:355–367
109. Wang D, Li A, Han H, Liu T, Yang Q (2018) A potent chitinase from Bacillus subtilis for the efficient bioconversion of chitin-containing wastes. Int J Biol Macromol 116:863–868
110. Wang T-N, Zhao M (2017) A simple strategy for extracellular production of CotA laccase in Escherichia coli and decolorization of simulated textile effluent by recombinant laccase. Appl Microbiol Biotechnol 101(2):685–696
111. Wang X, Yao B, Su X (2018) Linking enzymatic oxidative degradation of lignin to organics detoxification. Int J Mol Sci 19(11):3373
112. Wang Z, Xue M, Huang K, Liu Z (2011) Textile dyeing wastewater treatment. Adv Treat Text Effl 5:91–116
113. Wong JKH, Tan HK, Lau SY, Yap P-S, Danquah MK (2019) Potential and challenges of enzyme incorporated nanotechnology in dye wastewater treatment: a review. J Environ Chem Eng 7(4):103261
114. Wu XT, Ansari AR, Pang XX, Li HZ, Zhang ZW, Luo Y, Arshad M, Song H (2019) Visfatin plays a significant role in alleviating lipopolysaccharide-induced apoptosis and autophagy through PI3K/AKT signaling pathway during acute lung injury in mice. Arch Immunol Ther Exp 1(67):249–261
115. Xue F, Ya X, Tong Q, Xiu Y, Huang H (2018) Heterologous overexpression of Pseudomonas umsongensis halohydrin dehalogenase in Escherichia coli and its application in epoxide asymmetric ring opening reactions. Process Biochem 75:139–145

116. Yadav J, Bethea C, Reddy C (2000) Mineralization of trichloroethylene (TCE) by the white rot fungus Phanerochaete chrysosporium. Bull Environ Contam Toxicol 65(1):28
117. Zaini MS, Arshad M, Syed-Hassan SS (2023) Adsorption isotherm and kinetic study of methane on palm kernel shell-derived activated carbon. J Bioresour Bioprod 1;8(1):66–77
118. Zeng J, Lin X, Zhang J, Li X, Wong MH (2011) Oxidation of polycyclic aromatic hydrocarbons by the bacterial laccase CueO from E. coli. Appl Microbiol Biotechnol 89:1841–1849
119. Zhang J, Cao X, Xin Y, Xue S, Zhang W (2013) Purification and characterization of a dehalogenase from Pseudomonas stutzeri DEH130 isolated from the marine sponge Hymeniacidon perlevis. World J Microbiol Biotechnol 29:1791–1799
120. Zhuang M, Sanganyado E, Xu L, Zhu J, Li P, Liu W (2020) High throughput sediment DNA sequencing reveals azo dye degrading bacteria inhabit nearshore sediments. Microorganisms 8(2):233
121. Zu L, Li G, An T, Wong P-K (2012) Biodegradation kinetics and mechanism of 2, 4, 6-tribromophenol by Bacillus sp. GZT: a phenomenon of xenobiotic methylation during debromination. Biores Technol 110:153–159

The manufacturer's authorised representative in the EU is Springer Nature Customer Service Centre GmbH, Europaplatz 3, 69115 Heidelberg, Germany. If you have any concerns regarding our products, please contact ProductSafety@springernature.com

Printed and bound by CPI Group (UK) Ltd, Croydon, CR0 4YY
01/10/2026
02225914-0013